VOLUME ONE HUNDRED AND THIRTY SEVEN

ADVANCES IN HETEROCYCLIC CHEMISTRY

Heterocyclic Mesomeric Betaines and Mesoionic Compounds

EDITORIAL ADVISORY BOARD

VOLUME ONE HUNDRED AND THIRTY SEVEN

ADVANCES IN HETEROCYCLIC CHEMISTRY

Heterocyclic Mesomeric Betaines and Mesoionic Compounds

Editor

CHRISTOPHER A. RAMSDEN

Lennard-Jones Laboratories,
Keele University, Staffordshire,
United Kingdom

ELSEVIER

ACADEMIC PRESS
An imprint of Elsevier

Academic Press is an imprint of Elsevier
50 Hampshire Street, 5th Floor, Cambridge, MA 02139, United States
525 B Street, Suite 1650, San Diego, CA 92101, United States
The Boulevard, Langford Lane, Kidlington, Oxford OX5 1GB, United Kingdom
125 London Wall, London, EC2Y 5AS, United Kingdom

First edition 2022

ISBN: 978-0-323-98861-2
ISSN: 0065-2725

For information on all Academic Press publications
visit our website at https://www.elsevier.com/books-and-journals

Publisher: Zoe Kruze
Acquisitions Editor: Jason Mitchell
Developmental Editor: Jhon Michael Peñano
Production Project Manager: James Selvam
Cover Designer: Miles Hitchen

Typeset by STRAIVE, India

Contents

Contributors

Wim Dehaen
Molecular Design and Synthesis, Department of Chemistry, KU Leuven, Leuven, Belgium

Florea Dumitrascu
Center for Organic Chemistry, "C.D. Nenitzescu" Romanian Academy, Bucharest, Romania

Christopher G. Newton
The Research Laboratories, May & Baker Limited, Essex, United Kingdom

W. David Ollis
Department of Chemistry, The University of Sheffield, Sheffield, England

Tomas Opsomer
Molecular Design and Synthesis, Department of Chemistry, KU Leuven, Leuven, Belgium

Christopher A. Ramsden
Lennard-Jones Laboratories, School of Physical and Geographical Sciences, Keele University, Staffordshire, United Kingdom

Yingchun Wang
Molecular Design and Synthesis, Department of Chemistry, KU Leuven, Leuven, Belgium

Preface

Volume 137 of *Advances in Heterocyclic Chemistry* is devoted to the chemistry of heterocyclic mesomeric betaines (HMBs), which constitute an extensive family of heterocycles that can only be represented as resonance hybrids of dipolar structures. Chapter 1 (Christopher A. Ramsden, Keele University, UK) provides an overview of the common features of HMBs and the structural features that lead to the three fundamentally different subgroups, which are described as conjugated, cross-conjugated, and semi-conjugated HMBs.

Chapter 2 by Yingchun Wang, Tomas Opsomer, and Wim Dehaen (KU Leuven, Belgium) describes developments in the chemistry of 1,3a,6a-triazapentalenes, which are a family of bicyclic HMBs. Typically, these triazapentalenes undergo 1,3-dipolar cycloadditions and their physical properties make them of interest as fluorescent sensors and probes.

Chapters 3 and 4 by Florea Dumitrascu ("C.D. Nenitzescu" Romanian Academy Romania) and Christopher A. Ramsden (Keele University, UK) are devoted to developments in the chemistry of type A and type B mesoionic compounds during the period 1980–2020. A systematic coverage of ring structures from dioxoles to tetrazoles highlights 1,3-dipolar cycloadditions, N-heterocyclic carbene formation, and valence tautomerism, among other properties. Ring systems that have received little or no attention are also identified.

A special feature of this volume is the inclusion of two earlier reviews of mesoionic compounds, which are reproduced in Appendices 1 and 2. These are Mesoionic Compounds by W. David Ollis and Christopher A. Ramsden (*Adv. Heterocycl. Chem.* **1976**, *19*, 1) and Mesoionic Heterocycles (1976–1980) by Christopher G. Newton and Christopher A. Ramsden (*Tetrahedron* **1982**, *38*, 2965). The inclusion of these reprints provides comprehensive coverage of mesoionic chemistry up to 2020 in one volume.

CHRISTOPHER A. RAMSDEN
January 2022

CHAPTER ONE

Heterocyclic mesomeric betaines: An overview

Christopher A. Ramsden*
Lennard-Jones Laboratories, School of Physical and Geographical Sciences, Keele University, Staffordshire, United Kingdom
*Corresponding author: e-mail address: c.a.ramsden@keele.ac.uk

Contents

Abstract

The historical development of the concept of heterocyclic mesomeric betaines (HMBs) is briefly reviewed. The common features of heterocyclic mesomeric betaines are discussed with particular emphasis on the significance of molecular fragments associated with an odd number of conjugated atoms. The discussion demonstrates how the way that heteroatoms are connected to these fragments leads to three distinct types of heterocyclic mesomeric betaine: conjugated, cross-conjugated and semi-conjugated. The three types and their subclasses are systematically illustrated with examples from the literature.

Keywords: Mesomeric betaine, Cross-conjugation, Semi-conjugated, 1,3-Dipole, 1,4-Dipole, Mesoionic

Advances in Heterocyclic Chemistry, Volume 137
ISSN 0065-2725
https://doi.org/10.1016/bs.aihch.2021.10.001

1. Introduction and history

The foundation of the coal gas industry in the early 19th century provided building blocks for the rapid development of heterocyclic chemistry. Runge isolated pyrrole and aniline from coal tar in 1834 heralding the era of diazonium salts and the aniline dyes. In 1873 Fischer made phenylhydrazine by reduction of a diazonium salt, leading to his study of many hydrazine products including dithizone **1**. Fischer and Besthorn (1882)[1] showed that oxidation of dithizone **1** gave a product, named dehydrodithizone, which they formulated as the bisphenylazo structure **2** but which was later shown to have the mesomeric betaine structure **3**. This is regarded as the earliest example of the preparation of a heterocyclic mesomeric betaine and it is interesting to note that many of the reactions of dehydrothithizone **3** occur via the valence tautomer **2**.[2–4]

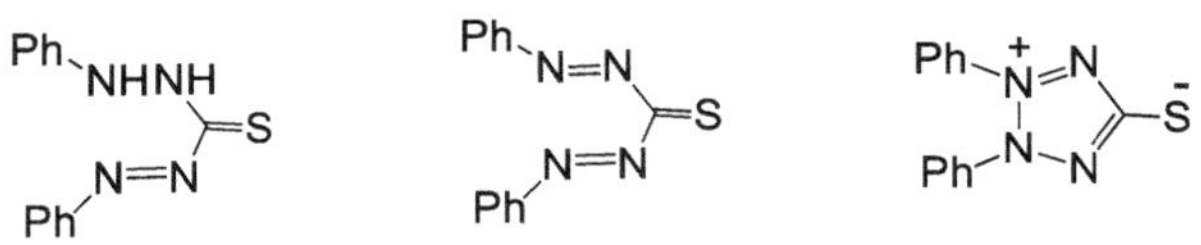

1 **2** **3**

Although Kekulé had demonstrated the structure of benzene in 1865, paving the way for the correct structural assignment of many heterocycles, the formulation of betaines remained an enigma. During a series of studies in the period 1895–1930, Busch and co-workers studied N^1-aryl-N^2-dithiocarboxyhydrazine salts, e.g., $PhNHNHCS_2K$, and related derivatives, and reported a number of novel products formulated as the bicyclic structures **4** (X or Y = O, S, NAr).[5] In an important and much overlooked 1938 contribution, Schönberg pointed out that the structures **4** do not conform to Bredt's rule (1924) and that the products are best represented as betaines **5**.[6] He also pointed out the possibility of isomers **5** and **6**, many of which have been subsequently characterized.[2] The molecules first prepared by Busch include the analytical reagent nitron **6** (R^1 = Ph, R^2 = H, X = Y = NPh),[7,8] which continues to be of notable interest in the 21st century due to recent discoveries of both N-heterocyclic carbene formation[9] and Blatter-type stable radical formation.[10] Later, in 1949, Baker and Ollis recognized the betaine structure of the sydnones **7** and coined the term mesoionic to describe the general class of heterocycles of this type.[2,11,12]

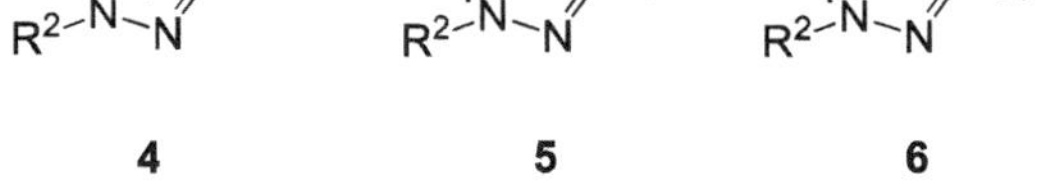

4 **5** **6**

An important development was the recognition that sydnones **7** contain an azomethine imine fragment that reacts as a 1,3-dipole with alkynes to give pyrazoles and carbon dioxide. This led to extensive studies of mesoionic compounds during the 1960s and 1970s with notable applications of 1,3-dipolar cycloadditions of the münchnones **8** by Huisgen and co-workers[13,14] and the isomünchnones **9** by Padwa and co-workers.[15,16]

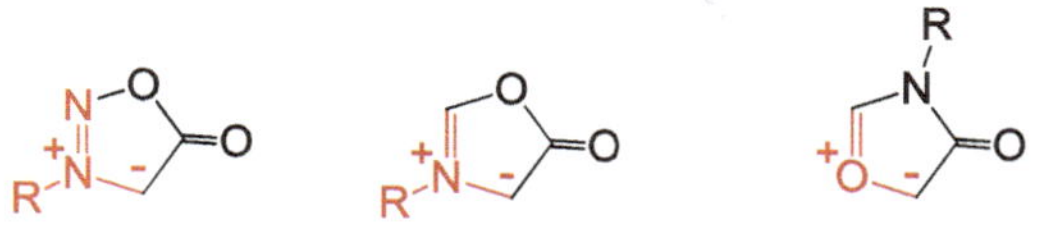

7 sydnones **8** münchnones **9** isomünchnones

Mesomeric betaines are not restricted to five-membered rings. The phthalazinium-3-olate derivatives **10** were described by Rowe in 1937,[17] and other early examples are known.[18] These six-membered heterocycles belong to a very large family of betaines which have common structural features and their known chemistry was reviewed in 1980.[18] Notable among these are the pyridinium-3-olates **11**. Katritzky and co-workers in the 1970s demonstrated that they undergo a wide variety of cycloadditions, dimerisations and rearrangements.[19]

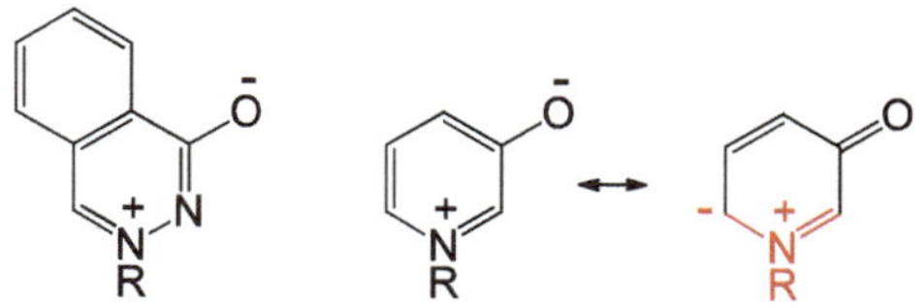

10 **11**

The structures **7–11** are all associated with a 1,3-dipole fragment and have the potential to participate in 1,3-dipolar cycloadditions. Collectively they belong to a very large class described as conjugated heterocyclic mesomeric betaines (HMBs).

In 1966/7 representatives of a new type of heterocyclic betaine were reported by Kheifets and co-workers[20] and by Prystaš and Šorm.[21] These are the 6-oxopyrimidinium-4-olates **12**. Although they are also HMBs, they have a very different electronic structure; the positive and negative charges are restricted to separate parts of the molecule. They were recognized as members of a discrete class of betaines containing cross-conjugated dipoles in 1974,[22] and are classified as cross-conjugated HMBs. A particularly significant difference is that they are associated with a 1,4-dipole fragment and have the potential to participate in 1,4-dipolar cycloadditions. This is readily seen by comparison with the 1,3-dipole fragment in the representative conjugated HMBs **13**, as shown in Fig. 1. During the 1970s and 1980s many new

Cross-conjugated HMBs **Conjugated HMBs**

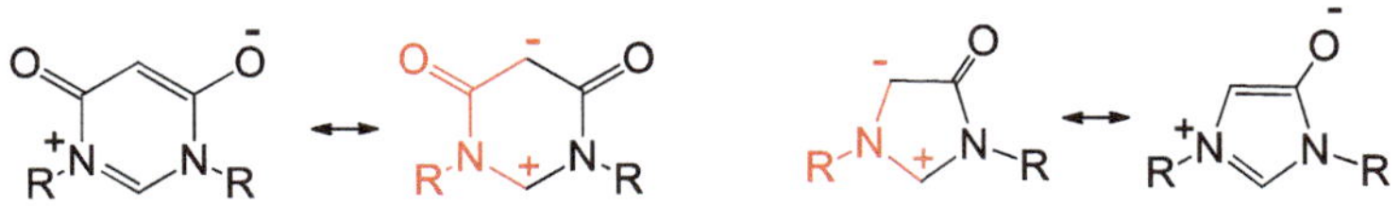

12 1,4-dipole 1,3-dipole **13**

Fig. 1 A comparison of heterocyclic mesomeric betaines containing 1,4-dipole and 1,3-dipole fragments.

examples of cross-conjugated HMBs were synthesized and their properties studied,[23] with notable contributions by the groups of Friedrichsen,[24] Kappe[25] and Potts.[26,27]

By the early 1980s it seemed that all HMBs could be classified as either conjugated HMBs or cross-conjugated HMBs and an overview of the subject classified HMBs in this way, with a separate category reserved for N-ylides.[23] And then, in 1984 Neugebauer, Fischer and Krieger described and fully characterized a deep blue crystalline compound which has the structure **14**.[28] These authors suggested that this molecule is "*a representative of a possibly large group of yet unknown betaines*." Indeed it is, and it is neither a conjugated HMB nor a cross-conjugated HMB, but so far no further examples have been forthcoming. In a systematic study of how 2π-donating heteroatoms are attached to the conjugated framework of HMBs, Ramsden showed that there are three discrete modes of connectivity corresponding to conjugated HMBs, cross-conjugated HMBs and a third group which he named semi-conjugated HMBs.[29,30] The molecule **14**, as far as we are aware, is the first and only example of a six-membered semi-conjugated HMB.

One hundred and forty years after Fischer and Besthorn reported the preparation of dehydrodithizone, HMBs can be recognized as belonging to three main classes (Fig. 2). To understand their properties it is important to understand how these classes differ. HMBs are many and varied. In what way do the isomeric rings **15**, **16** and **17** differ? What do the heteropentalene HMBs **18**, **19** and **20** have in common? To understand the nature of HMBs it is necessary to understand their common structural features and also the different modes of heteroatom connectivity that lead to the subdivisions of conjugated, cross-conjugated and semi-conjugated. These properties are explored in Sections 2–7.

A semi-conjugated HMB

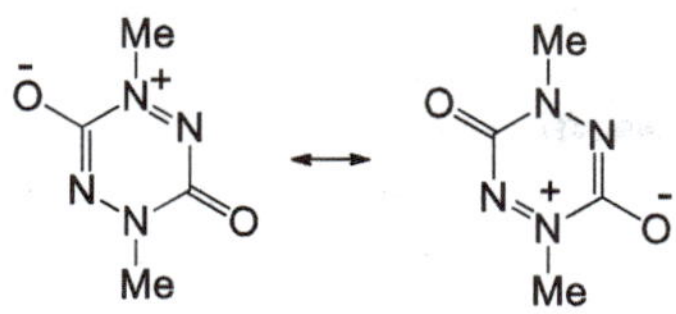

14

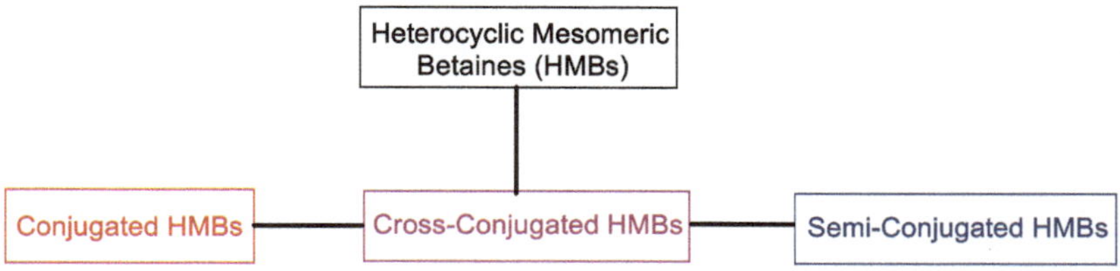

Fig. 2 The three main types of heterocyclic mesomeric betaine (HMB).

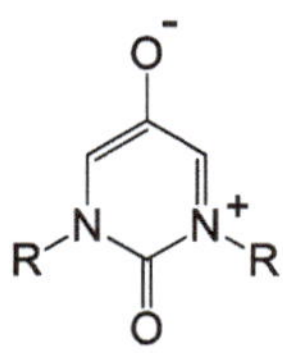

15

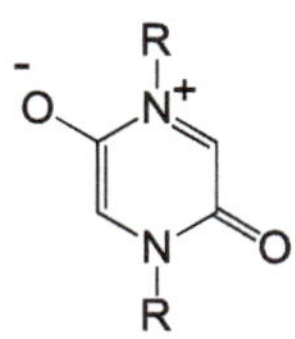

16

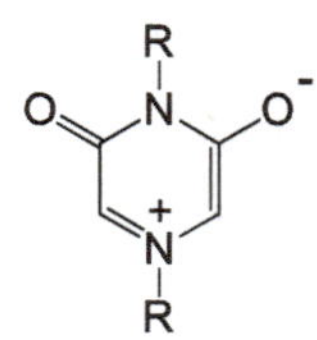

17

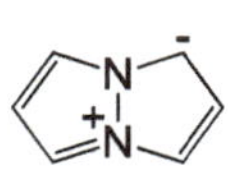

18

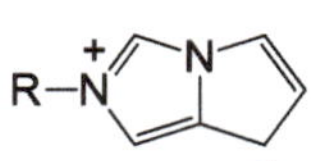

19

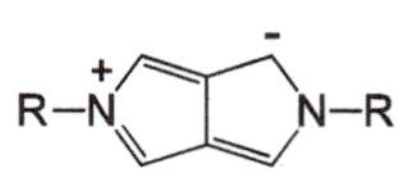

20

2. Heterocyclic mesomeric betaines with two 2π heteroatoms (Ẍ and Ÿ)—The common thread

This review is primarily concerned with heterocyclic mesomeric betaines (HMBs) that have two 2π-donating heteroatoms (Ẍ and Ÿ) in conjugation within the ring system, e.g., the nitrogen atoms in structures **15–20**. HMBs of this type can only be represented by dipolar structures because they contain pairs of conjugated fragments each with an odd number of atoms that are separated by the 2π-donating heteroatoms. HMBs with a single 2π-donating heteroatom, e.g., **10** and **11**, have different structural characteristics which have been discussed in detail elsewhere.[18] These features are briefly summarized in Section 7.

Odd fragments can only be represented as cations or anions (or radicals). Consider pyrazolo[1,2-*a*]pyrazole **21**. The structure contains two allyl fragments separated by nitrogens. Each odd fragment, shown as red and blue in resonance hybrids **22** and **23**, can only be represented as an anion or cation. The charges cannot cancel and the true structure is a resonance hybrid of many forms including **21**–**23**.

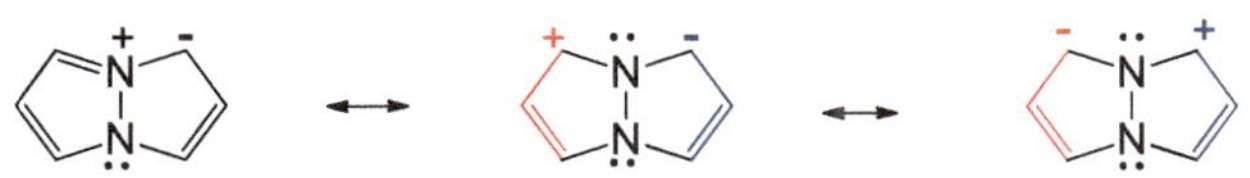

21 **22** **23**

A similar situation occurs with the structure of imidazolo[3,4-*a*]pyrroles represented by the resonance hybrids **24**–**26**. In this case there is an odd fragment (red) with one atom and an odd fragment (blue) with five conjugated atoms. Again, the structure is a hybrid of multiple dipolar forms.

Structures **21** and **24** are conjugated HMBs. However, the same structural features involving discrete odd fragments also occur in cross-conjugated and semi-conjugated HMBs as illustrated by structures **27** and **28**. This is a universal property of this class of HMBs which distinguishes them from neutral heterocycles that contain even conjugated systems, e.g., **29** and **30**. This does not preclude neutral heterocycles from being represented by dipolar structures, e.g., **31**, but they are not HMBs.

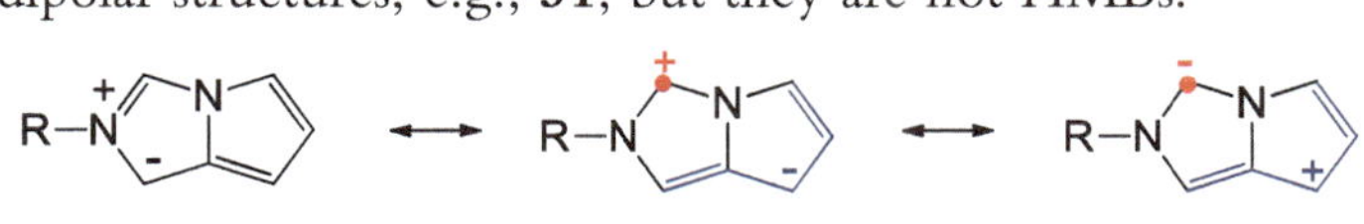

24 **25** **26**

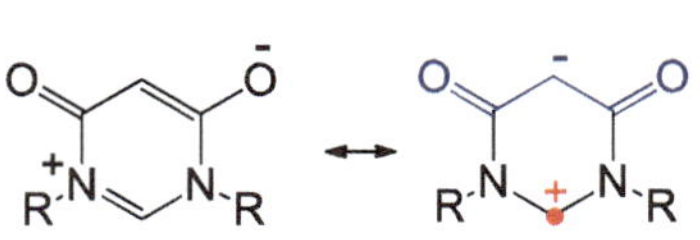

27

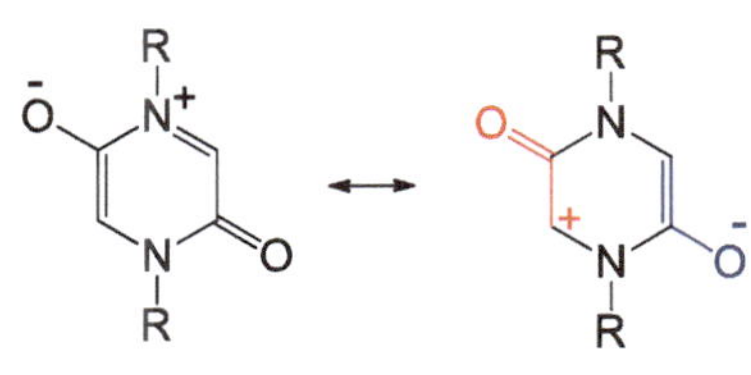

28

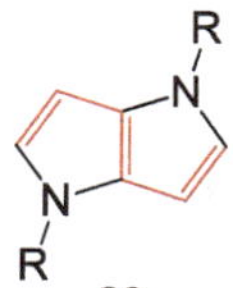

29

30 31

Having recognized this universal feature of these HMBs, there are significant general differences that distinguish the three types of HMB and this is related to the properties of odd conjugated chains. In particular, the atoms of these odd chains can be divided into two sets, starred (*) (the larger set) and unstarred (°), and atoms of each set are not bonded to each other. This is illustrated by the one-, three- and five-carbon chains shown in Fig. 3A. Importantly, the highest occupied molecular orbitals (HOMOs) of these chains are non-bonding molecular orbitals (NBMOs) that vanish on the unstarred atoms, as shown in Fig. 3B. Without delving too deeply into the bonding in HMBs, their properties are strongly influenced by whether the 2π-heteroatoms are attached to starred positions, where the NBMO has electron density (● or ○) or unstarred positions corresponding to a node.

Consider the structures **32–34** shown in Fig. 4. In structure **32** both nitrogen atoms are connected to starred positions of the odd fragments, and these positions are associated with electron density in the NMBO. This connectivity is characteristic of a conjugated HMB.

In the structure **33** both nitrogen atoms are connected to a starred position of one odd fragment (red) but to unstarred nodal positions of the other odd fragment (blue). This bonding is characteristic of a cross-conjugated HMB.

There remains one other general possibility. In structure **34** the nitrogen atoms are attached to starred positions of different odd fragments and to an unstarred position of the other fragment. This is characteristic of a semi-conjugated HMB.

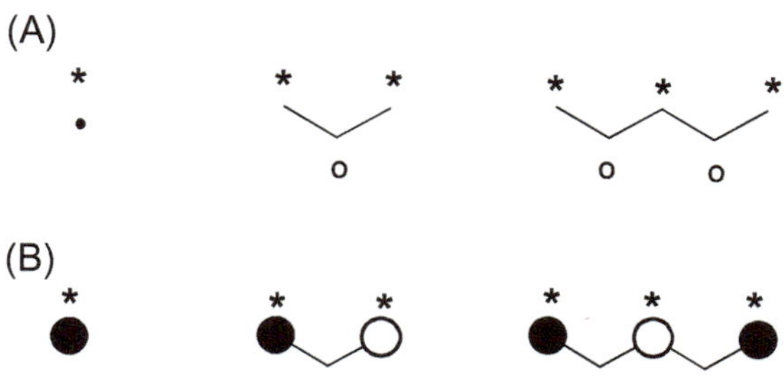

Fig. 3 Odd conjugated atoms showing (A) starred and unstarred positions and (B) the nature of associated non-bonding molecular orbitals (NBMOs).

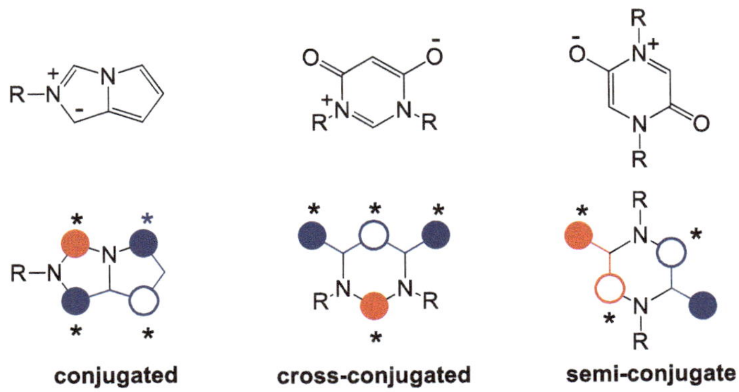

32 **33** **34**

Fig. 4 Odd conjugated fragments and associated NBMOs in representative HMBs.

Before summarizing the characteristics of the three types of HMB, there is one other situation that arises and this is illustrated by the pyrrolo[3,4-*c*] pyrroles **35** ↔ **36**. Superficially the heteroatoms appear to be bonded to a six-carbon even fragment, but this in fact is two cross-conjugated odd fragments linked by unstarred atoms **37**. The six atom fragment is derived from a non-Kekulé hydrocarbon which can only be represented by diradical or dipolar structures. These are illustrated by the 6- and 10-atom examples **38** and **39**. Cross-conjugated non-Kekulé systems are made up of odd-numbered components linked by unstarred atoms. The two odd fragments are independent (cross-conjugated). The HMBs **35** are in fact conjugated HMBs **37** in which the heteroatoms are both attached to starred positions of two cross-conjugated odd fragments. It is a special case of the odd fragments being unconnected, as in the previous examples.

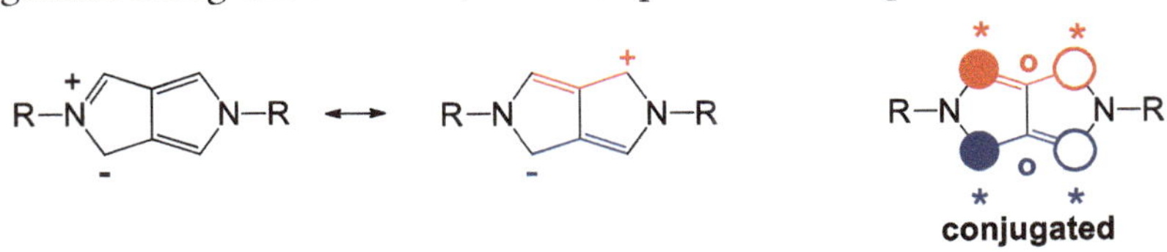

35 **36** **37**

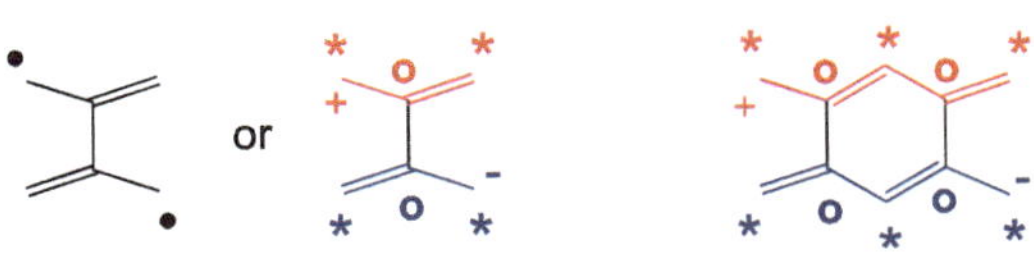

38 **39**

A second example of heterocycles containing a non-Kekulé fragment is the pyrazolium-4-carboxylates **40**. In this case both nitrogens are only attached to one odd fragment (red) and are completely isolated from the second (blue) **41**. These derivatives are therefore cross-conjugated HMBs.

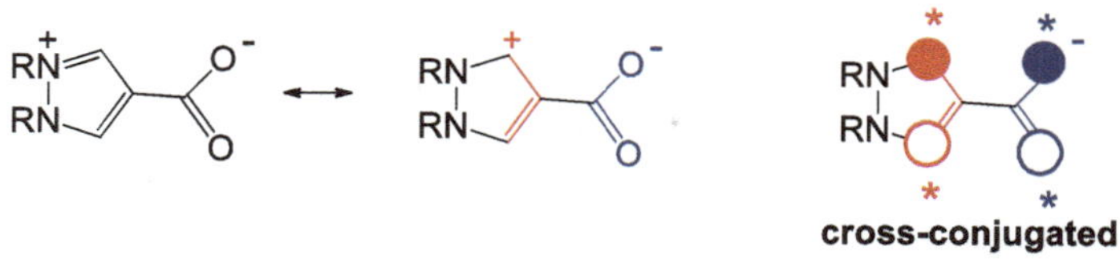

40 **41**

The general situation can be summarized as follows:

- HMBs are characterized by odd conjugated fragments that are either mutually isolated by intervening heteroatoms or are mutually isolated by cross-conjugation.
- In conjugated HMBs at least one heteroatom is attached to a starred position of both the odd fragments that constitute the HMB framework, e.g., **32** and **37**.
- In cross-conjugated HMBs heteroatoms are attached to starred positions of one odd fragment but neither heteroatom is attached to a starred position of the other odd fragment that constitutes the HMB framework, e.g., **33** and **41**.
- In semi-conjugated HMBs the two heteroatoms are each attach to different odd fragments by a starred position but not to a starred position of the other odd fragment, e.g., **34**.

3. Conjugated heterocyclic mesomeric betaines (Class 1)—The three types

In an analysis of the different ways that odd conjugated fragments can be connected to a pair of heteroatoms, seven different ways were identified.[29,30] These were represented by seven so-called connectivity matrices. This leads naturally to recognition of distinct classes of HMB. Conjugated HMBs (Class 1), cross-conjugated HMBs (Class 2) and semi-conjugated HMBs (Class 3). Class 1 and Class 2 each naturally divide into three subclasses: Class1A-C and Class 2A-C. Together with semi-conjugated HMBs (Class 3) (no subclasses), these make up the seven modes of heteroatom connectivity. In this review the differences between these classes and subclasses is illustrated by interactions between heteroatom lone pairs and NBMOs. An alternative view employing connectivity matrices can be found elsewhere.[30]

3.1 Conjugated HMBs—Subclass 1A

In Class 1A conjugated HMBs both of the 2π heteroatoms are flanked by starred atoms and both are the centre of 1,3-dipole fragments. This is illustrated by the heteropentalenes **42–44**.[31] The alternative 1,3-dipoles and the NBMOs are shown in Fig. 5.

The presence of 1,3-dipolar fragments does not guarantee participation in 1,3-dipolar cycloadditions, other factors are involved, but many examples are known.[31] In some cases both possible modes have been observed. For example, when the thieno[3,4-*c*]pyrrole **46** is reacted with fumaronitrile at 80 °C the azomethine ylide adduct **45** (kinetic product) is formed whereas at 110 °C the thiocarbonyl ylide adduct **47** (thermodynamic product) is obtained (Scheme 1).[32] Derivatives of the system **43** and **44** also form cycloaddition products at the alternative 1,3-dipole sites.[33,34]

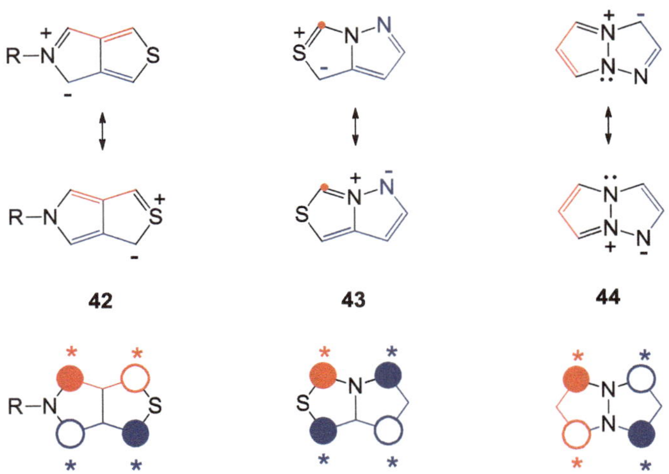

Fig. 5 Examples of the alternative 1,3-dipolar fragments in Class 1A HMBs.

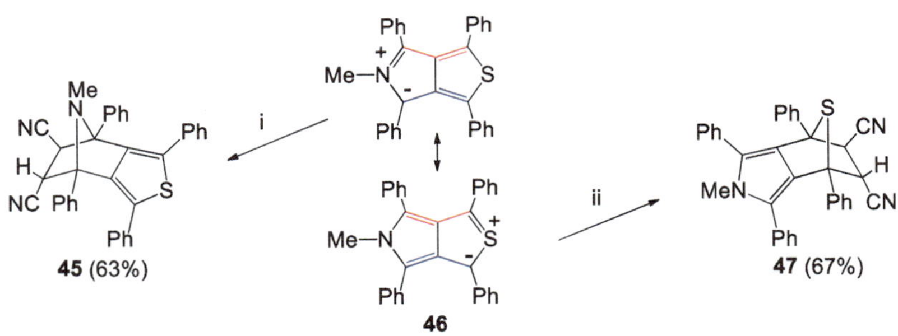

Reagents: i, fumaronitrile, benzene (80°); ii, fumaronitrile, toluene (110°).

Scheme 1 The alternative modes of 1,3-dipolar cycloaddition in a Class 1A HMB.

3.2 Conjugated HMBs—Subclass 1B

The best known examples of Class 1B conjugated HMBs are type A mesoionic rings typified by the münchnones **48**.[13,14] In these rings only one of the two heteroatoms is flanked by two starred positions and thus constitutes a 1,3-dipolar fragment. The second heteroatom is linked to one starred atom and one unstarred **49**. 1,3-Dipolar cycloaddition of type A mesoionic rings, including the münchnones **48**, have been studied extensively.[2,35]

48 münchnones

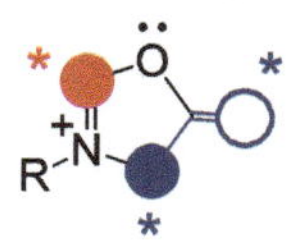

49

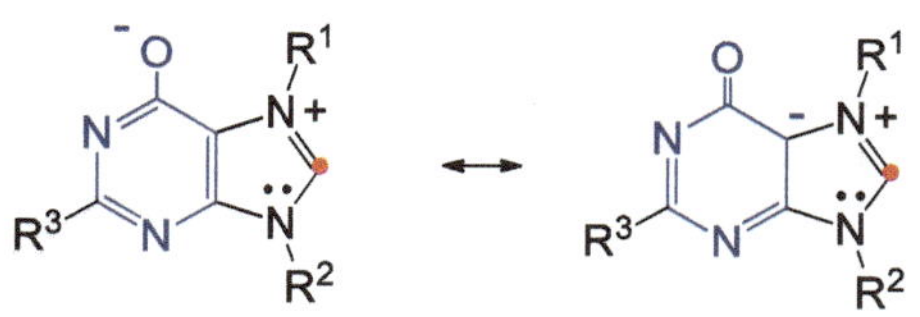

50

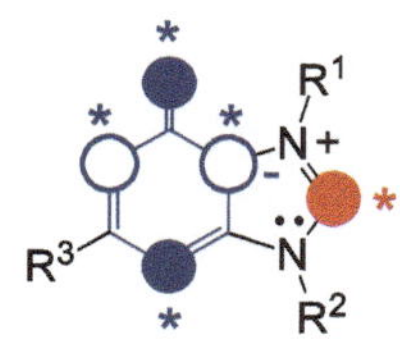

51

A second example is provided by the bicyclic heterocycles **50**. Again, only one of the 2π heteroatoms is flanked by two starred atoms **51**. In this case 1,3-dipolar cycloadditions by the azomethine ylide fragment is unlikely due to the ring fusion. Examples of this type include the natural product herbipolin **50** ($R^1 = R^2 = Me$, $R^3 = NH_2$) and the derivative **50** ($R^1 = Me$, $R^2 =$ D-ribosyl, $R^3 = NH_2$), which plays a role in the biochemistry of messenger RNA.[23] Other examples of this class are the polycyclic rings **52**[36,37] and **53**,[38] which are drawn to show the 1,3-dipole fragments.

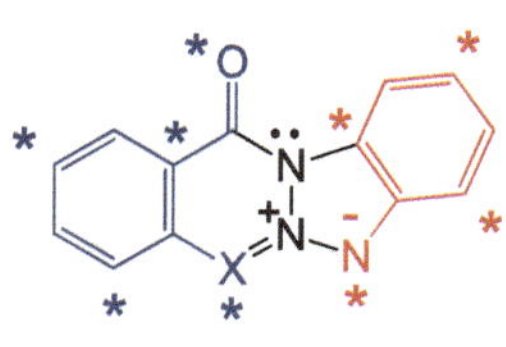

52 (X = CR, N)

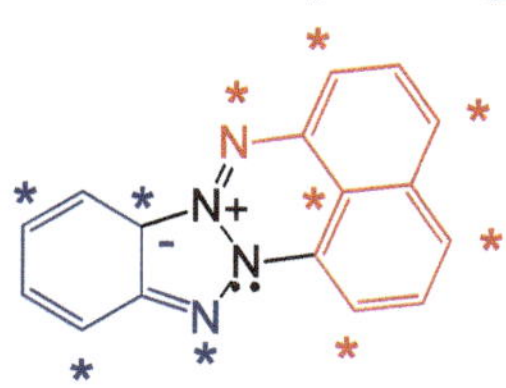

53

3.3 Conjugated HMBs—Subclass 1C

In Class 1C the second heteroatom is exclusively bonded to unstarred positions and examples of this type are quite rare. Examples of the ring system **54**, including the derivative **54** (Ar=Ph, R=SPh), have been prepared and participation of the azomethine ylide fragment (shown in structure **55**) in 1,3-dipolar cycloadditions has been demonstrated.[39,40]

54 **55**

4. Cross-conjugated heterocyclic mesomeric betaines (Class 2)—Another three types

Cross-conjugated HMBs (Class 2) are also associated with three modes of connection of the 2π heteroatoms and are sub-classified as Classes 2A, 2B and 2C.

4.1 Cross-conjugated HMBs—Subclass 2A

The best known examples of this class are heterocycles with the general structures **56–59**, where X or Y=RN, O, S, Se. Their chemistry has been reviewed,[23–25] and a common feature is their participation in 1,4-dipolar cycloadditions. The defining feature of these rings is the interaction of both the 2π heteroatoms with starred positions of one fragment but unstarred nodal positions of the second, as shown in structures **60** and **61**. Unlike conjugated HMBs, there is little interaction between the positive and negative fragments of the rings. This is nicely illustrated by the crystal structure of the tetraphenyl derivative **62** in which the N1-C6 and N3-C4 bonds are essentially single bonds (1.48 Å) and not involved in cyclic delocalisation. This observation is also consistent with the low calculated values of indexes of classical aromaticity for the 1,3-dimethyl derivative **56** (X=Y=NMe); the Bird index (I_6 22) and HOMA (0.230) values are both low and the aromatic stabilization energy (ASE) has been calculated to be only 3.8 kcal mol^{-1}. The calculated ASE value for the isomeric conjugated HMB **54** (Ar=Me, R=H) is 13.9 kcal mol^{-1}.[41]

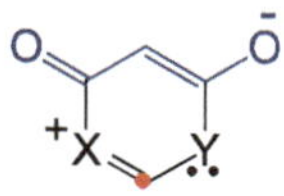

56

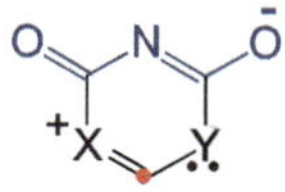

57

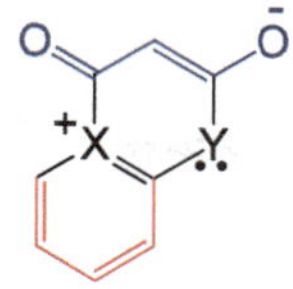

58

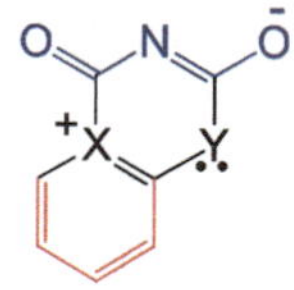

59

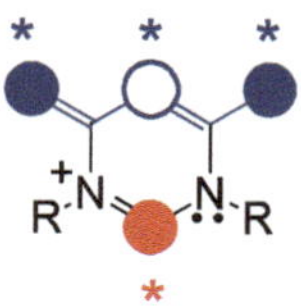

60

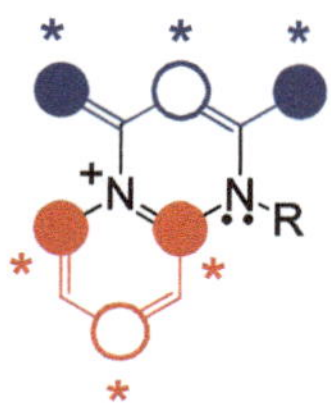

61

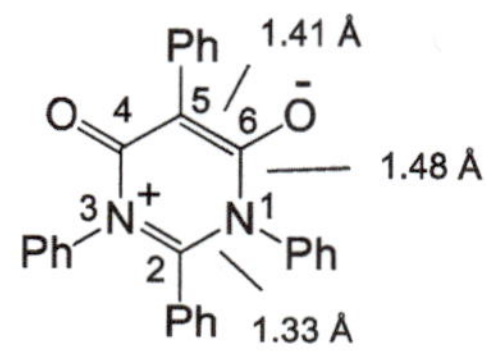

62

Further examples of cross-conjugated HMBs belonging to Class 2A are the structures **63**,[42] **64**[43–45] and **65**.[46] Note that in structure **65** both nitrogen atoms are attached to at least one starred atom of the cationic fragment.

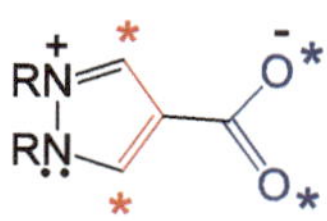

63

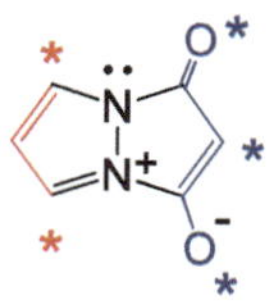

64

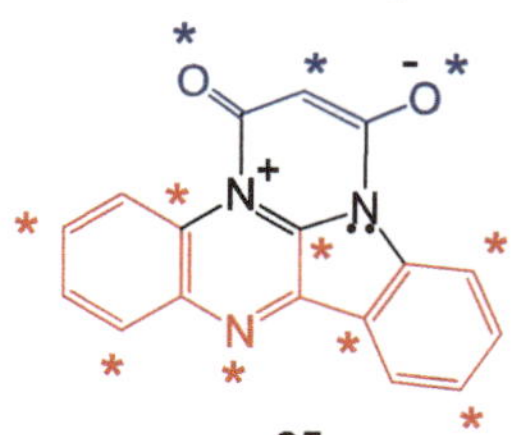

65

4.2 Cross-conjugated HMBs—Subclass 2B

Class 2B cross-conjugated HMBs have only one of the 2π heteroatoms attached to a starred atom of the cationic fragment. Examples are the structures **66** and **67**.[43] There is insufficient knowledge of this class to conclude whether their properties are significantly different to those of Class 2A cross-conjugated HMBs.

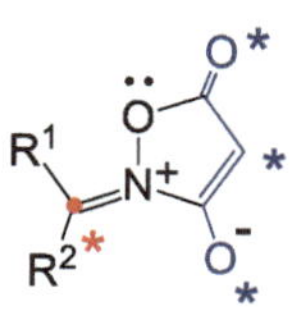

66

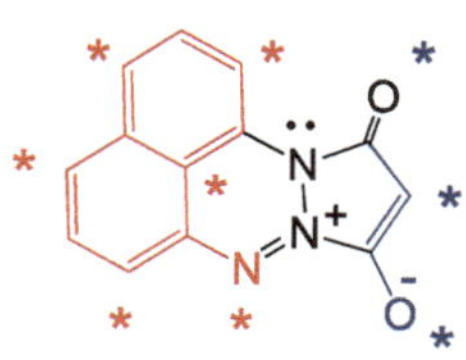

67

4.3 Cross-conjugated HMBs—Subclass 2C

As far as we are aware no Class 2C cross-conjugated HMBs have been prepared. The structure **68** is an illustrative example. In this class both 2π heteroatoms are exclusively connected to unstarred positions of both the anionic and cationic fragments as shown in structure **69**.

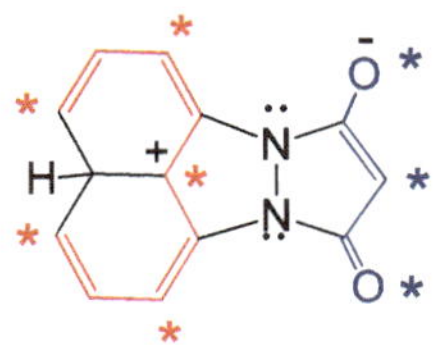

68

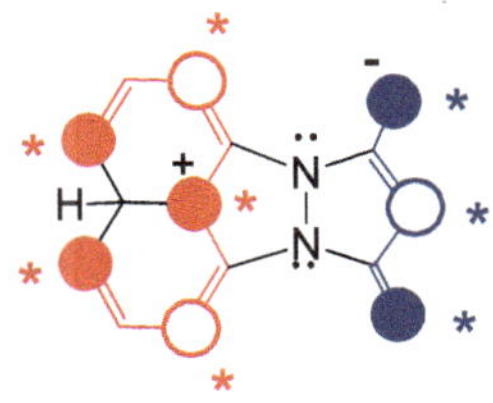

69

5. Semi-conjugated heterocyclic mesomeric betaines (Class 3)

Only one general type of semi-conjugated HMB occurs (Class 3). The only known example is the tetrazine **70** which has the heteroatom connectivity shown in structure **71**. It has been fully characterized, including an X-ray crystal structure **72**. The molecule is planar and symmetrical. It is interesting to note that the C–N bonds are intermediate between single and double bonds (compare structure **62**) whereas the CO and NN bonds approach double bond dimensions. The molecule readily forms a radical anion and protonates in acid.[28] Cycloadditions have not been investigated.

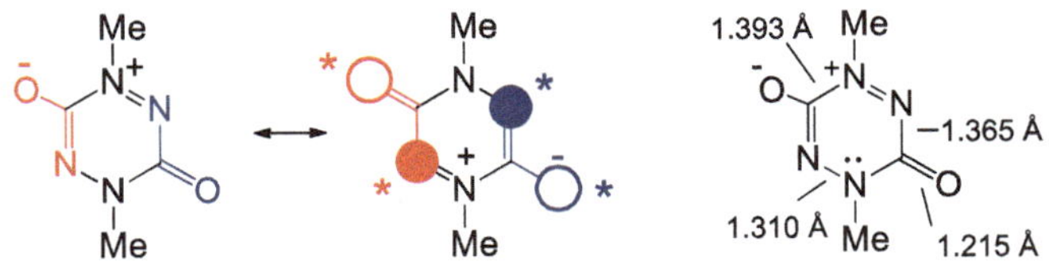

70 **71** **72**

Neugebauer, Fischer and Krieger recognized that structure **70** is "*a representative of a possibly large group of yet unknown betaines*,"[28] but further examples are yet to be described. The simplest examples would be the dioxopiperazine derivatives **73**. It is interesting to note that although many naturally occurring epidithiodioxopiperazines **74**, which are formally dithio adducts of HMBs **73**, have been investigated, none of these studies appear to have uncovered examples of the betaines **73**, either by accident or design.[47] Further examples would be the derivatives **75**.

73

74

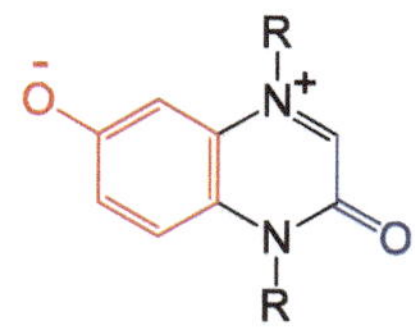

75

Fig. 6 shows calculated properties of the semi-conjugated HMBs **76** and **77** in the gas phase. In the diaza analogue **77** the CO bond lengths are shortened and the ring CN bonds lengthened relative to those in structure **76**. This is attributable to the electronegative ring nitrogens attracting greater electron density into the ring. This also reduces the aromatic stabilization energy (ASE)(Fig. 5) by increasing the antiaromatic character. The ASE of HMB **76** is greater than that of conjugated and cross-conjugated HMB isomers.[41] A more detailed discussion of the experimental and calculated properties of these ring systems has been published.[30]

76 **77**

ASE 18.1 kcal mol^{-1} 6.5 kcal mol^{-1}

Fig. 6 Calculated geometries and aromatic stabilization energies (ASE) of the HMB **76** and its diaza analogue **77**.[41]

6. The curse of pseudo-cross-conjugation

It would be convenient if a discussion of the types of HMBs ended here, but it cannot. There is a complication relating to cross-conjugation that is of little significance most of the time but, for a full understanding and rigorous classification of HMBs, cannot be ignored.

Cross-conjugate non-Kekulé structures were discussed in Section 2. The odd fragments are linked by unstarred positions and cationic and anionic fragments are mutually insulated as shown in structures **78a** and **78b**.

However, there is a second type of non-Kekulé structure in which two odd fragments are linked by an unstarred atom on one and a starred atom on the other. Again, only dipolar (or diradical) structures can be drawn. This is illustrated by structures **79a** and **79b**. In these cases resonance structures can be drawn in which the charges are transferred to the second fragment, e.g., **79c** and **79d**. Non-Kekulé structures of this type are referred to as pseudo-cross-conjugated.[30]

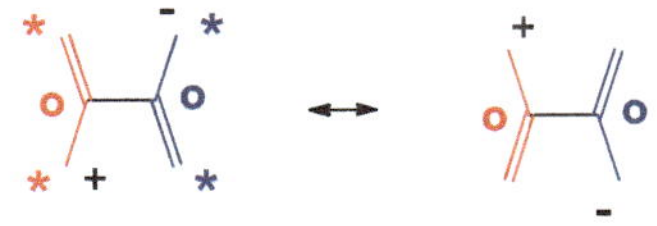

78a **78b**

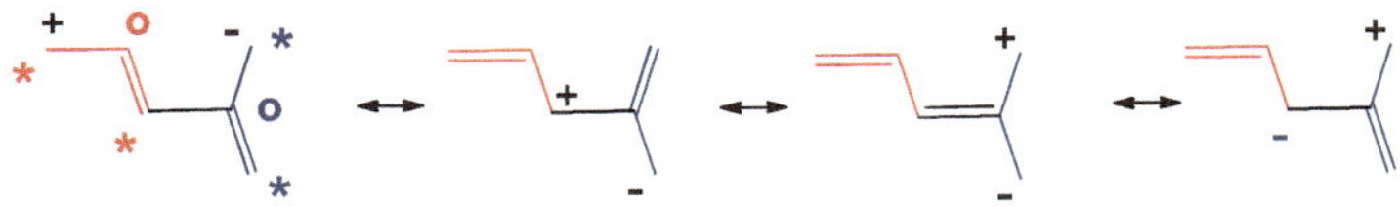

79a **79b** **79c** **79d**

The term pseudo-cross-conjugated HMB was coined by Ollis, Ramsden and Stanforth to describe HMBs containing this type of non-Kekulé fragment and a detailed valence bond analysis of examples of pseudo-cross-conjugated HMBs is included in a 1985 review of HMBs.[23] Although pseudo-cross-conjugated HMBs have similar properties to cross-conjugated HMBs, including charge separation, a consequence of the bonding is that some negative charge transfers to the positive fragment which is illustrated by resonance hybrid **79d**. For pseudo-cross-conjugated HMBs the positive and negative charges are effectively but not exclusively in separate fragments and this may have an influence on some of their properties.

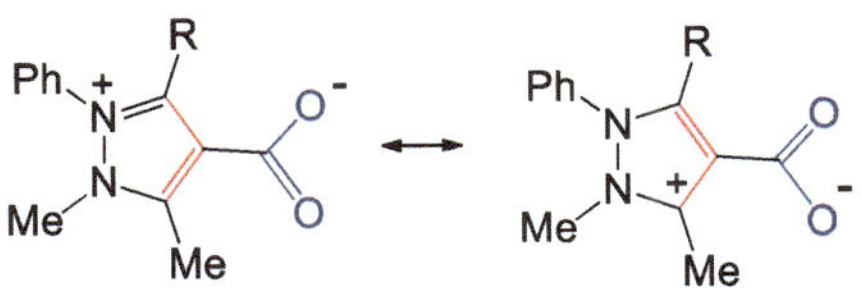

80a **80b**

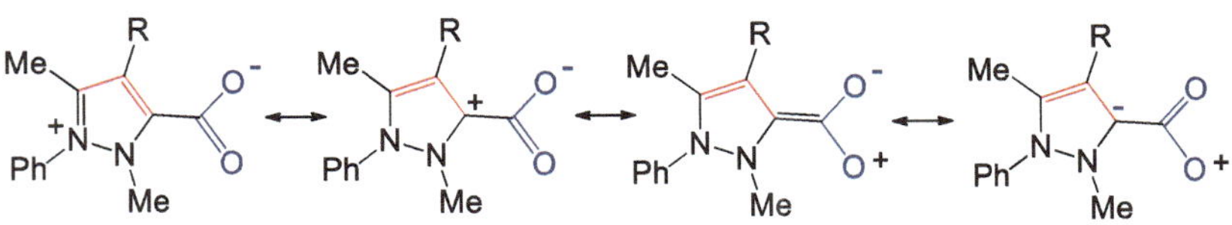

81a **81b** **81c** **81d**

Subtle differences between these two types of HMBs are illustrated by the work of Schmidt and coworkers who have studied a number of examples. Isomeric pyrazolium carboxylates **80** and **81** are of some interest. In both structures neither heteroatom (NR) is directly bonded to the blue fragment. The 4-carboxylates **80** are cross-conjugated HMBs whereas the 3-carboxylates **81** are pseudo-cross-conjugated HMBs. A Raman spectroscopy study of the isomers **80** and **81** (R=H) showed a slight interaction across the C3-CO_2 bond in the isomer **81** which was attributed to contributions of resonance hybrids of the type **81c, d**.[48]

The subtle difference between the isomeric pyrazolium carboxylates **80** and **81** is also illustrated by their ease of decarboxylation. The decarboxylation of the cross-conjugated HMBs **80** to the N-heterocyclic carbenes (NHCs) **82** has been observed in a mass spectrometer and the energy barrier to decarboxylation calculated to be 120 kJ mol^{-1}. In contrast, the pseudo-cross-conjugated HMBs decarboxylate to give the NHCs **83** under much milder condition (e.g., gentle heating in toluene) (Scheme 2).[49]

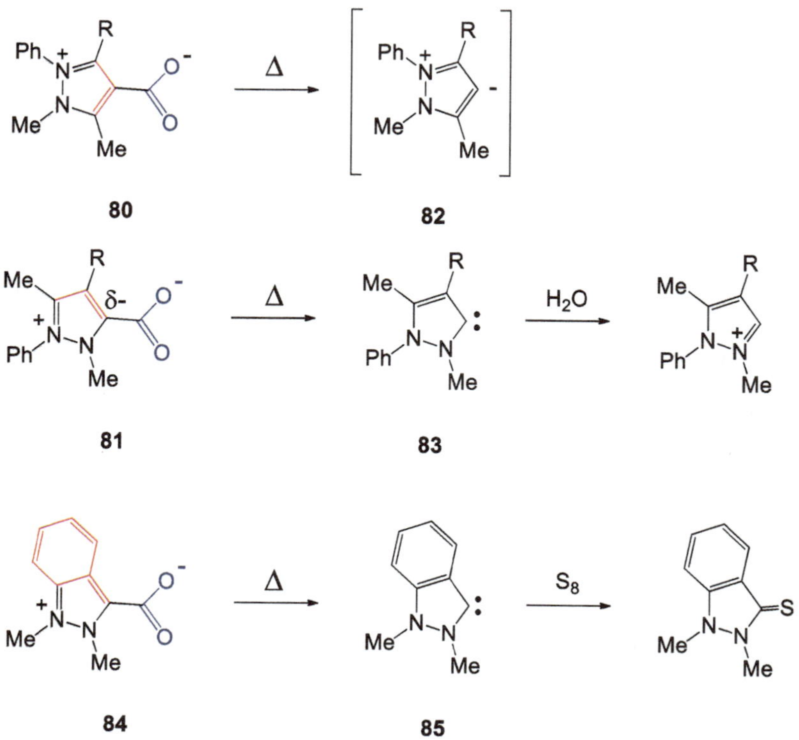

Scheme 2 Decarboxylation of cross-conjugated HMBs **80** and pseudo-cross-conjugated HMBs **81** and **84**.

In both types of HMB **80** and **81** the negative charge is associated with the highest occupied molecular orbital (HOMO) which is effectively restricted to the anionic fragments. In the pseudo-cross conjugated systems calculations suggest that a tiny amount of negative charge is on the C3 carbon atom of the ring, e.g., **81** (Scheme 2).[50] Thus, in pseudo-cross-conjugated molecules the separation between positive and negative segments is largely maintained with resonance hybrids **81a, b** being dominant. The bonds linking cross-conjugated fragments are often referred to as union bonds and are long, e.g., **62**. In the HMB **84** the union bond was found to be 1.519 Å with a torsion angle of 21.5°, consistent with charge separation characteristic of cross-conjugated systems.[51] Further details of pseudo-cross-conjugated HMBs and their relation to N-heterocyclic carbenes can be found in excellent reviews by the Schmidt group.[50,52]

Other examples of pseudo-cross-conjugated HMBs are the structures **86** and **87**. Potts and co-workers have made a theoretical and experimental study of several systems including the pseudo-cross-conjugated HMB **87** and compared its properties with the cross-conjugated HMB **88**. The experimental dipole moments of both molecules (**87**; 6.88 D and **88**; 6.42 D) are similar and consistent with a high degree of charge separation in both molecules. Spectroscopic and kinetic studies also support the view that both molecules have discrete anionic and cationic fragments and many of their properties can be attributed to ionic character and the weak nature of the union bonds.[26,27]

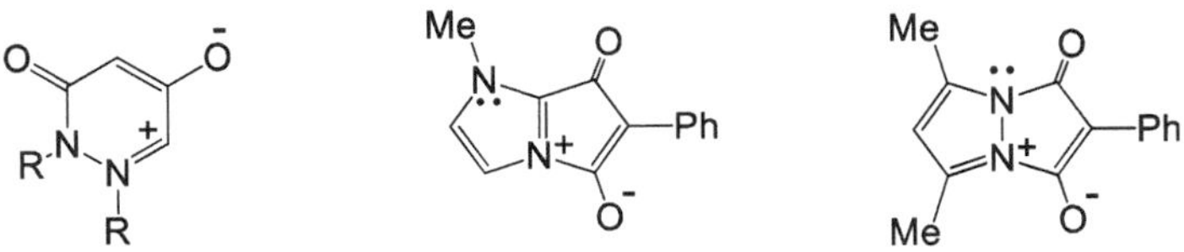

86 **87** **88**

In truth, pseudo-cross-conjugated HMBs are rare species. Based on the coverage in a 1985 review,[23] and with the exception of the molecules discussed above, few examples have been characterized or their reactions explored. It is important to recognize that they are different to cross-conjugated HMBs but for general practice the distinction is not critical.

Similar structural components can also give rise to the possible occurrence of pseudo-semi-conjugated HMBs. In fact type B mesoionic compounds are now regarded as pseudo-semi-conjugated HMBs,[30] and an analysis of their structure can be found in Chapter 4 of this volume.[4]

At the beginning of this section we referred to pseudo-cross-conjugation as a curse, primarily because of its potential to add confusion to an area which is sometimes lacking in structural discrimination and precise nomenclature. Nevertheless, it cannot be ignored and the preliminary classification of HMBs shown in Fig. 2 must be revised to include pseudo-cross-conjugation, as shown in Fig. 7.

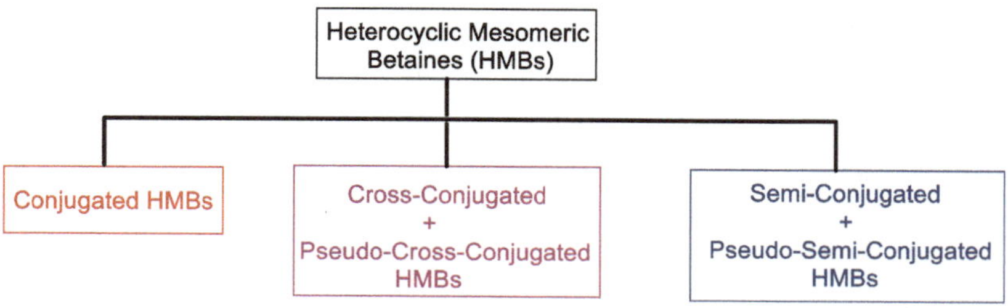

Fig. 7 The classification of Heterocyclic Mesomeric Betaines (HMBs).

7. Heterocyclic mesomeric betaines with one 2π heteroatom ($\ddot{X}$)

7.1 Conjugated HMBs

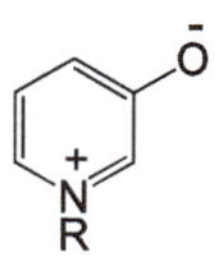

89

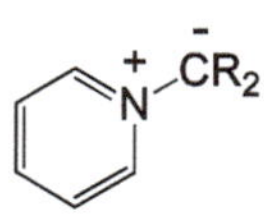

90

Many HMBs contain a single 2π heteroatom. In these molecules odd fragments cannot be isolated between a pair of heteroatoms and a different "common thread" defines them. Typical examples are the pyridinium-3-olates **89** and the N-ylides **90**, both of which are isoconjugate with the benzyl anion **91**.

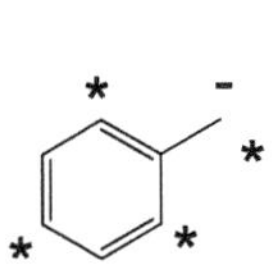

91

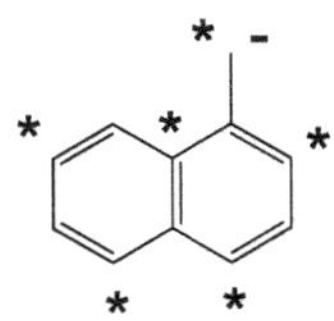

92

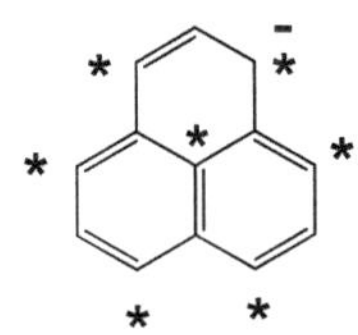

93

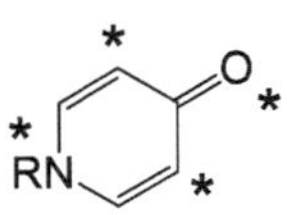

94

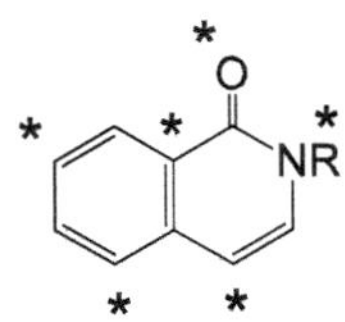

95

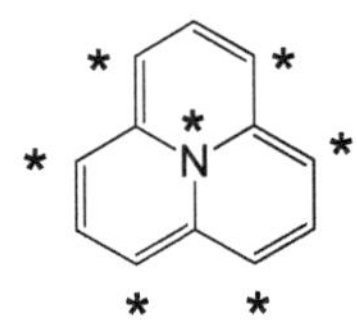

96

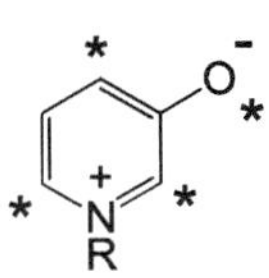

97

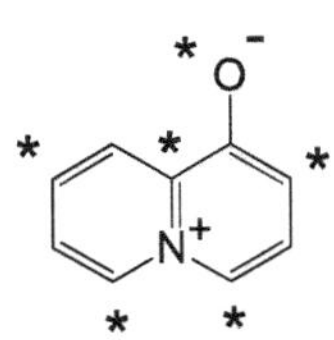

98

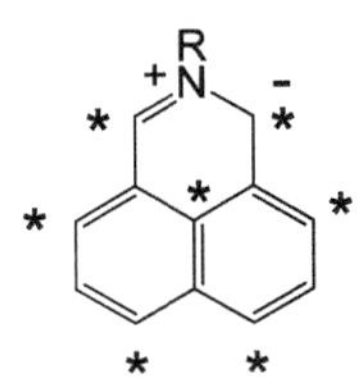

99

The relationship of structures **89** and **90** with the benzyl anion **91** is significant because this anion belongs to a large family, known as alternant hydrocarbon anions, in which the conjugated carbon atoms can be divided into starred and unstarred atoms such that pairs of starred or unstarred are not directly connected. This is demonstrated by the anions **91–93**. The starred atoms are always the larger set and a double bond must always link a starred and unstarred atom. In alternant hydrocarbon anions the starred atoms (n + 1) always exceed the unstarred atoms (n) by one. If we make an isoconjugate heterocycle by inserting a 2π heteroatom at a starred position, leaving equal numbers (n) of starred and unstarred atoms, they can all be linked by double bonds giving fully covalent structures, e.g., **94–96**. If the heteroatom is placed at an unstarred position the remaining n + 1 starred and n-1 unstarred atoms cannot be matched to form a fully covalent structure and only dipolar structures can be drawn, e.g., **97–99**.

Their relationship to alternant hydrocarbon anions defines an extensive family of HMBs. Furthermore, the 2π heteroatom is always flanked by starred positions which means that they are conjugated HMBs and associated with 1,3-dipoles. Since alternant hydrocarbon anions are associated with a

HOMO that is a non-bonding molecular orbital which is restricted to the starred position, the corresponding HMBs are associated with high energy HOMOs and associated reactivity.

The general properties of this large family of conjugated HMBs were discussed and the chemistry of know systems reviewed in 1980.[18] Further examples of this large family include the azaberbinone **100**,[53] and the alkaloid corrunine **101**.[54]

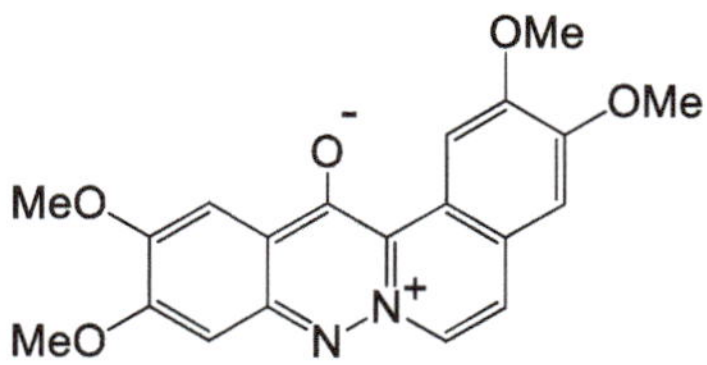

100

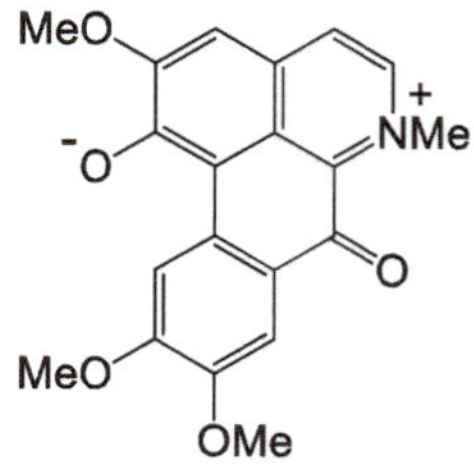

101

7.2 Cross-conjugated and pseudo-cross-conjugated HMBs

In contrast to the conjugated HMBs discussed in Section 7.1, cross-conjugated species are rare and have received little attention. The simplest example is the natural product trigonelline **102**. Examination of the resonance hybrid **103** reveals that the positive and negative fragments are both linked through unstarred positions and the charges are restricted to separate portions of the molecule. A similar situation arises with the pyridinium-3-phenolate **104**.[55] Quinolinium derived HMBs employing ethynyl spacers have also been prepared.[56]

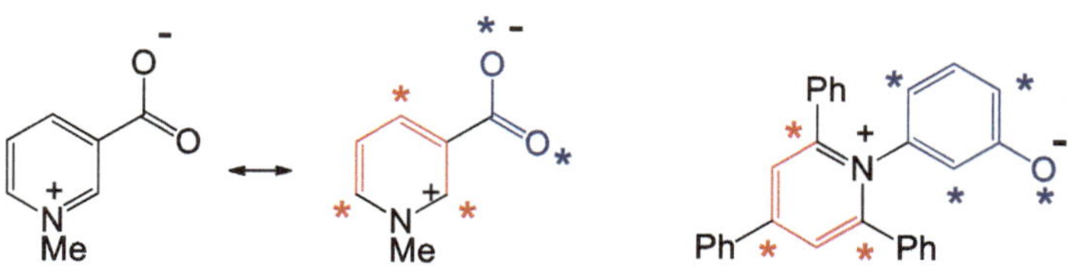

102 **103** **104**

Pseudo-cross-conjugated HMBs of this type are equally rare. Examples are the trigonelline isomer **105** and the derivatives **106**,[57] in which the pairs of odd fragments (red and blue) are pseudo-cross-conjugated.

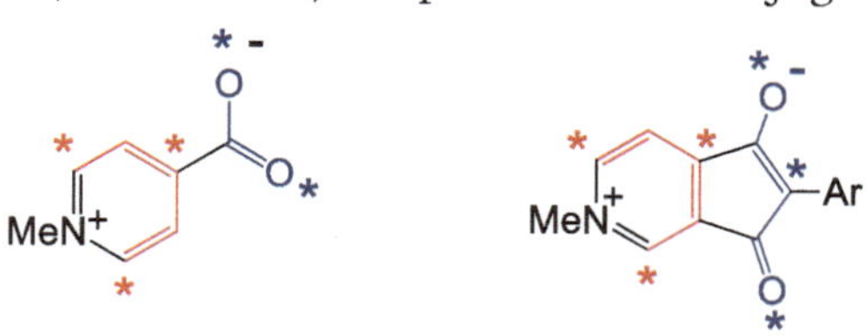

105 **106**

8. Conclusions

The systematic classification of HMBs, as described in the preceding sections, provides the opportunity of better understanding their properties and, in particular, similarities and differences in their structures and reactions. It also highlights areas that have not been explored.

We conclude this review with a brief analysis of another HMB that was first made in the 19th century. Identifying the class and subclass of individual HMBs needs a systematic approach and the recognition of the key structural fragments. The diphenylquinoxaphenazine **107** (5,7-DPQP) was first reported by Kehrmann and Bürgin in 1896.[58,59] More recently its properties have been investigated by Koutentis and coworkers and formulated as the" zwitterion" **107**.[60,61] This molecule is clearly a HMB—but what type? Inspection of the resonance hybrid **108** reveals two odd fragments (shown as red and blue) which are pseudo-cross-conjugated; the blue fragment is linked to the red fragment by unstarred positions. It can be seen from structure **108** that each of the 2π heteroatoms (PhN) is flanked by starred positions which clearly identifies it as a Class 1A conjugated HMB. Accordingly, resonance hybrids can be drawn showing each PhN as part of a 1,3-dipole, e.g., **109** and **110**.

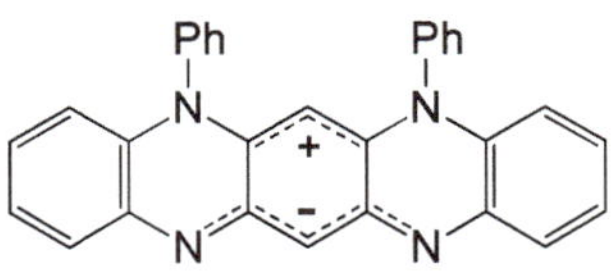

107

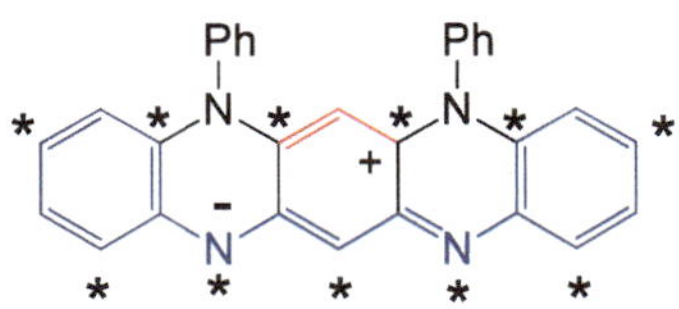

108

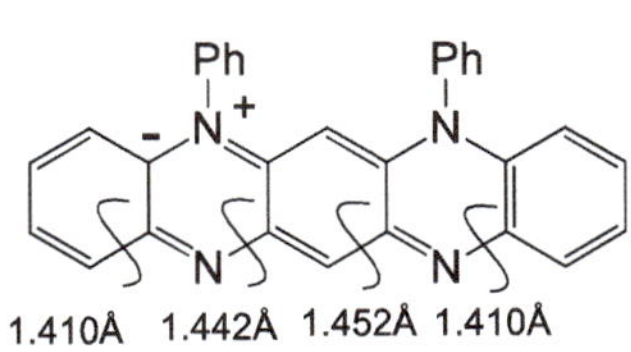

110

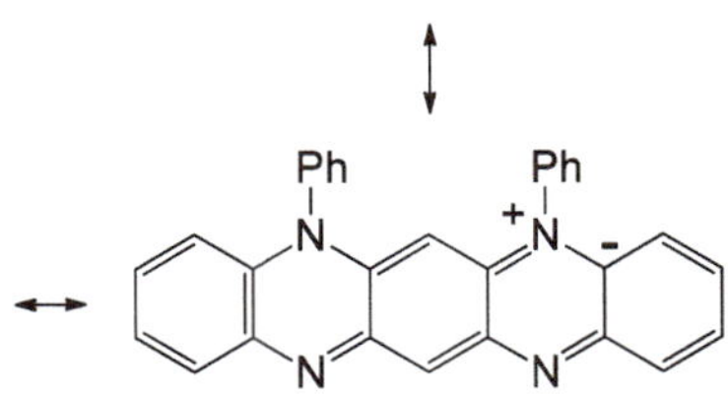

109

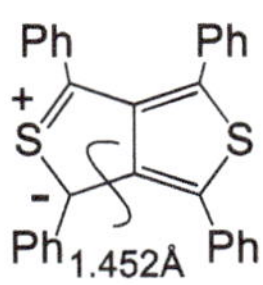

111

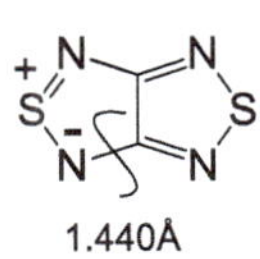

112

The pseudo-cross-conjugation results in minimal π bonding between the two odd fragments. The structure of 5,7-DPQP, determined by X-ray crystallography,[60] demonstrates that the union bonds are significantly lengthened as shown in structure **110**. Similar lengthening is observed for the union bonds in the thienothiophene **111**[62] and the thiadiazolothiadiazole **112**[31] in which the allylic fragments are cross-conjugated. Like 5,7-DPQP (**108**), the molecules **111** and **112** are Class 1A conjugated HMBs (Section 3.1) in accord with the general rules presented in Section 2.

We conclude with a reiteration of these rules.

- *Heterocyclic Mesomeric Betaines* (HMBs) with two 2π heteroatoms are characterized by odd conjugated fragments that are either mutually isolated by intervening 2π heteroatoms or are mutually isolated by cross-conjugation or pseudo-cross-conjugation.
- In *conjugated HMBs* at least one heteroatom is attached to a starred position of both the odd fragments that constitute the HMB framework.
- In *cross-conjugated and pseudo-cross-conjugated HMBs* heteroatoms are attached to starred positions of one odd fragment but neither heteroatom is attached to a starred position of the other odd fragment that constitutes the HMB framework.
- In *semi-conjugated HMBs* the two heteroatoms are each attached to different odd fragments by a starred position but not to a starred position of the other odd fragment.

References

1. Fischer, E.; Besthorn, E. *Justus Liebigs Ann. Chem.* **1882**, *212*, 316.
2. Ollis, W. D.; Ramsden, C. A. *Adv. Heterocycl. Chem.* **1976**, *19*, 1.
3. Oziminski, W. P.; Ramsden, C. A. *Struct. Chem.* **2021**, accepted for pub. STUC-D-21-00446R1.
4. Dumitrascu, F.; Ramsden, C. A. *Adv. Heterocycl. Chem.* **2022**, *137*.
5. Busch, M. *Ber. Dtsch. Chem. Ges.* **1905**, *38*, 856.
6. Schönberg, A. *J. Chem. Soc.* **1938**, 824.
7. Busch, M. *Ber. Dtsch. Chem. Ges.* **1905**, *38*, 861.
8. Cope, W. C.; Barab, J. *J. Am. Chem. Soc.* **1917**, *39*, 504.
9. Färber, C.; Leibold, M.; Bruhn, C.; Maurer, M.; Siemeling, U. *Chem. Commun.* **2012**, *48*, 227.
10. Grant, J. A.; Lu, Z.; Tucker, D. E.; Hockin, B. M.; Yufit, D. S.; Fox, M. A.; Kataky, R.; Chechik, V.; O'Donoghue, A. C. *Nat. Commun.* **2017**, *8*, 15088.
11. Baker, W.; Ollis, W. D. *J. Chem.Soc.* **1949**, 307.
12. Baker, W.; Ollis, W. D. *Quart. Rev.* **1957**, *11*, 15.
13. Huisgen, R. Aromaticity. *Chem. Soc. Spec. Publ.* **1967**, *21*, 51.
14. Reissig, H.-U.; Zimmer, R. *Angew. Chem. Int. Ed.* **2014**, *53*, 9708.
15. Padwa, A.; Sheehan, S. M.; Straub, C. S. *J. Org. Chem.* **1999**, *64*, 8648.
16. Padwa, A.; Bur, S. *Adv. Heterocycl. Chem.* **2016**, *119*, 241.
17. Rowe, F. M.; Adams, D. A. W.; Peters, A. T.; Gillam, A. E. *J. Chem.Soc.* **1937**, 90.

18. Ramsden, C. A. *Adv. Heterocycl. Chem.* **1980**, *26*, 1.
19. Dennis, N.; Katritzky, A. R.; Takeuchi, Y. *Angew. Chem. Int. Ed.* **1976**, *15*, 1.
20. Kheifets, G. M.; Khromov-Borisov, N. V.; Kol'tsov, A. I. *Zh. Org. Khim.* **1966**, *2*, 1516.
21. Prystaš, M.; Šorm, F. *Collect. Czech. Chem. Commun.* **1967**, *32*, 1298.
22. Ramsden, C. A. *J. Chem. Soc. Chem. Commun.* **1977**, 109.
23. Ollis, W. D.; Stanforth, S. P.; Ramsden, C. A. *Tetrahedron* **1985**, *41*, 2239.
24. Friedrichsen, W.; Kappe, T.; Böttcher, A. *Heterocycles* **1982**, *19*, 1083.
25. Kappe, T. *Heterocycles* **1984**, *21*, 358.
26. Potts, K. T.; Murphy, P. M.; Kuehnling, W. R. *J. Org. Chem.* **1988**, *53*, 2889.
27. Potts, K. T.; Murphy, P. M.; DeLuca, M. R.; Kuehnling, W. R. *J. Org. Chem.* **1988**, *53*, 2898.
28. Neugebauer, F. A.; Fischer, H.; Krieger, C. *Tetrahedron Lett.* **1984**, *25*, 629.
29. Ramsden, C. A. *Tetrahedron* **2013**, *69*, 4146.
30. Ramsden, C. A. *Prog. Heterocycl. Chem.* **2016**, *28*, 1.
31. Ramsden, C. A. *Tetrahedron* **1977**, *33*, 3203.
32. Potts, K. T.; McKeough, D. *J. Am. Chem. Soc.* **1974**, *96*, 4268.
33. Potts, K. T.; Marshall, J. L. *J. Org. Chem.* **1976**, *41*, 129.
34. Wang, Y.; Opsomer, T.; Dehaen, W. *Adv. Heterocycl. Chem.* **2022**, *137*.
35. Ramsden, C. A.; Dumitrascu, F. *Adv. Heterocycl. Chem.* **2022**, *137*.
36. Murray, A. W.; Vaughan, K. *Chem. Commun.* **1967**, 1282.
37. Morgan, D. O.; Ollis, W. D.; Stanforth, S. P. *Tetrahedron* **2000**, *56*, 5523.
38. Sieper, H.; Tavs, P. *Justus Liebigs Ann. Chem.* **1967**, *704*, 161.
39. Honzl, J.; Šorm, M.; Hanuš, V. *Tetrahedron* **1970**, *26*, 2305.
40. Tanaka, T.; Yamazaki, H.; Ohta, M. *Bull. Chem. Soc. Japan* **1977**, *50*, 1821.
41. Ramsden, C. A.; Oziminski, W. P. *Tetrahedron* **2014**, *70*, 7158.
42. Schmidt, A.; Wiechmann, S.; Freese, T. *Arkivoc* **2013**, (i), 424.
43. Potts, K. T.; Kanemasa, S.; Zvilichovsky, G. *J. Am. Chem. Soc.* **1980**, *102*, 3971.
44. Zvilichovsky, G.; David, M. *J. Org. Chem.* **1982**, *47*, 295.
45. Friedrichsen, W.; Böttcher, A.; Debaerdemaeker, T. *Heterocycles* **1983**, *20*, 23.
46. Kappe, T.; Lube, W. *Monatsh. Chem.* **1971**, *102*, 781.
47. Welch, T. R.; Williams, R. M. *Nat. Prod. Rep.* **2014**, *31*, 1376.
48. Ferrón, C. C.; Casado, J.; López Navarrete, J. T.; Dreger, A.; Schmidt, A.; Ramírez, F. J. *J. Raman Spectrosc.* **2009**, *40*, 238.
49. Schmidt, A.; Münster, N.; Dreger, A. *Angew. Chem. Int. Ed.* **2010**, *49*, 2790.
50. Schmidt, A.; Guan, Z. *Synthesis* **2012**, *44*, 3251.
51. Schmidt, A.; Snovydovych, B.; Habeck, T.; Dröttboom, P.; Gjikaj, M.; Adam, A. *Eur. J. Org. Chem.* **2007**, 4909.
52. Schmidt, A.; Wiechmann, S.; Otto, C. F. *Adv. Heterocycl. Chem.* **2016**, *119*, 143.
53. Cava, M. P.; Mitchell, M. J.; Hill, D. T. *J. Chem. Soc. D* **1970**, 1601b.
54. Ribas, I.; Sueiras, J.; Castedo, L. *Tetrahedron Lett.* **1971**, 3093.
55. Katritzky, A. R.; Ramsden, C. A.; Zakaria, Z.; Harlow, R. L.; Simonsen, S. H. *J. Chem. Soc. Perkin Trans. 1* **1980**, 1870.
56. Smeyanov, A.; Adams, J.; Hübner, E. G.; Schmidt, A. *Tetrahedron* **2017**, *73*, 3106.
57. Hrnčiar, P. *Chem. Zvesti* **1965**, *19*, 360.
58. Kehrmann, F.; Bürgin, H. *Chem. Ber.* **1896**, *29*, 1820.
59. Kehrmann, F.; Duret, A. *Chem. Ber.* **1898**, *31*, 2442.
60. Wudl, F.; Koutentis, P. A.; Weitz, A.; Ma, B.; Strassner, T.; Houk, K. N.; Khan, S. I. *Pure Appl. Chem.* **1999**, *71*, 295.
61. Ioannou, T. A.; Koutentis, P. A.; Krassos, H.; Loizou, G.; Re, D. L. *Org. Biomol. Chem.* **2012**, *10*, 1339.
62. Glick, M. D.; Cook, R. E. *Acta Crystallogr.* **1972**, *B28*, 1336.

CHAPTER TWO

Developments in the chemistry of 1,3a,6a-triazapentalenes and their fused analogs

Yingchun Wang†, Tomas Opsomer†, and Wim Dehaen*
Molecular Design and Synthesis, Department of Chemistry, KU Leuven, Leuven, Belgium
*Corresponding author: e-mail address: wim.dehaen@kuleuven.be

Contents

Abstract

The chemistry and applications of 1,3a,6a-triazapentalenes and their derivatives with fused benzo and heterocyclic rings are reviewed, with focus on their synthesis and reactivity. All the reports concerning this bicyclic heterocycle have appeared in the past six decades. The syntheses often rely on the intramolecular reaction of nitrenes and (fused) pyrazoles but also other cyclizations starting from (fused) pyrazole or (benzo)triazole

† Equal contribution.

Advances in Heterocyclic Chemistry, Volume 137
ISSN 0065-2725
https://doi.org/10.1016/bs.aihch.2021.10.002

heterocycles are possible. Typical reactions of triazapentalenes are cycloadditions (as 1,3-dipole), oxidations and electrophilic substitutions. In particular, the application of the triazapentalenes as fluorescent sensors and probes is highlighted.

Keywords: 1,3a,6a-Triazapentalenes, Pyrazole, 1,2,3-Triazole, Mesomeric betaines, Synthetic methodology, Nitrenes, Insertion, Organic fluorophores

1. Introduction

The 1,3a,6a-triazapentalenes (TAPs) and their fused analogs are a class of compounds with interesting structures, that can be seen as stable heteroderivatives of the anti-aromatic (8π-electron) pentalene. The extra electron pair of the aromatic 10π-electron system is provided by a conjugated nitrogen lone pair, making it isoelectronic with the aromatic pentalene dianion. At the same time, the TAPs are heterocyclic mesomeric betaines, meaning that they possess delocalized positive and negative charges, while none of the contributing structures are completely uncharged (Fig. 1).

Our own interest in TAPs stems from their interesting properties as novel low molecular mass fluorophores.[1,2] In a recent review,[3] we have focused on the bicyclic (non-(hetero)arene fused) TAPs and described their spectroscopic properties and use as fluorescent probes and sensors. This book chapter has a wider and different scope. It also includes the fused derivatives and involves a discussion of the reactivity of the betaine systems, e.g., in cycloaddition reactions. Conversely, for a more detailed discussion of the photophysical properties, especially of the non-fused TAPs, the interested reader is referred to the earlier review.[3]

There have been earlier literature studies on heterapentalenes[4] and mesoionic compounds,[5,6] including also information on the TAPs. Therefore, literature reports on TAPs previous to 1998 will only be treated

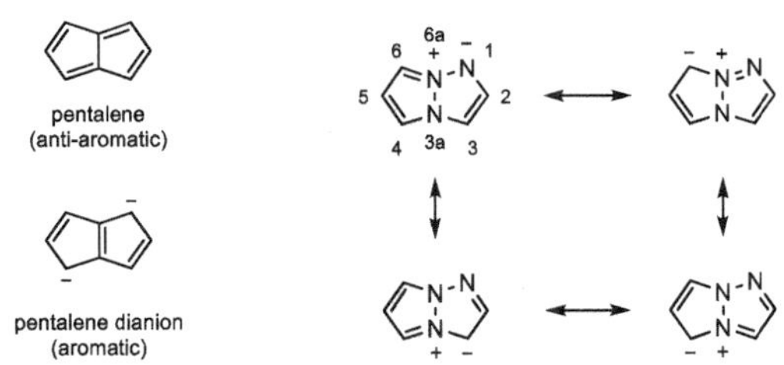

Fig. 1 Structure of pentalene, pentalene dianion and (numbering of) 1,3a,6a-triazapentalene (TAP).

in detail when it is relevant for the discussion. As far as we know, this is the first book chapter that is dedicated exclusively to the 1,3a,6a-triazapentalene system and the fused derivatives.

2. Bicyclic 1,3a,6a-triazapentalene

2.1 Synthesis of 1,3a,6a-triazapentalenes

The bicyclic TAPs have been obtained by applying a variety of synthetic pathways. Hirobe et al. were the first to report the bicyclic (non-fused) triazapentalenes in 1978.[7] After this initial report, nothing else was reported about this heterocyclic scaffold, until several decades later.[8] Namba et al. used the copper-catalyzed azide-alkyne cycloaddition (CuAAC) reaction for the synthesis of monocyclic 1,2,3-triazoles which were then cyclized to the TAPs. This resulted in a number of high impact reports throughout the past decade.[9–12] Additionally, other researchers, including our group, have been encouraged by these studies to investigate the TAP fluorophore and to develop alternative synthetic methodologies.[1,13]

2.1.1 Aminopyrazole-mediated syntheses

The first 1,3a,6a-triazapentalenes, reported by the Hirobe group, were synthesized via two different approaches from monocyclic pyrazoles (Scheme 1).[7] The first pathway led to the synthesis of 3-acetyl-2-methyl-TAP **2** (for numbering see Fig. 1). The 1-aminopyrazole **1** was condensed with 3-chloropentane-2, 4-dione upon heating in benzene solvent. In a following step, the TAP **2** could be deacetylated under acidic conditions to afford 2-methyl-TAP **3** as a colorless oil in almost quantitative yield (95%).

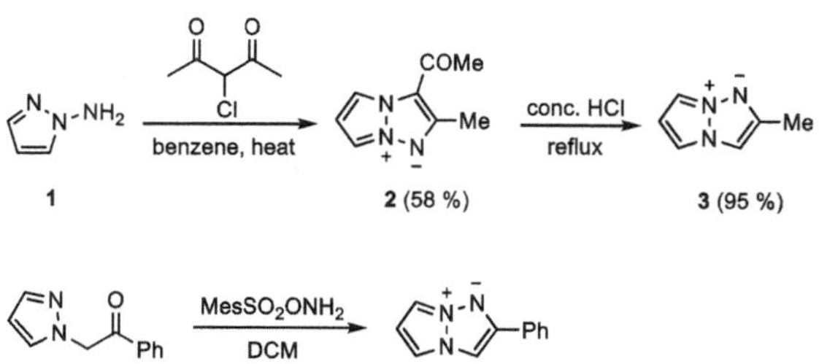

1 2 (58 %) 3 (95 %) 4 5 (66 %)

Scheme 1 Synthesis of 2-methyl- and 2-phenyl-TAP.[3]

In an alternative pathway, the 2-phenyl-TAP derivative **5** was synthesized under mild conditions through the introduction of nitrogen to

1-(phenacyl)pyrazole **4** with *O*-(mesitylenesulfonyl)hydroxylamine (MSH). Notably, the TAP phenyl derivative **5** could not be prepared from aminopyrazole **1** and phenacyl bromide. Interestingly, the deacetylated TAP **3** appeared to be sensitive to air and distinctly less stable compared to the 3-acyl derivative **2**. The 2-phenyl-TAP **5** was slightly more stable than its methyl analog **3**. From this initial study, it already became apparent that the TAP core is chemically sensitive and that this can be mitigated by a stabilizing effect of electron-withdrawing groups.

2.1.2 Triazole-first syntheses: CuAAC

Namba et al. prepared N1-substituted 1,2,3-triazoles by the very well-known copper-catalyzed azide acetylene cycloaddition ("click," CuAAC) reaction and realized their potential for the preparation of TAPs.[8,14] Thus, monosubstituted alkynes **6** were combined with 1-azidopropane building blocks **7** containing leaving groups (LG) at the 2- and 3-positions, in the presence of catalytic Cu(I) to afford 1,2,3-triazoles **8** as intermediate substrates. Cyclization of these 1,2,3-triazoles **8** readily occurred as a result of the basic environment via the substitution of one leaving group, followed by the elimination of the second one, causing aromatization (Scheme 2). In this case, the leaving groups on the azidopropanes **7** were trifluoromethanesulfonate (triflate) groups. A wide scope of TAP derivatives **9** with various functionalities at the 2-, 4-, and 5-positions were made available by changing the substituents on the alkynes and azido substrates.[8–11] However, the reaction conditions had to be adapted when 3-azido-2-methoxypropyl triflates were applied, as was the case for the synthesis 2,5-disubstituted TAPs.[9]

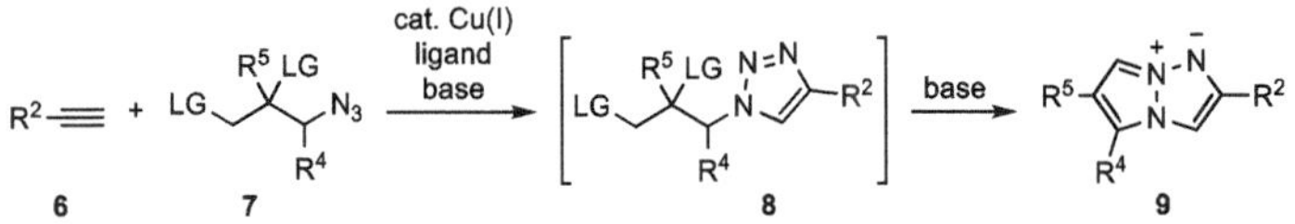

Scheme 2 CuAAC-based synthesis of substituted TAPs.[3]

The 2-substituted TAPs **9** (R^4, R^5 = H) were synthesized by the Namba group,[8,10] starting from terminal alkynes **6** and 3-azido-propane-1,2-bistriflate **7** (R^4, R^5 = H). The reactions are catalyzed by copper(I) iodide, occurring in the presence of bis[2-(*N,N*-dimethylamino)ethyl] ether (BDMAEE) ligand and triethylamine (TEA) base in THF solvent. It was observed that elimination of the second triflate did not occur in the absence of triethylamine, and that azidopropanes **7** with mesylate or tosylate groups only give the intermediate

1,2,3-triazoles **8**. An electron withdrawing aryl group at the C2 position leads to better reaction yields. Thus, the reaction with 4-nitrophenylacetylene afforded a 96% yield of the corresponding TAP **9**, while the TAP with a 4-methoxyphenyl substituent was only obtained in 56% yield. The parent unsubstituted triazapentalene **9** (R^2, R^4, R^5 = H) was obtained for the first time from trimethylsilylacetylene by CuAAC/cyclization and the direct desilylation of crude 2-trimethylsilyl-TAP intermediate.[8] It was noted that TAPs lacking an aryl substituent and also the electron-rich 4-methoxyphenyl-TAP gradually decomposed under UV irradiation.

The 2,5-disubstituted 1,3a,6a-triazapentalenes **12** were prepared analogously using (substituted aryl)ethynes **10** and 2-substituted 3-azido-2-methoxypropyl triflates **11**.[9] After the CuAAC, the cyclization of the triflate was done at reflux temperature. In this case, the aromatization failed when using triethylamine as the base. Also in the presence of the stronger base DBU, or in an acidic environment, elimination of the methoxy group, which is known not to be a very efficient leaving group, was unsuccessful. However, the elimination did occur by using the much stronger base potassium bis(trimethylsilyl)amide (KHMDS) at −78 °C, and thus 2,5-disubstituted TAPs **12** could be obtained (Scheme 3). Nearly all reaction yields were good, except for the examples with a phenyl group at the C5 position or with a 4-nitrophenyl and methyl substituent at positions 2 and 5, respectively. However, the latter derivative was prepared alternatively by using lithium diethylamide as the base, and all examples with a methyl or phenyl group at the C5 position were synthesized without heating to reflux temperature. This methodology was also applied to prepare TAP dimers and trimers.[15]

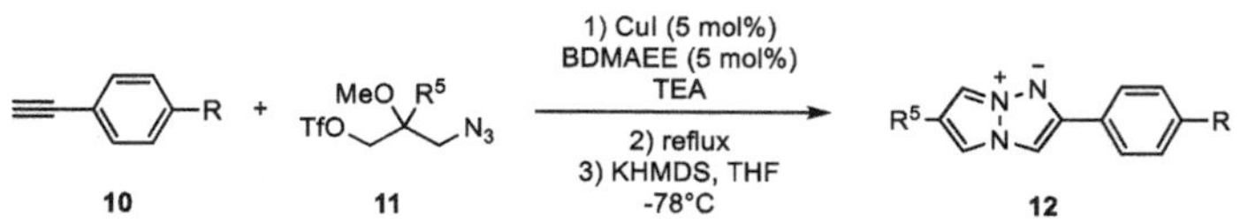

Scheme 3 Synthesis of 2,5-disubstituted TAPs.[3]

The 2,4-disubstituted TAP analogs (R^2 = Ph, R^4 = Me, Ph) were synthesized by the Namba group using the general procedure (Scheme 2).[8,11] The yields of the 2-aryl-4-methyl-TAP analogs were rather variable (19%–94%). On the other hand, the 2,4-diaryl-TAPs were obtained in moderate to high yields (70%–88%). The lower yields may be attributed to partial decomposition during chromatographic purification. For the 2,6-disubstituted TAPs, the general procedure starting from 3-azido-1-methylpropane-1,2-diyl bistriflates **7** was not possible since the latter are too unstable.[12] Based on a Payne-type rearrangement of 1-(oxiranylmethyl)-1,2,3-triazoles obtained by CuAAC,[16] the synthesis of 2,6-disubstituted 1,3a,6a-triazapentalenes **16** could be attained.[12] The intermediate **15** was formed by the triflic acid-mediated epoxide-opening reaction, and was subsequently subjected to acetylation and base-induced elimination (Scheme 4). Due to the poor stability of the 6-methyl-substituted TAPs **16**, all reaction yields were relatively low (14%–34%), while the more stable 6-(methoxycarbonyl)-TAP **18** was obtained in 40% yield. Nonetheless, the 6-phenyl-TAP was not obtained after various attempts.[12]

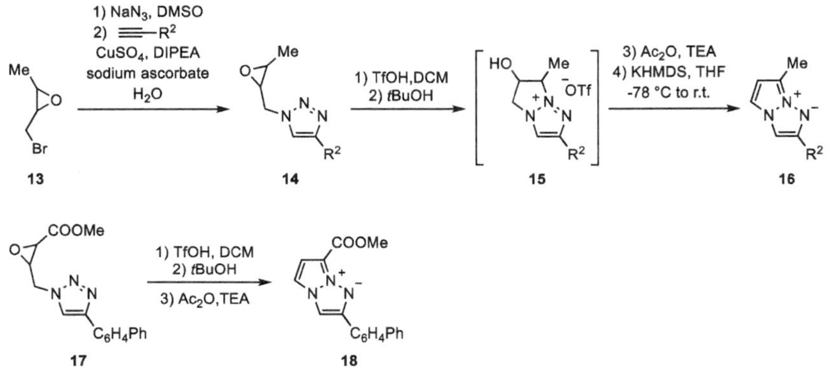

Scheme 4 Synthesis of 2,6-disubstituted TAPs.[3]

2.1.3 Triazolization of ketones and cyclization

In a series of articles starting in 2016, we described a very general synthesis of 1,2,3-triazoles from enolizable ketones, amines and 4-nitrophenyl azide as diazo transfer agent.[17,18] This metal-free organocatalytic method dubbed "triazolization" led to many different applications that were recently described in an account,[19] and an early application was indeed for the synthesis of 3-aryl-TAPs.[1] Thus, starting from commercially available 2,3-dihydroxypropylamine **19**, acetophenones **20** and the diazo transfer reagent, an acetic acid catalyzed triazolization reaction afforded 1-(2,3-dihydroxypropyl)-1,2,3-triazole **21**, that was converted with triflic anhydride and pyridine to the TAPs **22**. The substitution pattern of these TAPs is complementary to the ones obtained by the CuAAC route, but only 3-aryl groups with electron-withdrawing groups resulted in stable derivatives (Scheme 5).

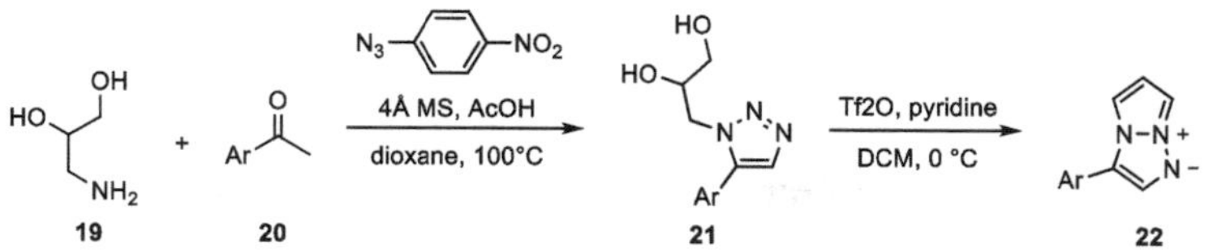

Scheme 5 Triazolization route towards 2-aryl-TAPs.[3]

2.1.4 Gold-catalyzed cyclization of propargyl-1,2,3-triazoles

N2-Substituted propargyl-1,2,3-triazoles **23** were prepared via the iron-catalyzed alkylation of *N*H-triazoles with propargyl alcohols,[20] and it was found that these triazoles can be further transformed to nonconjugated TAP derivatives with a stable Au–C bond via an intramolecular gold-catalyzed triazole-yne cyclization.[21] Later on, the same group demonstrated that this regioselective cyclization resulted in the formation of stable TAPs **26** in good to almost quantitative yields by using propargyl triazoles **23** bearing an electron-withdrawing group (Scheme 6).[13] The electron density distribution on the intermediate TAP-Au complexes **25** was changed by the electron withdrawing group. This reduces their basicity, causing protonation of the Au–C bond instead of the triazole ring. Protodeauration led to the desired gold catalyst turnover and afforded TAPs **26** having ketone and ester substituents. As before, the RCO groups caused high stability of the TAPs. Furthermore, starting from mono ester-substituted TAPs, 3-unsubstituted TAPs were obtained via LiOH-induced saponification/decarboxylation.[13]

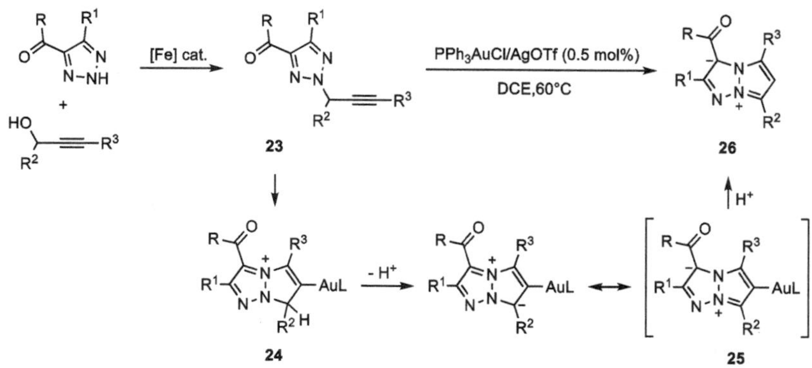

Scheme 6 Gold-catalyzed formation of 3-acylated TAPs.[3]

2.2 Reactivity of bicyclic 1,3a,6a-triazapentalenes

2.2.1 *Electrophilic substitutions*

The 2-substituted (R = Me, Ph) TAPs **3** and **5** are rather sensitive, but their acetylation and nitrosation occurs without problems at the C3 position, resulting in stable 3-acetyl and 3-nitroso derivatives **27–30** in high yields (Scheme 7). The successful electrophilic substitution reactions and the appearance of the carbonyl bands in the spectra of **27** and **28** at 1577–1585 cm^{-1} clearly demonstrated the electron-rich nature of the TAP core.[7] More recently, another example of acetylation of a 2-aryl-TAP was reported.[22]

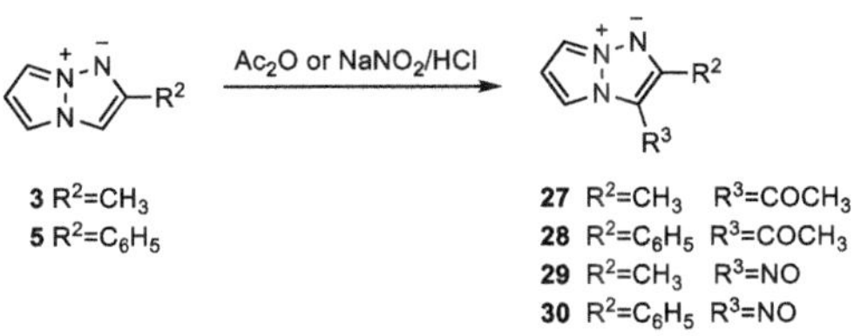

Scheme 7 Acylation and nitrosation of 2-substituted TAPs.

The 3-aryl-TAP **22** (Ar = 4-MeOOC-C_6H_4) was combined with benzenediazonium tetrafluoroborate in an attempt by our group to introduce a phenyl group after radical coupling (Scheme 8).[2] Due to the high reactivity of the triazapentalene towards electrophiles, azo-coupling was observed instead of CH-arylation, which was expected in view of our earlier work with the more electrophilic BODIPYs.[23] Another surprise was that under the reaction circumstances, a rearranged 2-aryl-4-phenylazo-TAP **32** was readily formed and isolated as the major product, accompanied by a 6-phenylazo-coupled 3-aryl-TAP **31** as the minor product. The structures of the products were confirmed using single-crystal X-ray diffraction, and a mechanism towards TAP **32** involving the formation of an open chain nitrilimine and alternative ring closure was proposed. In the presence of Hünig's base and an excess of diazonium salt, a second azo coupling on the intermediate 2-aryl-4-phenylazo-TAP **32** occurred at the C3 position

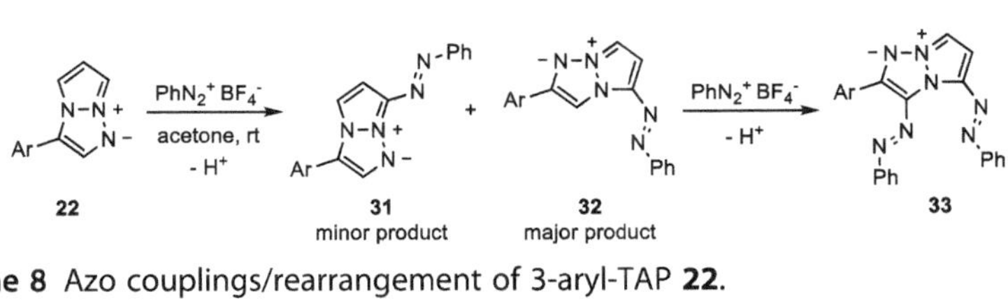

Scheme 8 Azo couplings/rearrangement of 3-aryl-TAP **22**.

(**33**; Scheme 8). The presence of base was required as the protonated form of TAP **32** would otherwise not react further. An attempt to reduce azo compound **32** with sodium dithionite led to cleavage of the TAP core via a nitrene pathway, rather than the formation of the amino derivative. In addition, hydrazinolysis of TAP **32** gave 3-phenylazo-1*H*-pyrazole and the corresponding NH-1,2,3-triazole.

2.2.2 Cycloadditions

Resonance structures of the TAPs often represent a 1,3-dipole of either the azomethine imine or azomethine ylide type. In view of the dipolar character of the TAPs, the possibility of 1,3-dipolar cycloaddition reactions is a relatively underexplored area. In fact, there is only a single early report in the literature.[7] Hirobe et al. reacted the 2-substituted TAPs **3** and **5** at room temperature with dimethyl acetylenedicarboxylate (DMAD). The 1:1 cycloadducts **34** that were formed regioselectively were characterized by NMR spectroscopy, and could be further oxidized with 2,3-dichloro-5,6-dicyano-1,4-benzoquinone (DDQ) to diazacycl[2.2.2]azines **35** (Scheme 9).

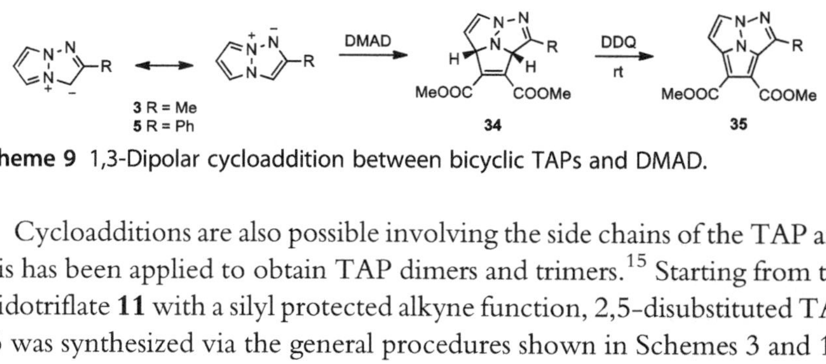

Scheme 9 1,3-Dipolar cycloaddition between bicyclic TAPs and DMAD.

Cycloadditions are also possible involving the side chains of the TAP and this has been applied to obtain TAP dimers and trimers.[15] Starting from the azidotriflate **11** with a silyl protected alkyne function, 2,5-disubstituted TAP **36** was synthesized via the general procedures shown in Schemes 3 and 10. After removing the *tert*-butyldimethylsilyl (TBS) group, the ethynyl substituted triazapentalene **36** was applied as the starting material together with one more equivalent of reagent **11** in a reiteration of this general

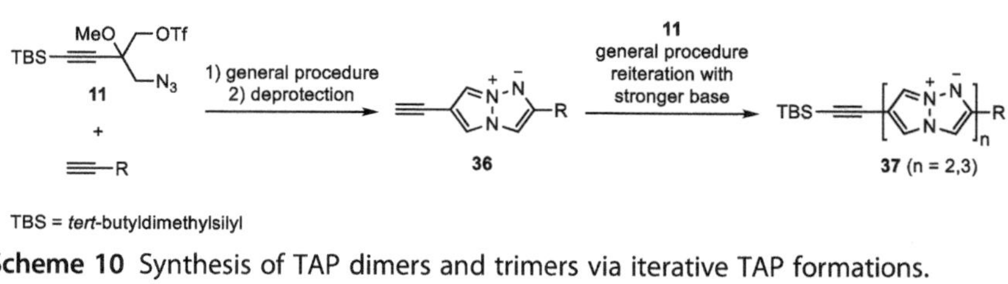

Scheme 10 Synthesis of TAP dimers and trimers via iterative TAP formations.

procedure, to develop the linearly bonded oligomeric systems **37** (Scheme 10). It should be noted that for the formation of the dimer, lithium diethylamide was required as a stronger base than KHDMS to induce elimination of the methoxy group. TAP dimers ($n=2$) with different substituents at the C2 position have been reported, including a 2-unsubstituted TAP dimer. Unfortunately, the trimers ($n=3$) are in general too unstable to obtain due to decomposition during the elimination reactions needed for TAP formation or during column chromatography. Only a 2-alkyl-TAP trimer could be prepared, by using lithium pyrrolidide in the final step.

2.3 Applications of bicyclic 1,3a,6a-triazapentalenes

The fact that many TAPs have been found to be highly fluorescent has led to many photophysical studies and applications in sensors and probes.

For the details of the photophysical properties of bicyclic TAPs we can refer to the earlier review.[3] The parent unsubstituted triazapentalene **9** (R^2, R^4, $R^5=H$) has λ_{max}(absorption) of 288 nm and λ_{max}(emission) of 389 nm in dichloromethane, with an unimpressive quantum yield of fluorescence (Φ_f) of 0.017. Arylated TAP derivatives in particular are more interesting with absorption and emission maxima for 2-aryl-TAPs between 326–466 nm and 413–632 nm, respectively, and Φ_f ranging from 0.03 to 0.44, with all of these values also measured in dichloromethane solvent.[8,10] The 3-aryl-TAP **22** (Ar = 4-MeOOC-C_6H_4) is among the most highly fluorescent TAP dyes with $\Phi_f=0.79$ in dichloromethane.[1] Electron-withdrawing substituents on the aryl groups show the expected bathochromic shifts in absorption and emission wavelengths, and increase the Stokes shift and the overall chemical- and photostability of the dyes. Therefore, they are often present in the TAPs used in applications. When the electron-withdrawing group is directly conjugated to the TAP, as in derivatives **26**, the Φ_f is lower.[13] In general, the Φ_f of TAP dyes is lower in more polar solvents, while the position of absorption and especially fluorescence maxima may be strongly affected by the solvent, with longer emission wavelengths in aprotic solvents.[8,10]

Also for the applications as sensors and probes, the detailed information can be found in the earlier review.[3]

We can list the application of a derivative of 2-aryl-TAP **9** ($R^2=$ 2-cyano-4-methoxycarbonylphenyl, R^4, $R^5=H$) as a fluorescent probe of low toxicity for live cell imaging.[10] The ester function of this TAP dye was converted to an amide with glycine ethyl ester or tripeptide Gly-Pro-Leu to obtain the labeled derivatives. These have the same fluorescence

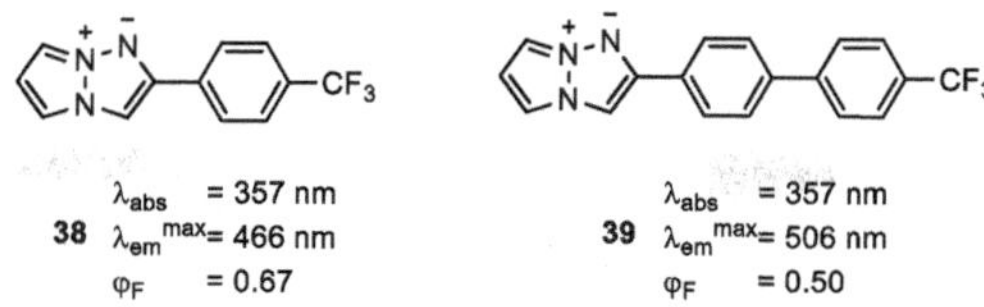

Fig. 2 Bifunctional fluorescent TAP probes **38** and **39**.

maxima in dichloromethane at 567 nm with $\Phi_f = 0.37$ and 0.24, respectively, while fluorescence measurements in water show decreased Stokes shifts and quantum yields of fluorescence. A similar amide coupling of a 2-aryl-3-acetyl-TAP derivative, via the aryl group, led to a tris(TAP) analog of the siderophore enterobactin, that was shown to be a selective and highly sensitive fluorescence-quenching sensor for iron(III).[22]

Another application was the use of 2-(4-biphenyl)-TAP as a fluorescent probe to observe cellular differentiation processes in living cells. Again, the nontoxity of the TAP dye was proven, and there was no effect on the cellular differentiation.[24]

2-Aryl-TAP derivatives **38** and **39** (Fig. 2) have been prepared as analogs of biphenyl-type kinesin spindle protein (KSP) inhibitors. They were then investigated as bifunctional fluorescent probes.[25] Both TAPs indeed showed inhibitory activity against KSP ATPase, with derivative **39** being more potent (IC_{50} 6.8 μM). The partial colocalization in cultured cells of compound **39** with KSP, combined with its inhibitory activity, demonstrated the possible use of the TAP fluorophore as a probe for visualization of bioactive substances and their targets.[25]

Photophysical data have been determined in dichloromethane solvent.[3] Most recently, TAP **40** functionalized with a Michael acceptor was applied as a thiol-specific fluorescent labeling reagent. Various thiols add to **40** in the presence of tetramethylguanidine (TMG) or other bases via thiol Michael additions to obtain sulfides **41** (Scheme 11). TAP **40** was nonfluorescent due to the conjugation of the enone at the C2 position. After the addition of the thiol group, the fluorescence of compound **41** was turned on while also exhibiting a hypsochromic shift compared to precursor **40**. This conjugation can also be used for labeling of water-soluble peptides in a phosphate

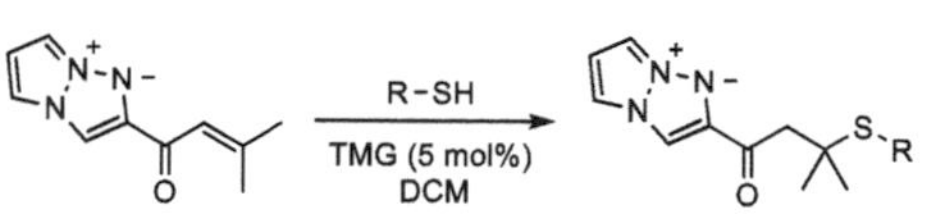

40 **41**

Scheme 11 Michael addition of thiols to acceptor TAP **40**.

buffer (pH 7) or for linking to captopril, a cystein-containing angiotensin converting enzyme (ACE) inhibitor. Advantageously, the TAP **40** did not have any background fluorescence and did not show any cytotoxicity.[26]

3. Monobenzo-fused 1,3a,6a-triazapentalenes

Of the three different possible isomers of monobenzo-fused TAPs (Fig. 3), only derivatives of the first isomer **42** (the pyrazolo[1,2-*a*]benzotriazole) have been reported, probably due to the availability of functionalized pyrazoles or benzotriazoles as starting materials for intramolecular cyclization reactions. These TAPs are rather stable due to the benzo-fusion, but interesting reactivity has been reported. Since these compounds **42** are nonfluorescent, few applications have been described.

3.1 Synthesis of monobenzo-fused 1,3a,6a-triazapentalenes

3.1.1 Pyrazole-first syntheses

Synthesis of the benzo-fused 1,3a,6a-triazapentalene **44** was first described by Lynch and Hung who reported the Cadogan deoxygenation/cyclization of 1-(2-nitrophenyl)-1*H*-pyrazole **43** ($R^1 = R^2 = R^3 = H$) in the presence of triethyl phosphite at 160 °C (Scheme 12).[27] The reaction involves a nitrene intermediate. Deoxygenation starting from a nitroso compound and pyrolysis of the corresponding azide were attempted as alternatives for the nitrene formation. Unfortunately, the product **44** ($R^1 = R^2 = R^3 = H$) was not observed since non-cyclized bispyrazoles deriving from alternative nitrene chemistry, for example as azo-, azoxy- or hydrazo-compounds, were formed during the reaction. This report was indeed the first one in the literature mentioning a 1,3a,6a-triazapentalene derivative.[27]

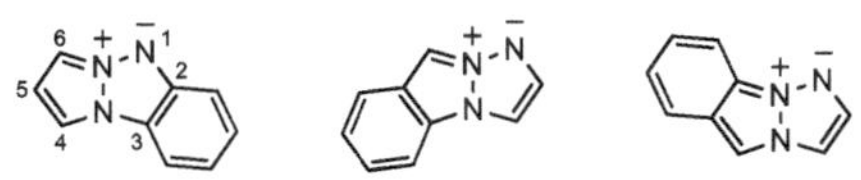

42

Fig. 3 Different possible isomers of monobenzo-fused TAPs.

43 **44**

Scheme 12 Deoxygenation of 1-(2-nitrophenyl)pyrazoles towards monobenzo-fused TAPs.

Subsequently, Harder and Kauer started from substituted 1-(2-nitrophenyl)pyrazoles **43** and triethyl phosphite. The reagents were first refluxed in xylene, followed by distillation at low pressure to remove the volatiles. The substituted TAPs **44** were then obtained after filtration and recrystallization. The substituents (R^1, R^2, R^3) used were mainly halogen-, cyano-, and aryl groups (Scheme 12).[28]

The competitive cyclization of the higher energy singlet nitrenes and triplet nitrenes with radical character was studied by the Meth-Cohn group based on reactions of 1-(2-nitrenophenyl)pyrazole intermediates (Scheme 13).[29,30] Deoxygenation of (2-nitrophenyl)pyrazoles **45** with triethyl phosphite (TEP) demonstrated that yields of TAPs **46** could be improved by the introduction of an electron-donating methyl group on the pyrazole ring, and/or an electron-withdrawing group on the phenyl ring (Table 1, **a–d**). These results verified the electrophilic nature of the singlet nitrene and supported formation of triazapentalenes **46** derived from the singlet nitrene. Variation of substituents on the phenyl ring of 3,5-dimethyl-1-(2-nitrenophenyl)pyrazole again showed that introduction of a halogen *para* to the nitrene could increase the yields of triazapentalenes **46** (Table 1, entries **e**, **h**, **i**).[29,30] However, the significance of this effect was argued by Bettinetti and co-workers, who isolated TAP **46e** in a yield of 57% after a reaction of 24 h under similar conditions.[31] They ascribed the low yield of 6% observed by Meth-Cohn *et al.* to the excessive reaction time. Nevertheless, it is beyond dispute that the electron-donating groups dimethylamine and methoxy *para* to the nitrene completely deactivate the singlet state and exclusively result in pyrazoloquinoxalines **47** and aminophenylpyrazoles **48** via the intermediate triplet nitrene (Table 1, entries **f**, **g**).[29,30] Although the effect of substituents at the *meta*-position was less straightforward, it was clear that the formation of singlet nitrenes was enhanced (Table 1, entries **j–m**). The reaction yields of **46h** in TEP, cumene and bromobenzene as the solvents were 66%, 62% and 50% yield, respectively.[30]

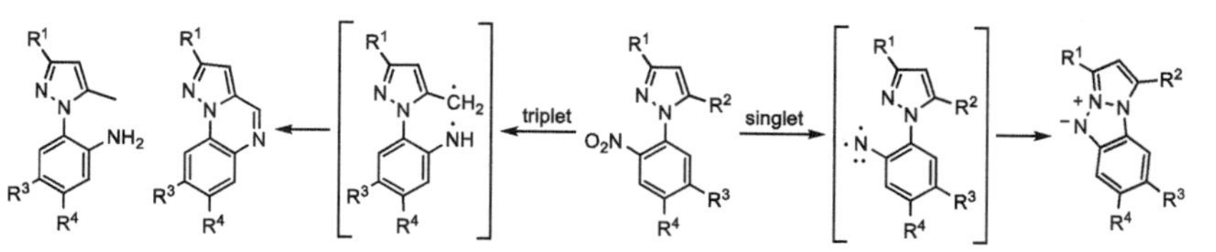

Scheme 13 Alternative reactions of singlet and triplet nitrenes in the deoxygenation of (2-nitrophenyl)pyrazoles **45** with triethyl phosphite (TEP).

Table 1 The reductive cyclization of (2-nitrophenyl)pyrazole derivatives and reaction yields.

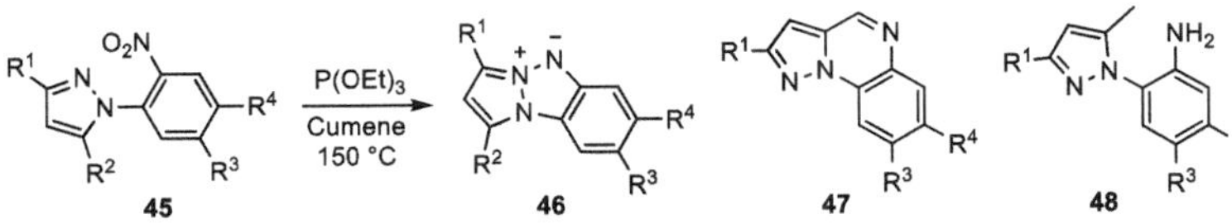

					Yield (%)		
Entry/Comp.	**R¹**	**R²**	**R³**	**R⁴**	**46**	**47**	**48**
a	H	H	H	H	15	–[a]	–[a]
b	H	H	Cl	H	21	–[a]	–[a]
c	Me	H	H	H	29	–[a]	–[a]
d	Me	H	Cl	H	65	–[a]	–[a]
e	Me	Me	H	H	6	–[b]	5
f	Me	Me	OMe	H	0	23	32
g	Me	Me	NMe_2	H	0	47	19
h	Me	Me	Cl	H	62	16	11
i	Me	Me	Br	H	70	6	2
j	Me	Me	H	Cl	57	0	–[b]
k	Me	Me	H	CF_3	63	0	–[b]
l	Me	Me	H	NO_2	26	0	–[b]
m	Me	Me	H	Me	63	0	–[b]

[a]Not applicable.
[b]Trace.

This implies that the solvent effect on the intermediate singlet nitrene such as the heavy atom effect was small. It is worth noting that Bettinetti et al. proposed the occurrence of electron transfer from the pyrazole N2 nitrogen to the nitrene as the underlying mechanism, which would be much faster compared to intersystem crossing.[31]

On the contrary, the effect of solvents on the formation of triplet nitrenes was more pronounced in the thermolysis of 1-(2-azidophenyl)-3,5-dimethylpyrazoles (Table 2).[29,30] Due to the heavy atom effect present in bromobenzene, and deactivation of the singlet nitrene by lone-pair interaction with diglyme, the ratio of triplet-derived products was increased (Table 2, entries 1–3 and 7–8). All thermolysis reactions were carried out

Table 2 Thermolysis and photolysis of 1-(2-azidophenyl)-3,5-dimethylpyrazoles **49** and reaction yields.

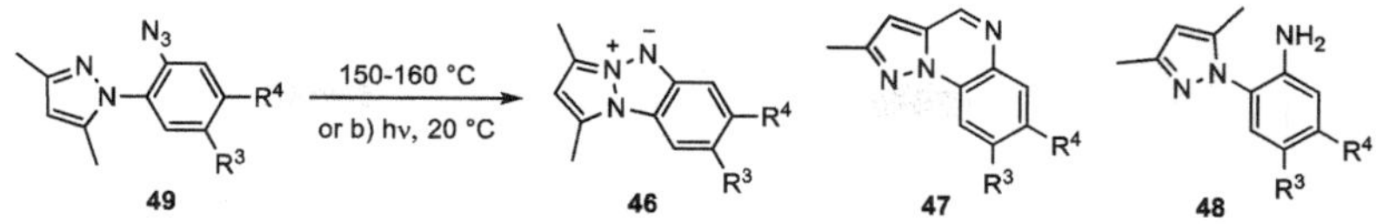

					Yield (%)		
Entry	**Comp.**	**R⁴**	**R³**	**Solvent**	**46**	**47**	**48**
1	**h**	H	Cl	Cumene	39	5	2
2				Diglyme	–[b]	9	84
3				PhBr	8	6	21
4[a]				DCM	28	6	1
5[a]				DCM-pyrene	42	3	–[b]
6[a]				PhAc	–[b]	48	43
7	**i**	H	Br	Cumene	26	11	11
8				PhBr	4	8	2
9[a]				DCM	–[b]	21	24
10[a]				DCM-pyrene	10	9	15
11[a]				PhAc	0	23	11
12	**g**	H	NMe_2	Cumene	0	44	41
13[a]				DCM	0	12	19
14[a]				DCM-pyrene	0	8	15
15[a]				PhAc	0	50	38
16	**k**	CF_3	H	Decalin	67	0	0
17[a]				DCM	33	0	13
18[a]				DCM-pyrene	15	0	44
19[a]				PhAc	0	41	22

[a]Photolysis of various 1-(2-azidophenyl)-3,5-dimethylpyrazoles.
[b]Trace.

at 150–160 °C. The effect of the substituents roughly showed the same trends as for deoxygenation of (2-nitrophenyl)pyrazoles **45**. An electron-withdrawing group on the phenyl ring increased the electrophilic nature of singlet nitrene and enhanced the yield of the singlet-derived products

(Table 2, entries 1,7,16), although this is again not a prerequisite according to Bettinetti et al.[32] By contrast, a dimethylamine group decreased the electrophilic character and eliminated the formation of the singlet-derived TAP **46g** (Table 2, entry 12).[29,30,32] It was concluded that the deoxygenation of 1-(2-nitrophenyl)pyrazoles with TEP is the preferred route for the preparation of monobenzo-fused TAPs.

Next, the photolysis of 1-(2-azidophenyl)-3,5-dimethylpyrazoles **49** was investigated by the Meth-Cohn group in different solvents.[29,30] The results of photolysis in acetophenone, a triplet sensitizer, showed that the solvent eliminated all singlet-derived compounds (Table 2, entries 6,11,15,19). On the other hand, when pyrene, a singlet photosensitizer, was used in combination with dichloromethane, the yields of the singlet-derived TAPs **46h** and **46i** were clearly increased. Again, photolysis of 1-(2-azidophenyl)pyrazole **49g** with a dimethylamine substituent only resulted in triplet-derived products. Although dimerized azo compounds **53** are not presented in Table 2, it should be noted that Bettinetti and co-workers, who also studied the thermolysis and photolysis of 1-(2-azidophenyl)-3,5-dimethylpyrazoles **49**, often observed the azo compound as the main product.[32]

The effect of temperature on the photolysis of 1-(2-azidophenyl)-3, 5-dimethylpyrazole **49e** was investigated by Bettinetti et al.[33] Based on results from irradiation of **49e** in several solvents at room temperature (Table 3, entries 1–4), ethanol and 2-methylpentane (MP) were chosen as solvents for the further study. The triazapentalene **46e** resulting from singlet nitrene cyclization was obtained in relative high yields when the temperature was 200 K or higher, both in ethanol and 2-methylpentane (Table 3, entries 5–7 and 11–12). At lower temperature, the yields of triazapentalene **46e** were decreased (Table 3, entries 8–10 and 14) or the product was not formed (Table 3, entry 15), and products derived from triplet nitrenes became the only products below 100 K.

To investigate the effect of a *para*-nitro substituent on the cyclization, photo-decomposition of 1-(2-azido-5-nitrophenyl)-3,5-dimethyl-1*H*-pyrazole **50a** was studied in ACN and ethanol (Table 4).[34] At low concentrations, the main product of photolysis with a 15 W lamp in both solvents was the triazapentalene **51a** (Table 4, entries 1 and 6). When increasing the concentration and using a 200 W lamp, the yield of TAP **51a** was only mildly affected (Table 4, entries 1, 2 and 6, 7). In the presence of Michler's ketone or diethylamine, the reduction of the nitrene to the amine is more favored (Table 4, entries 4, 5 and 8, 9). Photolysis at low temperature (90 K) resulted in amine **52a** as the main compound while TAP **51a** was only

Table 3 Effect of temperature and solvent on the formation of TAP **46e** by photolysis of 1-(2-azidophenyl)-3,5-dimethylpyrazole **49e**.

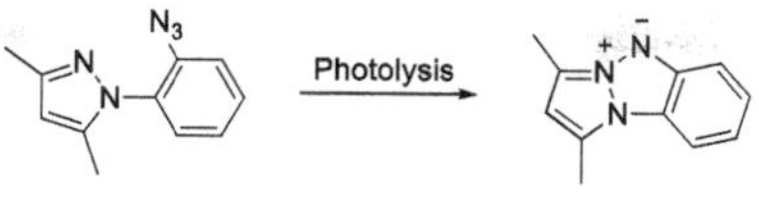

Entry	Temperature (K)	Solvent	Yield (%)
1	298	EtOH	66[a]
2	298	ACN	58[a]
3	298	MP[b]	53[a]
4	298	MP[b], Et_2NH[c]	7[a]
5	297	EtOH	72[b]
6	270	EtOH	85[d]
7	230	EtOH	82[d]
8	180	EtOH	60[d]
9	140	EtOH	27[d]
10	100	EtOH	2[d]
11	295	MP[b]	56[d]
12	240	MP[b]	59[d]
13	200	MP[b]	33[d]
14	160	MP[b]	9[d]
15	90/77	MP[b]	–[e]

[a]30 Min irradiation; Isolated yields from column chromatography.
[b]2-Methylpentane (MP).
[c]0.06 M.
[d]8 Min irradiation; HPLC monitoring.
[e]Not observed.

observed in trace amounts. Although the nitro-group increases the electrophilic character of the nitrenophenyl, the relative yields showed that intermolecular trapping by nucleophiles, such as ethanol or diethylamine (DEA), remained the less favored process.[34]

Bettinetti et al. further investigated the effect of *para*-substituents on the singlet and triplet nitrenes.[35] As shown in Table 5, the photodecomposition of azido compounds **50** at 295 K gives the corresponding TAPs **51** as the main products via intramolecular trapping of the singlet nitrene, except in

Table 4 Products and reaction yields from irradiation of 1-(2-azido-5-nitrophenyl)-3,5-dimethylpyrazole **50a** at room temperature.

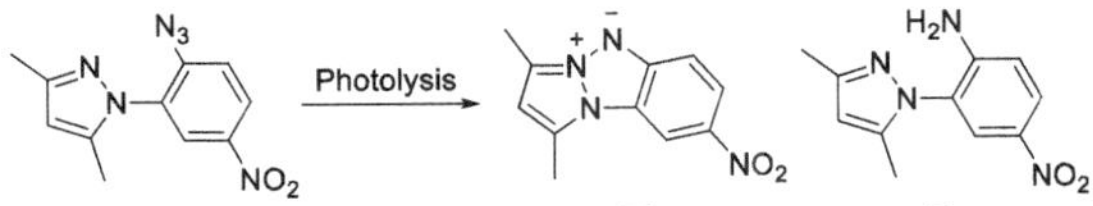

Entry	Solvent	Concentration	Lamp additive	Yield (%) 51a	52a
1	Acetonitrile (ACN)	1×10^{-4}	–[a]	37	4
2		4×10^{-3}	–[b]	41	3
3		1×10^{-4}	EtOH, 1 M[a]	38	4
4		4×10^{-3}	MK[c], $9*10^{-3}$ M[b]	5	38
5		1×10^{-4}	DEA[d], 0.1 M[a]	32	36
6	Ethanol	1×10^{-4}	–[a]	45	15
7		4×10^{-3}	–[b]	32	12
8		4×10^{-3}	MK, $9*10^{-3}$ M[b]	–[e]	42
9		1×10^{-4}	DEA[d], 0.1 M[a]	36	31

[a]Low-pressure mercury arc (15 W) inserted below the cell.
[b]Focused high-pressure mercury arc (200 W).
[c]Michler's ketone (MK, 4,4′-bis(dimethylamino)benzophenone) as sensitizer.
[d]DEA = diethanolamine.
[e]Not formed.

the case of **50g** and **50h**. For **50c** and **50e** equivalent amounts of amine were formed via the triplet nitrene under the same conditions. In contrast, azo-compounds **53** were the main products of photolysis at 90 K in ethanol. As an exception, the dimethylamino derivative **50h** gave amine **52** as the major product. The reactions of **50b**, **50f** and **50h** were also analyzed with DEA as an external nucleophile. Except for the example with a dimethylamine substituent, the singlet nitrene was in equilibrium with a dihydroazepine which could be trapped by DEA to from 3*H*-azepines. It was found that introducing an electron-withdrawing group had little effect on this equilibrium, whereas in the case of the dimethylamine group intramolecular trapping of the singlet nitrene was suppressed. Additionally, the electron-donating group dimethylamine largely enhances the energy of the triplet nitrene, making it part of the equilibrium with the singlet nitrene. Reactions in the presence of oxygen gave nitroso and nitro compound for all

Table 5 Products and reaction yields of the photolysis of 1-(2-azidophenyl)-3,5-dimethylpyrazoles **50** at 295 K and 90 K in ethanol.

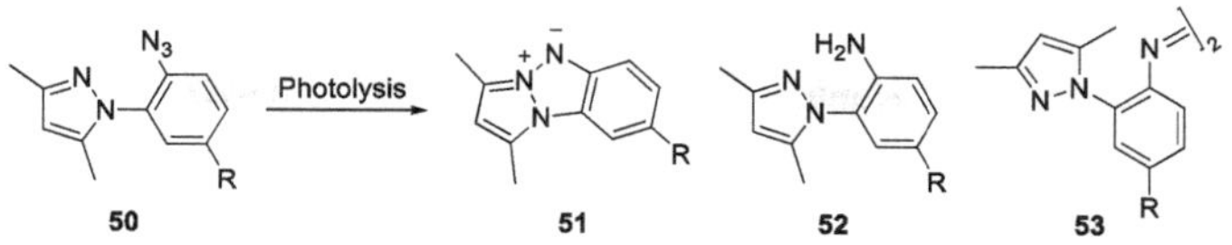

		Yield (%)[a]		
Azide	R	51	52	53
b	H	72 (1)	–[b] (4)	–[c] (41)
c	Me	32 (–)[c]	24 (6)	–[c] (36)
d	Cl	64 (–)[b]	–[c] (15)	7 (45)
e	F	27 (–)[b]	28 (15)	–[c] (64)
f	CF_3	66 (11)	1 (17)	–[c] (44)
g	OMe	–[d]	3 (10)	1 (31)
h	NMe_2	–[d]	33 (29)	5 (5)

[a]Yields in brackets result from irradiation at 90 K in ethanol glass followed by "fast heating", the other yields result from irradiation at 295 K.
[b]Trace (≤1%).
[c]Not observed.
[d]**51** not obtained at 295 K and 90 K.

derivativs. Further studies on the different products resulting from photodecomposition of **50** involving also computational results, were reported in 2005 by the same group.[36]

McNab et al. studied thermolysis in the gas phase by flash vacuum pyrolysis (FVP) of 1-(2-azidophenyl)pyrazole derivatives without methyl groups on the pyrazole ring.[37] FVP of the unsubstituted 1-(2-azidophenyl)pyrazole gave mainly **46a** at 300 °C, while at higher temperatures pyrazolo[1,5-*a*] benzimidazole (C–H insertion) and quinoxalines are also formed. At 700 °C the quinoxaline was the only product formed. The presence of a methoxy group *para* to the azide almost exclusively led to the corresponding pyrazolobenzimidazole via CH insertion of the triplet nitrene.[37]

A special case of the Cadogan cyclization of (2-nitrophenyl)pyrazoles is given by dithienyl derivative **54** after heating at reflux in triethyl phosphite. The expected benzotriazapentalene **55** is formed together with the unexpected 2-(2-thienyl)pyrazolo[1,5-*a*]pyrrolo[2,1-*c*]quinoxaline **56**, in 14%

and 27% yield, respectively. Apparently, in the latter case the nitrene inserts into the thiophene rather than on the pyrazole nitrogen, and sulfur is extruded (Scheme 14).[38]

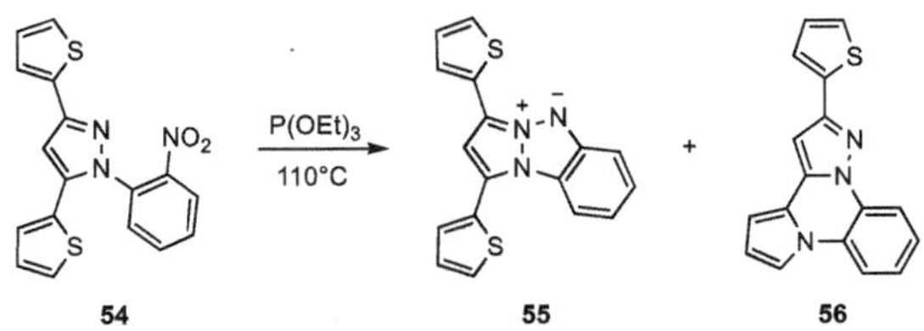

Scheme 14 Expected and unexpected product from the Cadogan cyclization of *N*-(2-nitrophenyl)dithienylpyrazole.

3.1.2 Triazole-first syntheses

3.1.2.1 Pummerer-type annulation pathway

A novel pathway to synthesize trisubstituted monobenzo-fused TAP derivatives via a Pummerer-type intramolecular cyclization was reported by Kim et al. (Scheme 15).[39] Starting from the 1-benzylbenzotriazole **57** treated with *n*-BuLi in THF at −78 °C, diastereomeric mixtures of oxiranes were obtained after the addition of 1-aryl-2-chloroethanones in a Darzens-like condensation reaction. The intermediate oxirane substrates were ring opened with 1.5 equivalents *n*-BuLi at −78 °C, followed by acidification, affording the (*E*)- and (*Z*)-alcohols **58**. Next, the transformation from alcohols **58** to sulfoxides **59** was carried out via oxidation of sulfides obtained from the corresponding allylic chlorides. Treatment of sulfoxides **59** with trifluoroacetic anhydride (TFAA) in THF at room temperature, gave the final TAPs **60** in good yield as a result of a Pummerer reaction. In only one case (Ar$=$R^1$=$Ph, R^2$=$$CH_2CH_2OH$) was the corresponding triazapentalene **60** obtained in a lower (34%) yield.[39]

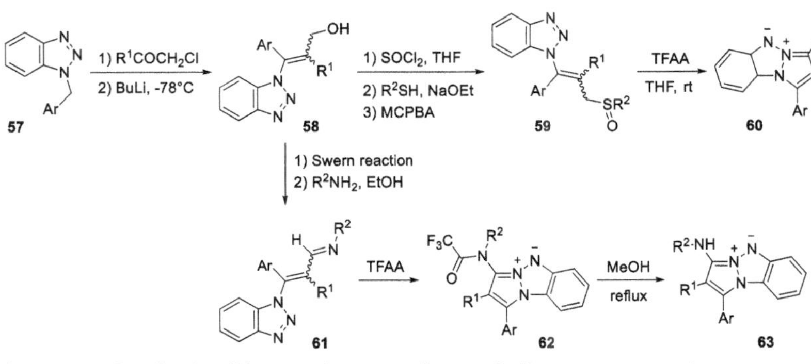

Scheme 15 Synthesis of benzotriazapentalenes via Pummerer reaction.

In order to explore new analogs of benzotriazapentalenes, aminobenzotriazapentalenes **63** were synthesized by the same research group, using a similar strategy (Scheme 16).[40] The conversion of alcohols **58** into *α,β*-unsaturated imines **61** was followed by cyclization in the presence of TFAA at room temperature. As a result of the formation of an N–C bond, benzotriazapentalenes **62** were obtained. Next, deacylation in methanol at reflux was applied to give amino substituted compounds **63**. It has to be noted that not all the amides **62** (R^2 = alkyl) can undergo deacylation to give arylamino-triazapentalenes **62**. In addition, both **62** and **63** slowly decomposed in the air.[40]

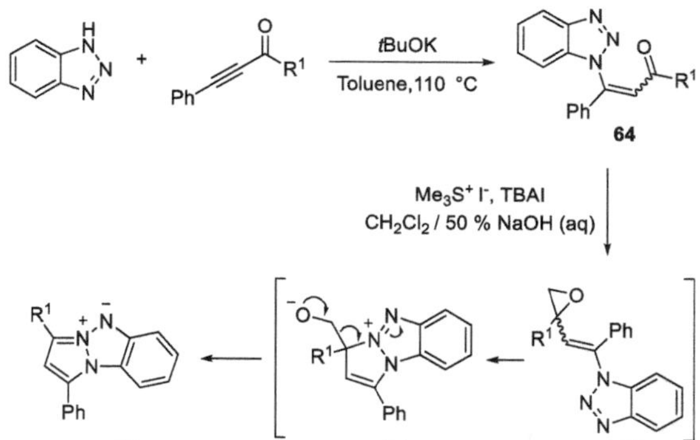

Scheme 16 Synthesis of benzotriazapentalenes **67**.

3.1.2.2 β-Benzotriazolyl substituted unsaturated ketone-mediated cyclization

Starting from ynones and benzotriazole, aza-Michael addition in the presence of *t*-BuOK in toluene as the solvent provided *β*-benzotriazolyl substituted unsaturated ketones **64**. These intermediate substrates were converted into 2-oxiranyl-vinyl benzotriazoles **65** by treatment with trimethylsulfonium iodide in DCM and 50% aqueous NaOH in the presence of tetrabutylammonium iodide (TBAI). Next, benzotriazapentalenes **67** were formed by an initial epoxide-opening reaction, followed by fragmentation of formaldehyde from intermediate **66** (Scheme 16). BenzoTAPs **67** were only formed when R^1 was an aryl group without a strong electron-withdrawing substituent, such as nitro, and the yields were very variable (Table 6).[41]

Table 6 Reaction yields of benzotriazapentalenes **67.**

Comp.	R^1	Yield (%)
a	4-CH_3-C_6H_4	71
b	4-Cl-C_6H_4	8
c	2-thienyl	42
d	4-MeO-C_6H_4	42

3.1.2.3 Gold-catalyzed cyclization of propargyl-benzotriazoles

Ballesteros et al. reported an efficient method to synthesize monobenzo-TAPs **69** (Scheme 17).[42] Different from the *5-endo-dig* cyclization starting with the N2-substituted propargyltriazoles (2014CC7303),[13] compounds **69** were directly obtained from 1-propargyl-1*H*-benzotriazoles **68** with catalytic amounts of *i*PrAuNTf$_2$. The synthesis of compound **69a** was first carried out in DCE at 70°C for 4h. By comparison, the reported yield of 78% was much lower than the 99% obtained in acetonitrile (ACN) at room temperature for 30min. Thus, ACN was chosen as the solvent in these reactions. For the preparation of **69c**, heating up to 60°C was required.

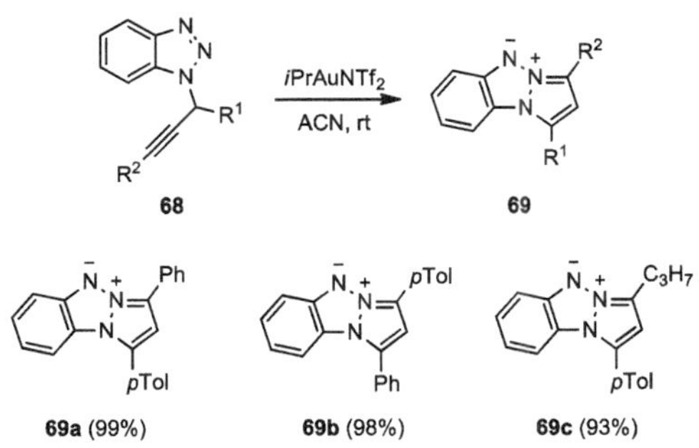

Scheme 17 Gold-catalyzed synthesis of benzotriazapentalenes **69.**

3.2 Reactivity of monobenzo-fused 1,3a,6a-triazapentalenes

3.2.1 Cycloaddition reactions

The Bettinetti group studied addition reactions of compound **46e** with dimethyl acetylenedicarboxylate (DMAD) under various conditions (Scheme 18). When the reaction was done in aprotic anhydrous solvents, such as CCl_4, $CHCl_3$ and C_6H_6, cycloadduct **71** was obtained in quantitative yield (NMR analysis) after cyclization of an aza-Michael adduct **70**. Because decomposition occurred during work-up, the structure was only

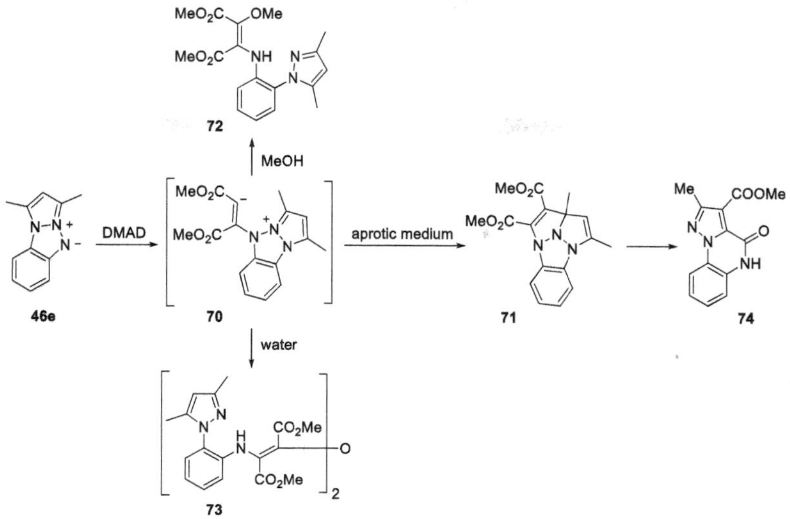

Scheme 18 The cycloaddition of benzotriazapentalene **46e** and DMAD.

confirmed by NMR analysis. Stable enamine **72** was successfully isolated in 65% yield when the reaction was carried out in methanol as a protic trapping agent. On the other hand, the ether **73** was produced in the presence of water. The zwitterionic intermediate **70**, which could be trapped by a protic solvent, was responsible for the formation of the enamine and the ether rather than the cycloaddition products. The decomposition of **71** lead to a complicated reaction mixture from which a major compound **74** was isolated in 37% yield, of which the formation was rationalized by a radical process followed by propyne extrusion.

Ballesteros et al. reported efficient gold-catalyzed reactions of benzo-fused triazapentalenes with an alkyne (Scheme 19).[42] Starting from the benzo-fused TAPs **69a,b** and phenylacetylene, the gold-catalyzed reaction in the presence of methanesulfonic acid as an additive, followed by treatment with sodium hydroxide solution, provided *ortho*-imidazolylpyrazolylbenzene

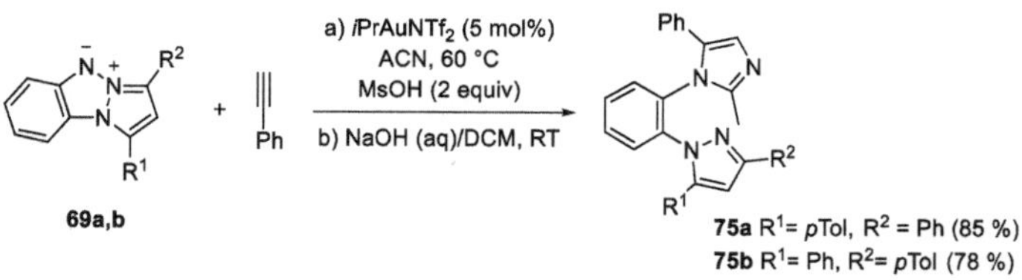

Scheme 19 The gold-catalyzed reaction of benzo-fused TAPs with phenylacetylene.

derivatives **75** in high yields. The structure of compound **75a** was confirmed by X-ray diffraction analysis.

Further investigation on benzo-fused TAPs **69** and ynamides reported by Ballesteros et al. provided 7-pyrazolylindoles **76** which were formed in the presence of a gold catalyst in DCE at 70 °C for 6 h (Scheme 20).[43] Differently from the previous reaction with phenylacetylene, these reactions could easily occur in the absence of methanesulfonic acid. Most attempted reactions gave excellent yields, regardless of whether R^3 included an electron-donating or -withdrawing group. Even for R^3 being an ester group, the yield was 80%. In addition, starting with ynamides derived from 2-oxazolidinone, the 7-pyrazolylindoles **76** were obtained mostly in good yields.

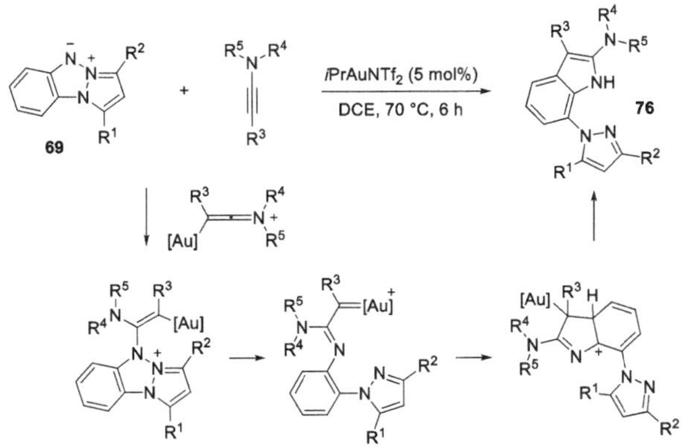

Scheme 20 Gold-catalyzed reactions of benzo-fused TAPs and ynamides.

3.2.2 Ring-opening reactions

Irradiation of TAPs **46a**, **46e** and **46h** was carried out by Bettinetti et al., affording nitrene-derived products (Scheme 21).[44] The photolysis of compound **46a** in benzene with or without di-*n*-propylamine (DPA) yielded the azo-product **77a** as the main compound. The irradiation of compounds **46e** and **46h** resulted in a mixture of products **47** and **48**. For TAP **46e**, the solvent had a mild impact on the yield of compound **47e**. Furthermore, the

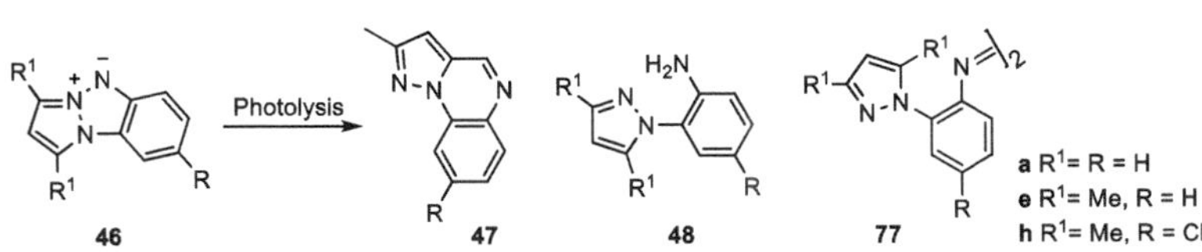

Scheme 21 The photolysis of compounds **46a**, **46e** and **46h**.

presence of DPA reduced the yields of products **47e** and **58e**. Interestingly, the reaction of TAP **46h** in benzene, regardless of the presence of the DPA, gave higher yields for both products **47h** and **48h** compared to reaction in *i*PrOH.[44]

On the other hand, photolysis of benzotriazapentalenes **46a** and **46e** by irradiation at 365 nm in argon matrices at 12 K, or in solution at 190 K gave a ring closure reaction to triazasemibullvalenes **78**, and these strained compounds were characterized spectroscopically (NMR, IR). The reaction is reversible and **78** goes back to **46** by heating or irradiation at 313 nm. Theoretical calculations were used to confirm the spectra and to rationalize the mechanism (Scheme 22).[45]

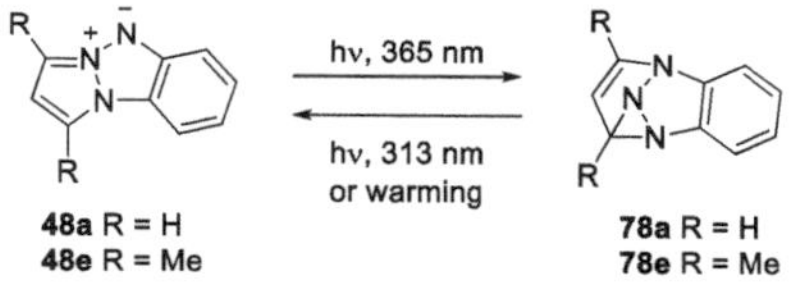

Scheme 22 Formation of triazasemibullvalenes **78** by irradiation of benzo-fused TAPs **46a** and **46e**.

3.2.3 Oxidative ring-opening reactions

The dye-sensitized (Rose Bengal, R. B.) photo-oxidation of compounds **46a** (R = H) and **46e** (R = Me) were also studied by the Bettinetti group (Scheme 23).[46,47] Two sets of compounds (**81**, **82** and **86**, **87**) were obtained in aprotic solvents via intermediates **80** and **85** arising from the attack of singlet oxygen and ring-opening of the adducts **79** and **84**. Surprisingly, the irradiation of compounds **46a** (R = H) and **46e** (R = Me) in methanol gave each

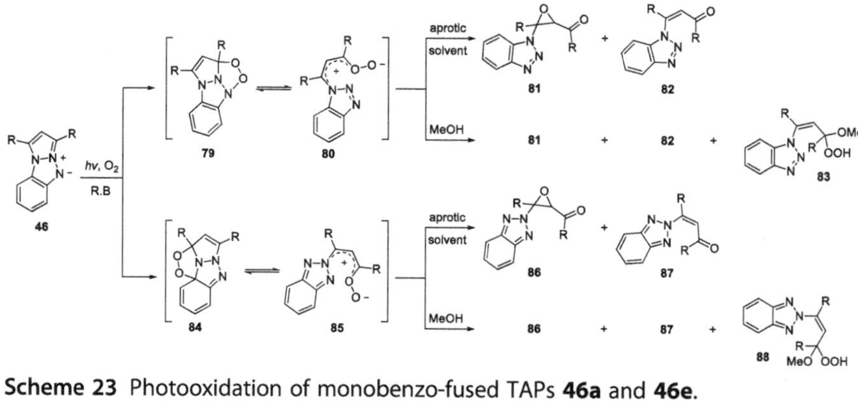

Scheme 23 Photooxidation of monobenzo-fused TAPs **46a** and **46e**.

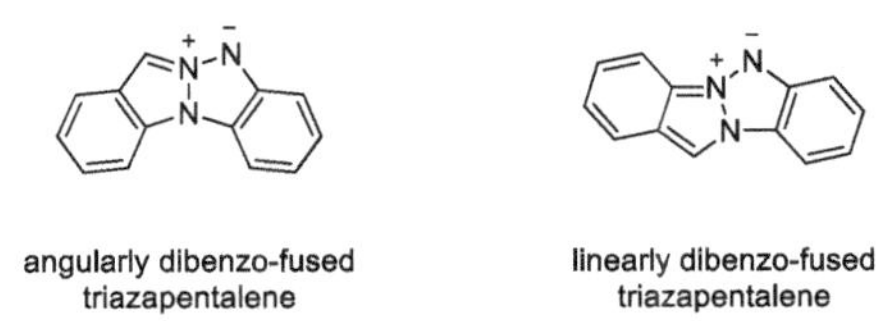

Fig. 4 Two isomeric dibenzotriazapentalenes.

two methoxy hydroperoxides **83** and **88** as major products due to trapping of intermediates **80** and **85** by methanol. These were accompanied by the aldehydes/ketones (**82** and **87**), as minor products.[46,47] Addition of compounds like trimethylsilyl phenyl ketone (TSPK), which might intercept carbonyl oxides, in aprotic solvents had little influence on the yields of oxiranes **81** and **86** (R = Me) but clearly diminished the yield of unsaturated ketones **82** and **87** (R = Me) in acetonitrile.[48]

4. Dibenzo-fused 1,3a,6a-triazapentalenes

Two possibilities exist to fuse two benzene rings to bicyclic triazapentalene, leading to an angular isomer and a linear isomer (Fig. 4). Most publications are about derivatives of the angular isomer. In this section, we will also discuss doubly fused TAPs with another carbocyclic ring including saturated rings.

4.1 Synthesis of dibenzo-fused 1,3a,6a-triazapentalenes

In 1971, the synthesis of dibenzo-TAP **90a** ($R^1 = R^2 = R^3 = H$) via the reductive cyclization of 1-(2-nitrophenyl)-1*H*-indazole **89a** ($R^1 = R^2 = R^3 = H$) was reported by Tsuge and Samura (Scheme 24A).[49] Tetracyclic TAP **90a**

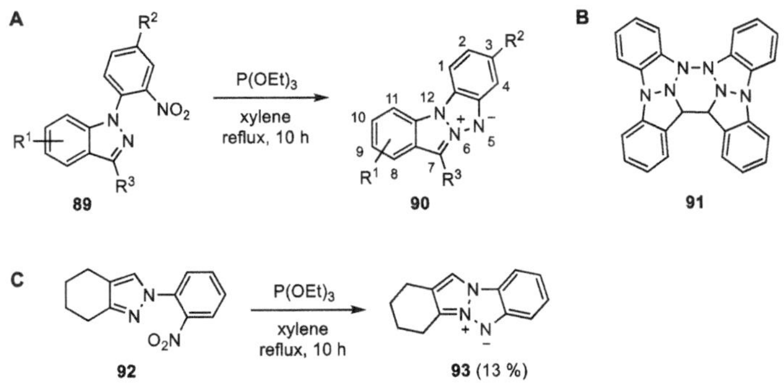

Scheme 24 (A) Reductive cyclization towards and numbering of dibenzo-fused TAPs **90**; (B) homocoupling product **91**; and, (C) synthesis of the linear tetrahydro derivative **93**.

was obtained in 30% yield, accompanied by homocoupling compound **91** in 31% yield. By photolysis with a 30 W mercury lamp (with pyrex filter) of the corresponding azidophenylindazole **94** in benzene at room temperature, triazapentalene **90** was obtained in 23% yield, while homocoupling azo compound **95** was collected as a side product in 20% yield (Scheme 25). In order to increase the stability of the unsubstituted TAP **90a**, triazapentalenes **90b–g** with various substituents were prepared via the reductive cyclization from **89** (Table 7). However, similar or lower yields were obtained compared to TAP **90a** ($R^1 = R^2 = R^3 = H$).[50] Similarly, linearly cycloalkyl-fused triazapentalene **93** was obtained from 2-(2-nitrophenyl)-tetrahydro-2*H*-indazole **92** in 13% yield.[30,51] The angularly fused structure was first erroneously reported by Tsuge and Samura as a product of 1-(2-nitrophenyl)-tetrahydro-1*H*-indazole, but this was rectified by the Meth-Cohn group a few years later.[30] Attempts to make the angularly fused structure from 1-(2-nitrophenyl)-tetrahydro-1*H*-indazole or 1-(2-azidophenyl)-tetrahydro-1*H*-indazole were unsuccessful.[30]

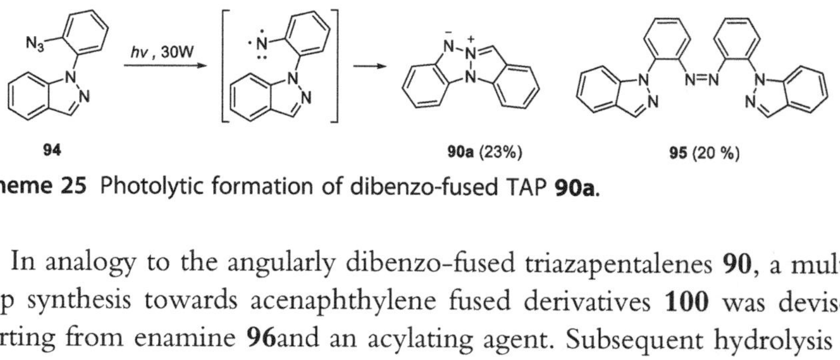

94 90a (23%) 95 (20 %)

Scheme 25 Photolytic formation of dibenzo-fused TAP **90a**.

In analogy to the angularly dibenzo-fused triazapentalenes **90**, a multistep synthesis towards acenaphthylene fused derivatives **100** was devised starting from enamine **96**and an acylating agent. Subsequent hydrolysis of

Table 7 Structures and yields of dibenzo-triazapentalenes **90** obtained by deoxygenation of 1-(2-nitrophenyl)-1*H*-indazoles **89**.

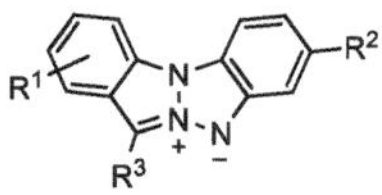	R^1					
	Me					Cl
	C8	C9	C10	H	H	C8
Comp.	**90b**	**90c**	**90d**	**90e**	**90f**	**90g**
R^2	H	H	H	H	Me	H
R^3	H	H	H	Me	H	H
Yield (%)	33	11	24	32	0	32

the acyl compounds **97** gave the acenaphthenones **98**. The latter were condensed with 2-nitrophenylhydrazine under acidic conditions, followed by Cadogan cyclization of the resulting fused pyrazole **99** in the presence of triethyl phosphite in *m*-xylene at reflux. The phenyl and methyl substituted benzoacenaphtho-fused triazapentalenes **100** were obtained in 71% (R = Ph) and 53% (R = Me) yield, respectively (Scheme 26).[52]

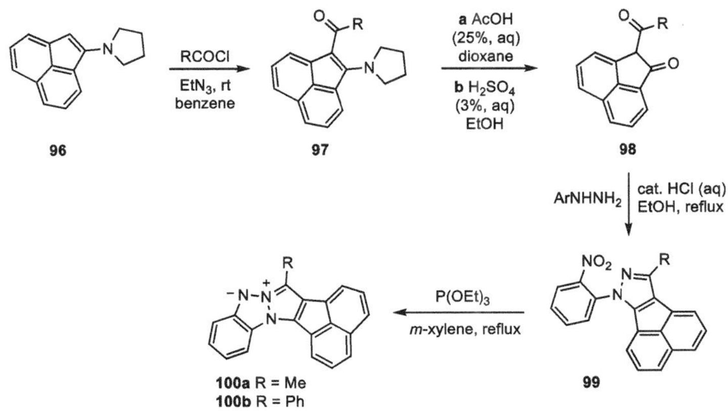

Scheme 26 Synthesis of benzoacenaphtho-fused triazapentalenes **100**.

Simamura et al. reported an efficient method to synthesize 2-(*o*-benzoylphenyl)-2*H*-benzotriazole **102** via the reduction of 2-(*o*-nitrophenyl) azobenzophenone **101** with triethyl phosphite at 160–170 °C. The intermediate benzotriazole **102** could be further transformed to linear dibenzo-fused triazapentalene **103** by increasing the temperature to 180–200 °C (Scheme 27).[53]

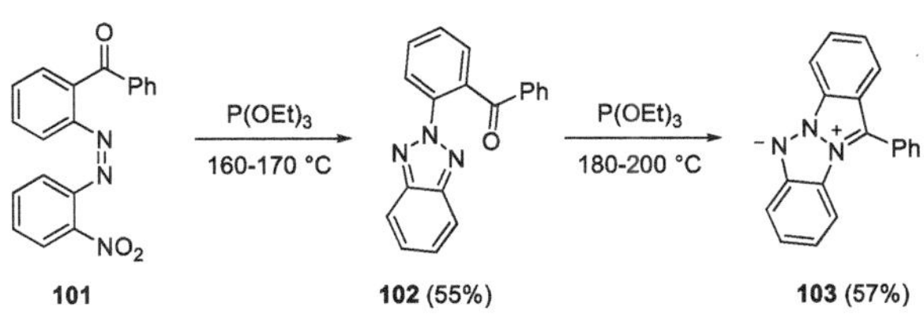

Scheme 27 Reductive cyclization of 2-*o*-nitrophenylazobenzophenone.

Nunn and Rowell reported linearly and angularly cycloalkyl-fused TAPs **105** and **107** via reductive deoxygenation of the corresponding (2-nitrophenyl)pyrazoles **104** and **106** (Scheme 28).[54] All products were obtained in fair yields.

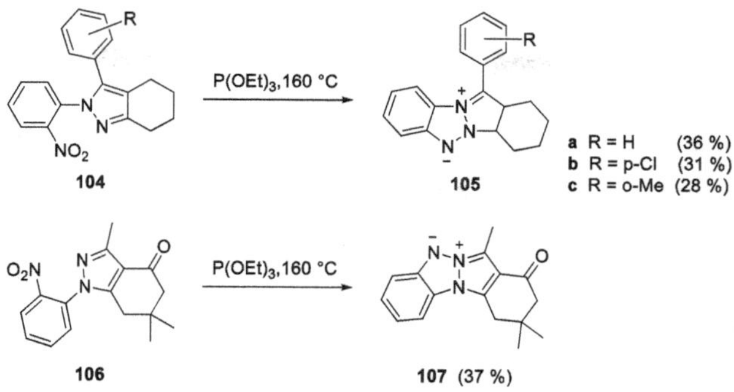

Scheme 28 Reductive deoxygenation towards cycloalkyl fused TAPs.

4.2 Reactivity of dibenzo-fused 1,3a,6a-triazapentalenes

4.2.1 Electrophilic substitutions

After the development of dibenzo-fused TAPs **90b** by Tsuge and Samura in 1971,[49] various electrophilic substitution reactions were investigated.[55] A first reaction with several acids was applied to 8-methyl dibenzo-fused TAP **90b**. Interestingly, the protonation reaction with 70% $HClO_4$ and 42% HBF_4 easily occurred at room temperature in methanol (Scheme 29), affording colorless compounds **108** in 80% and 99% yield, respectively.

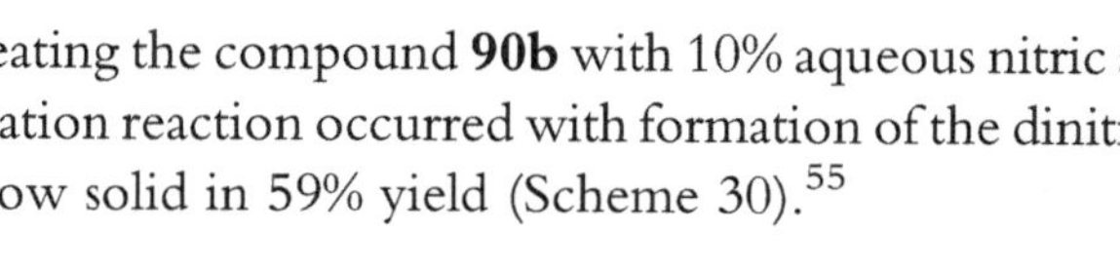

Scheme 29 Protonation of dibenzo-fused triazapentalene **90b**.

When treating the compound **90b** with 10% aqueous nitric acid at 20 °C for 3 h, a nitration reaction occurred with formation of the dinitro derivative **109** as a yellow solid in 59% yield (Scheme 30).[55]

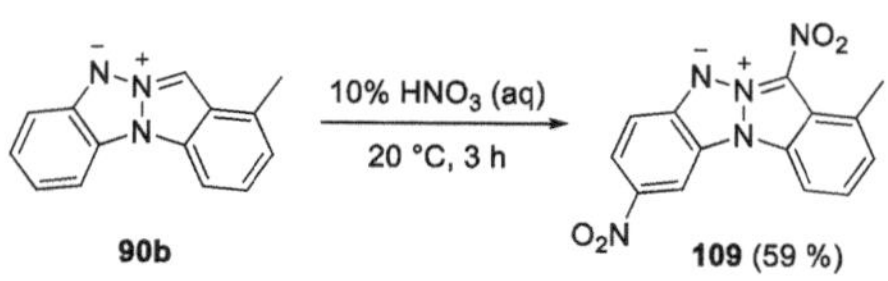

Scheme 30 The nitration of dibenzo-fused triazapentalene **90b**.

As the electrophilic substitution of compounds **90** occurred smoothly, a Vilsmeier reaction was carried out with $POCl_3$ in DMF at 0 °C

(Scheme 31).[55] As expected, the corresponding aldehydes **110** were obtained. The 7-acetyl derivative was produced in 93% yield via the reaction of compound **90b** and DMA/$POCl_3$ at room temperature for 12 h.

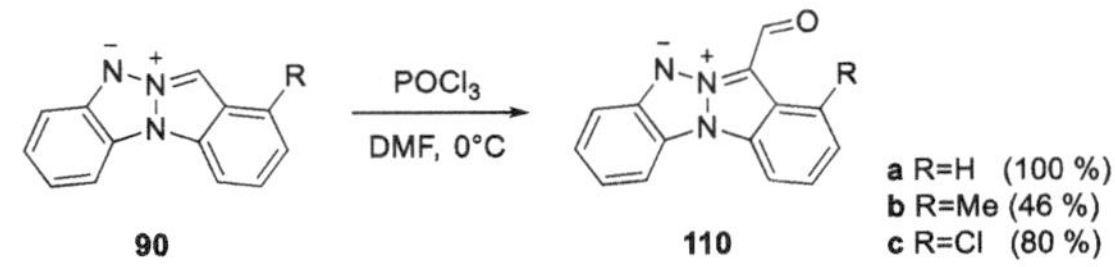

Scheme 31 Vilsmeier reactions of dibenzo-fused triazapentalenes **90**.

The Wittig and Knoevenagel condensation reactions were applied to aldehyde derivative **110a** with benzylidene(triphenyl)phosphorane and diethyl malonate reagents to obtain the corresponding alkenes **111** and **112** (Fig. 5) in 45% and 47% yield, respectively.[55]

Acylation of dibenzo-fused TAPs **90** with trifluoroacetic anhydride (TFAA) or 4-nitrobenzoyl chloride in CCl_4 at room temperature resulted in formation of the corresponding 7-trifluoroacetyl or *p*-nitrobenzoyl derivatives **113** in good to excellent yields (Table 8).[55]

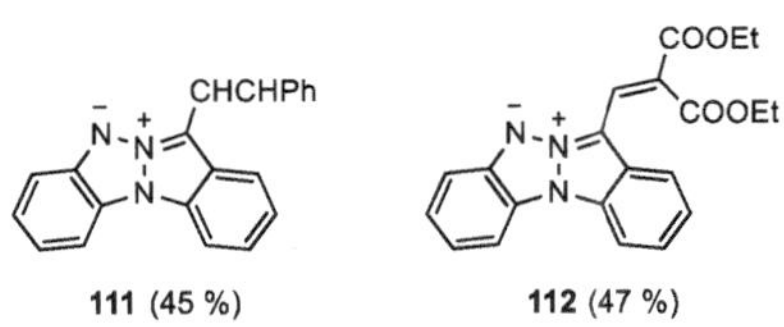

Fig. 5 The structure of compounds **111** and **112**.

Table 8 Acylation and reaction yields of compounds **113**

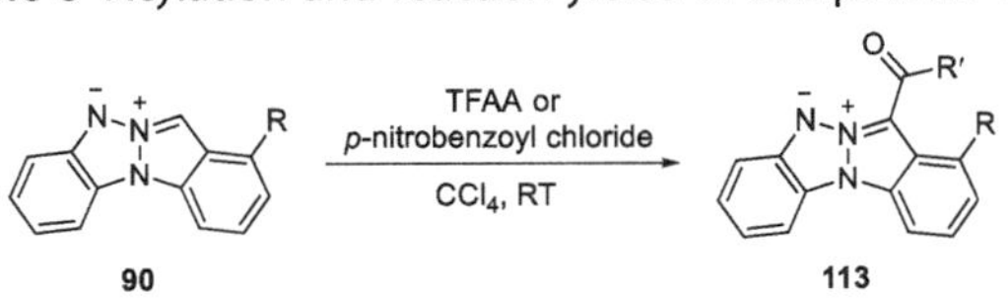

	R′ (Yield, %)	
Comp. 113	**CF_3**	**4-$C_6H_4NO_2$**
R=H	80	91
R=Me	83	90
R=Cl	76	95

4.2.2 Cycloaddition reactions

The 1,3-dipolar cycloaddition reactions of azomethine-like dibenzo-fused triazapentalenes **90** with acetylenes were studied by the Tsuge group (Scheme 32).[56,57] When a solution of TAP **90b** was treated with symmetrical acetylenes in refluxing ethanol (with catalytic sulfuric acid) or xylene, the cycloadducts **114** (R=R')were obtained as the major compound, accompanied by Michael-adducts **115** as the minor products for the acetylenes containing ester functions. The structures of the products were confirmed by spectral data. Moreover, the results of bromination of the minor product showed that the Michael-adduct **115** was indeed formed. In the reaction of **90e** with DMAD, the cycloadduct was formed in a high yield (90%) while the Michael adduct was not obtained.[56] Unsymmetrical terminal acetylenes like phenylacetylene (R=H, R'=Ph) underwent cycloaddition regioselectively. Interestingly, 1-chloro-4-ethynylbenzene and phenylacetylene add in opposite directions. On the other hand, the internal alkyne methyl phenylpropiolate (R=Ph, R'=COOMe) gave a mixture of regioisomeric adducts **114**. When **90b** was heated in DCM with two equivalents ofDMAD, the diazepine derivative **116** was formed in 60% yield.[57] Theoretical calculations (Hückel frontier molecular orbital) were used to rationalize the cycloaddition behavior.[58]

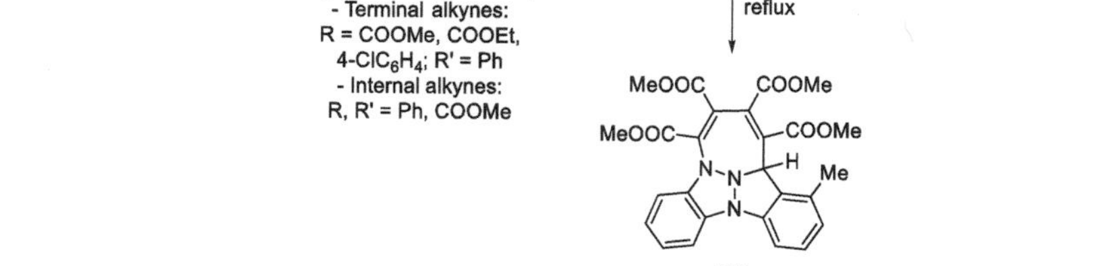

116

Scheme 32 The 1,3-dipolar cycloaddition of dibenzo-fused TAPs and acetylenes.

Analogously, the reactions of **100a,b** with DMAD at higher temperature gave mixtures of the cycloadducts **117** and pyrazinopyrimidines **118**; the yields strongly depend on the reaction condition. The structures of these products were confirmed by X-ray crystallographic analysis and a mechanism was formulated (Scheme 33).[59] TAP **100b** is clearly more reactive, and the methyl substituent of **118b** underwent a double Michael addition when excess of DMAD was added.

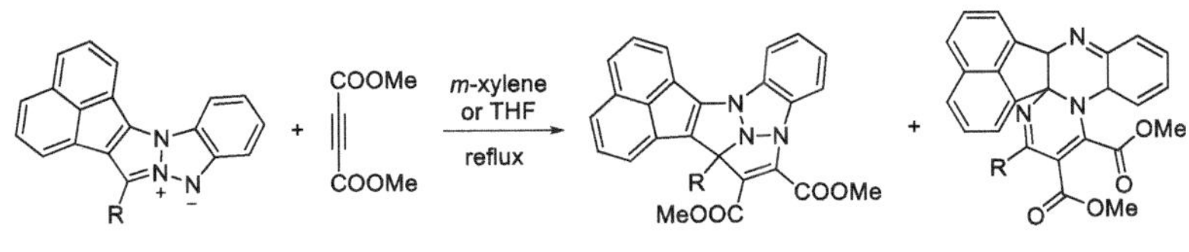

100a R = Ph
100b R = Me

117

118

Scheme 33 Cycloaddition of DMAD to benzoacenaphtho-fused TAPs.

5. Heteroaryl-fused 1,3a,6a-triazapentalenes

In recent years, several heterocycle-fused TAPs have been described with fluorescent properties, and applications in sensors and probes. The fused heterocycles investigated are mainly electron-poor pyridine and diazines.

5.1 Synthesis of heteroaryl-fused 1,3a,6a-triazapentalenes

5.1.1 Nitrene-mediated cyclizations

The thermal and photodecomposition of 2-azidophenazines was studied by the Bettinetti group.[60–62] The results from the photodecomposition of 2-azido-1-(3,5-dimethylpyrazol-1-yl)phenazine **119** in various solvents showed that the TAP **120** were produced by intramolecular trapping of the singlet nitrene in unreactive media (Table 9, entries 1–5).[61,62] In ethanol, the aminophenazine resulting from the triplet nitrene was formed as the major product, while in other solvents this was just a minor product. Singlet nitrenes were also intercepted by excess nucleophiles, and thus could not be trapped by the pyrazole nitrogen atom to form phenazine-fused TAPs (Table 9, entries 5–7).

Pyridine-fused TAPs **122** (Table 10), prepared by deoxygenation of nitro(pyrazolyl)pyridines **121** in the presence of triethyl phosphite under microwave irradiation, were reported by Guillaumet et al. in 2008

Table 9 Irradiation of compound **119** and reaction yields.

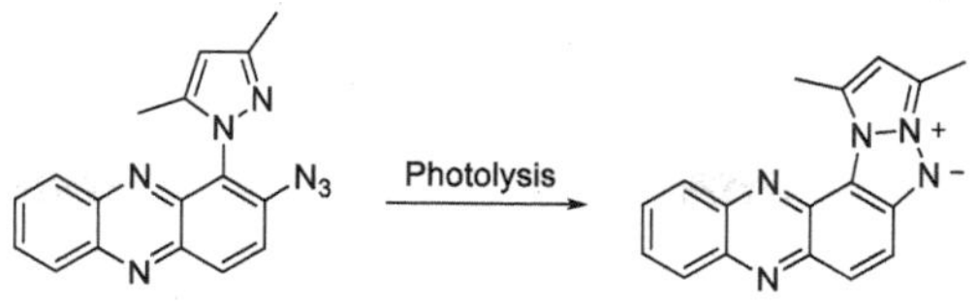

119 120

Entry	Solvent[a]	Yield (%)
1	Cyclohexane[b]	73
2	Tetralin[b]	71
3	ACN[b]	41
4	Benzene	64
5	Ethanol	14
6	Ethanol, 2M TFA	2
7	Ethanol, 2M EtONa	22

[a]Solution degassed by boiling and purged with Ar; Irradiated with a medium pressure mercury arc (Helios Italquartz, 125 W) through a Pyrex filter, unless otherwise specified.
[b]Solution degassed by boiling and cooled under a stream of nitrogen; Irradiated with a medium-pressure, water-cooled, mercury lamp (Hanau TQ 160) through a Pyrex filter.

Table 10 Reaction yields of the deoxygenation of nitro(pyrazolyl)pyridines **121** towards pyrido-fused TAPs **122**.

Comp. 122	R^1	R^2	R^3	Yields(%)
a (Y=N, X=Z=C)	H	H	H	85
b (Y=N, X=Z=C)	Me	Me	H	92
c (Y=N, X=Z=C)	H	CF_3	H	82
d (X=N, Y=Z=C)	H	H	H	20
e (Y=N, X=Z=C)	H	H	OMe	–[a]

[a]Not formed.

(Scheme 34).[63] The nitro(pyrazolyl)pyridines **121** were accessible via the nucleophilic aromatic substitution of chloropyridines with pyrazoles. Interestingly, although the deoxygenation reaction was only to a minor extent affected by the substituents on the pyrazole core, modification of the nitropyridine had a large effect on the reaction outcome. In the case

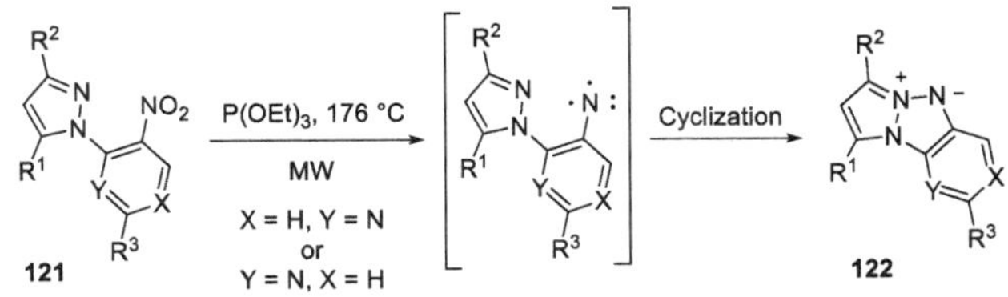

Scheme 34 Microwave-promoted cyclization of nitro(pyrazolyl)pyridines **121**.

of 6-methoxy-3-nitro-2-(1*H*-pyrazol-1-yl)pyridine **121e**, instead of N–N bond formation between the pyrazole nitrogen and the nitrene, degradation products and the corresponding phosphoramidate were formed. Degradation was also observed in the deoxygenation of **121d**, even though the temperature was lowered to 105 °C. As a result, the alternatively fused pyrido-TAP **122d** was only obtained in 20%.

Later in 2009, the same group prepared a library of pyrazine-fused TAPs **124** by thermolysis of 2-azido-3-(pyrazol-1-yl)pyrazines **123**.[64] These intermediate compounds can be formed by nucleophilic aromatic substitution starting from nitropyrazoles and 2-azido-3-chloropyrazine. Next, cyclization of **123** easily occurred in dichlorobenzene at 165 °C (Table 11). The TAP structures **122** and **124** were confirmed by NMR spectroscopy.[65]

To further investigate the fused TAPs, various pyrazino-fused TAPs **124** were synthesized under similar conditions in good to excellent yields.[66] In addition, two patents with much extra information have also appeared on this chemistry.[67,68]

Table 11 The synthesis of heterocyclic triazapentalenes and reaction yields of the thermolysis of azido(pyrazolyl)pyrazines **123** towards pyrazino-fused TAPs **124**.

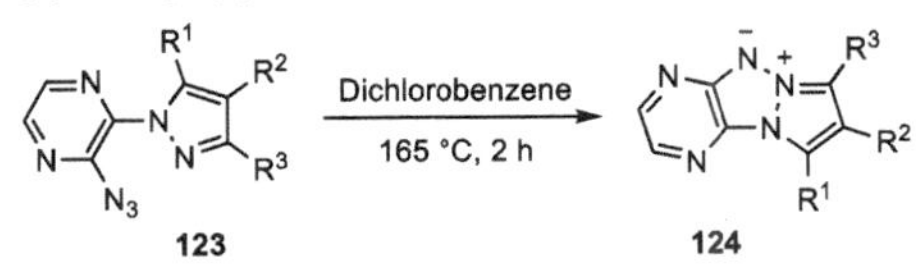

Comp.	R^1	R^2	R^3	Yield (%)
124a	H	H	H	85
124b	H	H	NO_2	97
124c	H	NO_2	H	91
124d	H	NO_2	NO_2	63

5.1.2 Oxidative intramolecular N–N bond formation

In 2020, Suzenet et al. reported an efficient method to synthesize heteroaryl-fused TAPs **124,127–131b** via the direct formation of an N–N bond in the presence of hypervalent iodine (Scheme 35).[69] Starting from halogenated heterocyclic amines and pyrazole derivatives, coupling reactions or nucleophilic aromatic substitution provided intermediate heteroaromatic amines **125**. Alternatively, 4-pyrazolyl-5-aminopyridines (**125**) could also be prepared from iodopyridinones via chlorination with $POCl_3$, substitution with pyrazole and finally a copper-catalyzed amination with aqueous ammonia. Next, oxidative cyclization of the amines readily occurred in a basic environment at room temperature in the presence of two equivalents of $PhI(OAc)_2$, liberating iodobenzene and acetic acid. This is unlikely to involve a radical process. The reaction yields were variable. In general for the diazine rings, the presence of an electron-withdrawing group on the pyrazine ring was beneficial for the reaction yield. However, high yields were also obtained without the presence of an electron-withdrawing group in the case of a pyrimidine with the amine located at the α-position. This showed that the position of the amine also influences the reaction outcome. In contrast, in the case of pyridine derivatives of **125**, the π-deficient character of the pyridine ring is probably too low to stabilize the iminoiodinane intermediate **126**, eliminating its reactivity in the absence of an electron-withdrawing group. For diazines, the influence of substituent on the pyrazole ring was studied. The presence of an electron-donating group had only a minor effect on the yield, while a trifluoromethyl (CF_3)

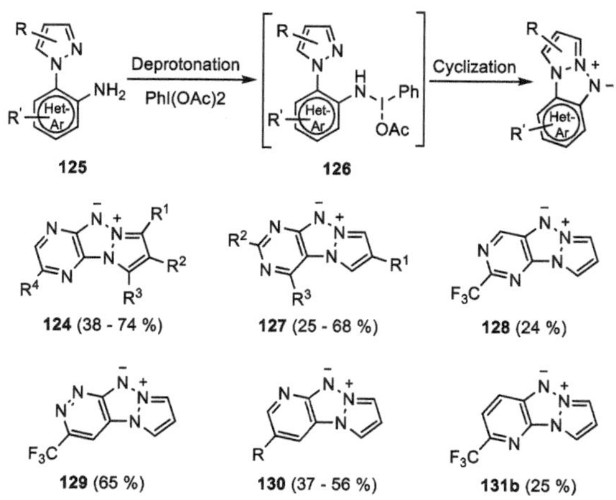

Scheme 35 Heteroaryl-fused TAPs via oxidative intramolecular N–N bond formation.

substituent on the pyrazole significantly lowered the yield. By comparison, the presence of a functional group on the pyrazole ring had little influence on the reaction yield when a CF_3-group was present on the pyrazine ring.

5.2 Reactivity of heteroaryl-fused 1,3a,6a-triazapentalenes

5.2.1 *Oxidative ring-opening reactions*

The first photo-oxidation reactions of triazapentalenes **120** and **133** were reported by the Bettinetti group in 1980 (Scheme 36).[70] The irradiation of phenazine-fused dimethyl-TAP **120** gave epoxy-ketone **132** in high yield (over 80%) in acetonitrile (ACN) both with and without sensitizer. In the presence of Rose Bengal (R. B.) as the sensitizer, irradiation of compound **133** obtained from cyclization of the corresponding 1-pyrazolyl-2-nitrene resulted in α,β-unsaturated aldehyde **134** in 42% yield. TAPs **120** and **133** could be reduced with Pd/C and hydrogen in benzene.[71] Although cleavage of the N–N bond could be expected, the bridging double bond between the outer quinoxaline and TAP moieties was reduced selectively in high yields. Photo-oxidation of the reduced products in the presence of R. B. gave the corresponding epoxy-ketone (from **120**) and unsaturated aldehyde (from **133**).[70]

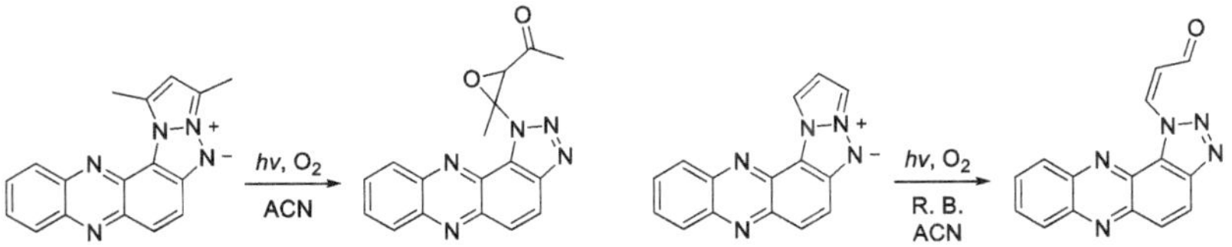

120 132 133 134

Scheme 36 Photo-oxidation of phenazine-fused TAPs **120** and **133**.

5.2.2 *Cycloaddition reactions*

Pentaazacyclic compounds **120** and **133** reacted with dimethyl acetylenedicarboxylate (DMAD) to give two different adducts **135** and **136** for the dimethyl derivative **120**, in 67% and 20% yield, respectively, and 77% yield of only one adduct **135** for the unsubstituted derivative **133** (Scheme 37).[72]

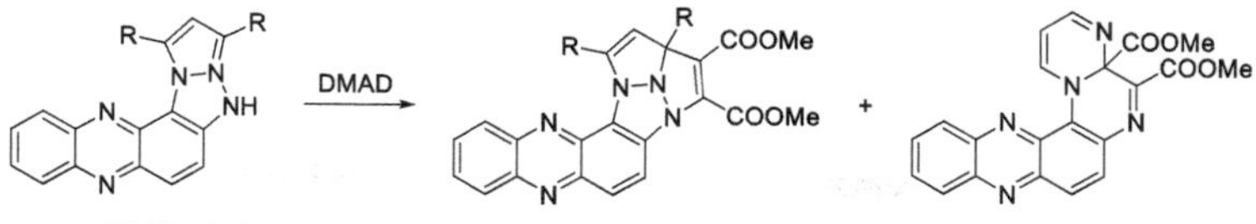

Scheme 37 1,3-Dipolar cycloaddition of phenazine-fused TAPs **120** and **133** with DMAD.

5.2.3 Suzuki reactions

Starting from the pyrazino-fused iodo- and bromo-TAPs **124k** and **124l**, Suzuki coupling reactions provided (hetero)aryl-substituted triazapentalenes **124t–z** in excellent yields under reflux conditions or in the microwave at 150 °C (Scheme 38).[66]

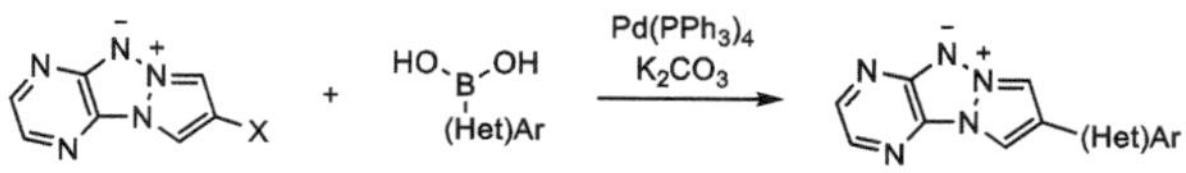

Scheme 38 Suzuki reaction starting from iodo- and bromo-TAPs **124k** and **124l**.

5.3 Applications of heteroaryl-fused 1,3a,6a-triazapentalenes

The hetero-aryl fused TAP show fluorescent properties that are of interest for potential future applications. The effect of the nitrogen atoms in the heteroaryl moiety on the fluorescent properties has been studied by Suzenet et al. (Table 12).[66] It is clear that the presence of nitrogen atoms in the aromatic ring drastically increases the quantum yield of fluorescence, and the number of nitrogens and position of aza substitution affect the fluorescence wavelengths. By comparison, adding one or more nitrogen atoms leads to a bathochromic shift (Table 12, entries **46a**, **131a**, **124a**). However, by changing the position of the extra nitrogen atom (from pyrazine to pyrimidine), a hypsochromic shift was observed (Table 12, entry **127a**). Furthermore, although compounds **124a** and **127a** gave the same quantum yield in DMSO, the molar extinction coefficient of compound **127a** was drastically decreased.[66]

In addition, the effect of substituents on the azole moiety on the fluorescent properties was studied (Table 13).[66] The presence of an electron-donating group on the azole moiety (compounds **124e**, **124h**) leads to a red shift of the emission wavelength and reduction of the quantum yield

Table 12 Fluorescence properties of different (hetero)aryl-fused TAPs in DMSO.[66]

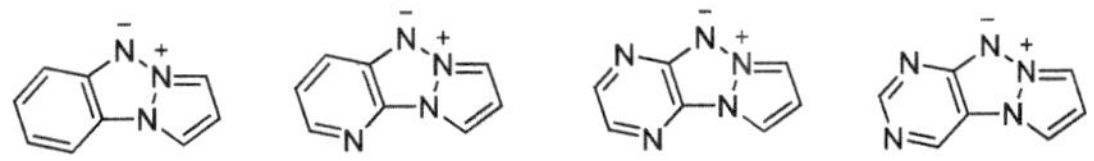

46a 131a 124a 127a

Entry	$\lambda_{abs,max}$ (nm)	$\lambda_{em,max}$ (nm)	Stokes shift (nm)	ε ($dm^3\ mol^{-1}\ cm^{-1}$)	φ_F (%)
48a	370	397	27	12,700	≤0.1
131a	387	446	59	18,700	2
124a	418	516	98	15,500	15
127a	362	440	78	1200	15

Table 13 Fluorescent properties and reaction yields of pyrazino-fused TAPs in DCM.[66]

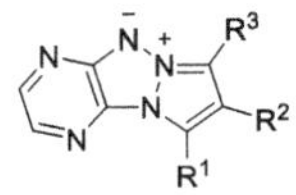

124	$\lambda_{abs,max}$ (nm)	$\lambda_{em,max}$ (nm)	Stokes shift (nm)	ε ($dm^3\ mol^{-1}\ cm^{-1}$)	φ_F (%)
124a ($R^1 = R^2 = R^3 = H$)	415 (418)[a]	494 (516)[a]	79 (98)[a]	13,000 (15,500)[a]	42 (15)[a]
124e ($R^3 = OMe$, $R^1 = R^2 = H$)	436[a]	529[a]	93[a]	13,200[a]	3[a]
124f ($R^3 = CF_3$, $R^1 = R^2 = H$)	403	464	61	14,706	3
124g ($R^3 = CO_2Et$, $R^1 = R^2 = H$)	416	473	57	24,157	2
124h ($R^1 = R^3 = H$, $R^2 = OMe$)	410 (415)[a]	512 (521)[a]	102 (106)[a]	7105 (13,700)[a]	8 (3)[a]
124i ($R^1 = R^3 = H$, $R^2 = CF_3$)	409	467	58	14,911	52
124j ($R^1 = R^3 = H$, $R^2 = CO_2Et$)	418	473	55	11,470	73
124k ($R^1 = R^3 = H$, $R^2 = I$)	414	483	69	16,443	27
124l ($R^1 = R^3 = H$, $R^2 = Br$)	412	487	75	19,300	27

[a]In DMSO.

of fluorescence. In contrast, an electron-withdrawing group causes a blue shift. Although an electron-withdrawing group in one position (**124f**, **124g**) resulted in a decreased quantum yield, the same substituents in another position **(124i, 124j)** increased quantum yields. In addition, in one case, the presence of an ester group (**124g**) highly increased the molar extinction coefficient. For compounds **124k** and **124l**, the presence of a halogen atom produced a reduction of the quantum yield with a slight effect on the other fluorescent properties. Further data on the fluorescence properties of heteroaryl fused triazapentalenes are found in Tables 14–16.

It is clear that the quantum yields of the diazino-fused TAPs are higher than the pyridino-fused TAPs.[66] The substituents on heteroaryl-fused TAPs have little impact on the wavelength and Stokes shift; however, a strong effect on the molar extinction coefficient and quantum yield is observed.

Table 14 Fluorescent properties and reaction yields of pyrazino-fused TAPs **124** in DCM.[69]

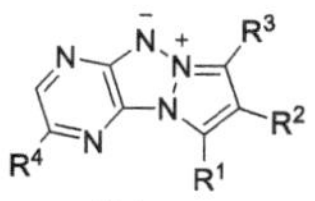

124

	$\lambda_{abs,}$ max (nm)	$\lambda_{em,}$ max (nm)	Stokes shift (nm)	ε ($dm^3\ mol^{-1}\ cm^{-1}$)	φ_F (%)
a ($R^1 = R^2 = R^3 = R^4 = H$)	415	494	79	13,000	42
h ($R^1 = R^3 = R^4 = H$, $R^2 = OMe$)	410	512	102	7100	8
m ($R^1 = R^2 = R^3 = H$, $R^4 = CF_3$)	415	516	101	6700	13
n ($R^1 = R^2 = R^3 = H$, $R^4 = Br$)	434	418	84	13,200	23
o ($R^1 = R^3 = R^4 = H$, $R^2 = Me$)	416	507	91	12,100	29
p ($R^1 = R^3 = Me$, $R^2 = R^4 = H$)	424	529	105	8900	10
q ($R^1 = R^3 = H$, $R^4 = R^2 = CF_3$)	412	489	77	9100	47
r ($R^1 = R^3 = H$, $R^4 = CF_3$, $R^2 = OMe$)	411	536	125	13,600	1
s ($R^1 = R^3 = H$, $R^4 = CF_3$, $R^2 = Br$)	413	507	94	13,800	21

Table 15 Fluorescence properties and reaction yields of pyrimidino-fused TAPs **127** in DCM.[69]

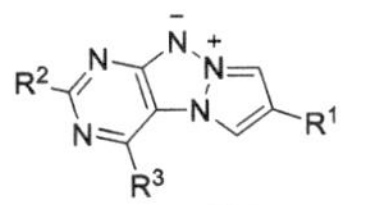

127

	$\lambda_{abs,max}$ (nm)	$\lambda_{em,max}$ (nm)	Stokes shift (nm)	ε ($dm^3\ mol^{-1}\ cm^{-1}$)	φ_F (%)
a ($R^1 = R^2 = R^3 = H$)	364	428	64	8600	2
b ($R^1 = R^2 = H$, $R^3 = CF_3$)	382	476	92	12,000	13
c ($R^1 = R^3 = H$, $R^2 = CF_3$)	371	452	81	3900	28
d ($R^1 = R^3 = H$, $R^2 = OMe$)	353	421	68	12,900	1
e ($R^1 = Me$, $R^2 = R^3 = H$)	364	434	70	7900	1
f ($R^1 = OMe$, $R^2 = R^3 = H$)	360	439	79	10,200	≤1
g ($R^1 = CF_3$, $R^2 = R^3 = H$)	360	410	50	13,200	≤1

Table 16 Fluorescence properties and reaction yields of heteroaryl-fused TAPs **128–131** in DCM.[69]

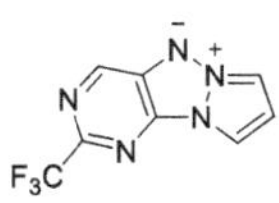

128 (24 %)

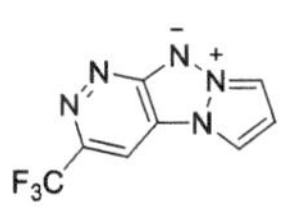

129 (65 %)

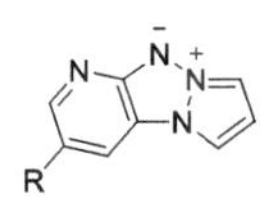

130 (37 - 56 %)

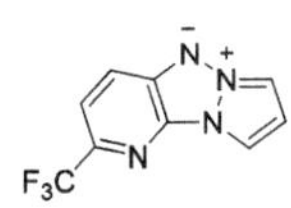

131b (25 %)

Entry	$\lambda_{abs,max}$ (nm)	$\lambda_{em,max}$ (nm)	Stokes shift (nm)	ε ($dm^3\ mol^{-1}\ cm^{-1}$)	φ_F (%)
128	396	470	74	7900	2
129	410	507	97	4300	2
130a (R = H)	–[a]	–[a]	–[a]	–[a]	–[a]
130b (R = CF_3)	395	436	41	22,400	4
130c (R = CN)	396	438	42	18,300	9
131a (R = H)	–[a]	–[a]	–[a]	–[a]	–[a]
131b	394	444	49	15,800	3

[a]No data available.

Compared to pyrazino-fused TAP **124**, the pyrido- and pyrimido-fused TAPs (**131** and **127**, respectively) showed an extensive hypsochromic shift and a smaller Stokes shift. Relatively high quantum yields were obtained for pyrazino-fused TAPs derivatives, i.e., 42% (**124a**) and 47% (**124q**) in DCM.

The fluorescence data of the conjugated pyrazino-fused TAPs **124t–z** prepared by Suzuki reaction showed that an electron-donating group is detrimental for the quantum yield, while also being beneficial to the Stokes shift and the molar extinction coefficient. In contrast, compounds containing an electron-withdrawing group **124w**, **x** and *N*-heteroaryl group **124y** show relatively high quantum yields (Table 17).

The first application of heteroaryl-fused TAPs was as a fluorescent probe for imaging of living cells (2019CC7776).[66] Epifluorescence microscopy experiments were carried out on HeLa cells with pyrazino-fused TAP **124w** (Fig. 6). The observed results showed that the dye **124w** could be a novel fluorescent probe with a good photostability in living cells.

Another application was based on bromo- and iodo substituted pyrazino-fused TAPs (as for instance **124l**), which were transformed into alkynyl-substituted TAPs via a Sonogashira cross-coupling reaction. A carboxylic

Table 17 Fluorescent properties and reaction yields of compounds **124t–z**.

(Het)Aryl

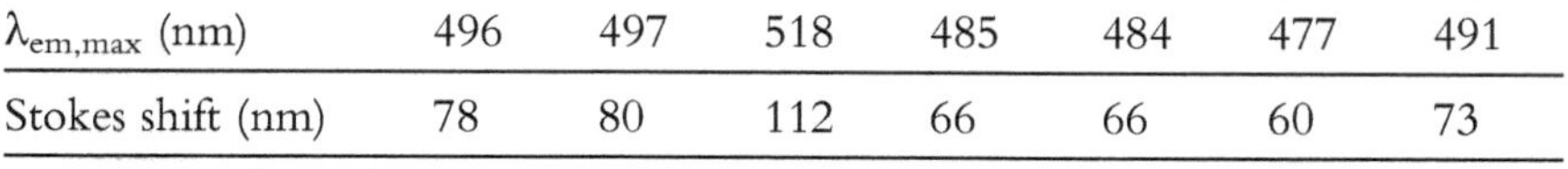

Comp. 124	t	u	v	w	x	y	z
$\lambda_{abs,max}$ (nm)	418	417	406	419	418	417	418
$\lambda_{em,max}$ (nm)	496	497	518	485	484	477	491
Stokes shift (nm)	78	80	112	66	66	60	73
ε ($dm^3\ mol^{-1}\ cm^{-1}$)	13,781	18,185	16,151	14,200	14,711	11,500	15,085
φ_F (%)	45	29	4	48	52	51	38

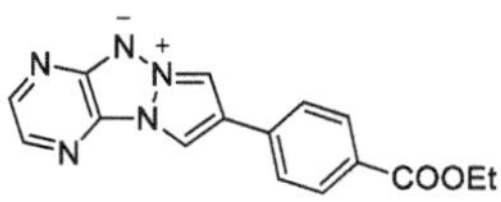

Fig. 6 The structure of fluorescence probe **124w**.

acid-functionalized derivative **137** was prepared from the coupling with methyl-5-hexynoate followed by hydrolysis in basic environment (Scheme **39**).[73] In a next study, TAP **137** was incorporated into a layered

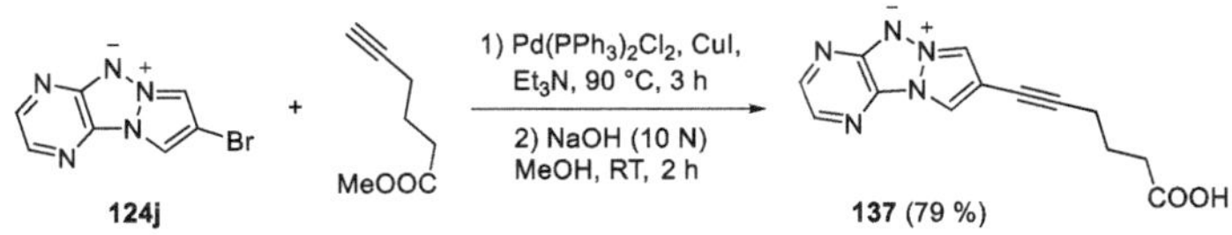

Scheme 39 Synthesis of alkynyl-substituted pyrazino-fused TAP **137**.

double hydroxide (LDH) via a direct co-precipitation method.[74] These dye co-intercalated LDH powders barely showed fluorescence (*ca.* 2%) under irradiation with UV or blue excitation wavelength, while a 12% quantum yield (λ_{ex} 320 nm) was observed when the system was used as a paste. When the dye co-intercalated LDH powder was dispersed into polymers via a compatible LDH host matrix, not enough fluorescence was produced. However, poly(ethylene oxide) combined with the dye co-intercalated LDH paste showed a 12% quantum yield (λ_{ex} 315 nm). Notably, compound **137** in powder form exhibited 6% to 11% quantum yield (λ_{ex} 315 nm). This was the first time the solid fluorescent properties of fused triazapentalenes had been reported.[74]

The probes **138** and **139** (Fig. 7) were designed for the quantification of nitroreductase, which may be overexpressed in regions of tumor hypoxia. Fluorescence was turned on after reduction of the nitro group, and this could be well correlated to hypoxia in the highly metastatic human melanoma A2058 cell line. The low toxicity, high photostability, and efficient uptake in cells are other important features of these novel probes.[75]

6. Miscellaneous products having nonconjugated 1,3a,6a-triazapentalene structures

In the literature, there are some TAP derivatives that are non-aromatic because their conjugation is interrupted. We will not discuss these compounds in detail, but a number of reports will be mentioned. Thus, Potts

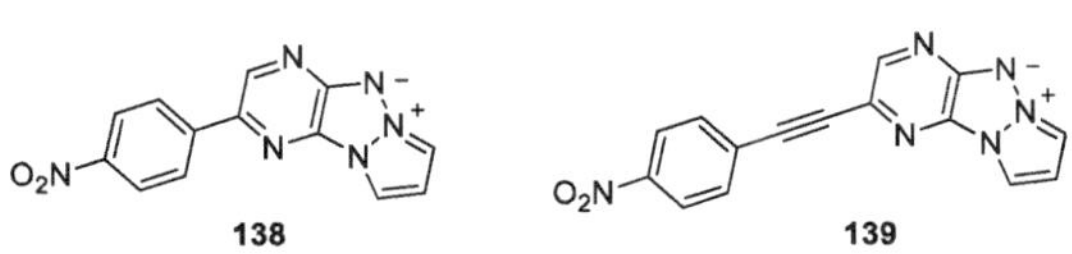

Fig. 7 The structure of probes **138** and **139**.

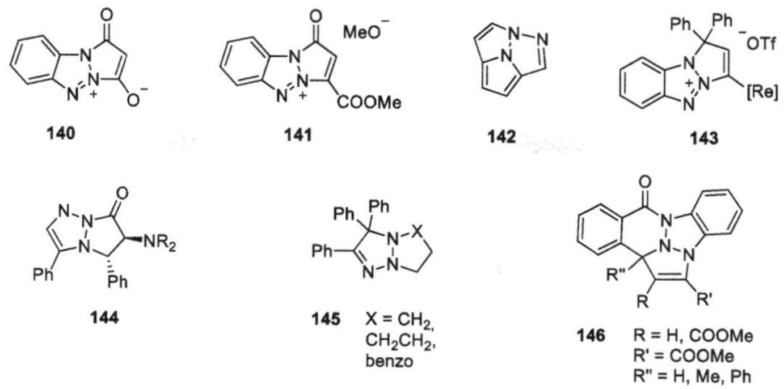

Fig. 8 Non-aromatic TAP derivatives.

reported a cross-conjugate betaine derivative **140** of the benzo-fused TAP, which was prepared from benzotriazole and phenyl(chlorocarbonyl) ketene.[76] The condensation of benzotriazole and DMAD was claimed to generate methoxide salt **141** (Fig. 8), but in our hands this unlikely compound was not possible to reproduce according to the reported procedure.[77] Triazaacepentalene **142** was described in a theoretical paper as a non-benzenoid aromatic system, together with other azaacepentalenes, but this compound has not been described experimentally.[78] Benzotriazole reacted with a rhenium(I) allenylidene complex ([Re] = [(triphos)$(CO)_2$Re], triphos = $MeC(CH_2PPh_2)_3$) to give the diphenyl substituted dihydrobenzo-TAP **143**.[79] The addition of cyanide anion onto azomethine imines gave phenyl substituted oxotetrahydro-TAPs **144**.[80,81] The reaction of 1-azido-3-iodopropane with 1,2,2-triphenyl vinyl Grignard reagent afforded reduced TAP **145**.[82] 1,3-Dipolar cycloaddition of mesomeric betaines to acetylenes led to the pentacyclic TAP derivatives **146** (Fig. 8).[83]

7. Conclusions

In the last decade, there have been several reports on triazapentalene (TAP) derivatives, fused or not, with the main goal of studying their photophysical properties or using them as sensors or probes. Traditional methods like nitrene insertions, with the nitrene resulting from nitro deoxygenation, thermal or photochemical azide decomposition, or amine oxidation, were used. The scope of the fluorescent triazapentalene derivatives was enlarged by palladium-catalyzed functionalization or electrophilic substitution reactions, although still many other possibilities remain to be investigated. In recent years, there have been no new reports on cycloaddition reactions of triazapentalenes, so this is a somewhat underexplored area, considering the many new derivatives that have been prepared. Given the small size of the triazapentalene and its unique structure, it may be expected that this system will continue to be investigated. However, in the case of the bicyclic non-fused TAPs, the inherent sensitivity of the TAP core will most likely remain a limiting factor.

References

1. Verbelen, B.; Dehaen, W. *Org. Lett.* **2016**, *18*, 6412–6415.
2. Wang, Y.; Opsomer, T.; Van Meervelt, L.; Dehaen, W. *J. Org. Chem.* **2020**, 9434–9439.
3. Wang, Y.; Opsomer, T.; Dehaen, W. *Chemosensors* **2021**, *9*, 16.
4. Ramsden, C. A. *Tetrahedron* **1977**, *33*, 3193–3202.
5. Ollis, W. D.; Stanforth, S. P.; Ramsden, C. A. *Tetrahedron* **1985**, *41*, 2239–2329.
6. Rodina, L. L.; Kolberg, A.; Schulze, B. *Heterocycles* **1998**, *49*, 587–618.
7. Koga, H.; Hirobe, M.; Okamoto, T. *Tetrahedron Lett.* **1978**, *19*, 1291–1294.
8. Namba, K.; Osawa, A.; Ishizaka, S.; Kitamura, N.; Tanino, K. *J. Am. Chem. Soc.* **2011**, *133*, 11466–11469.
9. Namba, K.; Mera, A.; Osawa, A.; Sakuda, E.; Kitamura, N.; Tanino, K. *Org. Lett.* **2012**, *14*, 5554–5557.
10. Namba, K.; Osawa, A.; Nakayama, A.; Mera, A.; Tano, F.; Chuman, Y.; Sakuda, E.; Taketsugu, T.; Sakaguchi, K.; Kitamura, N.; Tanino, K. *Chem. Sci.* **2015**, *6*, 1083–1093.
11. Nakayama, A.; Nishio, S.; Otani, A.; Mera, A.; Osawa, A.; Tanino, K.; Namba, K. *Chem. Pharm. Bull.* **2016**, *64*, 830–837.
12. Mera, A.; Ito, M.; Nakayama, A.; Namba, K. *Chem. Lett.* **2017**, *46*, 539–542.
13. Cai, R.; Wang, D.; Chen, Y.; Yan, W.; Geise, N. R.; Sharma, S.; Li, H.; Petersen, J. L.; Li, M.; Shi, X. *Chem. Commun.* **2014**, *50*, 7303–7305.
14. Namba, K.; Osawa, A.; Tanino, K. WO 2012121356, 2012.
15. Ito, M.; Mera, A.; Mashimo, T.; Seki, T.; Karanjit, S.; Ohashi, E.; Nakayama, A.; Kitamura, K.; Hamura, T.; Ito, H.; Namba, K. *Chem. A Eur. J.* **2018**, *24*, 17727–17733.
16. Osawa, A.; Mera, A.; Namba, K.; Tanino, K. *Synlett* **2013**, *24*, 207–210.
17. Thomas, J.; Jana, S.; John, J.; Liekens, S.; Dehaen, W. *Chem. Commun.* **2016**, *52*, 2885–2888.
18. Thomas, J.; Jana, S.; Liekens, S.; Dehaen, W. *Chem. Commun.* **2016**, *52*, 9236–9239.
19. Prakash, R.; Opsomer, T.; Dehaen, W. *Chem. Rec.* **2021**, *21*, 376–385.
20. Yan, W.; Wang, Q.; Chen, Y.; Petersen, J. L.; Shi, X. *Org. Lett.* **2010**, *12*, 3308–3311.
21. Chen, Y.; Wang, D.; Petersen, J. L.; Akhmedov, N. G.; Shi, X. *Chem. Commun.* **2010**, *46*, 6147–6149.
22. Hayashi, T.; Osawa, A.; Watanabe, T.; Murata, Y.; Nakayama, A.; Namba, K. *Tetrahedron Lett.* **2017**, *58*, 1961–1964.
23. Verbelen, B.; Boodts, S.; Hofkens, J.; Boens, N.; Dehaen, W. *Angew. Chem. Int. Ed.* **2015**, *54*, 4612–4616.
24. Kamada, R.; Tano, F.; Kudoh, F.; Kimura, N.; Chuman, Y.; Osawa, A.; Namba, K.; Tanino, K.; Sakaguchi, K. *PLoS One* **2016**, *11*, e0160625.
25. Sawada, J.; Osawa, A.; Takeuchi, T.; Kaneda, M.; Oishi, S.; Fujii, N.; Asai, A.; Tanino, K.; Namba, K. *Bioorg. Med. Chem. Lett.* **2016**, *26*, 5765–5769.
26. Nakayama, A.; Otani, A.; Inokuma, T.; Tsuji, D.; Mukaiyama, H.; Nakayama, A.; Itoh, K.; Otaka, A.; Tanino, K.; Namba, K. *Commun. Chem.* **2020**, *3*, 6.
27. Lynch, B. M.; Hung, Y.-Y. *J. Heterocycl. Chem.* **1965**, *3*, 218–219.
28. Harder, R. J.; Kauer, J. C. US 3262944, 1966.

29. McRobbie, I. M.; Meth-Cohn, O.; Suschitzky, H. *Tetrahedron Lett.* **1976**, *17*, 925–928.
30. Lindley, J. M.; McRobbie, I. M.; Meth-Cohn, O.; Suschitzky, H. *J. Chem. Soc. Perkin Trans. 1* **1980**, 982–994.
31. Albini, A.; Bettinetti, G. F.; Minoli, G. *Chem. Lett.* **1981**, *10*, 331–334.
32. Albini, A.; Bettinetti, G.; Minoli, G. *Heterocycles* **1995**, *40*, 597.
33. Albini, A.; Bettinetti, G.; Minoli, G. *J. Am. Chem. Soc.* **1997**, *119*, 7308–7315.
34. Albini, A.; Bettinetti, G.; Minoli, G. *J. Chem. Soc. Perkin Trans. 2* **1999**, (12), 2803–2807.
35. Albini, A.; Bettinetti, G.; Minoli, G. *J. Am. Chem. Soc.* **1999**, *121*, 3104–3113.
36. Carra, C.; Bally, T.; Albini, A. *J. Am. Chem. Soc.* **2005**, *127*, 5552–5562.
37. Clark, B. A. J.; McNab, H.; Sommerville, C. C. *Chem. Comm.* **1996**, 1211.
38. Harrington, R. W.; Stanforth, S. P. *Tetrahedron Lett.* **2012**, *53*, 2111–2113.
39. Kim, T.; Kim, K.; Park, Y. J. *Eur. J. Org. Chem.* **2002**, *2002*, 493–502.
40. Choi, Y.-A.; Kim, K.; Park, Y. J. *Tetrahedron Lett.* **2003**, *44*, 7507–7511.
41. Katritzky, A. R.; Hur, D.; Kirichenko, K.; Ji, Y.; Steel, P. J. *ARKIVOC* **2003**, *2004*, 109–121.
42. González, J.; Santamaría, J.; Suárez-Sobrino, Á. L.; Ballesteros, A. *Adv. Synth. Catal.* **2016**, *358*, 1398–1403.
43. Allegue, D.; González, J.; Fernández, S.; Santamaría, J.; Ballesteros, A. *Adv. Synth. Catal.* **2019**, *361*, 758–768.
44. Albini, A.; Bettinetti, G.; Minoli, G. *J. Am. Chem. Soc.* **1991**, *113*, 6928–6934.
45. Carra, C.; Bally, T.; Jenny, T. A.; Albini, A. *Photochem. Photobiol. Sci.* **2002**, *1*, 38–44.
46. Albini, A.; Bettinetti, G. F.; Minoli, G.; Vasconi, G. *J. Chem. Soc. Chem. Commun.* **1981**, *17*. 1089b–1091.
47. Albini, A.; Bettinetti, G. F.; Minoli, G. *J. Org. Chem.* **1983**, *48*, 1080–1083.
48. Albini, A.; Bettinetti, G.; Minoli, G. *J. Chem. Soc. Perkin Trans. 1* **1983**, 581–585.
49. Tsuge, O.; Samura, H. *J. Heterocycl. Chem.* **1971**, *8*, 707–710.
50. Tsuge, O.; Samura, H. *Org. Prep. Proced. Int.* **1972**, *4*, 273–281.
51. Tsuge, O.; Samura, H. *Org. Prep. Proced. Int.* **1974**, *6*, 161–167.
52. Tsuge, O.; Hatta, T.; Kojima, H.; Miyahara, N.; Sugaya, M. *J. Heterocycl. Chem.* **1994**, *31*, 1283–1287.
53. Lee, J. H.; Matsumoto, A.; Yoshida, M.; Simamura, O. *Chem. Lett.* **1974**, 951–954.
54. Nunn, A. J.; Rowell, F. J. *J. Chem. Soc. Perkin Trans. 1* **1975**, 7, 629–631.
55. Tsuge, O.; Samura, H. *Chem. Lett.* **1973**, 175–180.
56. Tsuge, O.; Samura, H. *Tetrahedron Lett.* **1973**, *14*, 597–600.
57. Tsuge, O.; Samura, H. *Heterocycles* **1974**, 27.
58. Mok, K.-L.; Nye, M. J. *J. Chem. Soc. Perkin Trans. 1* **1975**, (18), 1810–1818.
59. Tsuge, O.; Hatta, T.; Sugaya, M.; Yunoki, E.; Gau, H.; Kakehi, A. *Heterocycles* **1995**, 209.
60. Albini, A.; Bettinetti, G. F.; Minoli, G. *J. Chem. Soc. Perkin Trans. 1* **1981**, 4–7.
61. Albini, A.; Bettinetti, G.; Minoli, G. *J. Org. Chem.* **1987**, *52*, 1245–1251.
62. Albini, A. *Gazz. Chim. Ital.* **1988**, *118*, 61.
63. Nyffenegger, C.; Pasquinet, E.; Suzenet, F.; Poullain, D.; Jarry, C.; Léger, J.-M.; Guillaumet, G. *Tetrahedron* **2008**, *64*, 9567–9573.
64. Nyffenegger, C.; Pasquinet, E.; Suzenet, F.; Poullain, D.; Guillaumet, G. *Synlett* **2009**, *2009*, 1318–1320.
65. Palmas, P.; Nyffenegger, C.; Pasquinet, E.; Guillaumet, G. *Magn. Reson. Chem.* **2009**, *47*, 752–756.
66. Sirbu, D.; Diharce, J.; Martinić, I.; Chopin, N.; Eliseeva, S. V.; Guillaumet, G.; Petoud, S.; Bonnet, P.; Suzenet, F. *Chem. Commun.* **2019**, *55*, 7776–7779.
67. Suzenet, F.; Sirbu, D.; Bonnet, P.; Guillaumet, G. WO 2017/013135, 2017.
68. Suzenet, F.; Sirbu, D.; Guillaumet, G.; Bonnet, P. FR 3039153, 2017.

69. Daniel, M.; Hiebel, M.-A.; Guillaumet, G.; Pasquinet, E.; Suzenet, F. *Chem. A Eur. J.* **2020**, *26*, 1525–1529.
70. Albini, A.; Bettinetti, G. F.; Minoli, G.; Pietra, S. *J. Chem. Soc. Perkin Trans. 1* **1980**, 2904–2908.
71. Albini, A.; Bettinetti, G. F.; Minoli, G. *J. Chem. Soc. Perkin Trans. 1* **1981**, 1821–1825.
72. Albini, A.; Bettinetti, G.; Minoli, G.; Oberti, R. *J. Chem. Res. Synop.* **1980**, 404–405.
73. Sirbu, D.; Chopin, N.; Martinić, I.; Ndiaye, M.; Eliseeva, S. V.; Hiebel, M.-A.; Petoud, S.; Suzenet, F. *Int. J. Mol. Sci.* **2021**, *22*, 6645.
74. Legentil, P.; Chadeyron, G.; Therias, S.; Chopin, N.; Sirbu, D.; Suzenet, F.; Leroux, F. *Appl. Clay Sci.* **2020**, *189*, 105561.
75. Janczy-Cempa, E.; Mazuryk, O.; Sirbu, D.; Chopin, N.; Żarnik, M.; Zastawna, M.; Colas, C.; Hiebel, M.-A.; Suzenet, F.; Brindell, M. *Sens. Actuators B* **2021**, *346*, 130504.
76. Potts, K. T.; Murphy, P. M.; Kuehnling, W. R. *J. Org. Chem.* **1988**, *53*, 2889–2898.
77. Valizadeh, H.; Shomali, A.; Gholipour, H. *J. Heterocycl. Chem.* **2011**, *48*, 1440–1444.
78. Mascal, M. *J. Org. Chem.* **2007**, *72*, 4323–4327.
79. Mantovani, N.; Bergamini, P.; Marchi, A.; Marvelli, L.; Rossi, R.; Bertolasi, V.; Ferretti, V.; de los Rios, I.; Peruzzini, M. *Organometallics* **2006**, *25*, 416–426.
80. Svete, J.; Prešeren, A.; Stanovnik, B.; Golič, L.; Golič-Grdadolnik, S. *J. Heterocycl. Chem.* **1997**, *34*, 1323–1328.
81. Chuang, T.-H.; Sharpless, K. B. *Helv. Chim. Acta* **2000**, *83*, 1734–1743.
82. Suleymanov, A. A.; Ruggi, A.; Planes, O. M.; Chauvin, A.-S.; Scopelliti, R.; Fadaei Tirani, F.; Sienkiewicz, A.; Fabrizio, A.; Corminboeuf, C.; Severin, K. *Chem. A Eur. J.* **2019**, *25*, 6718–6721.
83. Morgan, D. O.; Ollis, W. D.; Stanforth, S. P. *Tetrahedron* **2000**, *56*, 5523–5534.

CHAPTER THREE

Type A mesoionic compounds (1980–2020)

Christopher A. Ramsden[a,*] and Florea Dumitrascu[b]

[a]Lennard-Jones Laboratories, School of Physical and Geographical Sciences, Keele University, Staffordshire, United Kingdom
[b]Center for Organic Chemistry, "C.D. Nenitzescu" Romanian Academy, Bucharest, Romania
[*]Corresponding author: e-mail address: c.a.ramsden@keele.ac.uk

Contents

Advances in Heterocyclic Chemistry, Volume 137
ISSN 0065-2725
https://doi.org/10.1016/bs.aihch.2021.09.003

Abstract

Type A mesoionic compounds are of particular interest due to their participation in synthetically useful 1,3-dipolar cycloaddition reactions. This review covers the chemistry of type A mesoionic rings published in the period 1980–2020 together with some work published in the first half of 2021. Type A mesoionic rings are conjugated heterocyclic mesomeric betaines (HMBs). Their relationship to other heterocyclic HMBs, including cross-conjugated HMBs, semi-conjugated HMBs and type B mesoionic rings, is discussed together with general aspects of their structure and properties. This is followed by a discussion of the chemistry of all the known type A ring systems that has been published since 1980.

Keywords: Mesoionic, 1,3-Dipoles, Mesomeric betaines, Cycloadditions, N-heterocyclic carbenes, Münchnones, Isomünchnones, Thioisomünchnones

1. Introduction

The first report of a mesoionic compound (dehydrodithizone **1**) was in 1882 by Fischer and Besthorn and the mesoionic literature from its origins in 1882 to September 1975 was comprehensively reviewed by Ollis and Ramsden.[1] The original survey was subsequently updated for the period 1976–1980 by Newton and Ramsden.[2] These earlier reviews are reproduced in Appendices 1 and 2 of this volume. This chapter covers the literature of type A mesoionic compounds for the period 1980 to December 2020. For a full coverage of the type A mesoionic literature this review should be used in combination with the two previous reviews. Type B mesoionic compounds are reviewed in the following chapter. Work described in the earlier reviews is not re-cited except, where appropriate, to place more recent results in context or to provide continuity of coverage. The chemistry of sydnones **2** and sydnone imines **3** is not included, except for discussions of general structural properties and reactions (Section 4), since these compounds were reviewed in Advances in Heterocyclic Chemistry in 2020 by Cherepanov and Moiseev,[3] and elsewhere.[4] During the period 1980–2020 there have been several reviews covering specific mesoionic derivatives or particular properties and these are referenced in the appropriate subsections. A general review by Gupta, Kumar and Gupta was published in 1999.[5] The chemistry of bioactive mesoionic heterocycles was reviewed in 2009.[6]

Mesoionic compounds cover a broad area of heterocyclic chemistry and, as with the previous reviews in this trilogy, it has been necessary to be selective in the choice of examples discussed in each section. However, we have attempted to be comprehensive in the citing of relevant references in the text so that readers can readily access the primary literature.

This chapter deviates from the two previous surveys in the representation of mesoionic rings. Formulae using a full circle with positive charge and an exocyclic negative charge are no longer regarded as useful and we subscribe to the use of mesomeric betaine structures with an exocyclic double bond, e.g., **1–3**, and the corresponding nomenclature, as used in the 1976 review.[1] Thus, sydnones **2** are named as 1,2,3-oxadiazol-5-ones. This is different from the current IUPAC nomenclature which is based on the historic betaine structures in which a positively charged aromatic ring is associated with an exocyclic negative charge, e.g., sydnones are named 1,2,3-oxazolium-5-olates. As a compromise, both names are given in the subheadings for individual rings in Section 5. The relevance of contributing mesomeric betaine resonance hybrids is discussed further in Section 4.

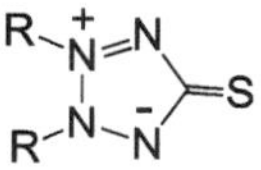

1

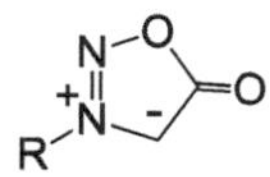

2

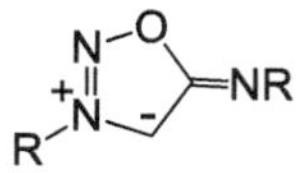

3

It should be noted that a subclass of N-heterocyclic carbenes is widely referred to as mesoionic due to their representation by dipolar structures. These interesting species are discussed elsewhere and are not covered by this review.[7,8]

2. Type A and type B mesoionic heterocycles

Mesoionic heterocycles are heterocyclic mesomeric betaines (HMBs), a class of heterocycles that is now recognized as large and diverse. When the term mesoionic was originally coined the extent of this class was not clear. Subsequently it was recognized that to retain the usefulness of the term mesoionic it was desirable to restrict its use to two types of five-membered HMB having the general structures **4** and **6**, and referred to as type A and type B.[1] This definition has been widely accepted. Detailed discussion of the history and definition of the term mesoionic can be found in an earlier review.[1] Both types of mesoionic ring have eight π electrons in conjugation. They differ in the origins of these π electrons.

In type A mesoionic rings the origins of the π electrons are shown in structure **4c** (Fig. 1) in which the atoms a and c each donate two electrons

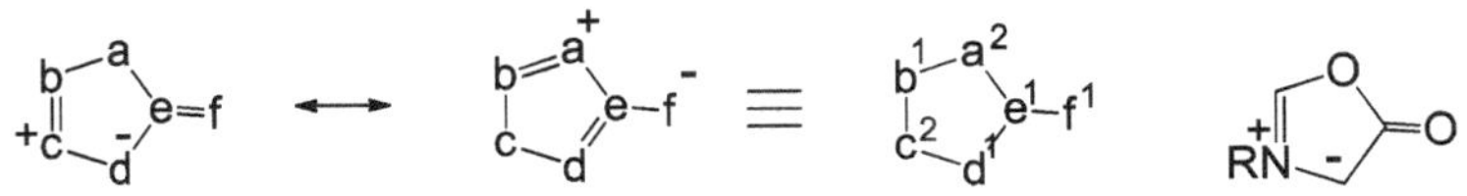

4a 4b 4c 5

Type A

Fig. 1 The general structure of type A mesoionic heterocycles **4**, exemplified by the münchnones **5**.

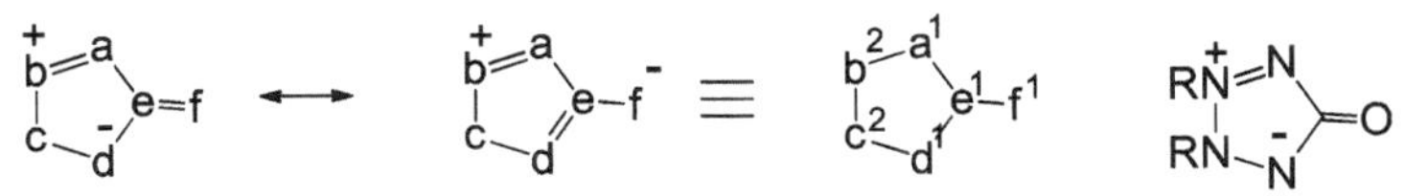

6a 6b 6c 7

Type B

Fig. 2 The general structure of type B mesoionic heterocycles **6**, exemplified by the tetrazol-5-ones **7**.

to the π conjugation. This is illustrated by the münchnones (1,3-oxazol-5-ones) **5**, which are named after the city where they were conceived.

In type B mesoionic rings the origins of the π electrons are shown in structure **6c** (Fig. 2). In this case two π electrons are each donated by the adjacent atoms b and c. Examples of type B mesoionic heterocycles are the tetrazolium-5-olates **7**. So far, no family of type B mesoionic rings is illustrious enough to have a common name.

3. The relationship of mesoionic heterocycles to other heterocyclic mesomeric betaines (HMBs)

Heterocyclic mesomeric betaines (HMBs) are the constituents of a large branch of heterocyclic chemistry. It is reasonable to ask the following questions: (1) what do they have in common? (2) why can they only be represented by dipolar structures? and (3) how do subgroups of HMBs (e.g., **11–13**, Fig. 3) differ from each other? These questions are answered in the following subsections.

3.1 Structural properties common to all HMBs

A molecule can only be represented by a dipolar structure if the conjugated π system contains either (i) two odd conjugated fragments separated by heteroatoms donating a lone pair or (ii) an even non-Kekulé component.

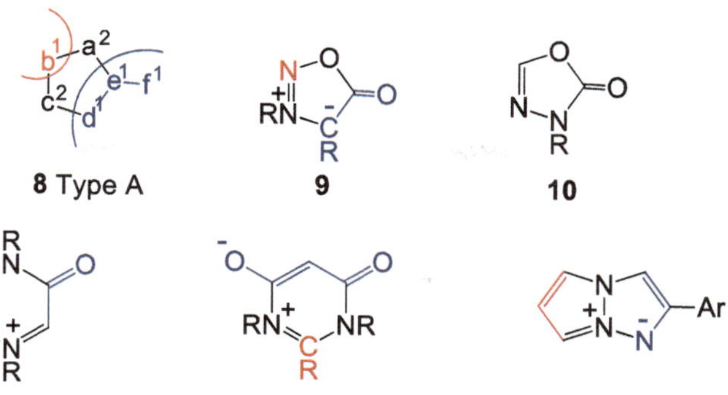

8 Type A **9** **10** **11** **12** **13**

Fig. 3 HMBs containing isolated odd conjugated fragments

These two possibilities are conveniently illustrated by mesoionic heterocycles of type A and type B.

(i) Consider the components of the generalized type A system as shown in structure **8** (Fig. 3). There are two odd-numbered fragments: b and d-e-f. These are mutually isolated by the heteroatoms a and c. Molecules can only be represented by fully covalent structures if all atoms can be paired to form double bonds, and this is not possible with isolated odd fragments. This is illustrated by the sydnones **9**; their isomers **10** contain even fragments and can be represented by covalent structures.

This is a general rule which, in addition to type A mesoionic molecules, defines many HMBs. Further examples are shown in structures **11–13**, where the isolated odd fragments are shown in red and blue (Fig. 3).

(ii) Consider the components of the generalized type B conjugation as shown in structure **14**. This contains a conjugated fragment a-e(d)-f that is associated with an even number of atoms but which, in spite of being an even number, cannot be represented by a fully covalent structure. This fragment is known as a non-Kekulé hydrocarbon, as shown in structure **15**. Non-Kekulé fragments can only be represented by dipolar (or diradical) structures. Compare this with the isomeric Kekulé hydrocarbon **16**. Heterocycles that contain an isolated non-Kekulé fragment cannot be represented by a fully covalent structure and are therefore HMBs. Type B mesoionic rings are good examples of this type as illustrated by the tetrazol-5-ones **17**.

Non-Kekulé fragments can be considered as a pair of cross-conjugated odd fragments a–e(d)–f and are therefore a special case of (i). Further examples, in which the cross-conjugated odd fragments of the non-Kekulé fragment are shown in red and blue, are the structures **18–20** (Fig. 4).

The whole HMB family is therefore defined by the two features described in (i) and (ii) above.

3.2 The different types of heterocyclic mesomeric betaine (HMB)

Having recognized the common structural features of HMBs (Section 3.1), it is important to distinguish discrete subgroups. Conjugated chains having an odd number of atoms [a – (b = c)$_n$] have special properties including a frontier orbital that vanishes on alternant atoms. These are known as non-bonding molecular orbitals (NBMOs).[9,10] For this reason, the positions that 2π ring heteroatoms (RN, O, S) are connected to the odd fragments or the non-Kekulé fragments leads to fundamentally different properties. The ways that the components of HMBs are assembled leads to three fundamentally different types that are described as conjugated, cross-conjugated and semi-conjugated HMBs,[11,12] as shown in Fig. 5. Type A mesoionic rings are conjugated HMBs and type B mesoionic rings are semi-conjugated HMBs.

Conjugated HMB are characterized by their association with one (or sometimes two) 1,3-dipole fragments and as a group they are associated with 1,3-dipolar cycloadditions. Type A mesoionic compounds clearly fall into this category and their cycloadditions continue to attract much synthetic interest. Many other conjugated HMBs are known including the pyridinium-3-olates **18** and related species,[13] and heteropentalene HMBs,[14] exemplified by structures **13** and **19.**

Cross-conjugated HMBs were first recognized as a separate class and described as cross-conjugated by Ramsden in 1977.[15] These HMBs have

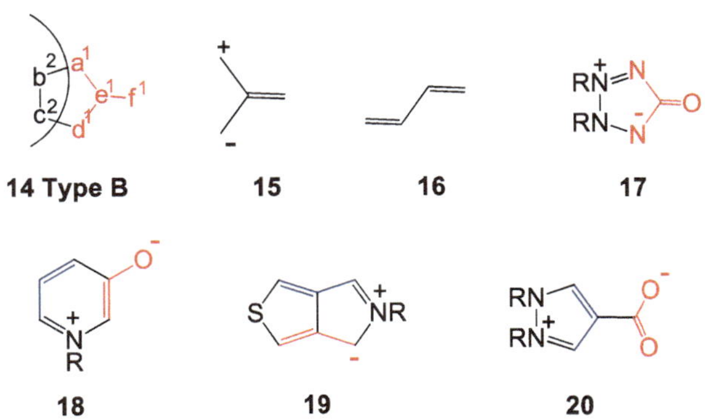

Fig. 4 HMBs containing non-Kekulé fragments.

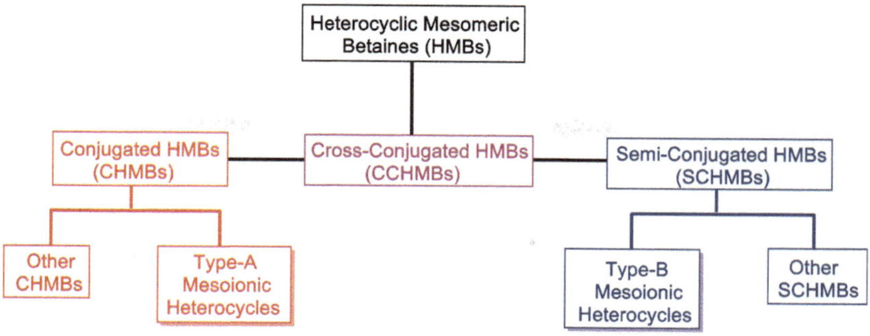

Fig. 5 The classification of HMBs.

structural[16] and chemical properties that are quite distinct from those of conjugated HMBs.[17] In particular, the positive and negative charges are restricted to separate portions of the molecule. This is illustrated by structures **12** and **20**. They are characterized by participation in 1,4-dipolar cycloadditions and these have found useful synthetic applications.[18,19]

Semi-conjugated HMBs have only recently (2013) been recognized as a discrete class of HMB,[11,12] and it is a very small class. Type B mesoionic heterocycles are now regarded as members of this group. In an earlier survey (1985) they were described as conjugated HMBs,[17] but this is unsatisfactory. They do undergo cycloadditions but they are not of the type that conjugated HMBs undergo. Their structure and properties are much more consistent with classification as semi-conjugated HMBs.[20] To the best of our knowledge, the only other example of a semi-conjugated HMB is the tetrazine derivative **21**, which has been fully characterized, including an X-ray structure.[21]

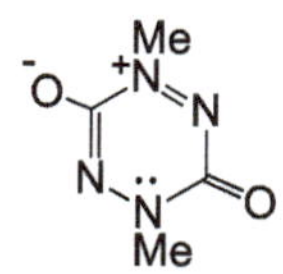

21

The types and relationships of HMBs shown in Fig. 5 is a simplification of the full picture.[11,12] A complete analysis requires a detailed consideration of the connectivity of the ring heteroatoms (RN, O, S) to the other components forming the HMB skeleton. There are seven possible connectivity matrices. This leads to three subclasses of conjugated HMBs (Class 1A–C), three subclasses of cross-conjugated HMBs (Class 2A–C) and one class of semi-conjugated HMB (Class 3).[11] The subject is further complicated by the fact that there are two subtlety different types of non-Kekulé hydrocarbon (cross-conjugated and pseudo-cross-conjugated). This leads to a further subdivision of cross-conjugated and semi-conjugated HMBs to include pseudo-cross-conjugated and pseudo-semi-conjugated HMBs. However, these finer points are of academic rather than practical interest and for every-day practice of heterocyclic chemistry Fig. 5 suffices. Nevertheless, a proper classification of new HMB species and a full appreciation of their properties does require an understanding of the detailed classification system. Anybody interested in the difference in chemical behavior between cross-conjugated and pseudo-cross-conjugated HMBs should refer to the experimental work of Schmidt and co-workers.[22,23]

4. Type A mesoionic systems—An overview

4.1 Known mesoionic systems of type A

If the atoms or groups a, b, c, d, e and f are selected from suitably substituted carbon, nitrogen, oxygen and sulfur atoms, 144 discrete mesoionic systems are described by the general formula **4** (Table 1).[1] In 1976, 44 type A ring systems were known. In 2020 this had increased to 68. In addition, nine systems containing selenium, phosphorus or arsenic have been reported bringing the total to 77. The known systems and the sections where they are discussed are listed in Table 1.

In surveying a subject it is important to assess what is not known as well as what is known. Table 1 allows the extraction of some interesting facts about known and unknown mesoionic rings.

Type A mesoionic rings contain a 1,3-dipole fragment. These rings can be regarded as being derived by union of a 1,3-dipole ($b{=}c^+ - d^-$) **22** with a heterocumulene (a=e=f) (Fig. 6) There are nine common 1,3-dipoles (b,c,d=O,S,N,C) and these are shown in Fig. 6. These can be combined with nine common heterocumulenes (e=C, a, f=O,S, N, C); each heterocumulene and 1,3-dipole can be combines in two or four ways depending upon symmetry and structure. The number of discrete mesoionic ring systems that can be derived from each 1,3-dipole and the nine heterocumulenes is shown in square brackets in Fig. 6. These add up to the 144 possible systems previously described.[1]

Examination of columns b, c and d of Table 1 reveals how many mesoionic rings with each 1,3-dipole are known and this is shown in round brackets in Fig. 6. It is interesting to note that most of the possible derivatives of mesoionic rings containing azomethine ylide, azonium imine and azomethine imine moieties have been reported. In contrast, as far as we are aware, no examples of mesoionic rings containing nitroso imine or

Table 1 Known mesoionic heterocycles of type A.

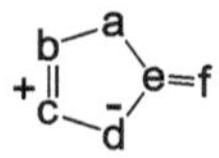

4

Atom or group

Parent system	Section	a	b	c	d	e	f
A. Dioxoles							
1,3-Dioxol-4-ones	5.1.1	O	CR	O	CR	C	O
B. Oxazoles							
1,3-Oxazol-4-ones	5.2.1	NR	CR	O	CR	C	O
1,3-Oxazol-4-imines	5.2.2	NR	CR	O	CR	C	NR
1,3-Oxazol-4-thiones	5.2.3	NR	CR	O	CR	C	S
1,3-Oxazol-5-ones	5.2.4	O	CR	NR	CR	C	O
1,3-Oxazol-5-imines	5.2.5	O	CR	NR	CR	C	NR
C. Oxathioles							
1,3-Oxathiol-4-ones	5.3.1	S	CR	O	CR	C	O
1,3-Oxathiol-4-thiones	5.3.2	S	CR	O	CR	C	S
1,3-Oxathiol-5-ones	5.3.3	O	CR	S	CR	C	O
D. Diazoles							
1,3-Diazol-4-ones	5.4.1	NR	CR	NR	CR	C	O
1,3-Diazol-4-imines	5.4.2	NR	CR	NR	CR	C	NR
1,3-Diazol-4-thiones	5.4.3	NR	CR	NR	CR	C	S
1,3-Diazol-4-enes	5.4.4	NR	CR	NR	CR	C	CR_2
1,3-Diazol-4-selones	5.4.5	NR	CR	NR	CR	C	Se
1,3-Diazol-4-phosphimines	5.4.6	NR	CR	NR	CR	C	PR
E. Thiazoles							
1,3-Thiazol-4-ones	5.5.1	NR	CR	S	CR	C	O
1,3-Thiazol-4-imines	5.5.2	NR	CR	S	CR	C	NR
1,3-Thiazol-4-thiones	5.5.3	NR	CR	S	CR	C	S
1,3-Thiazol-5-ones	5.5.4	S	CR	NR	CR	C	O
1,3-Thiazol-5-imines	5.5.5	S	CR	NR	CR	C	NR
1,3-Thiazol-5-thiones	5.5.6	S	CR	NR	CR	C	S
1,3-Thiazol-5-enes	5.5.7	S	CR	NR	CR	C	CR_2
F. Selenazoles							
1,3-Selenazol-4-ones	5.6.1	NR	CR	Se	CR	C	O
G. Dithioles							
1,3-Dithiol-4-ones	5.7.1	S	CR	S	CR	C	O
1,3-Dithiol-4-imines	5.7.2	S	CR	S	CR	C	NR
1,3-Dithiol-4-thiones	5.7.3	S	CR	S	CR	C	S
1,3-Dithiol-4-enes	5.7.4	S	CR	S	CR	C	CR_2

Continued

Table 1 Known mesoionic heterocycles of type A. —cont'd

		Atom or group					
Parent system	**Section**	**a**	**b**	**c**	**d**	**e**	**f**
H. Oxadiazoles							
1,3,4-Oxadiazol-2-ones	5.8.1	O	CR	NR	N	C	O
1,3,4-Oxadiazol-2-imines	5.8.2	O	CR	NR	N	C	NR
1,3,4-Oxadiazol-2-thiones	5.8.3	O	CR	NR	N	C	S
1,2,3-Oxadiazol-2-enes	5.8.4	O	CR	NR	N	C	CR_2
1,2,3-Oxadiazol-5-ones	5.8.5	O	N	NR	CR	C	O
1,2,3-Oxadiazol-5-imines	5.8.6	O	N	NR	CR	C	NR
I. Oxathiazoles							
1,3,2-Oxathiazol-5-ones	5.9.1	O	N	S	CR	C	O
J. Triazoles							
1,2,3-Triazol-4-ones	5.10.1	NR	N	NR	CR	C	O
1,2,3-Triazol-4-imines	5.10.2	NR	N	NR	CR	C	NR
1,2,3-Triazol-4-thiones	5.10.3	NR	N	NR	CR	C	S
1,2,3-Triazol-4-enes	5.10.4	NR	N	NR	CR	C	CR_2
1,2,3-Triazol-4-selones	5.10.5	NR	N	NR	CR	C	Se
1,2,4-Triazol-3-ones	5.10.6	NR	CR	NR	N	C	O
1,2,4-Triazol-3-imines	5.10.7	NR	CR	NR	N	C	NR
1,2,4-Triazol-3-thiones	5.10.8	NR	CR	NR	N	C	S
1,2,4-Triazol-3-enes	5.10.9	NR	CR	NR	N	C	CR_2
1,2,4-Triazol-3-selones	5.10.10	NR	CR	NR	N	C	Se
K. Thiadiazoles							
1,2,3-Thiadiazol-4-ones	5.11.1	NR	N	S	CR	C	O
1,2,3-Thiadiazol-4-imines	5.11.2	NR	N	S	CR	C	NR
1,2,3-Thiadiazol-4-thiones	5.11.3	NR	N	S	CR	C	S
1,2,3-Thiadiazol-5-ones	5.11.4	S	N	NR	CR	C	O
1,2,3-Thiadiazol-5-imines	5.11.5	S	N	NR	CR	C	NR
1,2,3-Thiadiazol-5-thiones	5.11.6	S	N	NR	CR	C	S
1,2,3-Thiadiazol-5-enes	5.11.7	S	N	NR	CR	C	CR_2
1,2,4-Thiadiazol-3-ones	5.11.8	NR	CR	S	N	C	O
1,2,4-Thiadiazol-3-imines	5.11.9	NR	CR	S	N	C	NR
1,3,4-Thiadiazol-2-ones	5.11.10	S	CR	NR	N	C	O
1,3,4-Thiadiazol-2-imines	5.11.11	S	CR	NR	N	C	NR
1,3,4-Thiadiazol-2-thiones	5.11.12	S	CR	NR	N	C	S
1,3,4-Thiadiazol-2-enes	5.11.13	S	CR	NR	N	C	CR_2
L. Dithiazoles							
1,3,2-Dithiazol-4-ones	5.12.1	S	N	S	CR	C	O
1,3,2-Dithiazol-4-imines	5.12.2	S	N	S	CR	C	NR
1,3,2-Dithiazol-4-thiones	5.12.3	S	N	S	CR	C	S

Table 1 Known mesoionic heterocycles of type A. —cont'd

Parent system	Section	a	b	c	d	e	f
		Atom or group					
M. Oxatriazoles							
1,2,3,4-Oxatriazol-5-ones	5.13.1	O	N	NR	N	C	O
1,2,3,4-Oxatriazol-5-imines	5.13.2	O	N	NR	N	C	NR
1,2,3,4-Oxatriazol-5-thiones	5.13.3	O	N	NR	N	C	S
N. Tetrazoles							
1,2,3,4-Tetrazol-5-ones	5.14.1	NR	N	NR	N	C	O
1,2,3,4-Tetrazol-5-imines	5.14.2	NR	N	NR	N	C	NR
1,2,3,4-Tetrazol-5-thiones	5.14.3	NR	N	NR	N	C	S
1,2,3,4-Tetrazol-5-enes	5.14.4	NR	N	NR	N	C	CR_2
O. Thiatriazoles							
1,2,3,4-Thiatriazol-5-ones	5.15.1	S	N	NR	N	C	O
1,2,3,4-Thiatriazol-5-imines	5.15.2	S	N	NR	N	C	NR
1,2,3,4-Thiatriazol-5-thiones	5.15.3	S	N	NR	N	C	S
1,2,3,4-Thiatriazol-5-enes	5.15.4	S	N	NR	N	C	CR_2
P. Dithiadiazoles							
1,3,2,4-Dithiadiazol-5-one	5.16.1	S	N	S	N	C	O
1,3,2,4-Dithiadiazol-5-imines	5.16.2	S	N	S	N	C	NR
Q. Diphosphadiazole							
1,3,2,4-Diphosphadiazol-4-ones	5.17.1	NR	P	NR	P	C	O
1,3,2,4-Diphosphadiazol-4-imines	5.17.2	NR	P	NR	P	C	NR
R. Arsenic derivatives							
1,2,3,4-Arsatriazol-5-imines	5.18	NR	N	NR	As	C	NR
1,3,2,4-Arsaphosphadiazol-5-imines	5.18	NR	P	NR	As	C	NR

carbonyl imine 1,3-dipoles have been described. Whether this is a problem with synthetic access or with stability remains to be seen. However, they do account for most of the unknown systems. Perhaps there is no interest in making them; it is a pity that they are not natural products.

4.2 Structural aspects of type A mesoionic rings

There is a consensus that the general structure **23**, in which a positive (a–b–c) and a negative (d–e–f) fragment are separated by longish bonds (a–e and c–d), generally summarizes the structures of many type A mesoionic rings.[24] These structural features are accommodated by the resonance hybrids

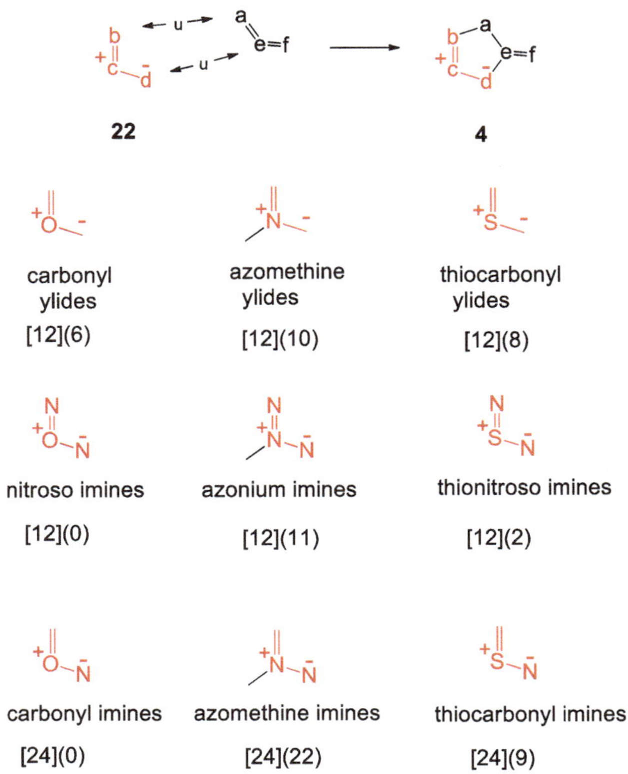

Total possible rings [144]; total known rings (68)

Fig. 6 The relationship between 1,3-dipoles **22** and type A mesoionic rings **4**. For each 1,3-dipole, square brackets [] show the number of possible systems and round brackets () the number of known systems incorporating each 1,3-dipole.

24–26, with structure **24** predominating.[24,25] A convenient feature of structures of the type **24** is that they are inherently consistent with 1,3-dipolar reactivity.

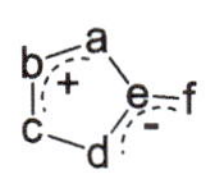

23

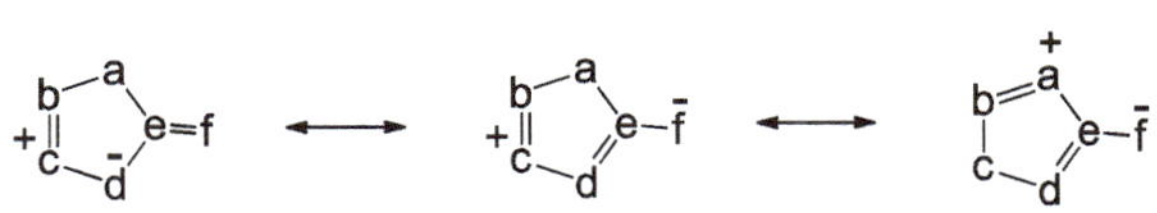

24 **25** **26**

This view is supported by X-ray crystallography and theoretical studies. An interesting feature of sydnone structures is angular distortion about the carbon atom bonded to the exocyclic oxygen atom. Fig. 7 shows typical bond angle distortion for an arylsydnone **27**,[26] and two similar rings **28** and **29**.[27,28] There are significant differences between the O—C=O angles (120°) and the C—C=O (135°) and N—C=O (132°) angles (Fig. 7). Thiessen and Hope attributed this distortion to a back donation of exocyclic oxygen lone pairs in the plane of the ring resulting in weakening of the ring C—O bond and increasing C=C=O (or N=C=O) character.[1] This anomeric effect also rationalizes the short exocyclic C=O bonds, which are comparable to ketone carbonyl groups in length. A Natural Bond Order (NBO) analysis of 1,3-dimethyl-1,3-diazol-4-one indicates a strong donor-acceptor interaction between one of the oxygen lone pairs with an antibonding orbital associated with the MeN—CO bond.[20]

Based on the crystals structures **30**,[29] **31**,[30] **32** (nitron)[31] and **33**,[32] the situation for imines is more complex (Fig. 8). In the oxatriazole derivative **30** and the thiadiazole **31** (and also 1,2,3,4-thiatriazol-5-imines,[33] Section 5.15.2), in which steric interactions are minimal, the nitrogen lone pair is *trans* to the C—Ẍ bond, and the angular distortion is much less, particularly in structure **30**. In nitron **32** and the diazole **33**, in which steric interaction impose the opposite configuration, the distortion is comparable to that in the oxygen analogues **27–29**. Calculations suggest that *cis* and *trans*

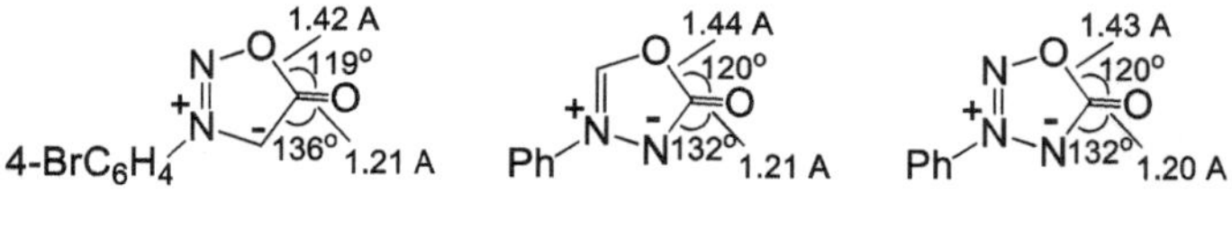

27 **28** **29**

Fig. 7 Bond angles and lengths for three mesoionic 5-ones determined by X-ray crystallography.

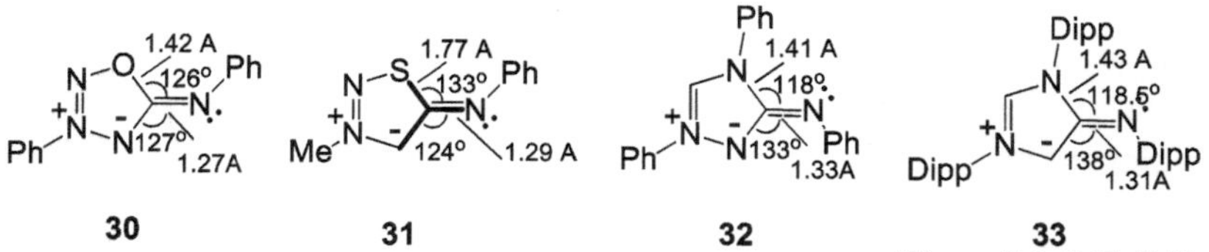

30 **31** **32** **33**

Dipp = 2,6-di-*i*PrC_6H_3

Fig. 8 Bond angles and lengths for four mesoionic imine rings determined by X-ray crystallography.

lone pairs have different effects on the geometry. In the imine **30**, a favorable anomeric interaction of the nitrogen lone pair (donor) *trans* to the C-Ö π* bond (acceptor) favors the observed configuration and geometry. In the derivatives **32** and **33** steric effects favor the opposite configuration and the absence of *trans* lone pairs and the influence of the lone pairs *cis* to the C_{Ar}-Nar bonds results in a different geometry about the C=NPh bonds.[34]

A number of other X-ray crystal structures of type A mesoionic rings have been reported since 1980 and these are referred to in the relevant sections. Although the configuration of exocyclic lone pairs has an influence on geometry, it is important to appreciate that other factors including the number and type of heteroatoms in the ring and substituent effects also influence ring geometry. As with other heterocycle, there is also some intrinsic asymmetry about the exocyclic double bonds.

4.3 The aromaticity of type A mesoionic rings

When the family of mesoionic compounds was first recognized, aromaticity and aromatic sextets were a hot topic and the suggestion that their stability is associated with a 6π cationic ring and an exocyclic anion represented by structure **34** seemed reasonable. Subsequently, X-ray crystallography (and other techniques) showed short e—f bonds and long a—e bonds, which are not consistent with an aromatic structure. The ring C—O bonds (1.41–1.43 Å) in structures **27–29** (Fig. 6) are much longer than in furan (1.36 Å). Structures of the general type **23**, in which there are implicit long a—e bonds and short e—f bonds, are more reasonable. Anomeric effects that account for these bond lengths are discussed in Section 4.2.

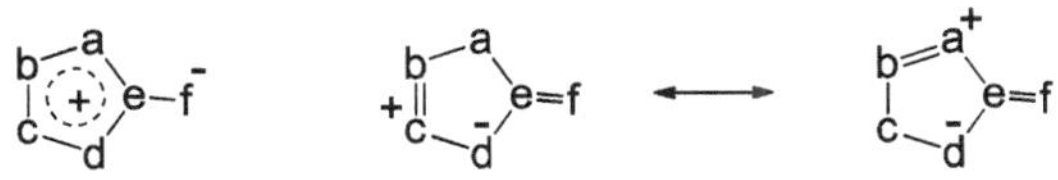

34 **35**

A detailed study by Miller and co-workers in 1998 concluded that type A mesoionic compounds are not aromatic.[35] This view is supported by a number of computational studies and aromaticity indices.[20,25] The Bird Index of classical aromaticity for 15 mesoionic derivatives shows some trends with variation of heteroatom.[36] Both the Bird and HOMA aromaticity indices suggest modest aromaticity with some increase on aza substitution of the ring.[20] However, a calculation of aromatic stabilization energy (ASE) of 1,3-dimethyl-1,3-diazol-4-one found an ASE value of $0.55\,kcal\,mol^{-1}$, suggesting very little stabilization due to cyclic conjugation.[20]

It is important to distinguish between classical and magnetic aromaticity which for a series of heterocyclic rings show different trends.[37] The NICS(1) magnetic aromaticity index for a series of mesoionic rings gives values in the range -6.6 ± 1.1 (benzene is -10.2).[20] In a separate study, the value for *N*-phenylsydnone was found to be -6.8.[38] These values suggest modest magnetic aromaticity.

4.4 General modes of reaction of type A mesoionic rings

4.4.1 1,3-Dipolar cycloadditions

Following the pioneering work of Huisgen and co-workers in the 1960s, 1,3-dipolar cycloaddition reactions have been a major aspect of mesoionic chemistry. Reactions with alkynes, alkenes and heterocumulenes, including intramolecular additions, have proved to be a rich source of novel heterocyclic systems. In addition to the sydnones, work on the münchnones (Section 5.2.4) and isomünchnones (Section 5.2.1) has been particularly prolific. A number of reviews focusing on this aspect of mesoionic chemistry have been published. These include comprehensive coverage of mesoionic oxazoles upto 2002,[39,40] and surveys of the application of mesoionic cycloadditions in the synthesis of alkaloids,[41,42] other heterocycles and natural products.[43] Other surveys include: (i) formation of pyrroles from mesoionic oxazoles[44]; (ii) the chemistry of isomünchnones and thioisomünchnones[45,46]; developments in sydnone chemistry[4,47]; and the exploration of electronic and steric control.[48] A 2020 review includes click-and-release reactions of mesoionic compounds and their applications in chemical biology.[49]

In general there are three common modes of reaction of mesoionic rings with 1,3-dipolarophiles. The most common outcome is reaction of the mesoionic ring **36** with an alkyne or alkyne equivalent leading to a cycloadduct **37** that can eliminate a heterocumulene (often CO_2) to give a five-membered heterocycle **40** (Scheme 1). This is particularly useful for the preparation of furans, thiophenes, pyrroles and pyrazoles. An alkyne equivalent is an alkene that can readily eliminate a small molecule. Ideally the reactions are regioselective giving predominately or exclusively one of the two regioisomers **37** or **38**. Much effort has been spent on developing regio- and stereoselective additions for control of the structures **37** or **38**. Conveniently, the use of an alkyne equivalent can sometimes give the opposite regioselectivity to the alkyne.

For some systems the cycloadduct **37** can lose the bridging group c leading to six-membered products **41**. A notable case is the isomünchnones where a bridging oxygen (c=O) can be lost as water leading to pyridone

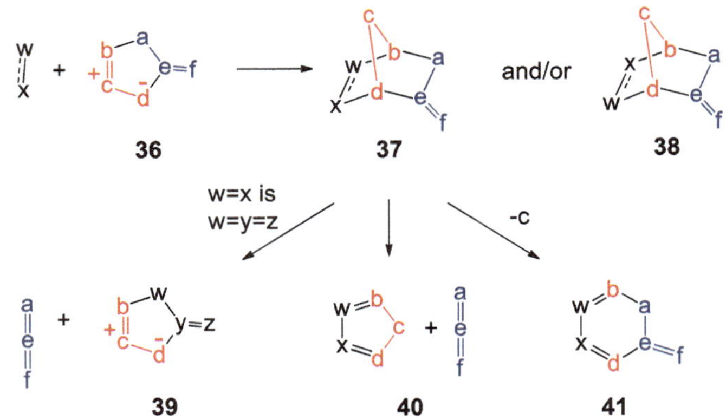

Scheme 1

derivatives. A bridging sulfur (c=S)[50,51] or selenium (c=Se)[52] can also be conveniently lost.

If the dipolarophile is a heterocumulene (w=y=z) the adduct **37** can eliminate another heterocumulene (usually carbon dioxide) to give a new mesoionic ring **39**. Münchnones have been used in this way to make several new mesoionic systems, although there is some computational evidence that these reaction occur *via* a dipolar intermediate rather than an adduct **37**.

A less common mode of reaction of the adducts is cleavage of the a—b bond to give a zwitterion **42** (Scheme 2). For some münchnone reactions these products are stable as carboxylates (Section 5.2.1.2.1). In some thioisomünchnone reactions proton transfer (**42** → **43**) gives amides (Section 5.5.1.2.1).

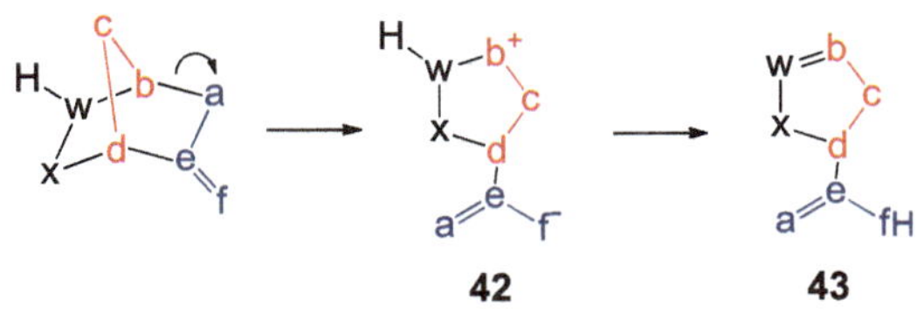

Scheme 2

Specific examples of the reactions summarized in Scheme 1 can be found in the relevant subsections in Section 5. Each ring system has its own characteristics. Reactivity tends to decrease with aza substitution of the ring.

Predicting the regioselectivity of the additions **36** → **37** or **38** is a desirable aim but frontier molecular orbital theory (FMO) has been found to be an unreliable predictor. A detailed computational study of reactivity and regioselectivity by Houk and co-workers found that the calculation of transition state (TS) interaction and distortion energies gives reliable predictions.[53]

4.4.2 N-Heterocyclic carbene formation

A remarkable new chapter in the chemistry of mesoionic rings is their formation of N-Heterocyclic Carbenes (NHCs) and anionic NHCs.

4.4.2.1 Equilibration with NHC tautomers

The equilibration of some mesoionic 1,3-diazol-4-imines (Section 5.4.2) and 1,2,4-triazol-3-imines (Section 5.10.7) **44** with a tautomeric NHC **45** has been reported (Scheme 3). Examples have been demonstrated by NMR studies and trapping to form thiones **46** and other adducts. For example, in solution *N*-alkyl diazolimines **44** (X=CR, Y=N-alkyl) are in equilibrium with the tautomeric carbenes **45**.[54,55]

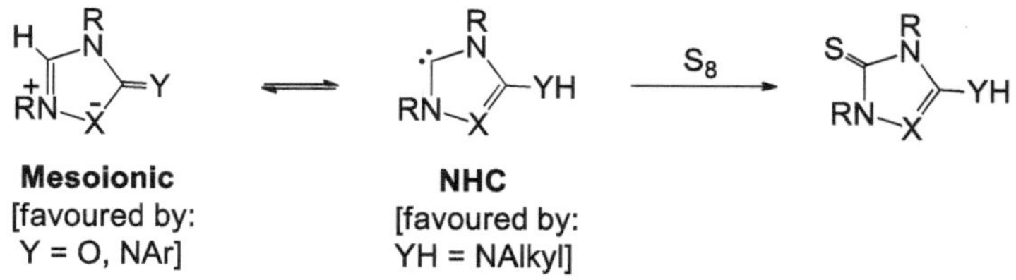

44 X = CR, N **45** X = CR, N **46**

Scheme 3

The position of equilibrium appears to be determined by the substituent effect stabilization energies (SESE) of the groups Y and YH (Scheme 3). The SESE is an estimate of the extra stabilization due to the groups Y and YH compared to reference structures.[56] For 1,2,4-triazoles **44** and **45** (X=N) the calculated SESEs for the mesoionic substituents Y are: O −4.8; NPh −3.8; NMe +0.02 kcal mol^{-1}. For the carbene substituents YH they are: OH +1.6; NHPh −0.9; NHMe −1.2 kcal mol^{-1}.[56] Based on these values it is clear why mesoionic rings with an exocyclic oxygen will not equilibrate with a carbene tautomer and why an exocyclic imine bearing an alkyl group most favors equilibration.

4.4.2.2 Formation of anionic NHCs

Protons on mesoionic rings can be removed by strong base to give anionic carbenes **48** (Scheme 4), which can trapped. For the 1,2,4-triazole carbenes

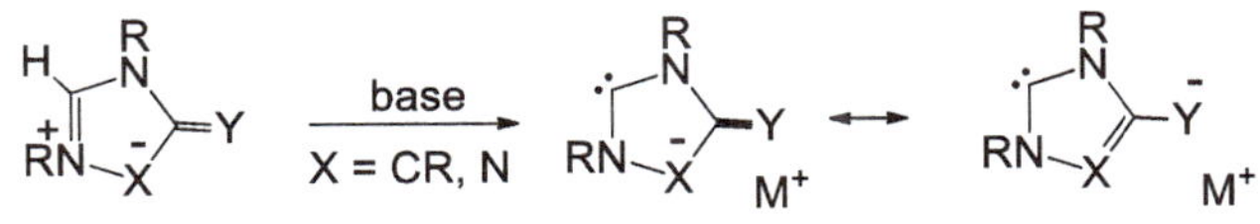

47 Mesoionic **48 Anionic NHC**

Scheme 4

48 (X=N) the calculated SESEs for Y are: O^- −8.0; PhN^- −6.5 kcal mol^{-1}, indicating that for mesoionic rings with an exocyclic O or NAr substituent the anionic carbene enjoys significant stabilization.[56] Examples include treatment of 1,3-diazol-4-ones **47** (X=CR, Y=O) (Section 5.3.1) with lithium bis(trimethyl)amide (LiHMDS).[57]

The ease of anionic NHC formation of mesoionic rings can be estimated using the CREF index, which is a calculated ionization energy expressed as an index.[58,59] Fig. 9 compares values; a lower index value indicates easier deprotonation. For 1,3-diazol-4-ones deprotonation is easier at the 2 position than the 5 position (**49** vs. **50**). Aza substitution facilitates carbene formation (**49** vs. **51**) and exocyclic NPh is comparable to exocyclic O (**51** vs. **52**). In accord with experiment, sydnones **53** have accessible anionic carbenes.[38] It is noteworthy that the well-known 2-unsubstituted 3-aryl-1,3,4-thiadiazol-5-thiones, e.g., **54**,[1] have a particularly low CREF value but, as far as we are aware, the formation of anionic carbenes by these species has never been investigated.

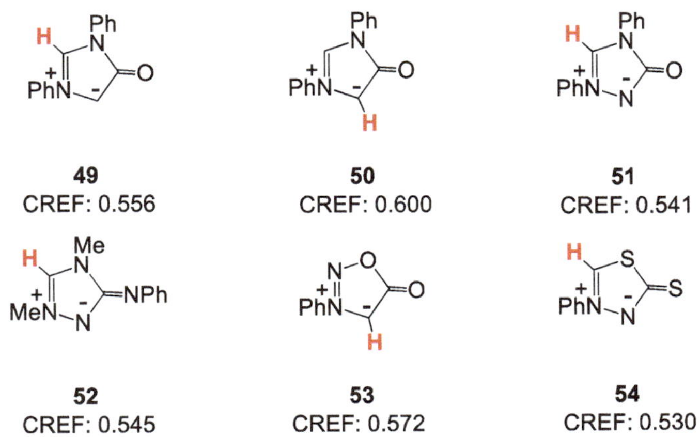

Fig. 9 CREF Index values for anionic heterocyclic carbene formation from mesoionic rings.

4.4.3 Valence tautomerism

Although it is not a common aspect of type A mesoionic chemistry, a reoccurring theme is the formation of products *via* the acyclic valence tautomers **55** (Scheme 5). Experimental evidence, provided by many research groups, takes the form of the formation of [2+2] cycloadducts **58** from mesoionic precursors **56**. In fact Hamaguchi and co-workers have provided evidence that 1,3-dioxol-4-ones are in rapid equilibrium with their ketene tautomer (Section 5.1.1).[60,61] Münchnones (Section 5.2.4), including work by the Huisgen and co-workers,[1] are also known to give cycloadducts of general structure **58**.

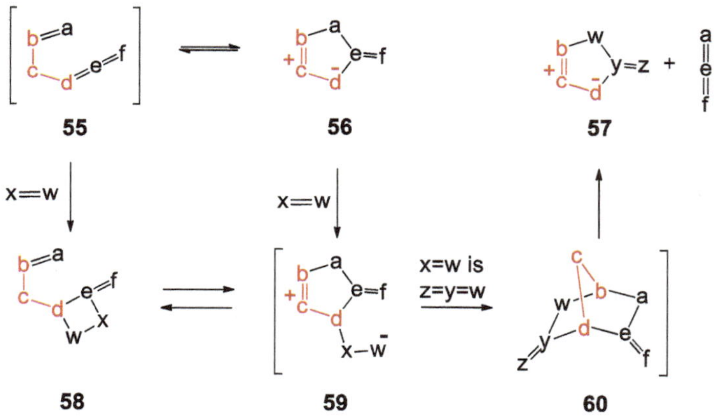

Scheme 5

However, not all products of type **58** are necessarily formed *via* the valence tautomers **55**. An alternative mechanism may be formation of zwitterionic intermediates **59** which cyclize to the observed four-membered rings **58**. In a study of the reactions of 1,3-dithiol-4-ones (Section 5.7.1) with phenyl isocyanate there is evidence that the products **58** are the kinetic products but longer reaction times give the thermodynamic products **57**. The latter can be envisaged as being formed by reversion to **59** and an alternative ring closure to **60** followed by elimination (Scheme 5).

The mode of reaction may depend on both the structure of the mesoionic ring **56** and the reagent (x=w), but there is insufficient evidence to make firmer conclusions about these modes of reaction.

4.4.4 Rearrangements

The general structure **61** is recognized as a good general representation of type A mesoionic rings. It is consistent with many properties including long a—e and c—d bonds and the size and direction of measured dipole moments.[62] It summarizes the contributions of resonance structures **62a–c**. Although resonance structure **62a** is probably the best single representation of mesoionic rings, this does not mean there is not significant negative charge on atom or group f, as represented by structures **62b** and **62c**. The short e—f bonds (Section 4.2) can to some extent be attributed to anomeric effects in the σ-bonding, which are largely independent of the π electron distribution.

As might be expected for structures **61**, stability is favored by atoms a and c that energetically favor the cationic fragment, i.e., NR > S > O. In contrast, stability is favored by groups d and f that stabilize negative charge, i.e., for atom f; S > O > NAr > NAlkyl > CR_2 and for atom d; N > CR. Computational studies support these empirical observations.[20,25] These atom effects are observed, for example, in the Dimroth-type rearrangements of a number of mesoionic rings (**62** → **63**) where the nature of a and f determine the most stable product.[63]

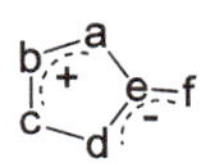

61

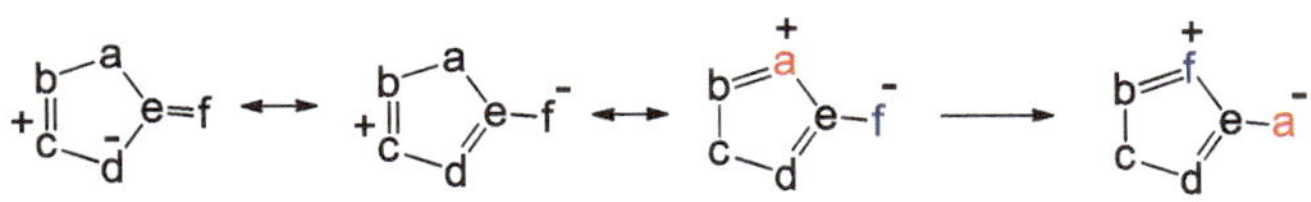

62a **62b** **62c** **63**

5. The chemistry of type A mesoionic compounds

5.1 Dioxoles

5.1.1 1,3-Dioxol-4-ones (1,3-dioxolium-4-olates) (64)

1,3-Dioxol-4-ones **64** are too unstable to be isolated but they can be trapped *in situ* by rapid 1,3-dipolar cycloadditions with reactive electron-deficient dipolarophiles. Hamaguchi and co-workers have provided strong evidence that the mesoionic rings **64** rapidly equilibrate with the ketene valence tautomers **65** (Scheme 6).[60,61]

An efficient method of generating the derivatives **64** in the presence of a trapping agent is by $Rh_2(OAc)_4$-catalyzed decomposition of diazoacetic anhydrides **66**. In the absence of a trapping agent an initial red color, attributable to initial formation of the mesoionic species **64**, rapidly fades. 1,3-Dipolar cycloaddition derived products arising from the mesoionic ring

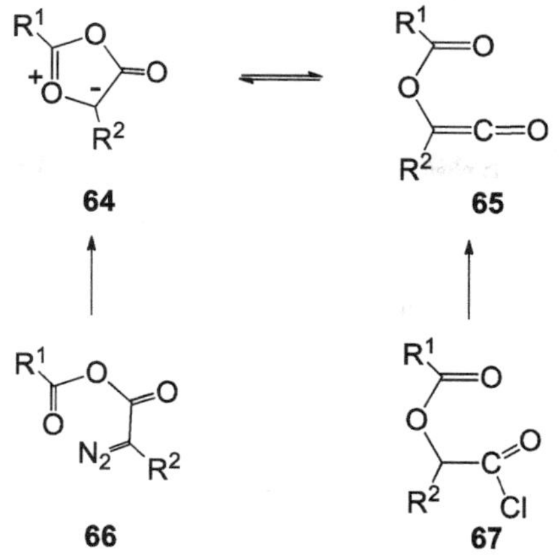

Scheme 6 Alternative routes to equilibrium mixtures of the valence tautomers **64** and **65**.

are also obtained by *in situ* generation of the acyloxyketenes **65** from α-acyloxyacetyl chlorides **67**. This provides strong evidence of the rapid equilibrium **64** ⇌ **65**.[60,61] PM3 calculations on the diphenyl isomers **64** and **65** ($R^1 = R^2 = Ph$) indicate that the mesoionic structure **64** is $9\,kcal\,mol^{-1}$ less stable than the ketene **65**.[64]

2-Methyl-5-phenyl-1,3-dioxolium-4-olate **64** ($R^1 = Me$, $R^2 = Ph$) has been generated *in situ* by thermal cyclodehydration of *O*-acetylmandelic acid.[65] 2-Trifluomethyl derivatives **64** ($R^1 = CF_3$) have been generated by treatment of α-hydroxycarboxylic acids with trifluoroacetic anhydride (TFAA).[66,67]

In extensive studies using a wide variety of dipolarophiles, Hamaguchi and co-workers have shown that the nature of cycloaddition products is determined by the reactivity of the reagent regardless of the method of generation of the equilibrium mixture **64** ⇌ **65** (Scheme 6). Using reactive electron-deficient 1,3-dipolarophiles, including azodicarboxylates, nitrosobenzenes, singlet oxygen, *N*-methylmaleimide, acenaphthylene, and acetylenedicarboxylates, fast reaction leads to formation of 1,3-dipolar cycloadducts of the mesoionic ring, even though it is a minor component of the equilibrium mixture.[61,68] This is illustrated by reactions of the 5-aryl derivatives **69** (Scheme 7). Reaction with DMAD gives the adducts **68** which undergo decarboxylation leading to furans.[65,69] Reaction with diethyl azodicarboxylate (DEAD) gives the unstable adducts **70** which ultimately give triacylbenzamidines.[70]

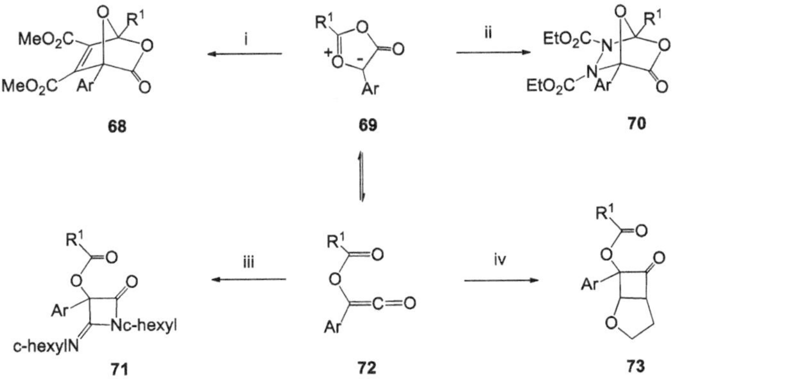

Reagents: i, dimethyl acetylenedicarboxylate (DMAD), C_6H_6, rt, 2h; ii, diethyl azodicarboxylate (DEAD), C_6H_6, 50 °C, 2h; iii, dicyclohexylcarbodiimide (DCC), C_6H_6, 50 °C, 1h; iv, 2,3-dihydrofuran, C_6H_6, 30 °C, 5h.

Scheme 7 The formation of either 1,3-diploar cycloadducts or [2+2] adducts from the equilibrium **69** ⇌ **72** depends on the reactivity of the reagent.

In contrast, less reactive reagents, such as 2,3-dihydrofuran, imines, carbodimides, and cyclopentadiene, add to the ketene component of the reaction mixture to form [2+2] adducts. This is illustrated by formation of the 2-imino-4-oxoazetidines **71** and the bicyclic adducts **73** (Scheme 7).[60] When the mesoionic derivatives **69** are generated from aryldiazoacetic anyhydrides **66** (R^2 = Ar) in the presence of methanol, methyl esters arising from addition to the ketene fragment are obtained in high yield.[60]

5.2 Oxazoles

5.2.1 1,3-Oxazol-4-ones (1,3-oxazolium-4-olates) (75)

Common name: isomünchnones

Mesoionic 1,3-oxazol-4-ones **75** are often referred to as isomünchnones because of their isomeric relationship to the well-known münchnones (1,3-oxazol-5-ones, Section 5.2.4). Since early studies in the 1970s, considerable new work on this system has been reported. This has been largely driven by their ready participation in 1,3-dipolar cycloadditions that lead to furan and pyridone derivatives of synthetic value. Early work was reviewed in 1994.[45]

5.2.1.1 Preparation

A convenient method of generating the derivatives **75** *in situ* is rhodium acetate ($Rh_2(OAc)_4$) catalyzed decomposition of the α-diazoimides **74** (Scheme 8).[71–74] Other catalyst have been used, including cupric acetylacetonate ($Cu_2(acac)_2$),[75] rhodium(II) perfluorobutyrate ($Rh_2(pfbm)_4$),[76,77]

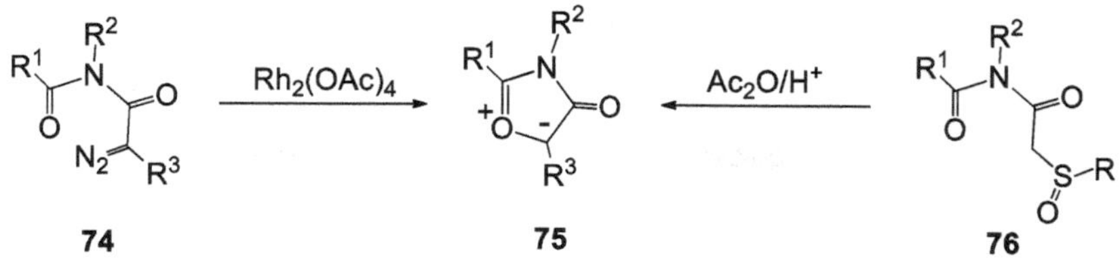

Scheme 8

rhodium(II) pivalate dimer ($Rh_2(Piv)_4$),[78,79] and Rh(II) octanoate dimer ($Rh_2(O_2C.C_7H_{15})_4$).[80] The choice of catalyst can be important. Padwa and co-workers have shown that $Rh_2(pfbm)_4$ favors five-membered isomünchnone formation but $Rh_2(AcO)_4$ can favor alternative six-membered cyclization pathways.[81–83] It should also be noted that Lautens and co-workers have demonstrated that Brϕnsted acids catalyze *in situ* generation of isomünchnones from diazoimides without the presence of a metal catalyst.[84]

An alternative method for *in situ* generation of the isomünchnone derivatives **75** (R^3 = SR) is Pummerer cyclodehydration of imidosulfoxides **76**. In these reactions initial formation of a thionium ion is followed by cyclization.[85–87] The use of trifluoroacetic anhydride (TFAA) for these reactions is best avoided since the initially formed mesoionic product reacts further to give oxazolidones and other products.[86]

The stable orange-red 2,3,5-triphenyl derivative **75** ($R^1 = R^2 = R^3$ = Ph) mp 175–177 °C has been obtained by treatment of *N*-benzoylphenylglyoxanilide (PhCO.CONPh.COPh) with hot triethyl phosphite.[88,89]

5.2.1.2 Cycloadditions

The *in situ* generation and trapping as 1,3-dipolar adducts has been extensively studied by Padwa and co-workers leading to access to the core skeletons of a wide variety of alkaloids, biologically important molecules and azapolycyclic systems.

5.2.1.2.1 Alkene [2+3]-cycloadditions A variety of alkene 1,3-dipolarophiles have been used including *N*-phenyl- and *N*-methyl maleimide,[74,88,90–97] maleic anhydride,[93] vinyl ketones,[90,92,93,96–98] vinyl sulfones,[86,91] vinyl ethers,[76,77,99–101] vinyl amides,[102] dialkyl ketene acetals,[90] methyl acrylate,[87,91,103,104] dimethyl maleate,[96,97] tetracyanoethylene,[93] 1,4-benzoquinones,[86,93,96,105] and buckminsterfullerene (C60).[106] Typical examples of these cycloadditions are shown in Scheme 9.

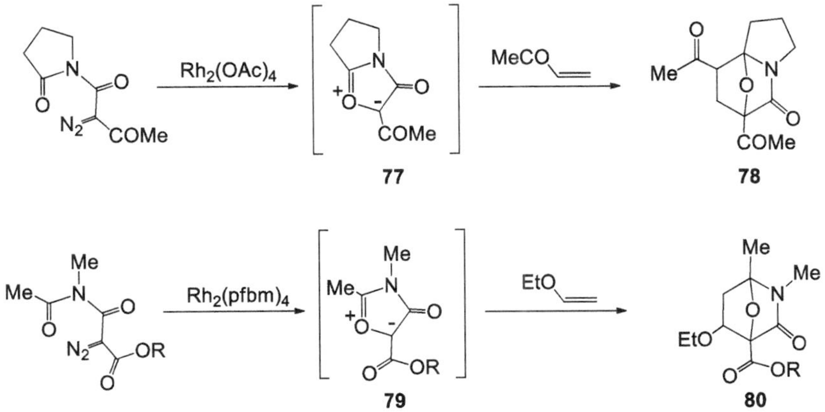

Scheme 9

Isomünchnones undergo addition with both electron-deficient and electron-rich alkenes and their regioselectivity is in accord with Frontier Molecular Orbital (FMO) theory.[90] However, the structure of the isomer formed depends on the nature of the dipolarophile (see Schemes 9 and 10). Methyl vinyl ketone gives the acid labile adduct **78** *via* the mesoionic intermediate **77**.[90] The facial selectivity of additions varies with ring substituents.[93] In a study of vinyl ether addition to the ester derivatives **79**, Austin and Savinov have shown that, in addition to regio and *endo*-selectivity, optimization of an attached chiral auxiliary gives diastereofacial selectivity in the adducts **80**.[77,100]

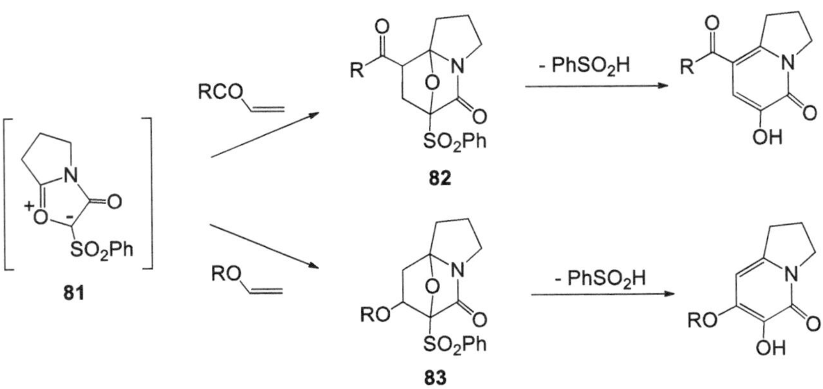

Scheme 10

The generation and trapping of 5-phenylsulfonyl isomünchnone derivatives, e.g., **81**, and facile elimination of phenylsulfinic acid ($PhSO_2H$)

from the unstable cycloadducts, e.g., **82** and **83**, provides a convenient route to 3-hydroxy-2-pyridone derivatives (Scheme 10).[91,107] This methodology provides convenient access to important nitrogen frameworks including indole alkaloid skeletons,[104] β-carbolines,[108] and antiviral drug analogues.[103] A 2(*H*)-pyridone library has been synthesized using this methodology.[109]

A synthetic strategy that has been usefully exploited is intramolecular 1,3-dipolar cycloaddition of an alkene function tethered to the isomünchnone ring. This approach is illustrated by the formation of the *endo* adduct **85** in 88% yield from the mesoionic precursor **84** (Scheme 11).[110] A number of other applications of intramolecular cycloadditions of isomünchnones have been reported.[19,71,73,78,80,85,92,110–125]

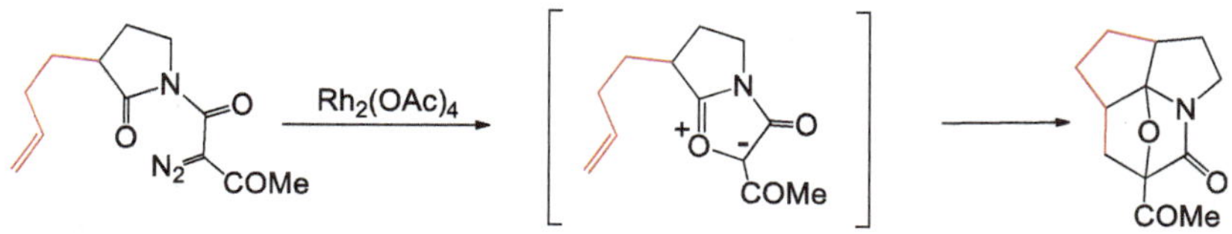

Scheme 11

The application of isomünchnone chemistry to synthetic strategies has been pioneered by Padwa and co-workers and some applications of these methods to the synthesis of natural products merit specific mention. These include the synthesis of (±)-vallesamidine,[19] (±)-lycopodine,[118] (±)-pumilotoxin C,[85] (±)-ipalbidine,[91,107] β-coniceine,[107] dielsiquinone,[87] (±)-costaclavine,[87] camptothecin,[101,124] (+)-tashiromine,[102] mappicine ketone (nothapodytine B),[103] and the ACE inhibitor (−)-A58365A.[107,126]

5.2.1.2.2 Alkyne [2+3]-cycloadditions Alkynes add to isomünchnones to give cycloadducts that readily undergo cycloreversion with loss of an isocyanate to give furans. This is a convenient route to furans and polymer-supported isomünchnones have been employed to generate combinatorial furan libraries.[83,127,128]

A number of reactions with dimethyl acetylenedicarboxylate (DMAD) have been reported.[73,74,83,90,94,95,97,127,128] In the example shown in Scheme 12, the initial adduct **86** releases a tethered isocyanate function **87** which is trapped to give the furan **88**.[90] However, in the seven-membered analogue the initial cycloadduct rearranges to a 3-hydroxy-2-pyridone.[129,130]

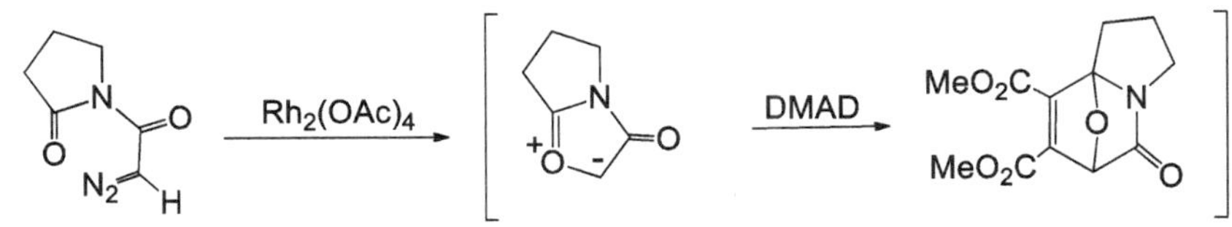

86

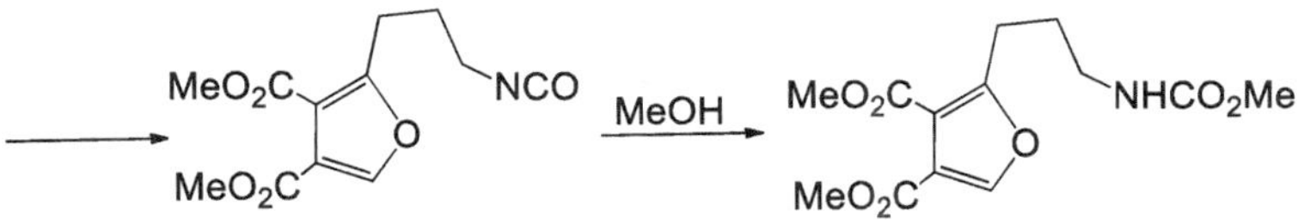

87 88

Scheme 12

Other alkynes that have been employed include diethyl acetylenedicarboxylate,[127] methyl and ethyl propiolate,[74,95,122,127] and dibenzoylacetylene.[127] Intramolecular reactions of isomünchnones with tethered alkyne substituents have been employed to prepare tetrahydrobenzo[*b*]furans in good yield. An interesting variation on this mode of reaction is the use of phosphaalkynes (RC≡P) to prepare 1,3-oxaphospholes.[72,131]

5.2.1.2.3 Aldehyde [2+3]-cycloadditions Examples of 1,3-dipolar cycloadditions of aldehydes were reported in an earlier review.[2] Harwood and co-workers have provided further examples that show high diastereofacial and high *exo*-selectivity and illustrated by the trapping of the isomünchnones **89** with benzaldehydes to form the adducts **90** (Scheme 13).[132]

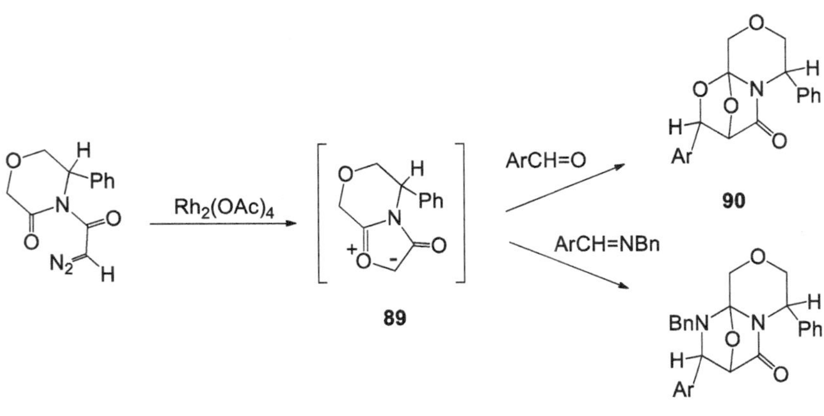

89 90 91

Scheme 13

5.2.1.2.4 Imine [2 + 3]-cycloadditions The Harwood group have also prepared the novel imine adducts **91** (Scheme 13). The structure of the *p*-methoxy derivative **91** (Ar = 4-$MeOC_6H_4$) was confirmed by X-ray crystallography.[133]

The [2 + 3]-addition of the C=N bond of 2,3,4-tri-*tert*-butylazete to give a cycloadduct in 91% yield has been reported.[134]

5.2.1.2.5 [4 + 3]-Cycloadditions Lautens and co-workers have shown that *in situ* generation of isomünchnones and *ortho*-quinone methides results in a formal [4 + 3] cycloaddition to give oxa-bridged oxazocine adducts in good yield (21 examples). This mode of reaction is illustrated by formation of the cycloadduct **92** (Scheme 14). A single diastereoisomer is formed and the reactants are generated simultaneously from phenol and diazoimide precursors by (+)-camphorsulfonic acid (CSA) catalysis without the need for a Rh(II) catalyst.[84]

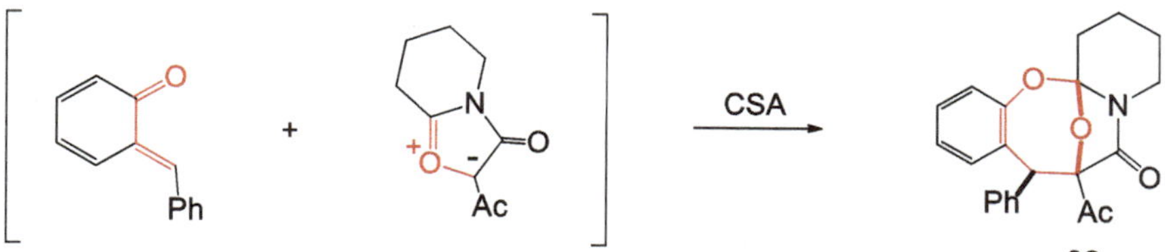

92

Scheme 14

In a separate study, Feng, Lin and co-workers reacted $Rh_2(Piv)_4$ generated isomünchnones with β,γ-unsaturated α-keto esters in the presence of a chiral *N,N'*-dioxide-Zn(II) complex to give oxa-bridged oxazocines in good yield and with excellent *ee* values (19 examples). This approach is illustrated by the formation of the cycloadduct **93** (Scheme 15).[79]

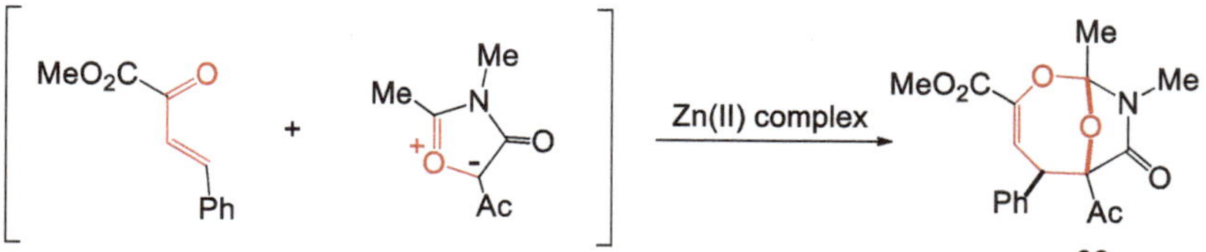

93

Scheme 15

It is interesting to note that in an earlier study, α,β-unsaturated amide derivatives (3-(2-alkenoyl)-2-oxazolidinones) in the presence of chiral

Lewis acids gave [3 + 2] cycloadducts.[102] The nature of the dipolarophile and catalyst both appear to be important in determining the periselectivity.

5.2.1.3 Reaction with nucleophiles

Isomünchnones are sensitive to water and alcohols. Water attacks the carbon at position 2 forming a hemiaminal, e.g., **94** which may cleave to form a hydroxy imide, e.g., **95**,[90,135,136] or in unconstrained derivatives an ester (Scheme 16).[88,113,137] For this reason it is important to avoid the presence of adventitious water when generating and trapping isomünchnones. Alcohols (and diols) form the stable hemiaminal products **96** and, when $n = 1$ or 2, the water adducts **96** (R = H) are also stable.[136]

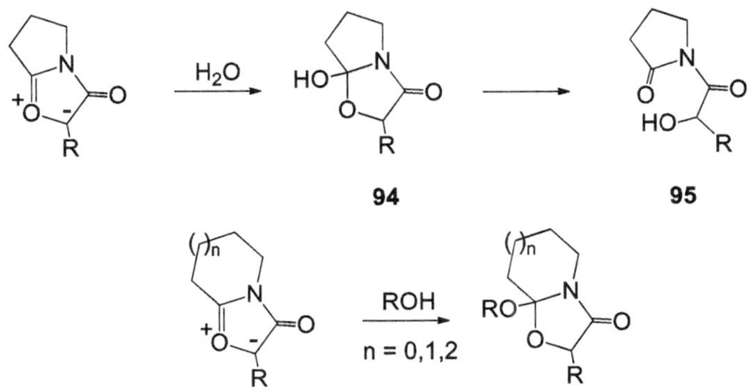

Scheme 16

Trifluoroacetic anhydride (TFAA) has been reported to add to isomünchnones to give oxazolidone derivatives.[86] Polymerization by nucleophilic attack of the exocyclic oxygen of an isomünchnone on its protoned form has been reported.[129,130]

5.2.2 1,3-Oxazol-4-imines (1,3-oxazolium-4-aminides) (97)

Common name: isomünchnone imines

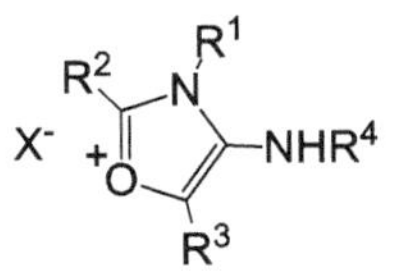

97

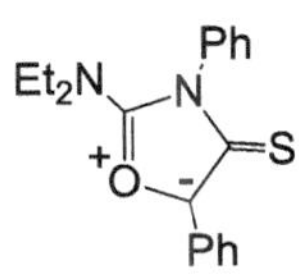

98

The salts **97** were described in 1969 but no new work has been reported.[1,2]

5.2.3 1,3-Oxazol-4-thiones (1,3-oxazolium-4-thiolates) (98)

The derivative **98** has been prepared and fully characterized by addition phenyl isothiocyanate to the corresponding 1,3-oxathiolium-4-olate **154** (R=Et) (Section 5.3.1).[138] This mesoionic ring system was previously unknown.

5.2.4 1,3-Oxazol-5-ones (1,3-oxazolium-5-olates) (100)

Common name: münchnones

Like the sydnones, the münchnones are named after the city of their origin. In the 1960s Huisgen and co-workers recognized the synthetic value of these reactive 1,3-dipoles and interest in their synthetic applications has continued to grow. Conveniently the münchnones can be generated *in situ* and trapped to give a wide variety of heterocycles and considerable new work on this system has been reported since 1980.

Several reviews of münchnones have appeared since 1980. These include chapters by Gribble,[139,140] and a short 2014 overview dedicated to Rolf Huisgen.[141] Some other reviews covering aspects of münchnone chemistry have appeared since 2000.[44,142]

5.2.4.1 Preparation

The original and widely used preparative method is cyclodehydration of *N*-acylamino acids **99** commonly using acetic anhydride as cyclizing agent (Scheme 17). Polymer bound münchnones have been generated using this method.[143,144] Other cyclizing agents that have been used include trifluoroacetic anhydride (TFAA),[145,146] ethyl chloroformate,[147] carbodiimides,[147–151] and isocyanates.[146] In general, münchnones are unstable, especially 4-unsubstitued derivatives, and are generated and used *in situ*. However, electron-withdrawing groups at the 4-position endow stability and a number of 4-trifluoroacetyl[146] and 4-imide[146] derivatives have been fully characterized.

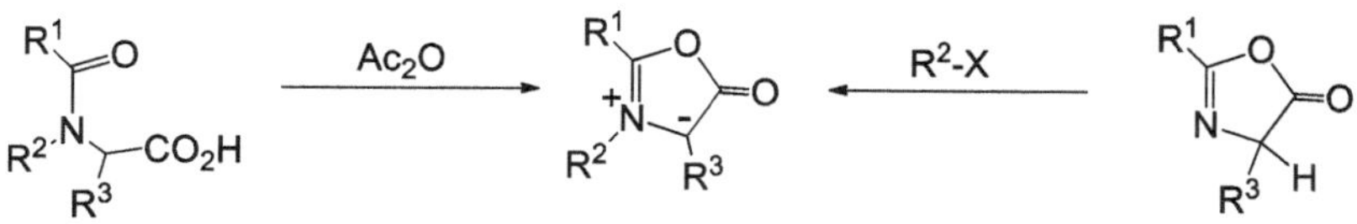

99 **100** **101**

Scheme 17

Sometimes the perchlorate salts of münchnones can be generated by treatment of the acids **99** with aqueous perchloric acid. The münchnone is then liberated in the presence of base.[152–154]

An alternative approach to *in situ* münchnone generation involves the treatment of azlactones **101** with strong alkylating agents, such as methyl trifluoromethanesulfonate or triethyloxonium tetrafluoroborate, in the presence of a base (**101** → **100**; R^2 = alkyl) (Scheme 17).[155] Trimethylsilyl chloride has been used to generate transient *N*-trimethylsilyl derivatives (**2**, $R^2 = Me_3Si$).[156,157]

A variation on the azlactone method is to use a metal catalyst to generate and trap N-metalated münchnones **103** (Scheme 18). For example, treatment of 2-phenylazlactones **102** (R^1 = Ph) with silver acetate (10 mol%) provides an efficient stereoselective route to pyrrolines *via* the silver derivatives **103** (R^1 = Ph, LM = AgOAc).[158] This approach has been extended to the use of Au(I)[159,160] and Cu(I)[161] catalysis and the *in situ* trapping of *N*-Au and *N*-Cu münchnones **103** (M = Au, Cu).

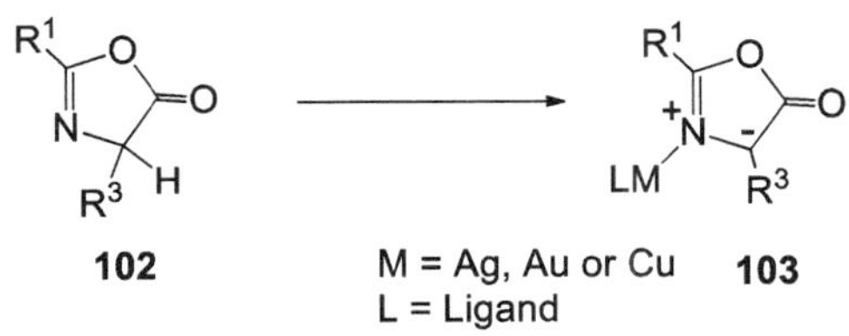

Scheme 18

A remarkable development has been the use of metal-catalyzed carbonyl insertions to generate münchnones (Scheme 19). In 2000, Merlic and co-workers showed that the acylamino chromium carbene complex **104** reacts with carbon monoxide to give the münchnone **100** ($R^1 = R^3$ = Ph, R^2 = Me) which can be isolated (27%) or trapped by dimethyl acetylenedicarboxylate (DMAD).[162] Arndtsen and co-workers have extensively studied the palladium-catalyzed coupling of imines with carbon monoxide. Acyliminium salts with $Pd_2(dba)_3$ reagents give the palladium

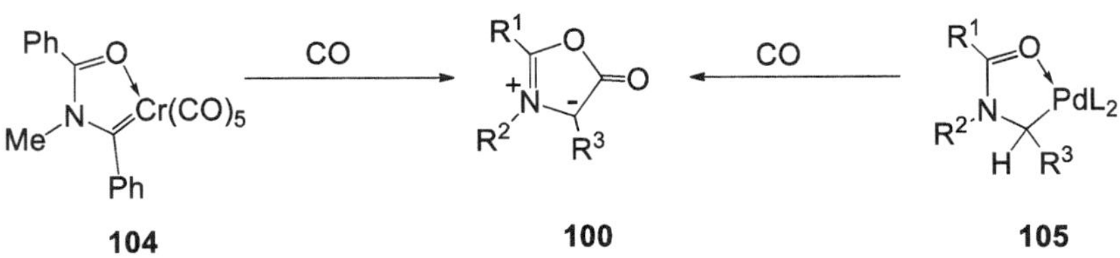

Scheme 19

complexes **105** which insert carbon monoxide forming münchnones.[163] Using this method, a number of münchnone derivatives **100** have been isolated in good yield.[164] Alternatively they can be trapped *in situ* by dipolarophiles giving imidazolines,[163,165] imidazoles[166] or pyrroles.[167–169]

5.2.4.2 Cycloadditions

5.2.4.2.1 Alkene [2 + 3]-cycloadditions Münchnones undergo 1,3-dipolar cycloadditions with alkenes to give cycloadducts that readily lose carbon dioxide to give a new 1,3-dipole that can be trapped. This is illustrated by the decarboxylation of the *N*-methylmaleimide adduct **106** and trapping by a second molecule of the maleimide to give the final product **107** as a mixture of stereoisomers (Scheme 20).[170,171]

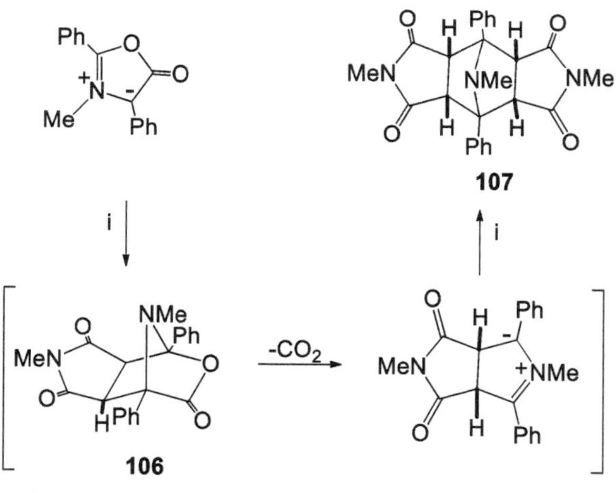

Scheme 20

A more synthetically useful approach is to use an "alkyne equivalent" as the dipolarophiles. The münchnones are electron-rich 1,3-dipoles and readily react with nitroalkenes. Typically the derivative **108** with 2-nitrostyrene gives the transient adducts **109** (only one regioisomer shown) which readily lose both carbon dioxide and nitrous acid to give a mixture of the isomeric pyrroles **110** and **111** in 81% yield (Scheme 21). This exemplifies a convenient route to pyrroles and, significantly, the major isomer **110** is the opposite regioisomer to that obtained using phenylacetylene as the dipolarophiles (i.e., **111**).[172] Although frontier molecular orbital (FMO) theory is successful in predicting the regiochemistry of many 1,3-dipolar cycloadditions,

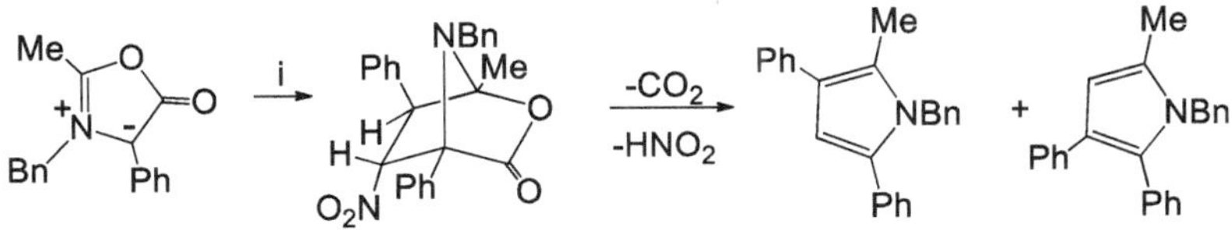

108 **109** **110** **111**

Reagent: i, 2-nitrostyrene

Scheme 21

it shows poor correlation with these nitroalkene reactions.[172,173] A theoretical study indicates that these reactions are controlled by steric and reactant-reorganization factors rather than by frontier orbital overlap.[172] A c-DFT study suggests an asynchronous mechanism with maximum charge transfer after the transition state.[174]

An important development is the use of eneamines as dipolarophiles in reactions with münchnones bearing electron-withdrawing 4-substituents, such as CF_3CO or TsN(H)CO. This is illustrated by the transformations **112** → **113** in which the pyrroles **113** are the major isomers (Scheme 22). As with the nitro group in nitroalkenes (Scheme 21), the amino function NR_2 acts as a convenient leaving group and these alkyne equivalents provide regioselectivity that complements other methods.

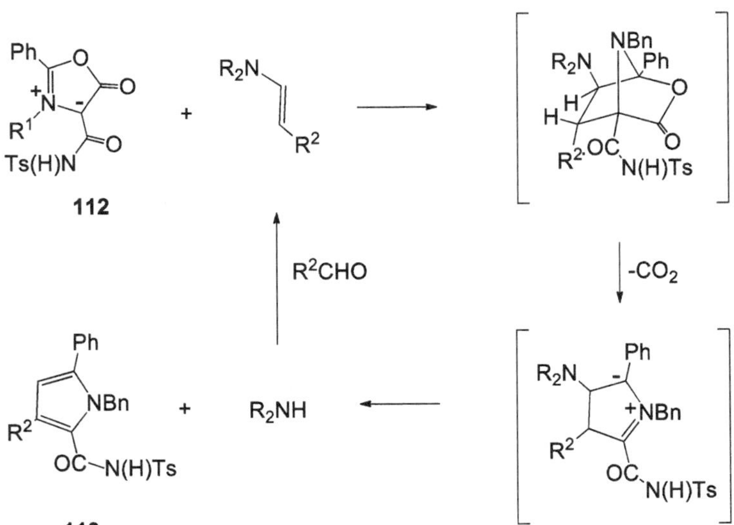

112

113

Scheme 22

In an ingenious variation, the amine R_2NH can recycle acting as a catalyst enabling benzaldehydes in toluene under reflux to be used as alkyne equivalents (Scheme 22).[175] The addition of trifluoroacetic acid is reported to enhance the yields of these enamine additions.[176]

Many other pyrrole derivatives have been prepared using münchnone cycloadditions of alkyne surrogates. These include studies employing nitroalkenes,[173,177–180] nitroheterocycles,[181–183] cyanocinnamates,[184] chloroalkenes,[152] acetoxyacrylic esters,[185] and vinylphosphonium salts.[186]

Other species reported to undergo C=C bond cycloaddition to münchnones include: fulvenes,[187] cyclopropenes,[188] cyclobutenes,[189,190] cyclobutenones,[191] cyclooctadiene and cyclooctatetraene,[192] naphthoquinone,[193] chromenones,[194–196] coumarins,[197] fullerenes,[161] and isothiazole 1,1-dioxides.[198,199]

Intramolecular cycloadditions of münchnones allow access to novel polycyclic systems and provide examples of stable primary adducts in which carbon dioxide is not eliminated. Padwa and co-workers investigated the intramolecular additions of *N*-(2-allyphenyl)- and *N*-(2-vinylphenyl) münchnones, generated *in situ* from the corresponding aminoacids. The regioselectivity of these additions is dependent on the ring substituents. For example, the 4-methyl derivatives **114** (R^1 = H, CF_3, R^2 = Me) give the adducts **115** (R^1 = H, CF_3, R^2 = Me) whereas the 4-phenyl derivatives **114** (R^1 = H, CF_3, R^2 = Ph) give predominantly the products **116** (R^1 = H, CF_3, R^2 = Ph) (Scheme 23).[200–202] Gataullin *et al.* have studied the intramolecular cycloadditions of a variety of 2-alkenylaniline derived münchnones.[147,203–206]

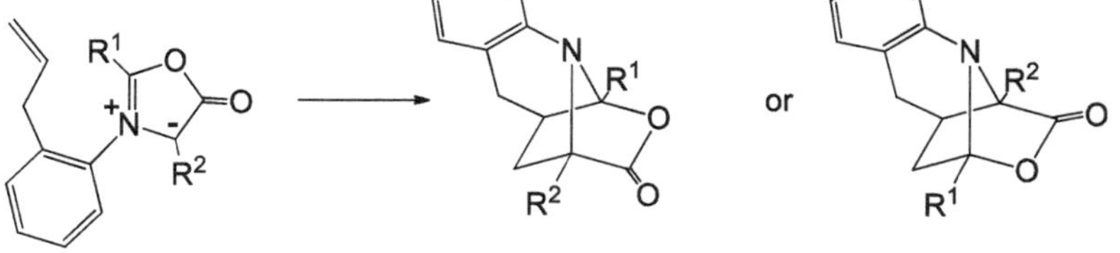

114 **115** **116**

Scheme 23

An interesting feature of the intramolecular 1,3-dipolar cycloadditions of 4-alkyl chain tethered alkenes, such as the derivative **117**, is that the primary adducts open to give stable zwitterions, e.g., **118** → **119** (Scheme 24). See also Scheme 2 in Section 4.4.1. These iminium carboxylates can be further

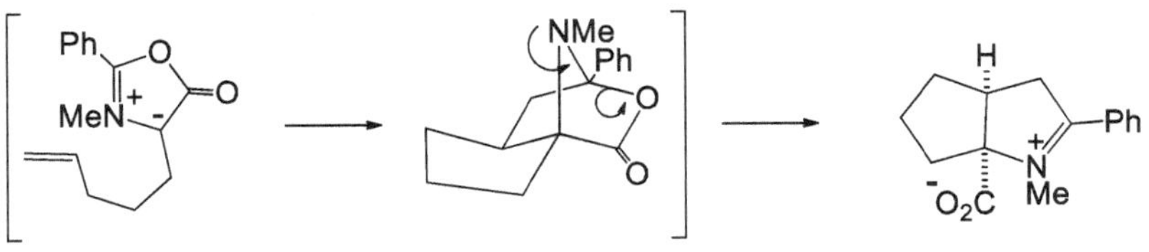

117 118 119

Scheme 24

manipulated to give synthetically useful products. Bélanger and co-workers have studied a number of examples of these 4-alkene-tethered reactions of münchnones.[207]

5.2.4.2.2 Alkyne [2+3]-cycloadditions Since the earliest studies by Huisgen,[1,2] 1,3-dipolar cycloaddition of alkynes to *in situ* generated münchnones has been recognized as an excellent route to pyrroles. This is illustrated by the preparation of the pyrrole intermediate **121** from münchnone precursor **120** in a synthesis of the antimitotic agent (−)-rhazinilam (Scheme 25).[208] Cyclization of the appropriate amino acid using acetic anhydride to generate the münchnone continues to be the most common methodology and many pyrroles of structural and biological interest have been made this way.[143,209–226] The reactions are usually regioselective and sometimes regiospecific. However, both steric and electronic properties influence the outcome and FMO theory is often not a good predictor of regioselectivity.[227]

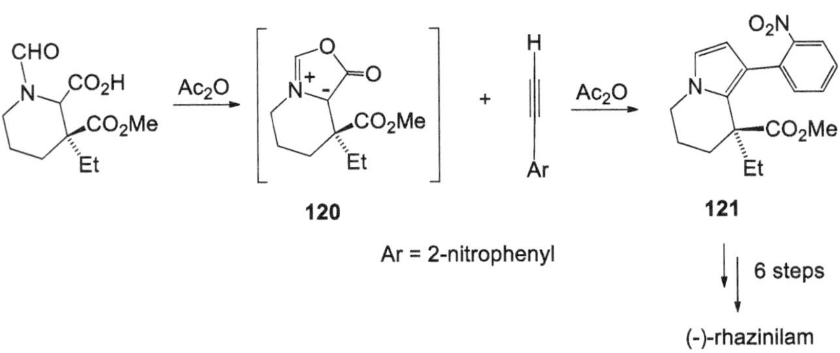

Scheme 25

Other methods that have been used in alkyne-münchnone routes to pyrroles include N-alkylation of azlactones **101** → **100** (Scheme 17)[155,228] and palladium-catalyzed carbonylation of amides, **105** → **100** (Scheme 19).[167–169]

The latter method is illustrated by the reactions of the 3-pyridyl derivatives **122** with carbon monoxide in the presence of a variety of alkynes to give the pyrroles **123** in good yields (Scheme 26).[167] Intramolecular alkyne additions have also been described using a variation of this method.[229]

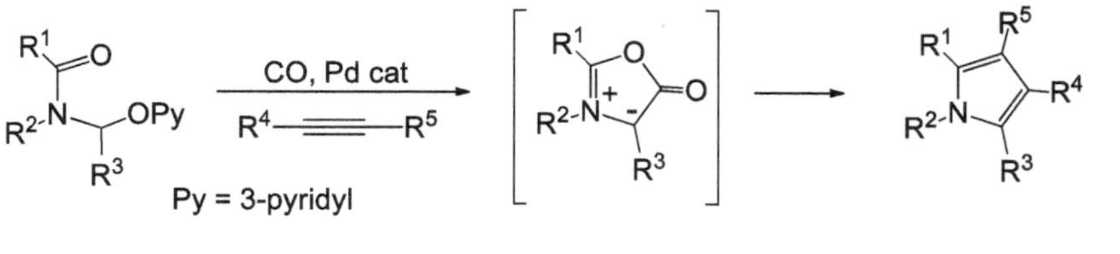

122 **123**

Scheme 26

Many studies have described the generation of bicyclic and polycyclic münchnones and their reactions with alkynes to give pyrrole-based structures, often of biological interest. Examples include the derivatives **124**,[230] **125**,[231] and **126**.[232–235] References to other studies are included here for completeness.[221,231,236–245]

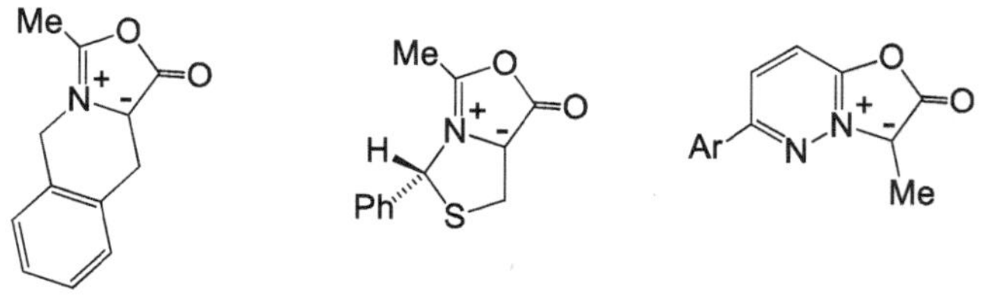

124 **125** **126**

Münchnones have been reacted with arynes to give isoindole derivatives or their adducts.[246,247]

5.2.4.2.3 Imine [2+3]-cycloadditions Two multicomponent syntheses of imidazolines employing 1,3-dipolar cycloaddition of imines to münchnones have been reported.

Tepe and co-workers have developed a one-pot diastereoselective route to the unsymmetrical imidazolines **128**. Treatment of the azlactones **127** with trimethylsilyl chloride (TMSCl) gives the *N*-silylmünchnones **129** and addition of imine leads to the products **128** (Scheme 27). The presence of a nucleophilic counterion is necessary to ensure an equilibrium between O- and N-silylation of the azlactone.[156,157]

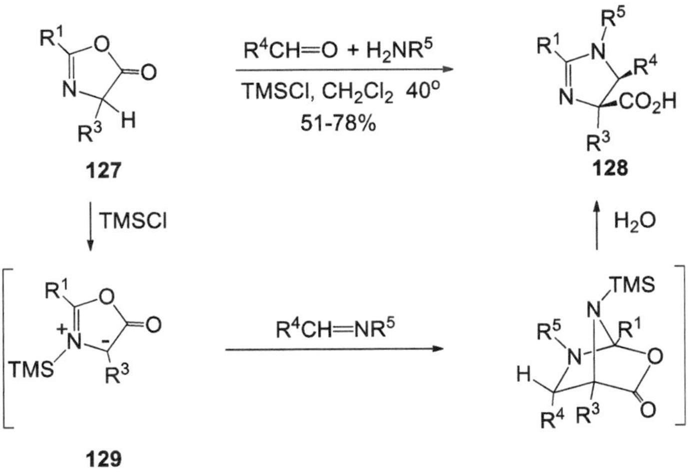

Scheme 27

In an alternative approach that gives the zwitterionic symmetrical imidazolines **130**, Arndtsen and co-workers have described a palladium-catalyzed procedure in which a münchnone intermediate **132** is generated by insertion of carbon monoxide into a palladium-amide complex **131**, formed from a mixture of acid chloride and imine (Scheme 28). A second molecule of imine then adds leading to the product **130**. In a variation of this method, a different imine can be incorporated at the cycloaddition stage leading to unsymmetrical products.[163,165]

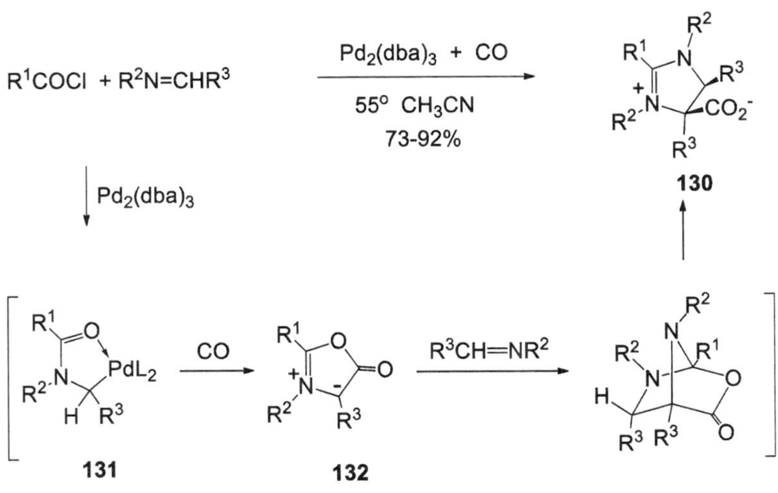

Scheme 28

The use of *N*-arenesulfonylimines as 1,3-dipolarophiles and subsequent loss of arenenesulfinic acid ($ArSO_2H$) and carbon dioxide from the resulting münchnone adduct gives access to imidazole derivatives **133**.[166,248–250] In an alternative approach, an imidazole derivative **133** ($R^4 = ArSO_2$) was obtained by addition of tosyl cyanide to a münchnone **134** (Scheme 29).[218]

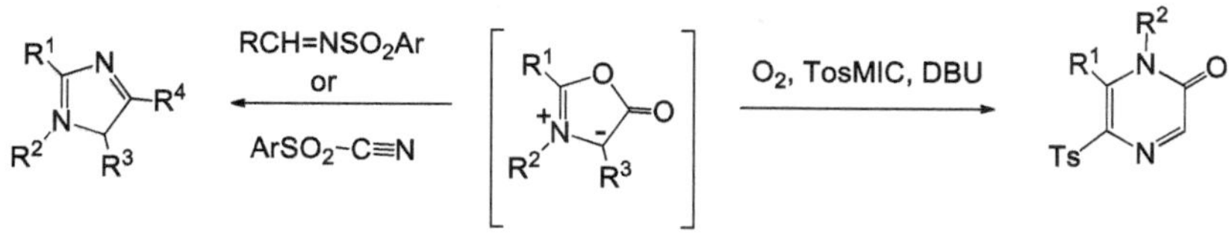

133 **134** **135**

Scheme 29

Reaction of 4-$COCF_3$ derivatives (**134**; $R^3 = COCF_3$) with (p-tolylsulfonyl)methyl isocyanide (TosMIC) in the presence of oxygen and a base gives 2(1*H*)-pyrazinones **135**. The oxygen in the product **135** originates in molecular oxygen.[251–253]

5.2.4.2.4 Diazene [2+3]-cycloadditions Diethyl azodicarboxylate (DEAD) or 4-phenyl-4*H*-1,2,4-triazoline-3,5-dione have been added to polymer-bound münchnones to give 3,5-disubstituted 1,2,4-triazoles.[144,254]

5.2.4.2.5 Other cycloadditions Münchnones react with heterocumulenes to give new type A mesoionic rings in which the "carbon dioxide" fragment is replaced by the heterocumulene, e.g., CS_2, ArNCS. These reactions are covered as preparative methods in the appropriate subsections but see particularly 1,3-thiazol-5-thiones (Section 5.5.6)[255,256]and 1,3-diazol-4-thiones (Section 5.4.3).[257]

3,5-Di-*t*-butyl-*o*-benzoquinone gives the [4+4] cycloadduct **136** with 2,4-diphenyl-3-methylmünchnone;[258] similar *o*-benzoquinone reactions have been reviewed.[17] La Rosa and co-workers have shown that the bicyclic münchnones **137** (X=O, S, SO_2; R=Ph, $PhSO_2$) with imines give the spirocyclic β-lactams **138**, in which an imine appears to have added to the ketene fragment of the acyclic valence tautomer, but this is not necessarily the mechanism.[242,250,259]

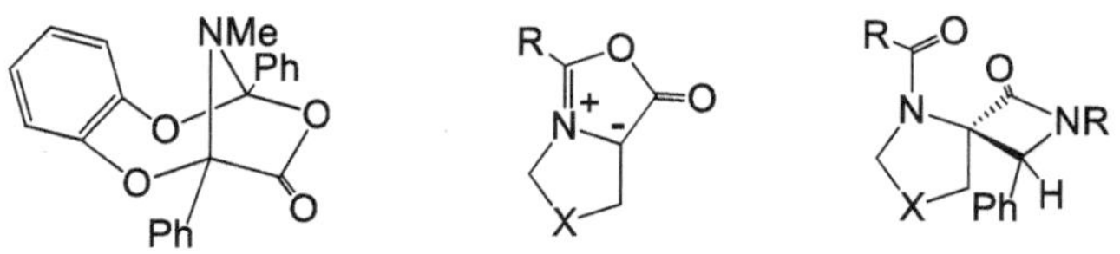

136 **137** **138**

5.2.4.2.6 Reaction with nucleophiles Kawase and co-workers have investigated the reactions of 4-trifluoroacetyl münchnones (**134**; $R^3 = COCF_3$) with a number of nucleophilic reagents. Reaction with phosphorus or sulfur ylides leads to 3-trifluoromethylpyrroles *via* initial nucleophilic attack at position 2 by the ylide anion, followed by a ring opening/ring closure sequence.[260,261] Treatment with amidines leads to 5-trifluoroacetyl imidazoles,[262,263] and a library of 200 imidazoles has been made using this methodology.[145] Reaction with ammonium acetate (or aqueous ammonia) leads to 3-hydroxyimidazolines which upon dehydration with $POCl_3$ give 4-perfluoroalkylimidazoles.[264–266] Tri-*n*-butylphosphine gives novel acyclic phosphorus zwitterions.[267] Reaction with hydrazines is more complex giving selectively 1,2,4-triazines, 5-pyrazolones or 3-hydroxypyrazoles depending upon solvent and temperature.[268,269] Hydroxylamine gives 1,2,4-oxadiazin-6-ols in high yield.[270]

5.2.5 1,3-Oxazol-5-imines (1,3-oxazolium-5-aminides) (141)

Common name: münchnone imines

The preparation of derivatives of this ring system was reported in previous reviews.[1,2] The trifluoroacetyl derivatives **141** ($R^4 = COCF_3$) can be converted to 4-lithio or 4-copper derivatives and reaction with electrophiles (RCHO, RCOCl) facilitates functionalization at the 4 position.[271]

A new method of generating this mesoionic ring system *in situ* involves reaction of an imine, an acid chlorides and an isocyanide in the presence of a base. The initially formed *N*-acyliminium salt **139** with isocyanide rapidly forms the ring **140** which with base gives the transient münchnone imine **141**. In the presence of an alkyne, cycloaddition gives an adduct **142** and elimination of isocyanate results in formation of substituted pyrroles **143** in high yield (Scheme 30). This simple, room temperature procedure provides an extremely convenient route to pyrroles.[272]

Polycyclic derivatives **141** ($R^1 = CF_3$, R^2-R^3 = isoquinoline) have been obtained as byproducts in multicomponent reactions of azines with isocyanides and TFAA.[273]

Salts of the münchnone imines **144** can be generated *in situ* by treatment of the nitriles **145** with strong acid. Reactions of the diphenyl derivatives with alkenes have been reinvestigated and the products are pyridones and not pyrroles as originally reported. Thus, the derivative **144** (Ar = Ph) with ethyl acrylate gives the pyridone **146** after initial formation of a 1,3-dipolar cycloadduct, rearrangement and elimination of aniline.[274]

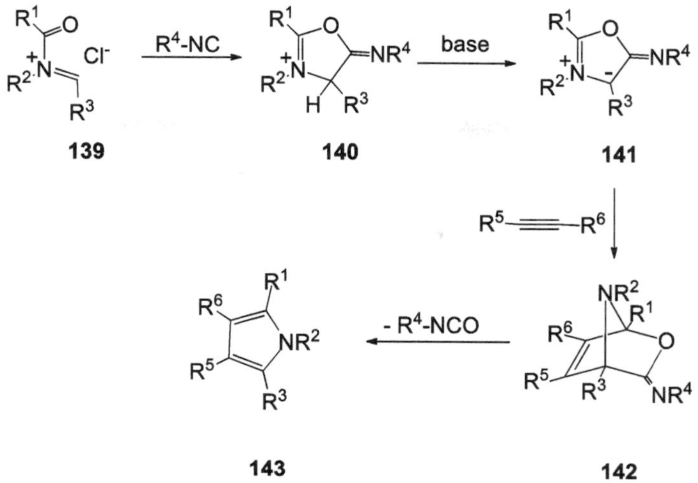

Scheme 30

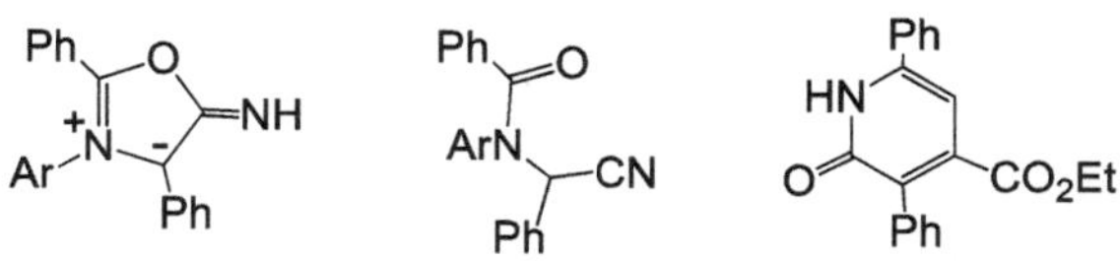

144 145 146

5.3 Oxathioles

5.3.1 1,3-Oxathiol-4-ones (1,3-oxathiolium-4-olates) (148)

This ring system has been studied by Gotthardt and co-workers who showed that cyclization of the acids **147** ($R^1 = SR$, NR_2, $R^2 = Ph$) using acetic anhydride or DCC gives *in situ* access to the oxathiol-4-ones **148** (Scheme 31).[2] Subsequently, they showed that when $R^2 = H$, the use of TFAA (or trichloroacetic anhydride) gives the stable 5-acyl derivatives **148** ($R^1 = SR$, NR_2, $R^2 = CF_3CO$, CCl_3CO) which can be isolated and fully characterized.[275–277]

The mesoionic 1,3-oxathiolium-4-olates **148** are carbonyl ylides and undergo 1,3-dipolar cycloadditions with alkynes to form furans. Reaction with DMAD gives the diesters **149** in good yield,[278] and symmetrical dibenzoylacetylene,[279] and bis(trifluoromethylthio)acetylene[280] also give the corresponding 3,4-disubstituted furans. Reactions of the 5-trifluoroacetyl derivatives (e.g., **151**) with methyl propiolate give mixtures of regioisomeric furans but reactions with electron-rich alkynes are regiospecific. The reaction

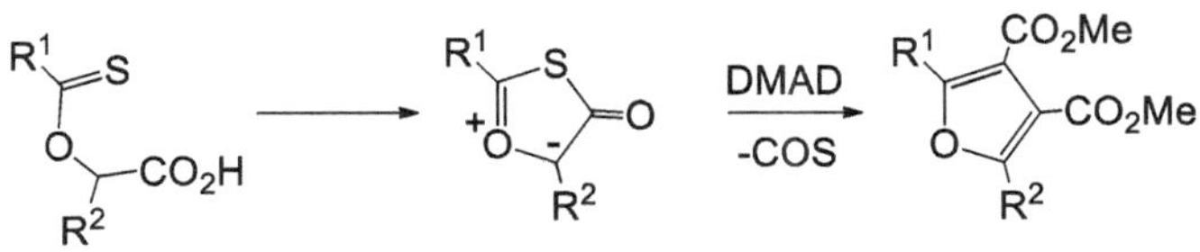

147 148 149

Scheme 31

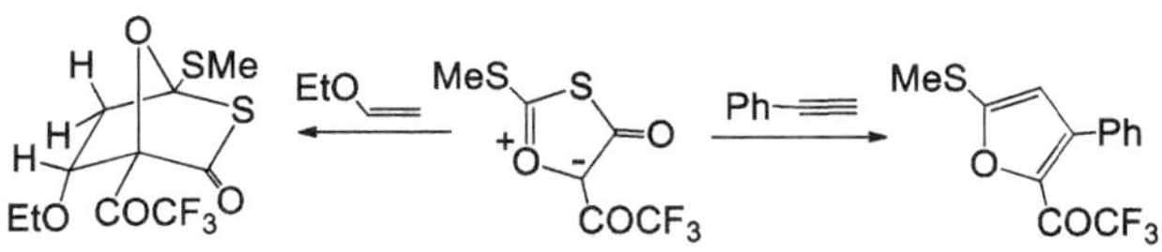

150 151 152

Scheme 32

of the derivative **151** with phenylacetylene to give the 3-phenylfuran **152** (80%) is typical (Scheme 32).

Alkenes react to give stable cycloadducts. This is exemplified by the formation of the adduct **150** (100%) using ethyl vinyl ether.[279] Reaction with heterocumulenes provides access to two new mesoionic ring systems (Scheme 33). Addition of carbon disulfide to the 5-phenyl derivatives **154** gives 1,3-oxathiol-4-thiones **153** (Section 5.3.2) in good yield. Similarly, phenyl isothiocyanate gives the 1,3-oxazol-4-thiones **155** (R = Et) (Section 5.2.3).[138]

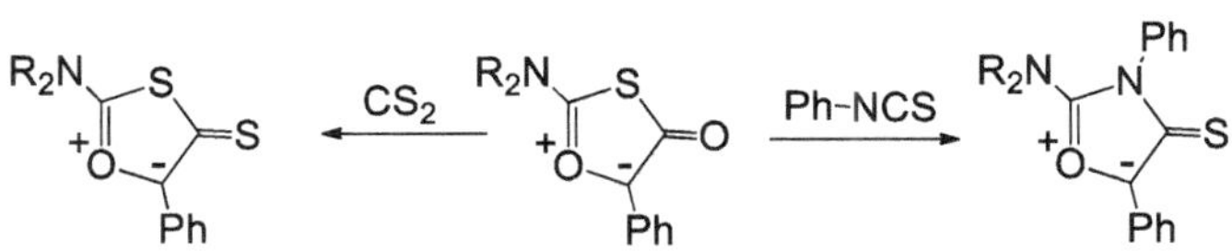

153 154 155

Scheme 33

Using the AM1 semi-empirical method, the equivalence of 1,3-oxathiol-4-ones, and other mesoionic rings, to polyenes in the design of materials for non-linear optics has been investigated.[281]

5.3.2 1,3-Oxathiol-4-thiones (1,3-oxathiolium-4-thiolates) (156)

2-Amino derivatives **156** have been prepared as stable crystalline compounds by addition of carbon disulfide to 1,3-oxathiol-4-ones **154** (Section 5.3.1). Treatment of the 2-dimethylamino derivative **156**

(R = Me) with methyl iodide gives the crystalline salt **157** (99%) which with acetylacetone gives the 1,4-oxathiafulvene **158**. 1,3-Dimethylbarbituric acid gives a similar product.[138]

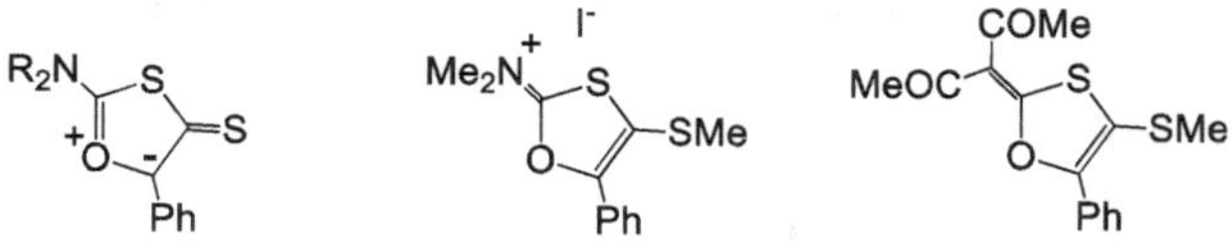

156 **157** **158**

5.3.3 1,3-Oxathiol-5-ones (1,3-oxathiolium-5-olates) (159)

Surprisingly little work has been done using this ring system since examples were first described.[2] Gribble and co-workers have exploited the reactivity of these thiocarbonyl ylide derivatives to prepare the hydrocarbon **161**. Cyclization of the acid **160** using diisopropylcarbodiimide (DIPC) and *in situ* trapping of the mesoionic ring **159** with cycloocta-1,5-diene (CODE) resulted in a tandem cycloaddition sequence giving the sulfide **162** in 32% yield. A two-step procedure *via* the sulfone then gave the product **161** (Scheme 34).[282]

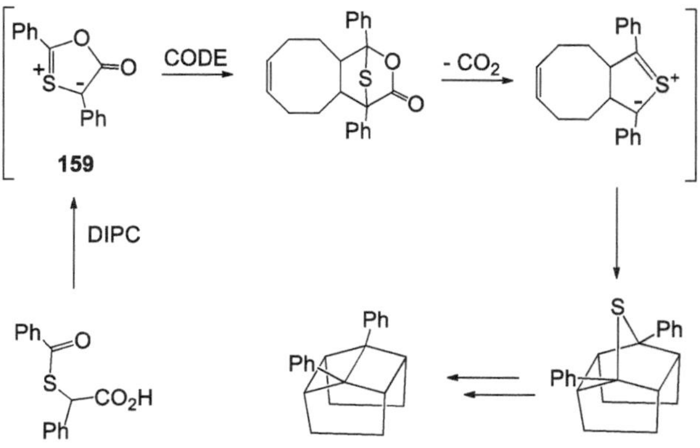

159 **160** **161** **162**

Scheme 34

5.4 Diazoles

5.4.1 1,3-Diazol-4-ones (1,3-diazolium-4-olates) (164)

5.4.1.1 Preparation

Derivatives of this ring system are stable crystalline compounds and several methods of preparation have been reviewed.[1,2] A recent development is the preparation of 2-unsubstituted derivatives **164** (Scheme 35), which are

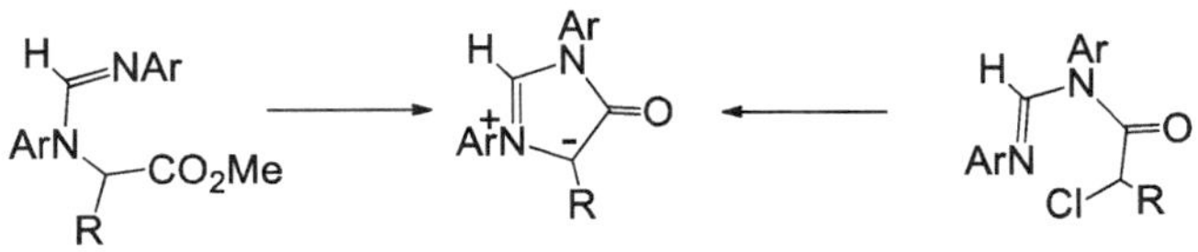

163 164 165

Scheme 35

precursors to N-heterocyclic carbene derivatives. *In situ* generation of the esters **163** from diarylformamidines and 2-bromo methyl esters in the presence of base at 120° gives the derivatives **164** in high yield.[57] The structure of the derivative **164** (Ar = 2,6-(iPr)$_2$C$_6$H$_3$, R = Et) has been determined by X-ray crystallography.[57] In an alternative approach, diarylformamidines and chloroacetyl chloride give the amidines **165** (R = H), which cyclize at 100°, and treatment of the resulting chlorides with base gives the diazol-4-ones **164** (R = H).[283,284]

Several new polycyclic derivatives have been prepared by cyclization of amidine carboxylic acid derivatives using acetic anhydride.[285–288]

Significant new preparative methods that do not employ amidine precursors have been reported. Treatment of the isocyanoacetamides **166** with arylsulfenyl chlorides gives the amides **167** which with base give the 2-arylthio derivatives **168** in good yield (Scheme 36). These products are reported to be stable as the NH tautomers and the X-ray structure of compound **168** (R = nPr, Ar = 2-NO$_2$C$_6$H$_4$) has been published, although the Cambridge Crystallographic Data Base (CCDC-1184397) shows the OH tautomer.[289]

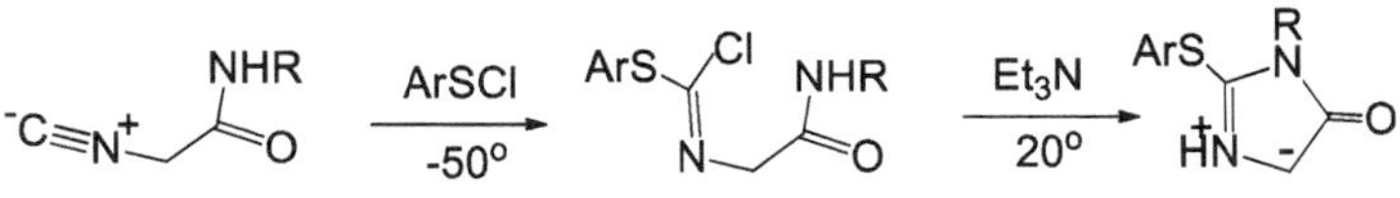

166 167 168

Scheme 36

Cyclization of the iminium chlorides **169**, formed *in situ* from the acid chloride of picolinic acid and an imine, gives the bicyclic derivatives **170** in high yield (Scheme 37). The crystal structure of one derivative has been reported.[290] In an alternative one-pot palladium-catalyzed process from the same laboratory, 2-bromopyridine, an imine and carbon monoxide give similar bicyclic derivatives **170**, which are conveniently trapped *in situ* (Scheme 37).[291]

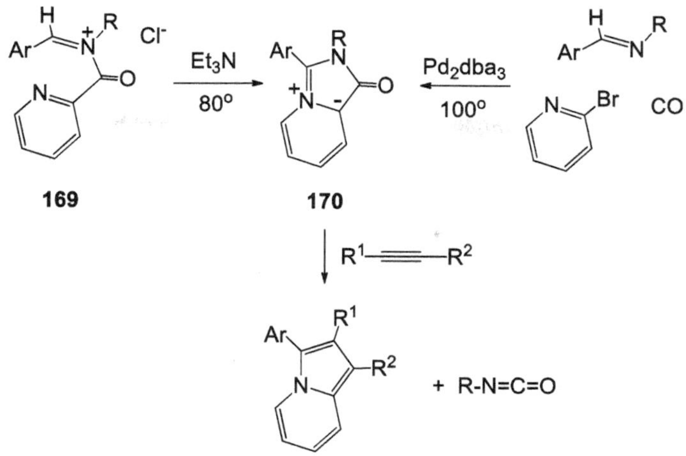

Scheme 37

Derivatives of the isomeric bicyclic system **171** have been prepared. Treatment of 2-anilinopyridine with chloroacetyl chloride in the presence of *N,N*-dimethylacetamide gives compound **171**. The structure of product **171**, which was incorrectly assigned by previous workers, has been confirmed by X-ray crystallography (Scheme 38). Displacement of the chloro substituent in derivative **171** by N- and O-nucleophiles gives sidechain modified products.[292] In an alternative approach, a number of derivatives **172** have been prepared by multicomponent reactions of isoquinoline with cyclohexyl isocyanide activated by fluorinated acid anhydrides. The crystal structures of the derivatives **172** (R = F, 2-CCl_3-1,2-dihydroisoquinolinyl) have been reported.[273] The derivatives **172** (R = F) are fluorescent and have been used to label biomolecules that contain an amino group.[293–295]

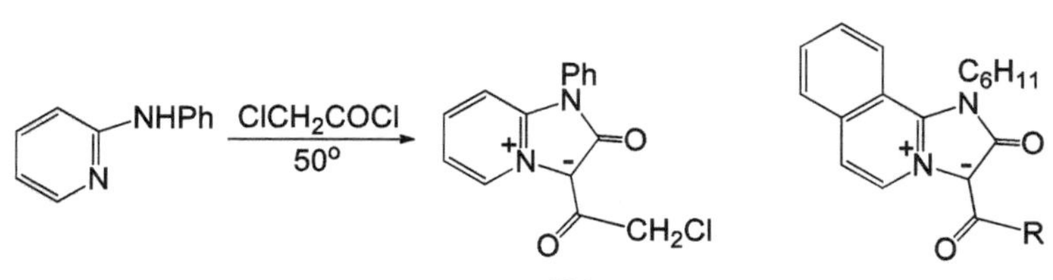

Scheme 38

Glycosyl analogues of the mesoionic imidazole nucleoside bredinin have been prepared using the silyl-Hilbert-Johnson method.[296] A limited number of derivatives of mesoionic 1,3-diazol-4-ones have been prepared by O-demethylation of the corresponding 4-methoxy-1,3-diazolium salts.[297,298]

5.4.1.2 Cycloadditions

5.4.1.2.1 [2 + 3]-cycloadditions Using the preparative methods shown in Scheme 37, the bicyclic diazol-4-ones **170** undergo 1,3-dipolar cycloadditions with alkynes *in situ* giving indolizines in good yield. Using these one-pot multicomponent methods a wide range of indolizine derivatives has been prepared from readily available starting materials.[290,291]

5.4.1.2.2 Other cycloadditions 1,3-Diazol-ones form [4 + 3] cycloadducts with *o*-benzoquinones and *o*-benzoquinone dimines.[299] Other type A mesoionic rings undergo similar reactions.[300]

5.4.1.3 Carbene formation

An important new development in the chemistry of type A mesoionic rings has been the generation of anionic carbenes and their trapping (see also Section 4.4.2). In 2009 Glorius and co-workers showed that treatment of the *N,N*-dimesityl (Mes) derivative **173** with lithium bis(trimethyl)amide (LiHMDS) gave a solution of the lithium salt of the anionic N-heterocyclic carbene (NHC) **174** which can be trapped by elemental sulfur to give the thione **175** (79%) (Scheme 39). Similar trapping with iridium reagents gives high yields of $Ir(CO)_2Cl$ complexes of the carbene **174**.[57]

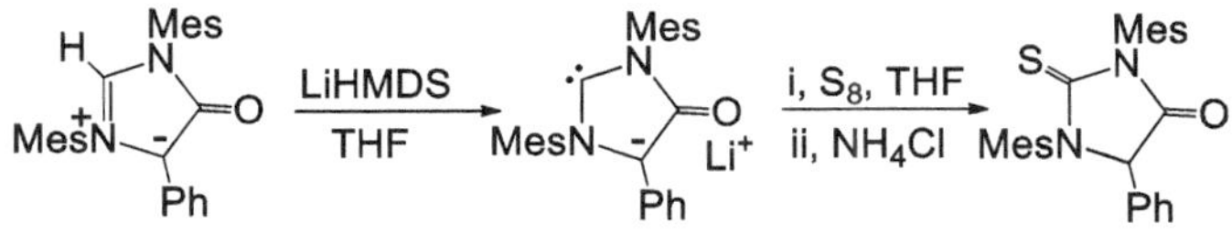

173 **174** **175**

Scheme 39

In a study of the 2,5-unsubstituted mesoionic ring, César, Lavigne and co-workers generated the NHC **177** in a similar procedure using LiHMDS. Trapping experiments gave the rhodium and copper complexes **176** (90%) and **178** (68%), as well as a thione (68%) analogous to **175** (Scheme 40).[283,284] Copper-carbene complexes have also been obtained starting from the 5-acetyl-1,3-dimesityl derivative.[301]

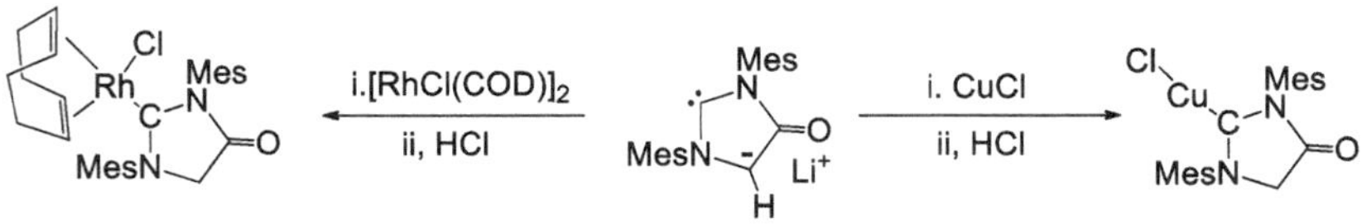

176 **177** **178**

Scheme 40

5.4.1.4 Computational studies

The aromatic stabilization energy (ASE) of the dimethyl derivative **179** (R^1 = H) has been calculated, using homodesmotic reaction schemes, and found to be 0.55 kcal mol^{-1}.[20] This low value is consistent with (i) other measures of classical aromaticity (Bird and HOMA index) and (ii) a ring system that readily undergoes 1,3-dipolar cycloadditions with alkynes.[302] The low classical aromaticity and high reactivity support the structural representation **179** rather than structure **180** which implies an aromatic sextet. The calculated NICS(1) (−5.8) and NICS(1)zz (−16.6) indexes indicate moderate magnetic aromaticity [benzene NICS(1) (−10.2) and NICS(1)zz (−29.3)].[20]

CREF index values for the acidity of the 2-proton in a series of 5-substituted derivatives **179** have been calculated (see also Section 4.4.2). The results, which measure the ease of deprotonation to form the carbenes **181**, show a significant correlation with both the Swain and Lupton field (F) and resonance (R) substituent constants for substituent R^1; this demonstrates that both σ and π substituent effects contribute to acidity. The 2-position is calculated to be more acidic than the 5-position.[303]

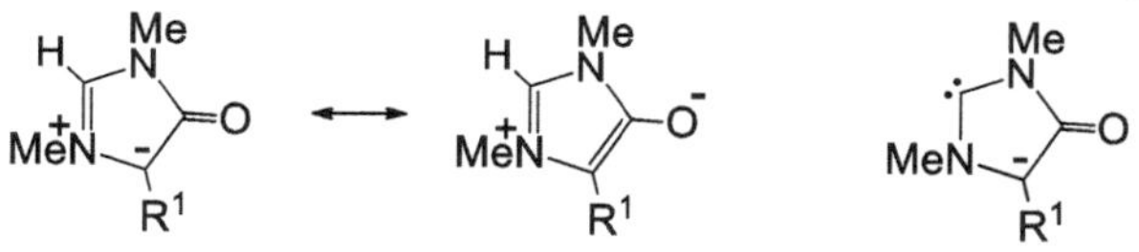

179 **180** **181**

5.4.2 1,3-Diazol-4-imines (1,3-diazolium-4-aminides) (183)

A fascinating development has been the preparation and isolation of stable mesoionic 1,3-diazol-4-imines by treatment of the imidazolium salts **182** with strong base. The *N*-isopropyl derivative **183** (Ar = mesityl, R = *i*Pr) appears to be the first isolable mesoionic imine derivative with an N-alkyl substituent; the structure was confirmed by X-ray crystallography. NMR studies show that this mesoionic derivative is in equilibrium with a small amount of the tautomeric NHC **184** (Ar = mesityl, R = *i*Pr). In THF solution at room temperature the minor NHC tautomer **184** can be trapped by carbon disufide or sulfur to give the adducts **186** (36%) and **187** (69%), respectively (Scheme 41).[54]

In a separate study, the triaryl derivative **183** (Ar = R = Dipp) was prepared in a similar manner, isolated and the structure confirmed by X-ray crystallography. For this N-aryl derivative, NMR studies confirmed that in solution only the mesoionic tautomer **183** (Ar = R = Dipp) is present and there is no evidence of equilibration with the NHC **184**

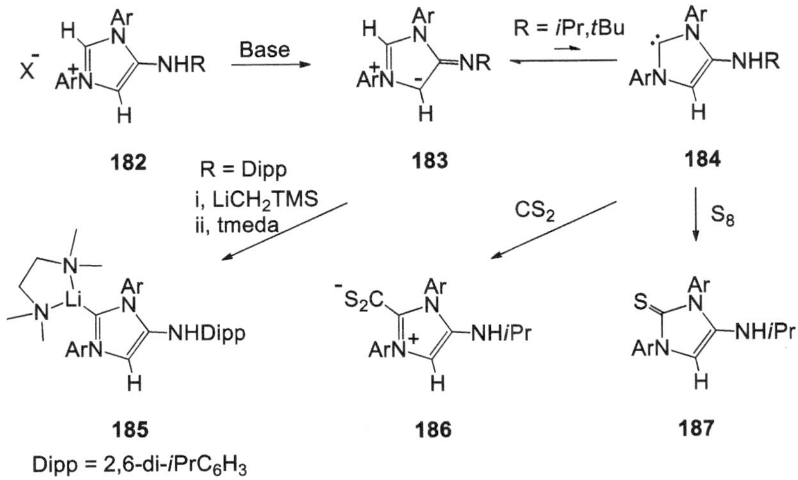

Scheme 41

(Ar=R=Dipp). This is consistent with the greater stabilizing effect, by extended conjugation, of N-aryl substituents in mesoionic imines.[32] However, NMR evidence demonstrates that the *N*-*t*Bu derivative **183** (Ar=Dipp, R= *t*Bu) is in equilibrium with the carbene form **184** in aromatic solvents at room temperature.[55] In fact, computational studies have shown that the substituent effect stabilization energies (SESE) for mesoionic imines (e.g., **183**) are =NPh> =NMe whereas the stabilization energies for the tautomeric NHCs (e.g., **184**) are —NHMe **>** —NHPh.[56] These differences in stabilization account for why the *N*-*i*Pr and *N*-*t*Bu species equilibrate whereas the *N*-Dipp derivative does not.

Anionic substituents are particularly good at stabilizing NHCs, e.g., replacing NHR in structure **184** by PhN^- or O^-: the calculated stabilization energies are $O^- > PhN^- >> NHPh$[56] This is illustrated by treatment of **183** (Ar=R=Dipp) with $LiCH_2TMS$ generating the anionic NHC which upon addition of tetramethyethylenediamine (tmeda) is trapped as the lithium adduct **185**.[32] In fact the anionic carbene (Ar=R=Dipp) was subsequently isolated and characterized as a salt in which there are discrete $[K(18\text{-crown-}6)]^+$ and anionic NHC ions.[55]

5.4.3 1,3-Diazol-4-thiones (1,3-diazolium-4-thiolates) (188)

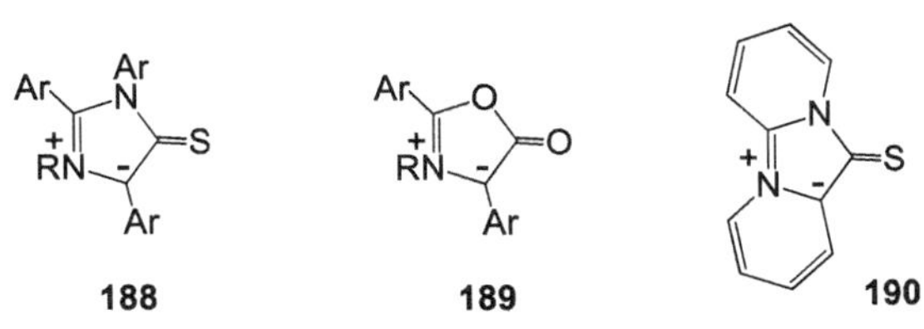

The preparation of 1,3-diazol-4-thiones from münchnones has been revisited. A number of 2,3,5-triaryl derivatives **188** have been prepared in moderate yield by addition of aryl isothiocyanates to the corresponding münchnones **189** under microwave conditions (140°) in DMF. DFT calculations suggest that these additions occur *via* dipolar intermediates rather than formal 1,3-dipolar cycloadducts.[257] Morel and co-workers have prepared a variety of monocyclic derivatives by the reaction of mesoionic 1,3-thiazol-5-imines with isothiocyanates.[304–306]

Full details of the preparation of the tricyclic compound **190** and derivatives from 2-picolines, pyridines and sulfur dioxide have been published. These compounds show interesting solvatochromic properties.[307]

5.4.4 1,3-Diazol-4-enes (1,3-diazolium-4-methylides) (191)

The orange/red derivatives **191** (X=CN, CO_2Et) were the first examples of 1,3-diazol-4-enes[298]; X-ray crystallography and nitrogen NMR studies have confirmed the structure of **191** (X=CN).[308]

A later study has shown that unsubstituted monocyclic ene derivatives can be made. Treatment of the diazolium iodide **193** with potassium bis(trimethyl)amide (KHMDS) gives the green (λ_{max} 664 nm) crystalline product **192** (59%) (Scheme 42). The structure has been confirmed by X-ray crystallography.[309]

The exocyclic carbon of the ene **192** is a Lewis base and slowly reacts with $B(C_6F_5)_3$ and $[Rh(\mu\text{-}Cl)(CO)_2]_2$ to form the adducts **194** and **195**, respectively. Deprotonation of an adjacent *i*Pr substituent leads to slow formation of the isomer **196** at room temperature (Scheme 42).[309]

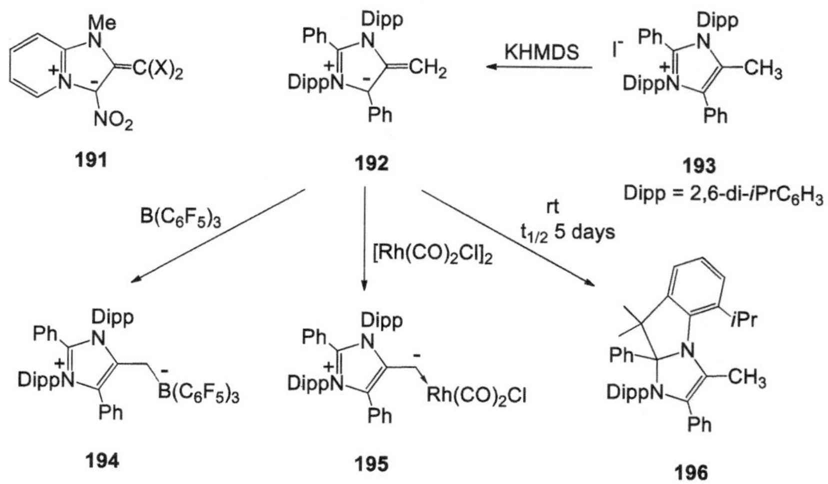

Scheme 42

5.4.5 1,3-Diazol-4-selones (1,3-diazolium-4-selenides) (198)

Treatment of the NHCs **197** (Ar=Ph, 4-$Me_2NC_6H_4$) with selenium powder gives the corresponding mesoionic 1,3-diazol-4-selones **198** (Scheme 43). These orange-red crystalline compounds obtained in 84% and 86% yield are examples of a ring system that has not previously been reported. Their structures have been confirmed by X-ray crystallography.[310] Two more examples have been reported.[311]

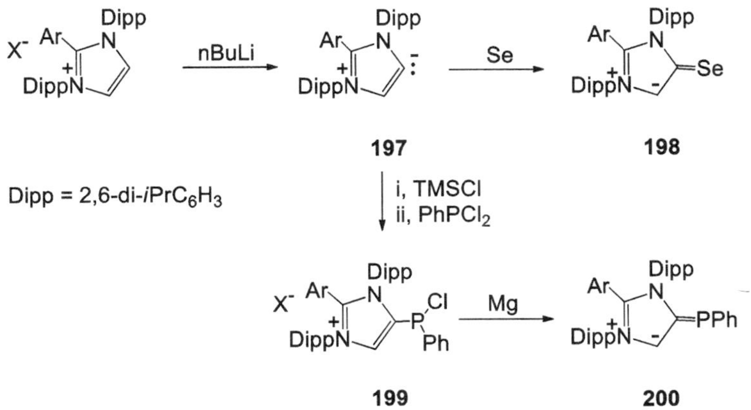

Scheme 43

5.4.6 1,3-Diazol-4-phosphimines (1,3-diazolium-4-phosphamidates) (200)

Sequential treatment of the NHCs **197** (Ar=Ph, 4-$Me_2NC_6H_4$) with trimethylsilyl chloride (TMSCl) and phenylphosphine dichloride ($PhPCl_2$) gives the salts **199** which with magnesium turnings give the mesoionic 1,3-diazol-4-phospimines **200** (Scheme 43). The structure of the green crystalline solid **200** (Ar=Ph) (mp 156 °C, yield 56%) was confirmed by X-ray crystallography. The exocyclic angles (C-C-P 135.5° and N-C-P 120.7°) are similar to those in the 1,3-diazol-4-imines **183**.[310]

5.5 Thiazoles

5.5.1 1,3-Thiazol-4-ones (1,3-thiazolium-4-olates) (201)

Common name: thioisomünchnones

1,3-Thiazolium-4-olates **201** are commonly referred to as thioisomünchnones because of their relationship to the isomünchnones (1,3-oxazolium-4-olates, Section 5.2.1). Some aspects of their chemistry were reviewed in 1994[45,312] and 2005.[46] Investigations of their chemistry have been driven by interest in applications of their cycloadditions.

5.5.1.1 Preparation

Convenient routes to these rings from thioamides were pioneered by Potts and co-workers in the 1970s and this work has been reviewed.[2,45] No significant new methods have appeared. The most convenient route is treatment of thioamides with bromoacetic acids followed by cyclization of the resulting thioacid using acetic anhydride (Scheme 44).[313] Use of thioureas provides access to 2-amino derivatives.[314] A convenient variation is to use the corresponding 2-haloacetyl chloride, which in the presence of base gives the product **201** in one step.[312,315,316] Diethyl bromomalonate has also been used as cyclizing reagent.[317]

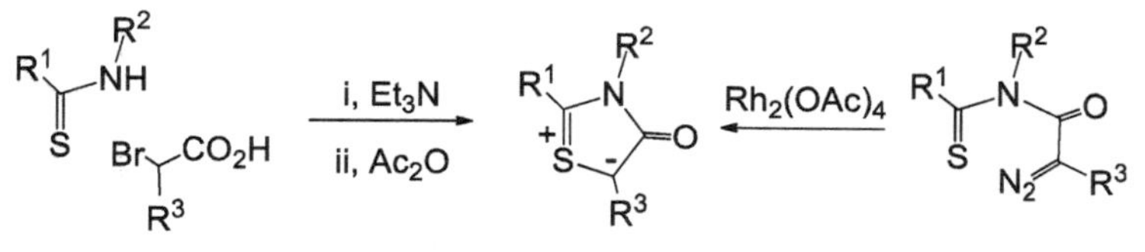

201 **202**

Scheme 44

Based on this well-established methodology, bicyclic derivatives have been prepared by cyclization of heterocyclic thioamides[318] or mercaptoazine derivatives.[319–325] Treatment of mercaptopyridines with 2,2-disulfonyl-oxirans gives pyridine-fused bicyclic derivatives.[326]

Rhodium acetate ($Rh_2(OAc)_4$) catalyzed decomposition of α-diazoamides **202** has been employed as an alternative preparative route,[125,327–329] but has not been widely used. Using tosylhydrazones as a transient carbene source, tricyclic mesoionic thiazolone derivatives have been obtained in good yield.[330] An unusual cyclization of dithioester derivatives has been used to generate bis- and tris-thioisomünchnones.[331]

5.5.1.2 Cycloadditions

5.5.1.2.1 Alkene [2+3]-cycloadditions Thioisomünchnones regioselectively react with alkenes to form stable 1,3-cycloadducts that readily lose hydrogen sulfide to give pyridone derivatives, e.g., **203** → **204** → **205** (Scheme 45). Sulfur loss may take place spontaneously, thermally or be facilitated by reagents such as *m*-chloroperbenzoic acid (MCPBA),[312]

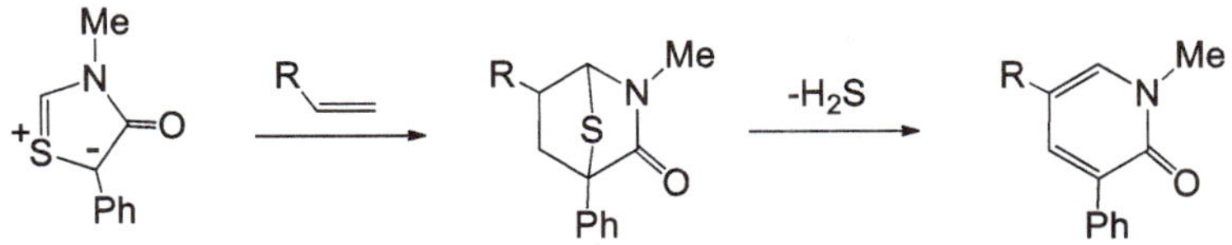

203 **204** **205**

Scheme 45

$Mo(CO)_6$,[312] Raney-Ni,[332,333] $Hg(OAc)_2$,[329,334,335] and silica gel.[92] In this way a number of polycyclic pyridone targets have been prepared.[92,329,332,336,337] Alkene dipolarophiles include N-substituted maleimides,[92,316,327,328,334,335,337–339] maleic anhydride,[340] dimethyl maleate,[329,341] dimethyl fumarate,[341] methyl acrylate,[334] phenyl vinyl sulfone,[312] vinyl ketones,[334,335,342] vinyl ethers,[341] acrylonitrile,[334,335,341] nitroalkenes,[343] benzoquinones,[335] fulvenes,[344] 1,2-diaza-1,3-butadienes,[345,346] and methylene triazolines.[333]

A novel feature of 2-aminothioisomünchnones **206** is their stereoselective reaction with nitroalkenes (Scheme 46). At room temperature the transient adducts **207** fragment to give 4,5-dihydrothiophenes **208** in high yield. This cycloadduct fragmentation, in which the heterocumulene (RN=C=O) fragment is retained as an amide, is unusual.[314,343] 1,2-Diaza-1,3-butadiene adducts,[345,346] and a fulvene adduct appear to undergoes a similar cleavage.[344] An intermediate betaine mechanism has been proposed (see also Section 5.5.1.2.3 for a similar fragmentation of aldehyde adducts)[340]; some münchnone adducts undergo a similar fragmentation to give zwitterionic products (Section 5.2.1.2.1).[207]

206 **207** **208**

Scheme 46

Intramolecular 1,3-dipolar cycloadditions employing alkenes tethered to the mesoionic 1,3-thiazol-4-one ring have been investigated by several groups and this provides access to novel polycyclic skeletons.[125,316,332,336,347] In some cases the cycloadducts are stable.[348]

Donar-acceptor interactions in the addition of alkenes to thioisomünchnones have been investigated using DFT methods and NBO analysis.[349]

5.5.1.2.2 Alkyne [2 + 3]-cycloadditions The addition reactions of mesoionic 1,3-thiazol-4-ones with alkynes are well established; the initial 1,3-dipolar cycloadduct fragments to give either a thiophene or a pyridone.[1,2] The outcome depends on reactant structure and conditions.[350] This is illustrated by the reactions of the 2-amino derivatives **209** with methyl propiolate (Scheme 47). If the aryl substituent is 4-$NO_2C_6H_4$, a mixture of the products **212** and **213** is obtained but for Ar = Ph, 4-$MeOC_6H_4$ the pyridones **212** are the exclusive products.[351] The reactions of the tungsten carbene complex **211** provide a further example; addition of alkyne **211** to 2,3,5-triaryl thioisomünchnones gives thiophenes but reaction with 3-alkyl-2,5-diaryl derivatives gives 2-pyridones.[352]

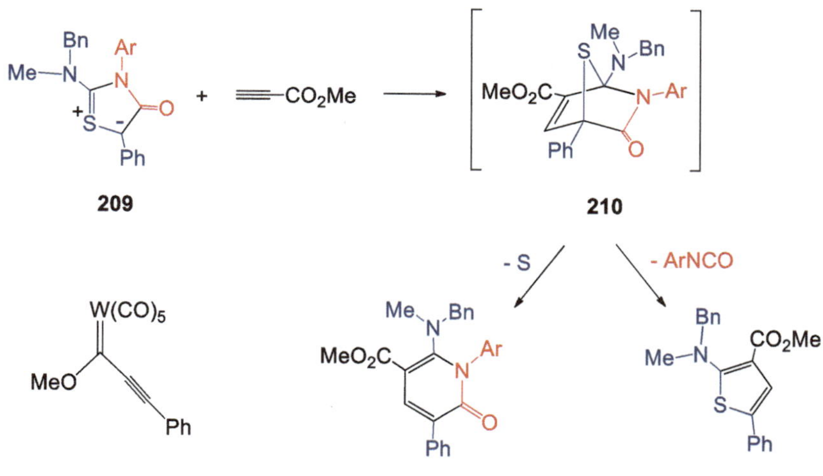

Scheme 47

Although both thiophene and pyridone products are possible, most recent studies report formation of 2-pyridone derivatives.[92,318,353–357] In some cases the cycloadduct can be isolated and subsequent desulfuration using $Hg(OAc)_2$ clearly favors 2-pyridone formation.[335] A computational study suggests that sulfur may be lost *via* thiirane intermediates.[51]

5.5.1.2.3 Aldehyde [2 + 3]-cycloadditions As part of an elegant series of studies of the cycloadditions and the stereoselectivities of 2-amino thioisomünchnones, Avalos and co-workers have shown that the derivatives **218** add alkyl aldehydes to give thiiranes **217** and/or β-lactams **219**. The intermediate cycloadducts are not isolated. The formation of the products **217** and **219** are rationalized by alternative fragmentations of the cycloadducts **215** to give the intermediates **214** or **216** and subsequent intramolecular cyclization and ring cleavage as shown in Scheme 48. The mode of fragmentation appears to be determined by the electronic character of the N-aryl substituent.[358–362] The fragmentation **215** → **216** is analogous to that observed for nitroalkene adducts (see Section 5.5.1.2.1, Scheme 46). It is worthy of note that isomünchnones have also been reported to undergo 1,3-dipolar cycloadditions with aldehydes.[2]

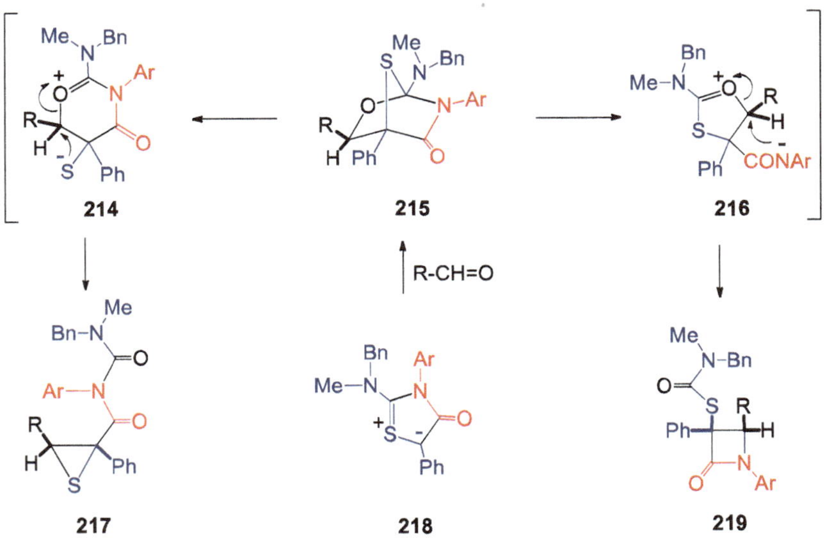

Scheme 48

5.5.1.2.4 Diazene [2+3]-cycloadditions Reaction of 2,5-diphenyl derivatives gives stable 1,3-dipolar cycloadducts with dialkyl azodicarboxylates or 4-phenyl-1,2,4-triazoline-3,5-dione (PTAD). The crystal structure of the diethyl adduct **220** has been reported. Heating in benzene gives the rearranged products **221**.[363,364]

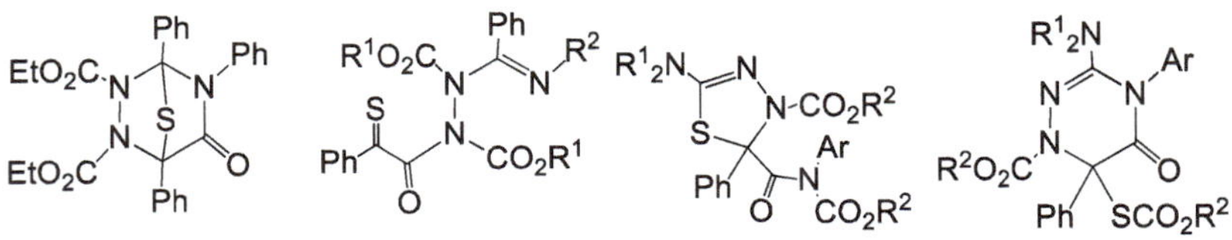

220 **221** **222** **223**

2-Amino derivatives give different results. The initial adducts have not been isolated and the 4-nitrophenyl derivative **218** (Ar = 4-$NO_2C_6H_4$) with azodicarboxylates under mild conditions gives 4,5-dihydro-1,3,4-thiadiazoles **222** (Ar = 4-$NO_2C_6H_4$) in 40–65% yield.[365] When electron-rich *N*-aryl substituents (Ar = Ph, 4-$MeOC_6H_4$) are employed, the products are the 1,2,4-triazine derivatives **223**.[366] This is a further example of how the outcome of thioisomünchnones cycloadditions is heavily dependent on the ring substituents. When azodicarboxamides are used as dipolarophiles on the same precursors, the products formed in moderate yield are acyclic thioureas. Plausible alternative fragmentation pathways of the initial cycloadducts have been proposed to account for the diverse products.[367]

5.5.1.2.5 Other cycloadditions

(i) *Thione addition.* Oxidation of the derivative **224** by molecular oxygen at ambient temperature gives the pseudo-dimer **225** (34%) with complete diastereoselectivity. Superficially this appears to be formed by cycloaddition of a thione but mechanistic studies suggest a complex cascade mechanism.[368]

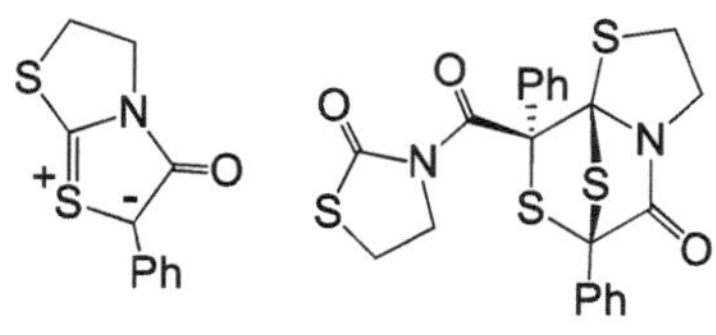

224 **225**

(ii) *Nitroso cycloaddition.* Triaryl derivatives react with nitrosobenzene to give cycloadducts which on standing give rearrangement products. 1-Chloro-1-nitrosocyclohexane gives adducts that on ethanolysis give thiazolones.[369]

(iii) *Heterocumulene cycloaddition.* A glucofuranose derivative reacts with arylisocyanates to give 6-oxopyridinium-4-olates. These cross-conjugated mesomeric betaines are formed by loss of sulfur from the 1,3-dipolar cycloadduct. The regioselectivity of these additions has been rationalized using computational methods.[50] In a study using arylisothiocyanates, tricyclic derivatives give the corresponding mesoionic 1,3-thiazol-4-thiones (Section 5.5.3) formed by loss of ArNCO from the cycloadduct. These products are accompanied by varying amounts of isomeric cross-conjugated mesomeric betaines, formed by loss of sulfur. The product structures and a theoretical study support stepwise mechanisms *via* polar intermediates.[370,371] Similar mechanistic conclusions have been shown for a similar study of monocyclic derivatives under microwave conditions.[372]

5.5.1.3 Tautomerism

2-Alkyl thioisomünchnones equilibrate with tautomeric 2-methylene-thiazolidin-4-ones. Typically the diphenyl derivative **226** reacts with aliphatic acid chlorides, *via* the tautomer **227**, to give an equilibrium mixture of the ketones **228** and **229** (Scheme 49). Aromatic acid chlorides give diketones. Isocyanates and isothiocyanates give amides and thioamides, respectively. Acryloyl chloride ($CH_2{=}CHCOCl$) reacts as an alkene and gives a 1,3-dipolar cycloadduct.[342,373,374]

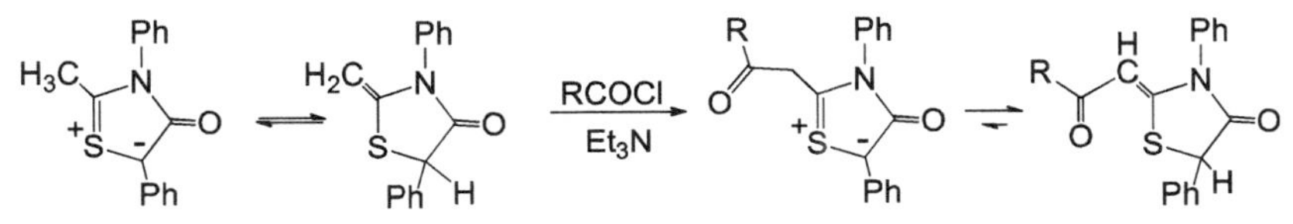

226 **227** **228** **229**

Scheme 49

An unusual ring-chain tautomeric equilibrium has been demonstrated for some N-bridged derivatives.[375]

5.5.1.4 Photochemistry

Earlier photochemical studies reported the formation of β-lactams.[2] The triphenyl derivative upon irradiation in the presence of a phosphine resulted in loss of sulfur and formation of a quinolinone (78%). This appears to proceed *via* an azetone which opens to a ketene followed by cyclization.[376,377] In a similar study in the presence of alcohols, the products were β-anilinocinnamate esters, possibly formed *via* similar ketene intermediates.[378]

5.5.2 1,3-Thiazol-4-imines (1,3-thiazolium-4-aminides) (230)

Cyclization of 2-(imidoylthio)carboximides using triphosgene gives the salts **231** (four examples) in good yield. The mesoionic derivative **230** ($R^1 = p$Tol, $R^2 =$ Ph) has been generated *in situ* from the corresponding salt using alumina-dispersed potassium fluoride as base. Reaction with DMAD gives the enamine **232** and not a 1,3-dipolar cycloadduct. A similar reaction with carbon disulphide gives a thiocarbamate.[379]

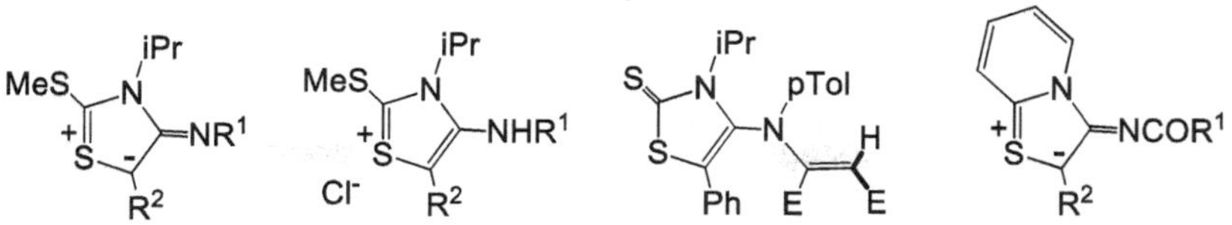

230 **231** **232** E = CO_2Me **233**

N-Acyl bicyclic derivatives **233** are conveniently prepared from (2-pyridylthio)acetamides or nitriles, and new examples have been reported.[321,380]

5.5.3 1,3-Thiazol-4-thiones (1,3-thiazolium-4-thiolates) (234)

2-Amino-3-aryl-1,3-thiazol-4-thiones **234** ($R^1 = NR_2$, $R^2 = Ar$) have been prepared by addition of arylisothiocyanates to mesoionic 1,3-dithiol-4-ones (Section 5.7.1).[381–383] Alternatively, 3-alkyl derivatives **234** ($R^1 = R^3 = Ph$, $R^2 = Me$, Et, Bn) have been made either by treatment of thiazolium tosylates **235** with sulfur and sodium hydride or by reaction of the tetrafluoroborates **236** with sodium hydrosulfide. In contrast to the reactions of thioisomünchnones with alkynes (Section 5.5.1.2.2), the 2,5-diphenyl-3-alkyl derivatives **234** ($R^1 = R^3 = Ph$, $R^2 = Me$, Et, Bn) give stable cycloadducts **237** with DMAD.[384]

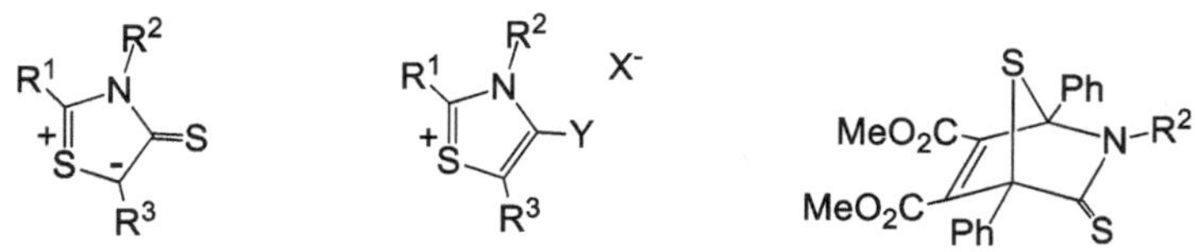

234 **235** Y = H, X = OTs **236** Y = OMe, X = BF_4 **237**

In a fascinating study, reactions of 2-amino-thioisomünchnones **239** ($R_2N = BnMeN$) with arylisothiocyanates have been found to give mixtures of the N-aryl products **238** (N-Ar^2) and **240** (N-Ar^1) (Scheme 50). The products **238**, which might be expected from conventional 1,3-dipolar cycloaddition, are accompanied by alternative products **240** in which the original aryl substituent (Ar^1) is retained but the exocyclic sulfur is introduced by the reagent. A detailed experimental and computational study revealed mechanisms in which betaine adducts **242** lead to either 1,3-dipolar cycloadducts **244** or thioazetidinediones **245**. Further reaction *via* the intermediates **241** and **243**, respectively, lead to the observed products. Good yields of mesoionic 1,3-thiazol-4-thiones **238** ($Ar^1 = Ar^2$), or mixtures (**238** + **240**), are obtained by heating in DMF at 100 °C or by microwave heating.[372] It is relevant to note that azetidinediones products, analogous

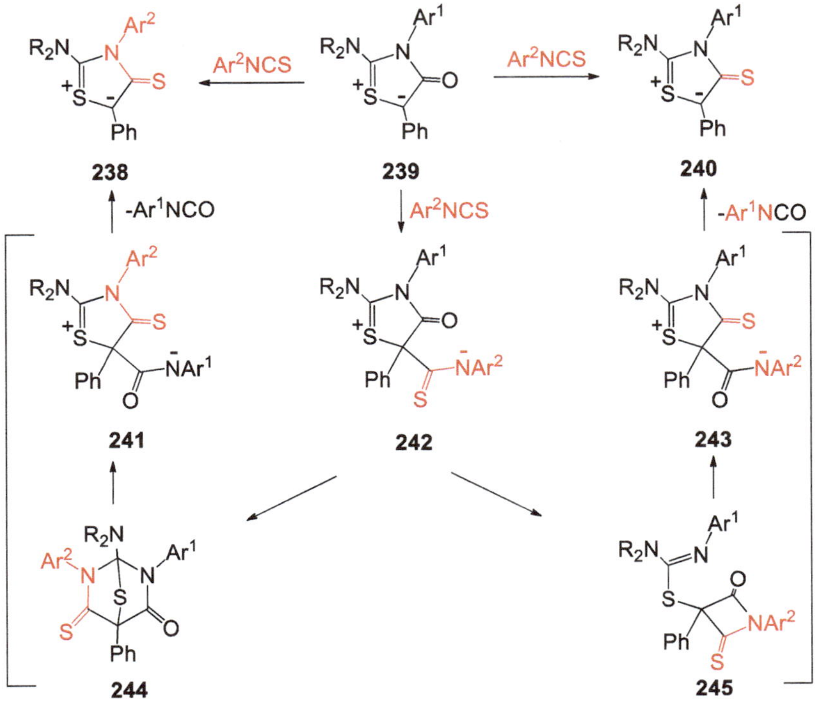

Scheme 50

to the intermediates **245**, have been isolated in reactions of mesoionic 1,3-dithiolones with PhNCO and longer reaction times give mesoionic 1,3-thiazolones (Section 5.7.1).

Similar thionation reactions of arylisothiocyanates forming bicyclic mesoionic 1,3-thiazol-4-thiones have been reported.[370,371]

5.5.4 1,3-Thiazol-5-ones (1,3-thiazolium-5-olates) (246)

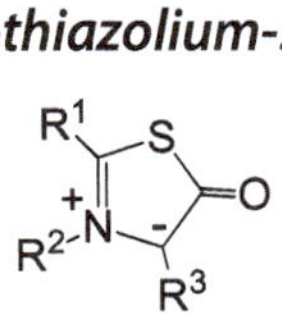

246

Little new work has been reported on this relatively well-known ring system.[2] A convenient route to 5-trifluoromethylthiazoles, involving treatment of *N*-thioacylprolines with TFAA/pyridine, proceeds *via* mesoionic 1,3-thiazol-5-one intermediates.[385]

Reaction of the 2,4-diphenyl-3-methyl derivative **246** ($R^1 = R^3 =$ Ph, $R^2 =$ Me) with benzocyclobutene gives a cycloadduct (33%) which upon heating in benzene loses COS to give 3-methyl-2,4-diphenyl-3*H*-3-benzazepine (87%).[386] The same derivative did not give any useful products when reacted with tropone.[387] Reaction with *o*-benzoquinones gives products formed *via* [4 + 4] cycloadducts.[388]

Calculations have identified mesoionic 1,3-thiazol-5-one rings connected by a polyenic bridge as having potentially useful photophysical properties.[389]

5.5.5 1,3-Thiazol-5-imines (1,3-thiazolium-5-aminides) (248)

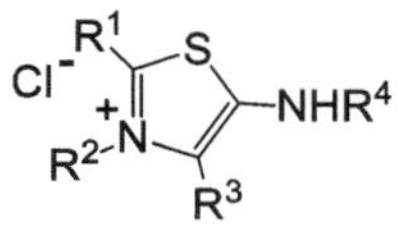

247

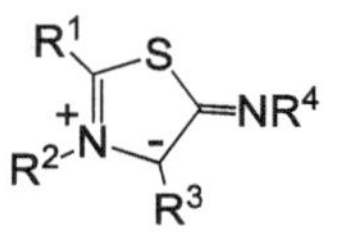

248

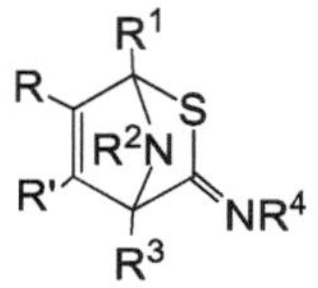

249

Three-component reactions of chlorothioformates (R^1-CSCl), aldimines (R^3-CH=NR^2) and isocyanides (R^4NC) provide convenient access to the chlorides **247** in good yield.[390] Treatment of these salts with base in the presence of alkynes rapidly gives pyrroles, formed by *in situ* generation of the mesoionic thiazolimines **248** followed by formation of the cycloadducts **249** and subsequent loss of isothiocyanate (R^4NCS). Alkenes were found to give products resulting from further reaction of the initial cycloadduct with retention of the isothiocyanate fragment.[391] Derivatives with an N-tethered alkene substituent have been shown to give a mixture of stable regioisomeric 1,3-dipolar cycloadducts.[392]

Reaction of *N*-*tert*-butyl derivatives **248** (R^4 = ^{t}Bu) with aryl- and alkylthioisocyanates gives mesoionic 1,3-diazol-4-thiones (Section 5.4.3), formed by regiospecific cycloaddition and subsequent *tert*-butylthioisocyanate loss. Interestingly, use of benzoylthioisocyanate leads to formation of stable *N*-benzoyl thiazolone derivatives **248** (R^4 = PhCO).[304] Phenylene-1,2-diisothiocyanate leads to bis products.[305] Treatment of salts **247** (R^1 = SPh, R^4 = tBu) with PhSH/Et_3N results in (i) Dimroth rearrangement to *N*-*t*butyl-1,3-diazol-5-thiones and (ii) reduction of the PhS substituent to give 2-unsubstituted final products.[306]

Bicyclic derivatives **250** have been made by treatment of 2-halo-*N*-phenacylpyridinium salts with KSCN followed by treatment with acid anhydrides [$(RCO)_2O$]. The structure of the derivative **250** (Ar = 4-$NO_2C_6H_4$, R = Me) has been determined by X-ray crystallography.[393] Similar products **250** (R = COPh) have been obtained by reaction of *N*-phenacyl heterocyclic salts with benzoylisothiocyanate.[394,395]

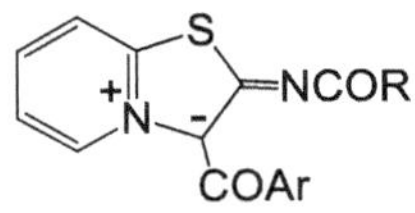

250

5.5.6 1,3-Thiazol-5-thiones (1,3-thiazolium-5-thiolates) (251)

The monocyclic derivatives **251** are conveniently prepared by reaction of *in situ* generated münchnones (Section 5.2.4) with carbon disulphide at room temperature.[255,396–398] Alternatively, the isolated münchnones can be heated with carbon disulphide in DMF.[256] The X-ray structures of three 2,4-diaryl-3-methyl derivatives have been reported.[396] Reaction of 2-halo (or mercapto) *N*-phenacylpyridinium salts with CS_2/Et_3N gives the bicyclic derivatives **252**.[399] In a similar manner, 2-bromoimidazolium salts give the derivatives **253**.[400] X-ray structures of representative examples of **252** and **253** have been reported.

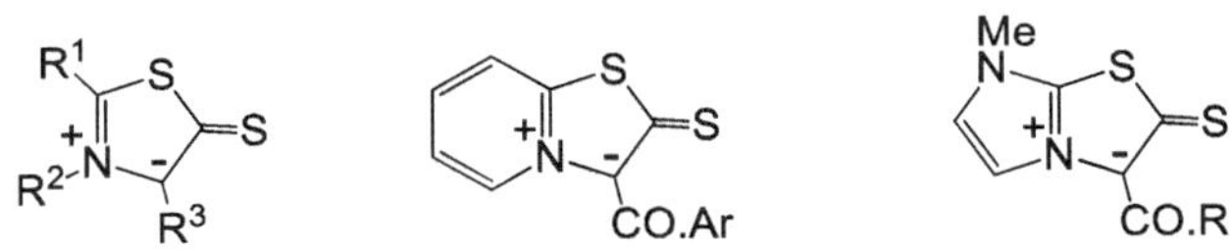

251 **252** **253**

In general, mesoionic 1,3-thiazol-5-thiones are stable compounds. 2,4-Diaryl-3-methyl derivatives have received significant attention due to (i) potential applications in non-linear optics[401–412] and (ii) potential biological applications.[413–417]

5.5.7 1,3-Thiazol-5-enes (1,3-thiazolium-5-methylides) (254)

The first examples of this mesoionic ring system have been reported. Treatment of O-acylated 1,3-thiazol-5-ones with the appropriate carbanion gives the crystalline derivatives **254** and **255** (R=H, Me) in variable yield.[418] The reaction of 2-bromo-1-(ethoxycarbonylmethyl)pyridinium bromide with CS_2/Et_3N gives the derivative **256** and not the expected mesoionic 1,3-thiazol-5-thione (e.g., **252**, Section 5.5.6). The structure **256** has been confirmed by X-ray crystallography.[419]

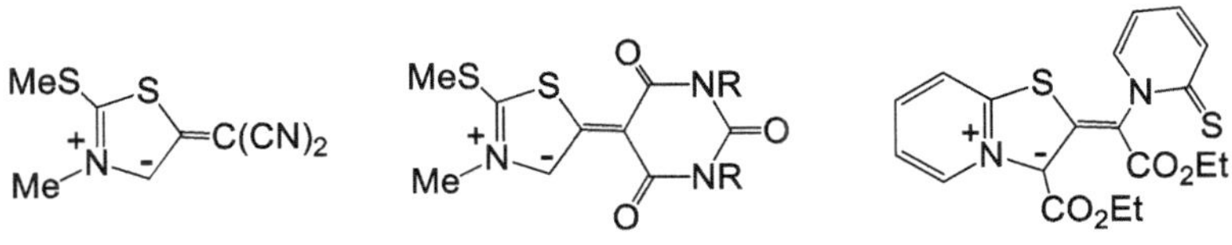

254 **255** **256**

5.6 Selenazoles

5.6.1 1,3-Selenazol-4-ones (1,3-selenazolium-4-olates) (*257*)

The preparation and some cycloadditions of 2,3,5-triaryl 1,3-selenazol-4-ones **257** have been reported previously.[2] With DMAD a 2-dimethylamino derivative gives the pyridone **259** (Scheme 51). Under the same conditions, the corresponding thiazolone gives the alternative thiophene product. Preferential pyridone formation is attributed to the weaker C—Se bonds in the cycloadduct **258**.[350]

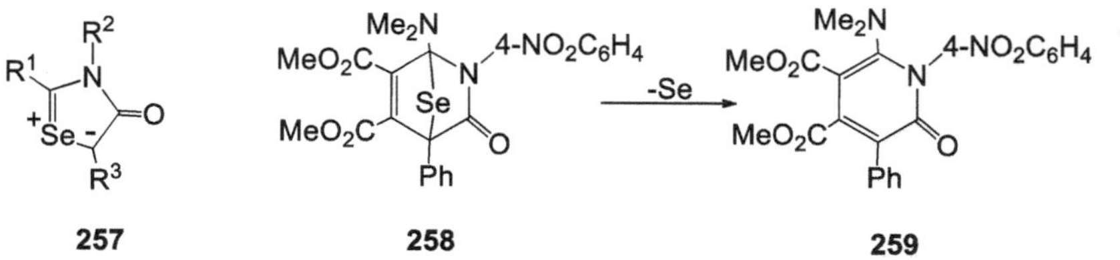

Scheme 51

5.7 Dithioles

5.7.1 1,3-Dithiol-4-ones (1,3-dithiolium-4-olates) (*261*)

This mesoionic system **261** is conveniently obtained by cyclization of the acids **260** (Scheme 52).[2,420,421] Using this method Taran and co-workers prepared twelve dark blue diaryl derivatives in moderate to good yield.[422]

Low yields (10–26%) of diaryl derivatives **261** (R^1 and R^2 = Ar) have been obtained by thermolysis of mesoionic 1,3,2-oxathiazol-5-ones **296** (Section 5.9.1). This reaction appears to proceed *via* a radical mechanism (Scheme 55).[423]

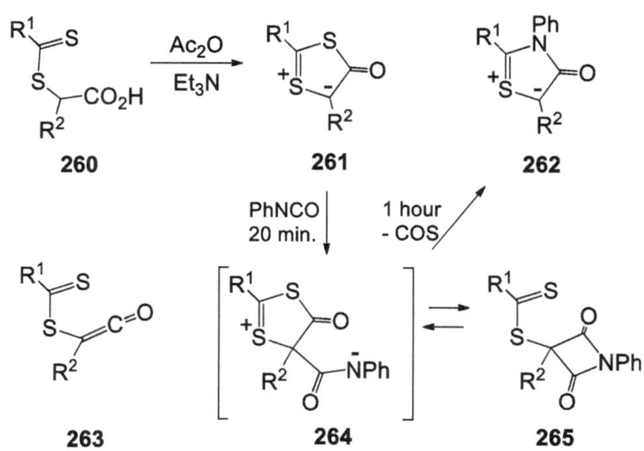

Scheme 52

1,3-Dipolar addition of alkynes gives thiophenes.[424–428] Reaction with *in situ* generated arynes gives good yields of benzo[*c*]thiophenes.[2,422] Reactions with cyclooctyne derivatives give the corresponding thiophenes in high yield and these strain-promoted reactions are of potential interest in the development of double ligation reactions as biological tools.[422]

In a manner analogous to thiophene formation, treatment of the diphenyl derivative with *tert*-butylphosphalkyne gives 2,5-diphenyl-1,3-thiaphosphole.[429] Similar products are formed using *P*-chloro-*C*-trimethylsilylphosphaalkenes as phosphalkyne equivalents.[430] Reaction with fulvenes gives stable [4+2] cycloadducts[431]; however, use of fulvenes bearing a bulky substituent leads to [4+6] addition.[344,432] Intramolecular addition of an alkene derivative also gives a stable cycloadduct.[424] Reaction with benzocyclopropene gives cycloadducts which either lose COS to give methylene bridged [9]-annulenes or, after oxidation, lose sulfur monoxide to give bridged [10]-annulenes.[433,434] Benzocyclobutene,[386] triphenylphosphirene,[435] and buckminsterfullerene (C_{60})[436] give 1,3-dipolar cycloadducts. Benzoquinones give [4+4] adducts.[437]

Treatment of 2-amino derivatives **261** ($R^1 = NR_2$) with heterocumulenes is a source of new mesoionic rings. Reaction with carbon disulphide gives mesoionic 1,3-dithol-4-thiones (Section 5.7.3)[438,439] and aryl isothiocyanates in hot benzene give mesoionic 1,3-thiazol-4-thiones (Section 5.5.3).[382] Reaction with phenyl isocyanate for 1 h gives mesoionic 1,3-thiazol-4-ones **262**. However, reaction for 20 min gives an azetidinedione **265** which on further heating gives the mesoionic product **262**. The azetidinedione **265** may be formed by [2+2] addition to the valence tautomer **263**.[381,440] Alternatively, formation may occur *via* the zwitterionic intermediate **264** which gives the kinetic product **265** or with longer reaction times gives the thermodynamic product **262** (Scheme 52).

Reduction by sodium borohydride in the presence of alkyl halides gives 4-alkylthiodithioles.[441] Diaryl derivatives react with cobalt reagents to give dithiolatocobalt complexes.[442] Photo-oxidation[443–445] and photodimerization[446] products have been reported.

5.7.2 1,3-Dithiol-4-imines (1,3-dithiolium-4-aminides) (266)

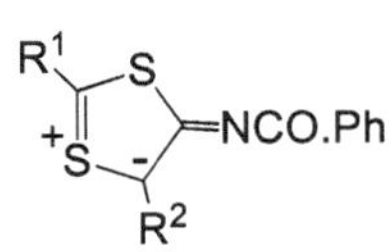

266

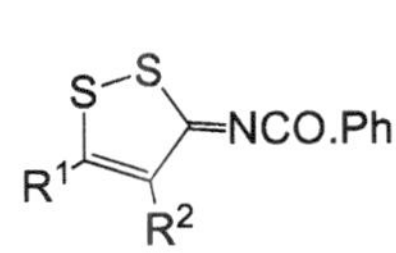

267

No new studies of this ring system have been reported. Further details of the photochemical rearrangement of *N*-benzoyl derivatives **266** to dithioles **267** have been published.[446]

5.7.3 1,3-Dithiol-4-thiones (1,3-dithiolium-4-thiolates) (268)

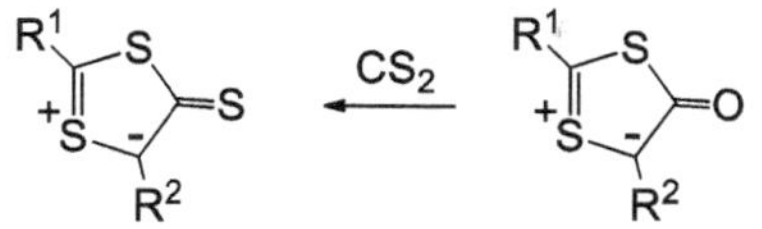

268 **269**

2-Amino derivatives **268** ($R^1 = NR_2$) have been obtained as orange solids by reaction of mesoionic 1,3-dithiol-4-ones **269** with carbon disulphide in boiling benzene.[381] The precursors **269** can be generated *in situ* from the acids **260** in the presence of CS_2.[447–450] The structures of three 2-amino-dithiolthiones **268** (R^1 = piperidino, morpholino, Me_2N) have been determined by X-ray crystallography.[451–453] There is one report of a reaction with an isothiocyanate giving a mesoionic 1,3-thiazol-4-thione (90% yield).[454]

2-Amino derivatives **270** are convenient intermediates for the synthesis of tetrathiofulvalenes (TTFs). Typically, alkylation gives the salts **271** which are reduced by $NaBH_4$/EtOH to the dithioles **272**. Coupling takes place under various conditions to give TTFs **273** (Scheme 53). Using this approach, and variations on the same theme, a wide variety of TTF derivative have been prepared.[438,439,447,448,455–470]

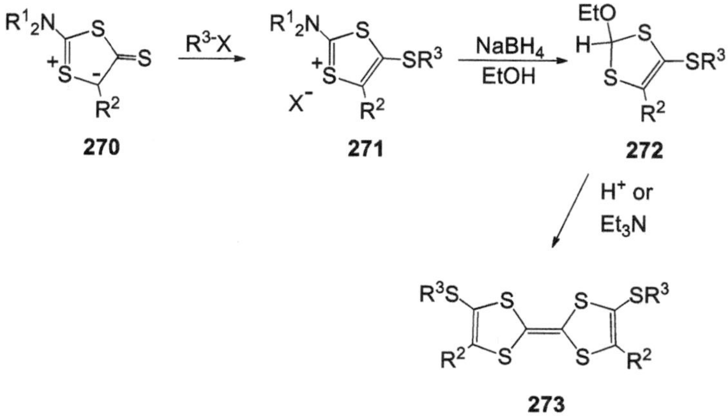

Scheme 53

5.7.4 1,3-Dithiol-4-enes (1,3-dithiolium-4-methylides) (274)

The first examples of this mesoionic ring system have been reported. Treatment of O-acylated 1,3-thiazol-5-ones with the appropriate carbanion gives the crystalline derivatives **274** and **275** (R'=H, Me) in variable yield.[418]

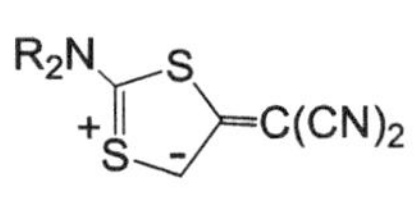

274

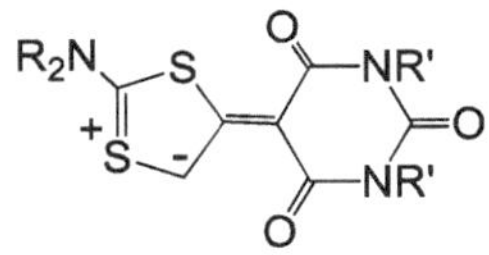

275

5.8 Oxadiazoles

5.8.1 1,3,4-Oxadiazol-2-ones (1,3,4-oxadiazolium-2-olates) (276)

Common name: Isosydnones

Isosydnones **276** are usually prepared by the reaction of *N*-acyl-*N*-(alkyl or aryl)hydrazines with phosgene; using this method several new derivatives have been reported,[471] including the 2-Me_2N derivative **276** (R^1 =Me_2N, R^2 =Me).[472] In a similar manner, the bicyclic derivative **277** has been prepared from the corresponding 1-aminopyrid-2-one and 1,1′-carbonyldiimidazole. In an alternative, higher yielding approach, thermal cyclization of the ureas **278** (250°) gives the bicyclic product **277** (82%).[473,474] In an unexpected reaction, the ureas **279** (X=Cl, Br) react with phosgene to give low yields of the tricyclic derivatives **280**.[475]

The 2-phenyl-3-methyl derivative **276** (R^1 =Ph, R^2 =Me) has been made (37%) by methylation (MeI) of the silver salt of 2-phenyl-1, 3,4-oxadiazol-5(4*H*)-one.[476]

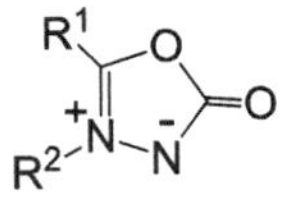

276

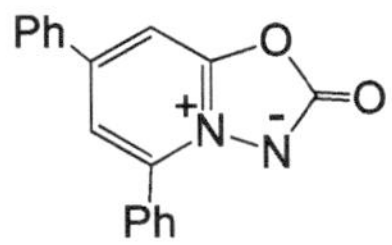

277

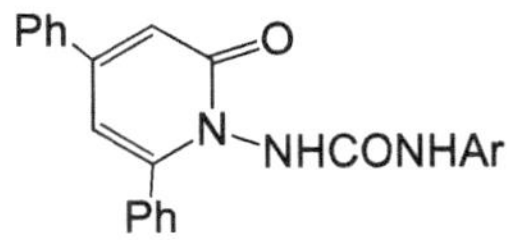

278

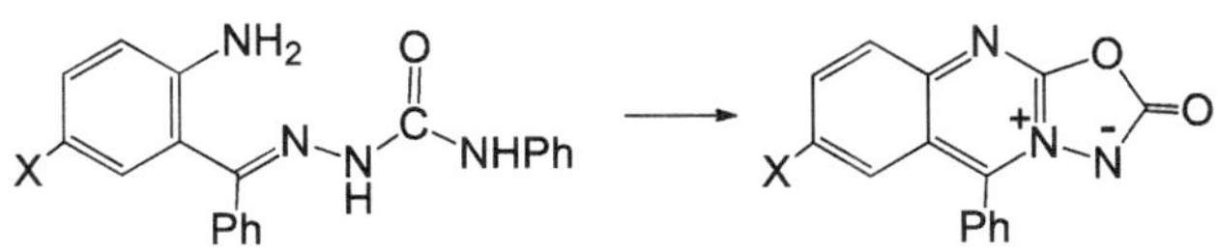

279 **280**

The bicyclic derivative **277** reacts with *N*-arylaminophosphoranes (ArN=PPh_3) to give mesoionic 1,3,4-oxadiazol-2-imines **283** (Section 5.8.2).[477,478] A number of comparative NMR studies have been reported, including ^{13}C-, ^{14}N-, ^{15}N- and ^{17}O-NMR.[479–481] A DFT computational study suggests that isosydnones will be more reactive than sydnones in orthogonal cycloadditions with strained cycloalkynes.[482]

5.8.2 1,3,4-Oxadiazol-2-imines (1,3,4-oxadiazolium-2-aminides) (281)

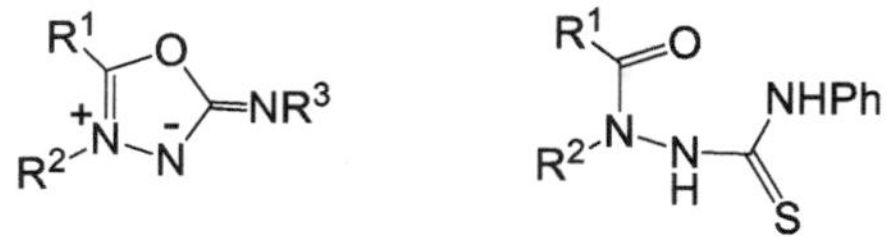

281 **282**

Monocyclic derivatives **281** have previously been made by reaction of aroylhydrazines with arylisocyanide dichlorides.[1] In a new approach, treatment of the acylthiosemicarbazides **282** with mercuric oxide gives *N*-phenyl derivatives **281** (R^3 = Ph).[471]

The bicyclic derivatives **283** are conveniently prepared by reaction of the oxadiazolones **277** (Section 5.8.1) with *N*-arylaminophosphoranes (ArN = PPh_3)[477,478] Alternatively, they can be made by treatment of the derivatives **284** with triphenylphosphine[474] or by reaction of the iminophosphorane **285** with isocyanates or isothiocyanates (Scheme 54).[474] The latter approach gave the *N*-Me derivative as a stable compound.[483]

In hot ethanolic K_2CO_3 the oxadiazolimines **283** undergo a Dimroth-type rearrangement to the isomeric triazolones **286** (Scheme 54).[478] Although this type of interconversion of mesoionic rings is well known (Section 4.4.4),[1] this is the first example of a 1,3,4-oxadiazol-2-imine to 1,2,4-triazol-3-one mesoionic rearrangement. Reaction of the imines **283** with amines results in ring opening to give the guanidines **287**, which can be recyclized to give mesoionic 1,2,4-triazol-3-imines (Section 5.10.7).[477]

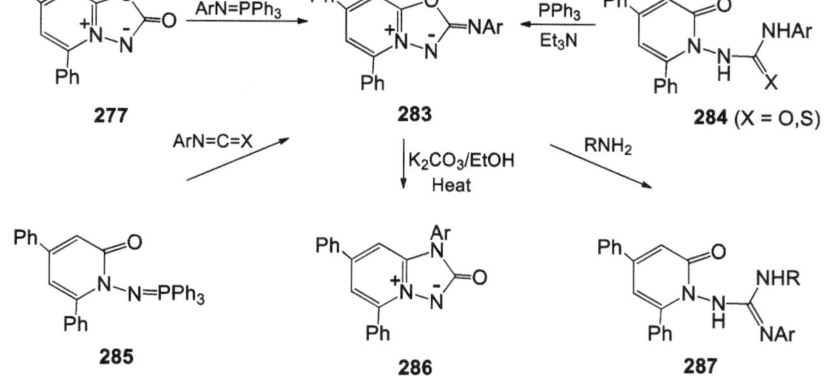

Scheme 54

5.8.3 1,3,4-Oxadiazol-2-thiones (1,3,4-oxadiazolium-2-thiolates) (288)

No new studies of the monocyclic derivatives **288** have been reported since the subject was reviewed in 1980.[2] The preparation of two bicyclic derivatives with general structure **289** has been described.[484]

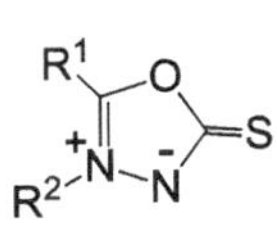

288

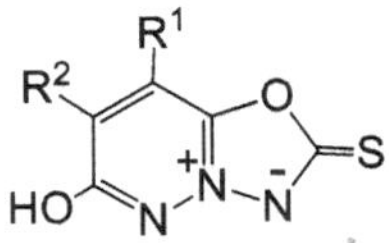

289

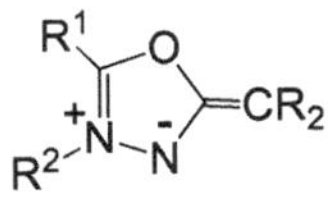

290

5.8.4 1,3,4-Oxadiazol-2-enes (1,3,4-oxadiazolium-2-methylides) (290)

As far as we are aware, no new work has been reported on this mesoionic system since the subject was reviewed in 1976.[1]

5.8.5 1,2,3-Oxadiazol-5-ones (1,2,3-oxadiazolium-5-olates) (291)

Common name: Sydnones

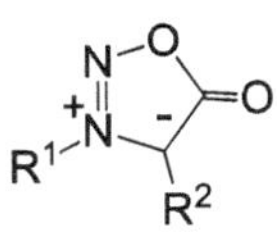

291

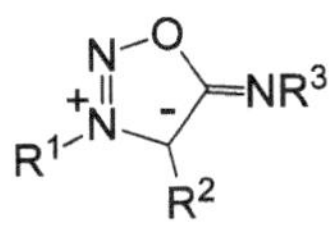

292

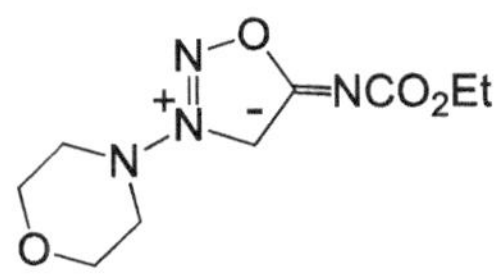

293 molsidomine

A review of the chemistry of sydnones **291** was published in Advances in Heterocyclic Chemistry in 2020,[3] and a discussion of sydnone chemistry is not included here. Other reviews of sydnone chemistry were published in 2006,[47] 2010,[4] and 2017.[485,486]

5.8.6 1,2,3-Oxadiazol-5-ones (1,2,3-oxadiazolium-5-olates) (292)

Common name: Sydnone imines

A review of the chemistry of sydnone imines **292** was published in Advances in Heterocyclic Chemistry in 2020,[3] and a discussion of sydnone imine chemistry is not included here. Much interest has been focused on molsidomine **293** which is a commercial antianginal agent.

5.8.7 1,2,3-Oxadiazol-5-thiones (1,2,3-oxadiazolium-5-thiolates) (294)

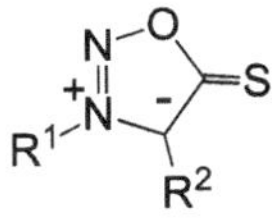

294

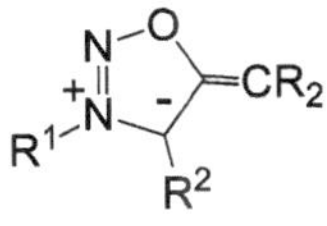

295

1,2,3-Oxadiazol-5-thiones **294** are known to rearrange to mesoionic 1,2,3-thiadiazol-5-ones **365** (Section 5.11.4) in ethanolic ammonia.[2] Electrochemical and ESR properties of pairs of these isomers have been studied. Novel conditions for the rearrangement of derivative **294** ($R^1 = 4\text{-}EtCO_2C_6H_4$, $R^2 = H$) were found to be sodium dithionite ($Na_2S_2O_4$) in hot aqueous ethanol.[487]

5.8.8 1,2,3-Oxadiazol-5-enes (1,2,3-oxadiazolium-5-methylides) (295)

The orange crystalline derivative **295** ($R^1 = Ph$, $R^2 = H$, $R = CN$) has been prepared (22% yield) by reaction of *N*-phenylsydnone with trifluoromethanesulfonic anhydride (Tf_2O) followed by treatment with malononitrile/Et_3N. This is the first example of this type of mesoionic heterocycle.[488,489] The ^{17}O-NMR spectrum has been compared with those of related species.[480] A novel indenide derivatives has also been reported.[490]

5.9 Oxathiazoles

5.9.1 1,3,2-Oxathiazol-5-ones (1,3,2-oxathiazolium-5-olates) (296)

Oxathiazolones **296** are prepared by nirosation of 1-mercaptoarylacetic acids **298** (R = Ar) and *in situ* cyclization of the *S*-nitroso product **297**. Reagents used include sodium nitrite/Ac_2O,[1] EtONO/Ac_2O,[491] EtONO/DCC,[492] and ^{i}BuONO/DCC.[493,494] Using this methodology a number of new aryl derivatives **296** have been prepared including the 4-phenyl-5-[^{18}O] derivative.[492]

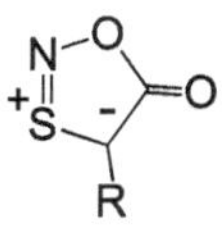

296

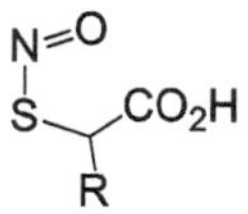

297

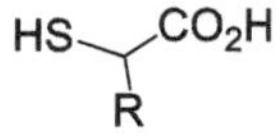

298

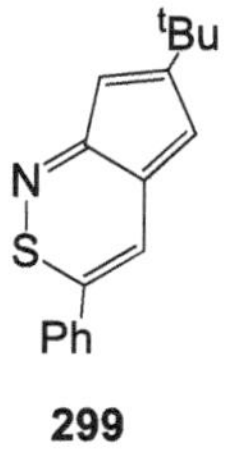

299

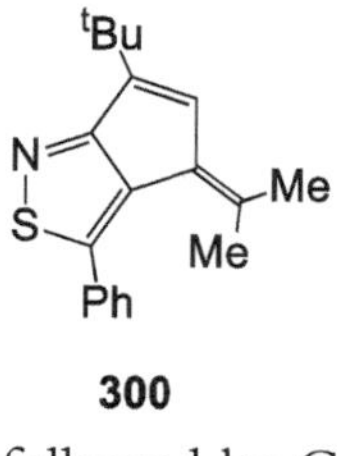

300

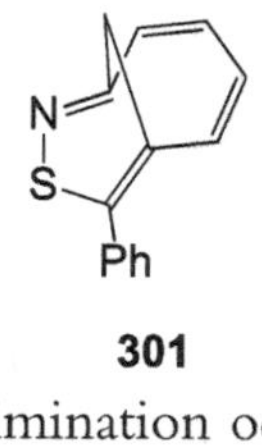

301

1,3-Dipolar cycloaddition followed by CO_2 elimination occurs with a variety of reagents. Thermal reaction with DMAD gives the expected isothiazole in good yield.[491] Reaction with 2-*tert*-butylfulvenes gives [4 + 2] or

[4 + 6] cycloaddition products depending on the fulvene 6-substituent. The 6-dimethylamino fulvene with the 4-phenyl compound **296** (R = Ph) gives the product **299** (32%) (isoelectronic with azulene) whereas the 6,6-dimethylfulvene gives the isothiazole **300** (19%).[344,431,432] Reaction with benzocyclopropene slowly gives the methanothiazonine **301** (32%).[495]

Cycloaddition of *tert*-butylphosphaalkyne (^{t}Bu-C≡P) gives the thiazaphosphole **302** (27%) which is accompanied by the isomer **303** (3%). Formation of the latter is attributed to thermal decarboxylation and rearrangement to form PhC≡N$^+$-S$^-$, which then cycloadds the phosphaalkyne.[429]

302 **303**

In the absence of 1,3-dipolarophiles, thermal decomposition (>110°) of 4-aryloxathiazol-5-ones **296** (R = Ar) occurs *via* a radical mechanism. Valence tautomerism to the ketene **304** and sequential loss of NO and CO appears to give the radicals **306** and **307** which combine to form mesoionic 2,5-diaryl-1,3-dithiol-4-ones **261** (10–26%) (Section 5.7.1).[423]

New photochemical studies have been reported. Irradiation leads to alternative fragmentations *via* the valence tautomers **304** and **305** (Scheme 55). Subsequent decomposition gives either the radicals **306** and **307** or the nitrile sulfide **308**, which lead to the observed products.[491,492,496] The generation of the NO-donating *S*-nitroso thiols **304** from a stable source, i.e., **296**, by light-induced ring opening of mesoionic oxathiazol-5-ones has been investigated.[494]

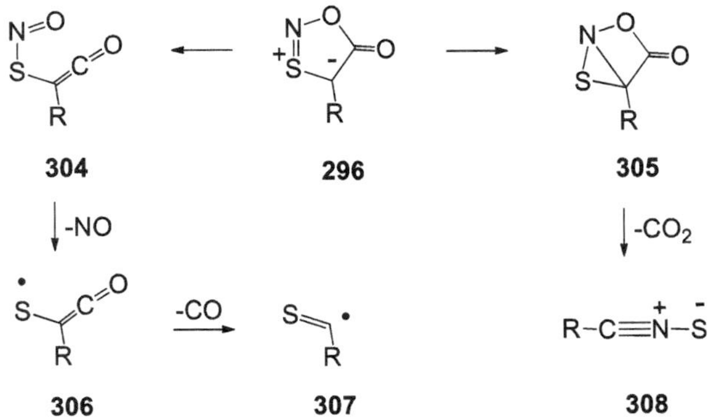

Scheme 55

Ring opening of mesoionic 4-aryloxathiazol-5-ones **296** and release of NO together with an S-centered radical also takes place under acidic conditions (pH 5). Potential biological applications of these NO-releasing properties have been investigated.[493] In related studies, the 4-phenyl derivative **296** (R = Ph) was found to be a glutathione S-transferase inhibitor with potential anticancer properties.[497]

5.10 Triazoles

5.10.1 1,2,3-Triazol-4-ones (1,2,3-triazolium-4-olates) (309)

Several routes to the mesoionic derivatives **309** have been described in previous reviews.[1,2] A 2008 review of condensed 1,2,3-triazoles includes a section on mesoionic 1,2,3-triazol-4-ones.[498]

An established synthetic route is cyclization of the triazenes **310** (Ac_2O/pyridine) which can be conveniently generated *in situ* from an α-aminoacid and a diazonium salt. Using this approach a number of new mono- and polycyclic derivatives have been prepared.[499–501] An alternative approach involves N-alkylation of the triazole sodium salts **311** and this has also been used to generate new derivatives.[502–509] In a number of these studies the initial monocyclic products **309** are further elaborated to give bicyclic derivatives.[510]

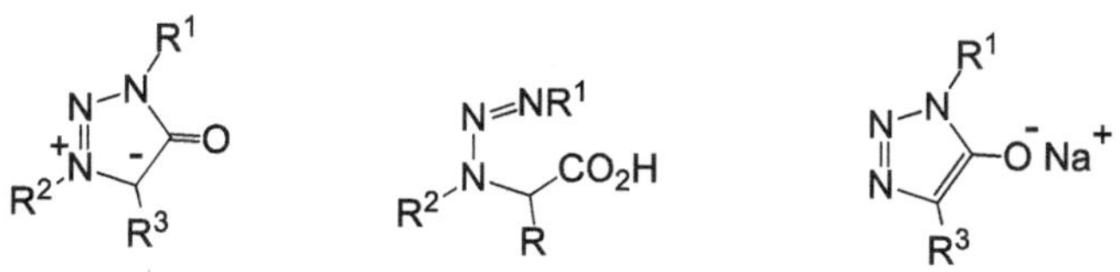

309 **310** **311**

Three interesting new routes to mesoionic 1,2,3-triazol-4-ones have been reported in the last decade. The derivatives **313** have been obtained upon treatment of the iodides **312** with CsOH. Higher yields and purity are obtained when the iodides **312** are first transformed into the copper complexes **314** (Ag_2O/CuCl) and then treated with CsOH (Scheme 56). The structure of the diphenyl derivative **313** ($R^1 = R^3 = Ph$) was confirmed by X-ray crystallography.[511]

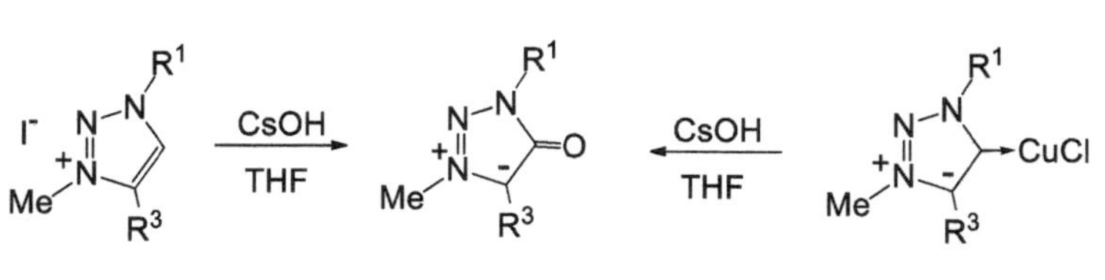

312 **313** **314**

Scheme 56

In a related transformation, the 2,6-diisopropylphenyl (Dipp) salt **315** on treatment with KOH/EtOH gives the mesoionic product **317** in quantitative yield. This reaction is believed to occur *via* the intermediate **316** (Scheme 57). The structure of compound **317** was confirmed by X-ray crystallography.[512]

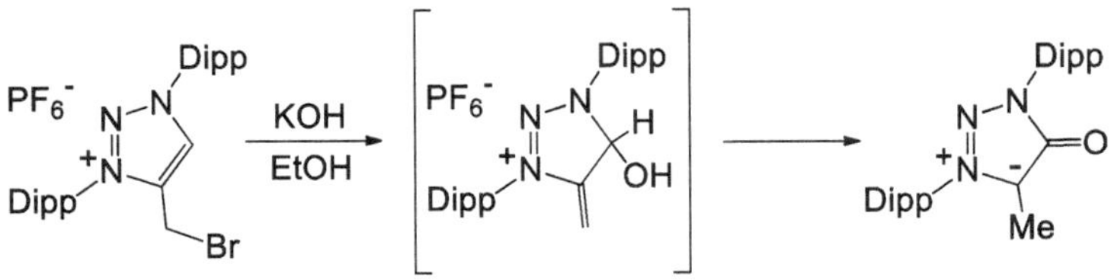

315 **316** **317**

Scheme 57

A novel route to isoquinolinium derivatives **319** involves the Rh-catalyzed annulation of 4-aryl-1,2,3-triazoles **318** and alkynes (Scheme 58). Using this methodology a wide range of derivatives in moderate to good yield have been prepared.[513]

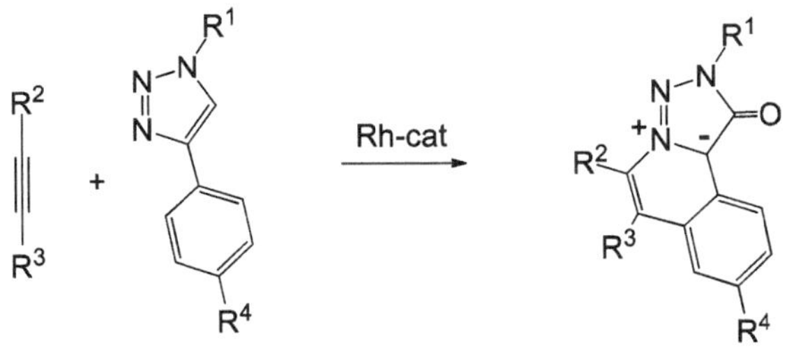

318 **319**

Scheme 58

The dipole moment of the 1-methyl-3-*p*-tolyl derivative **309** ($R^1 = p$-tolyl, R^2 = Me, R^3 = H) in benzene is 4.48 D.[514] The ^{13}C, ^{14}N and ^{15}N NMR spectra of representative examples of the mesoionic rings have been analyzed.[515–517]

Tricyclic derivatives have been found to be highly potent aryl hydrocarbon receptor (AhR) agonists.[518] Some fungicidal activity has been found in bicyclic derivatives.[510]

5.10.2 1,2,3-Triazol-4-imines (1,2,3-triazolium-4-aminides) (320)

No new preparative work on monocyclic mesoionic 1,2,3-triazol-4-imines **320** has been reported.[1] Bicyclic derivatives **321** (R = Ph, CH_2OH, X = H, Br) have been prepared by copper-catalyzed cyclization of the azide salts **322**, and their fluorescent properties investigated.[519]

Dipole moments of three aryl derivatives **320** have been measured and are in the range 4.7–4.9 D.[514] The X-ray structure of the *p*-tolyl derivative **323** (Ar = *p*-Tol) has been reported,[520] together with ^{13}C NMR spectra.[515] An interesting development is the use of the derivative **324** as an effective H-atom transfer (HAT) catalyst for photoredox C-H functionalization, apparently by formation of an amidyl radical that participates in HAT.[521]

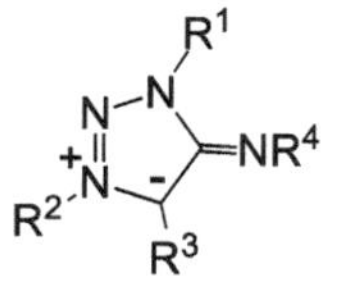
320

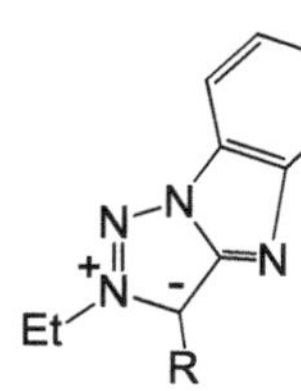
321

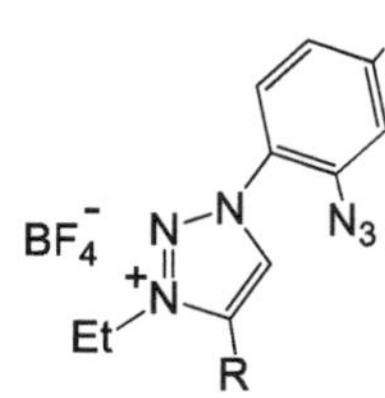
322

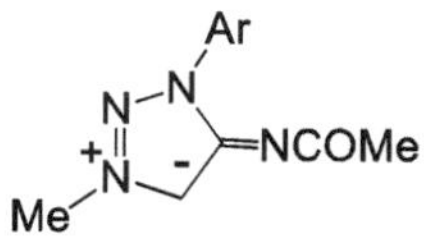
323

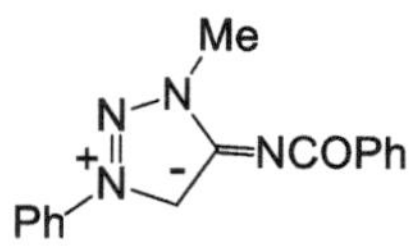
324

5.10.3 1,2,3-Triazol-4-thiones (1,2,3-triazolium-4-thiolates) (325)

The monocyclic derivative **325** (R^1 = mesityl, R^2 = Me, R^3 = Ph) has been prepared (80%) by reaction of the corresponding triazolium iodide with K_2CO_3/S_8 in methanol at 68°. The exocyclic sulfur forms complexes with zinc halide.[522] In a similar manner, the pincer ligands of type **326** and **327** (X = CH, N) have been prepared in good yield by treatment of the corresponding triazolium salts with K_2CO_3/S_8. The derivatives **327** (X = CH) form Pd *bis* pincer complexes with $PdCl_2$.[523,524]

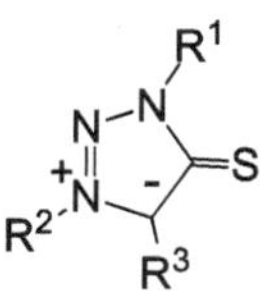
325

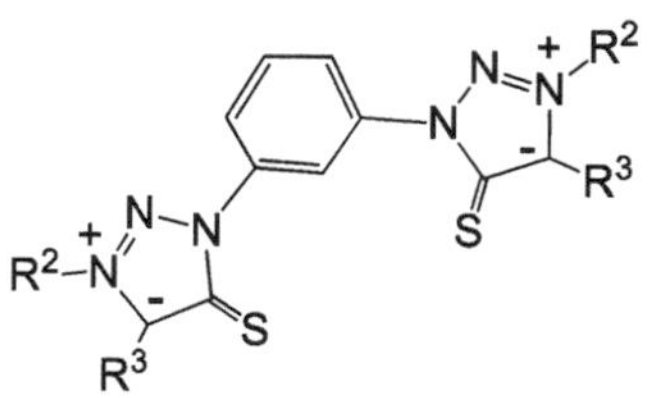
326

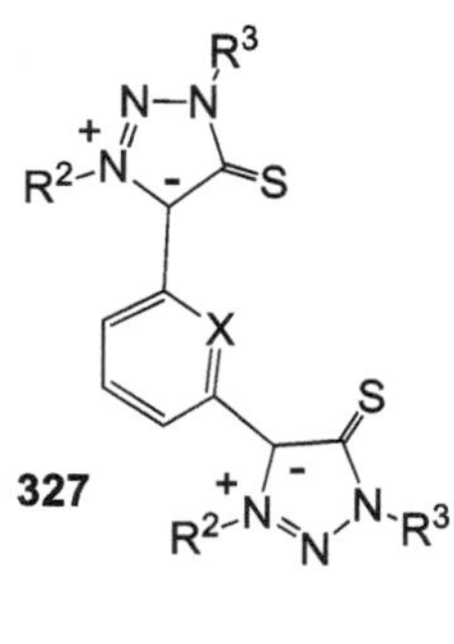
327

5.10.4 1,2,3-Triazol-4-enes (1,2,3-triazolium-4-methylides) (328)

This class of mesoionic ring was unknown until 2020 when three groups described the preparation of derivatives **328** by treatment of triazolium salts **329** with potassium bis(trimethyl)amide (KHMDS). Two groups prepared unsubstituted derivatives **328** (R^4 = H) in good yield. They showed that they are strong donor ligands forming coordination complexes with rhodium and boron (see also Section 5.4.4),[309] and that they are efficient catalysts for hydroboration reactions.[525] A third group has prepared a C-Ph derivative **328** (R^4 = Ph) and shown that with Lewis acids it forms zwitterionic kinetic products at the exocyclic position, which rearrange to thermodynamically more stable isomers.[526] All three studies confirmed the mesoionic structures by X-ray crystallography.

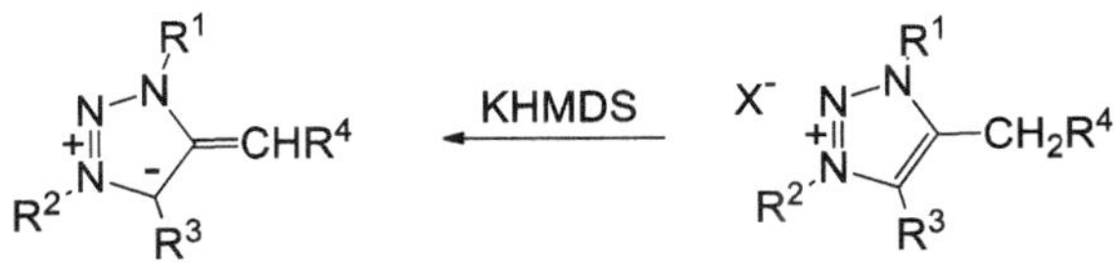

328 **329**

5.10.5 1,2,3-Triazol-4-selones (1,2,3-triazolium-4-selenolates) (330)

Treatment of 4-unsubstituted 1,2,3-triazolium salts with strong base (K_2CO_3, KHMDS or KO^{t}Bu) in the presence of selenium powder gives the mesoionic triazol-4-selones **330**.[522,527] X-ray structures of two derivatives have been reported.[522] The selenium acts as a ligand and forms metal complexes with zinc halides[522] and gold chloride (e.g. **331**).[527] Copper(II) oxidation gives diselenide salts.[528]

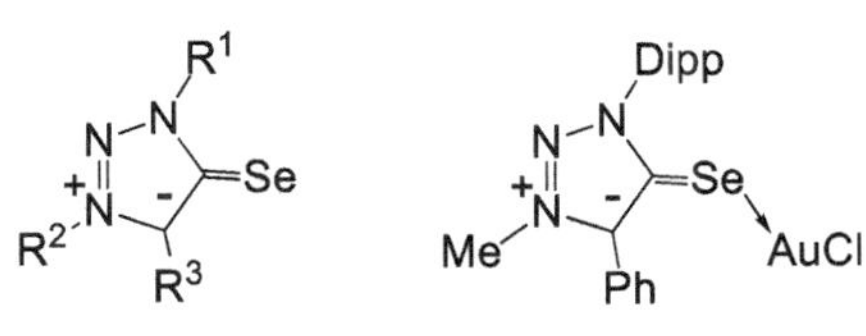

330 **331**

5.10.6 1,2,4-Triazol-3-ones (1,2,4-triazolium-3-olates) (332)

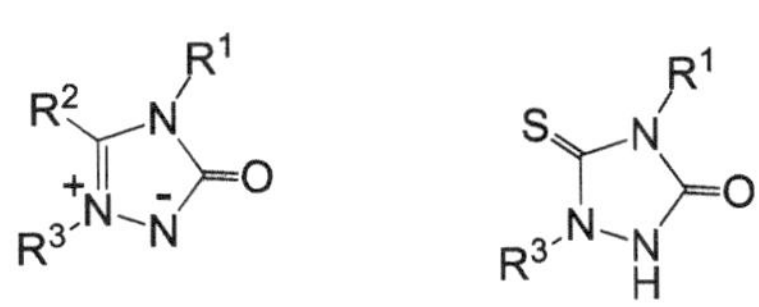

332 **333**

Mesoionic 1,2,4-triazol-3-ones (**332**) are well known and have been prepared by a variety of methods.[1,2] In new studies, monocyclic 5-thio derivatives **332** (R^1 = Ar, R^2 = SR) have been prepared by treatment of the 5-thioxo-triazolin-3-ones **333** with an alkyl iodide and base.[529] A photochemical reaction between *N,N*-dialkylcarbamoyl azides and carbodiimides gives 5-dialkylamino derivatives **332** (R^2 = R_2N) as minor products.[530] Methylation (MeI) of a 5-aryl-4-methyl-1,2,4-triazol-3-one gave the 1-methyl mesoionic derivative as the minor N-methylation product (7.6%).[531]

The bicyclic systems **334** are conveniently prepared by base catalyzed rearrangement of the isomeric mesoionic 1,3,4-oxadiazol-2-imines **283** (Section 5.8.2).[478,483] *N*-Alkylpyrimidinium derivatives **335** (R = Alkyl) have been prepared by an N-alkylation procedure and one structure confirmed by X-ray crystallography.[532] *N*-Carbonyl derivatives (R = COR') of the bicyclic system **336** have been prepared by N-acylation and used as acylating agents in peptide synthesis.[533,534]

A novel tricyclic derivative was obtained as a minor product (3.5%) in the photolysis of a quinazolinium-*N*-imide.[535] Four tricyclic derivatives have been obtained using the arylhydrazine/phosgene route.[536]

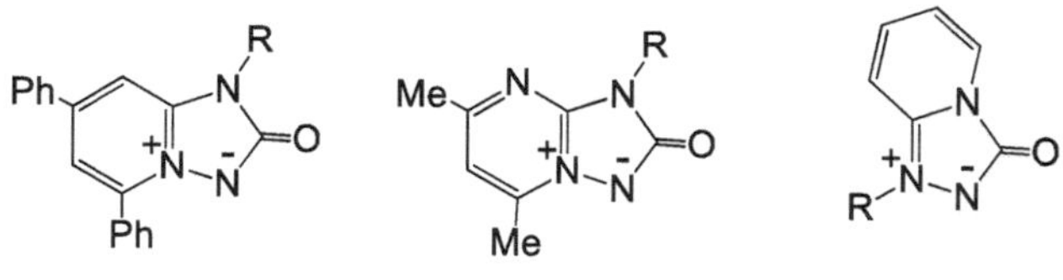

334 **335** **336**

A limited amount of new chemistry has been reported for this ring system. Benzyne addition to a 5-amino derivative **332** (R^2 = R_2N) gave a 3-aminoindazole after isocyanate elimination from the initially formed cycloadduct.[530] Reactions of representative derivatives **335** with DMAD and diphenylcyclopropenone have been reported.[537,538]

In a remarkable study, Kim and O'Shea obtained an almost quantitative yield of the mesoionic derivative **338** on addition of tetracyclopropylethylene (TCPE) to the 1,2,4-triazoline-3,5-dione **337** at room temperature. Heating resulted in formation of the diazetidine **339** (Scheme 59).[539]

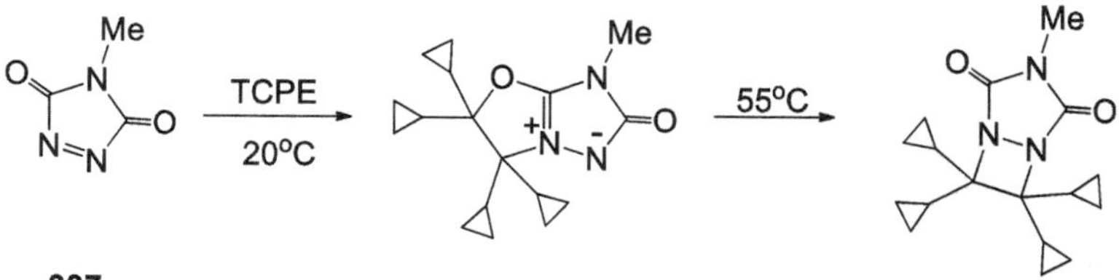

337 **338** **339**

Scheme 59

5.10.7 1,2,4-Triazol-3-imines (1,2,4-triazolium-3-aminides) (340)

A remarkable development has been the discovery by Siemeling and co-workers that nitron **341**, a commercially available analytical reagent first prepared by Busch in 1905,[540] readily equilibrates with an N-heterocyclic carbene (NHC) tautomer **342** in solution.[541] The preparation of mesoionic 1,2,4-triazol-3-imine derivatives **340** is discussed Section 5.10.7.1 and novel chemical transformation of nitron and related derivatives are summarized in Section 5.10.7.2.

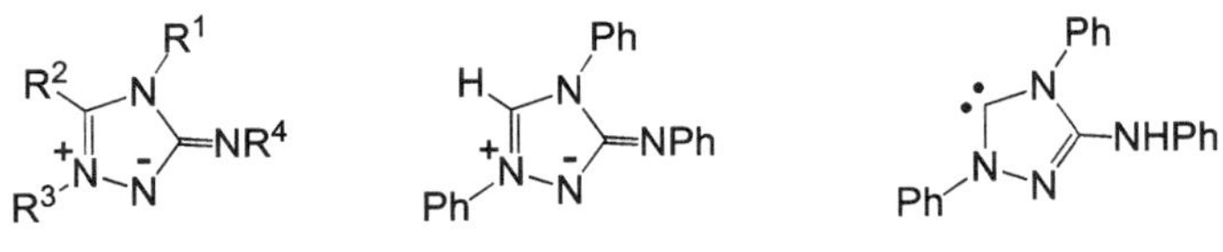

340 **341** nitron **342**

5.10.7.1 Preparation

An improved procedure for preparing nitron **341**, which gives higher yields and a purer product, has been described,[542] and several analogues have been made using this method.[543] The N-^{t}Bu derivative **340** (R^1 = Me, R^2 = H, $R^3 = R^4$ = ^{t}Bu) has been prepared by N-methylation ($Me_3O^+BF_4^-$) of the corresponding 1,2,4-triazole and deprotonation (^{t}BuO$^-$K$^+$) of the resulting salt; the structure has been confirmed by X-ray crystallography.[544] An interesting aspect of this structure is its stability (yellow microcrystals, 91% yield) since mesoionic exocyclic imines usually require an *N*-aryl or *N*-acyl substituent for stability.

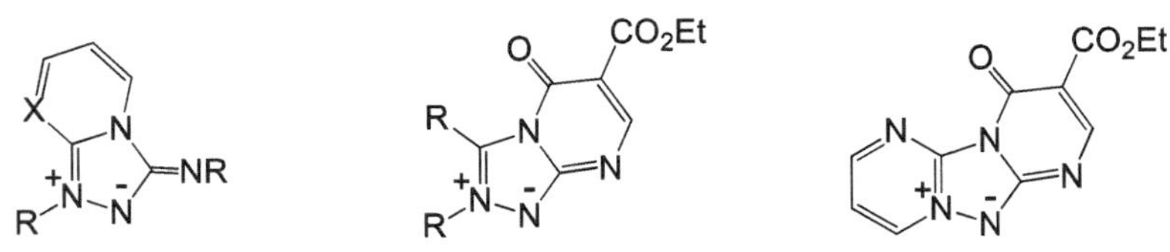

343 (X = N, CH) **344** **345**

Other monocyclic derivatives that have been reported include exocyclic =N—COR[545–547] and =N—N(R)COR[548] derivatives. A variety of bi- and tricyclic derivatives has been reported using cyclocondensation, N-alkylation and deprotonation methodology. Typical polycyclic structures studied include the ring systems **343**,[549–552] **344**,[553,554] **345**,[555,556] and closely related structures.[557–559]

The ^{1}H, ^{13}C, ^{14}N and ^{15}N NMR spectra of nitron have been reported.[560] Semi-empirical[561] and DFT[562] theoretical studies have been published.

5.10.7.2 Reactions of nitron and related species

In detailed studies, Siemeling and co-workers have shown that nitron **341** and an N^{t}Bu analogue **346** (R^1 = Me, R^2 = R^3 = ^{t}Bu) are in equilibrium with their NHC tautomers **347**. The amount of NHC tautomer is below the NMR detectable limit but it is readily trapped to give adducts in good yield (Scheme 60). Typically, sulfur in THF at room temperature gives the thione adducts **348**.[541,544,563] Similarly, carbon disulphide or [Rh(μ-Cl)$(CO)_2]_2$ give products of the type **349** and **350**, respectively.[541,544] Other NHC complexes are formed by ruthenium,[564] and by copper, silver and gold, e.g., **351**.[565] Other workers have studied Ni^{II}, Ru^{II}, Rh^{III}, Ir^{I}, Ir^{III} and Pd^{II} complexes formed by nitron.[566,567] Nitron spontaneously self-assembles carbene-based monolayers on Pt and Au sufaces.[568] The nucleophilic exocyclic nitrogen atom of nitron reacts with 1-CF_3-prop-2-yne iminium salts but the nucleophilic NHC tautomer reacts with 1-CF_3-prop-2-yne imines.[569]

The mechanism of formation of the carbenes **347** is not clear. Nitron is quite a strong base (pK_{aH} 10.34)[570] but probably not strong enough to self-generate an anionic NHC. It is possible that a trace of nitron salt may

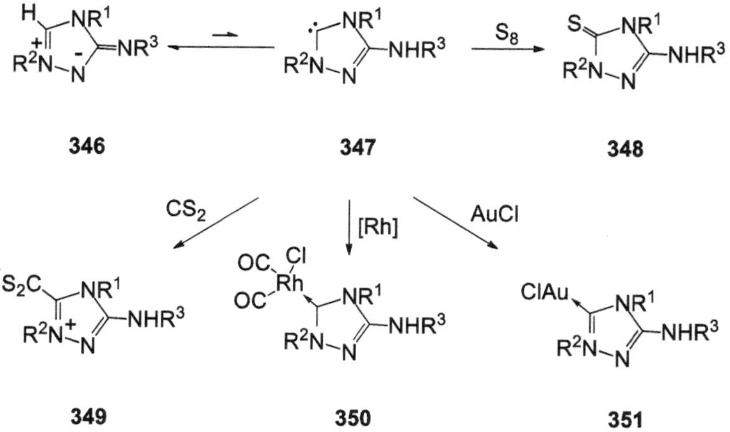

Scheme 60

facilitate deprototonation and be self perpetuating but this is speculation.[58] Gas phase DFT calculations show that the carbenes **347** are less stable than the mesoionic isomers **346** with the N^{t}Bu derivative **347** (R^1 = Me, R^2 = R^3 = ^{t}Bu) ($\Delta G°$ 4.7 kcal mol^{-1}) being lower in relative energy than the nitron tautomer **347** (R^1 = R^2 = R^3 = Ph) ($\Delta G°$ 5.7 kcal mol^{-1}).[544] This is consistent with calculated substituent effect stabilization energies (SESE) which show that for mesoionic imines (e.g., **346**) substituent stabilization is in the order =NAr > =NAlk whereas for the tautomeric NHCs (e.g., **347**) substituent stabilization is reversed (—NHAlk > —NHAr).[56] N-Aryl groups favor the mesoionic structure whereas N-alkyl groups favor the NHC.

Another recently discovered property of this old chemical is its ready formation of a Blatter-type stable radical in good yield. In wet acetonitrile nitron (and analogues) undergoes aerial oxidation to give the fully characterized radical **352** in 58% yield. The mechanism shown in Scheme 61 has been proposed.[543]

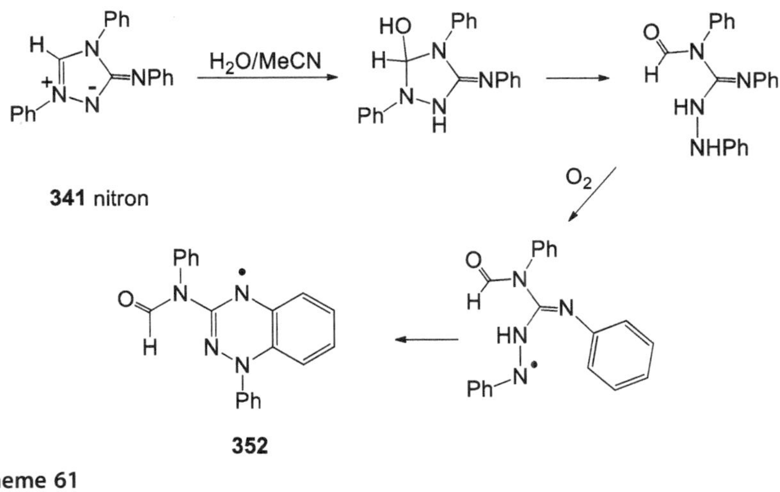

Scheme 61

5.10.7.3 Applications

Some interest continues to be focused on the analytical applications of nitron **341**.[571–573]

5.10.8 1,2,4-Triazol-3-thiones (1,2,4-triazolium-3-thiolates) (353)

Mesoionic 1,2,4-triazol-3-thiones **353** are well known and preparative methods have been summarized.[1] The main approaches are based on cyclization of 4-acyl-1,4-disubstituted thiosemicarbazides **354** or reaction of *N*-aminoamidines **355** (X=NR[1]) or *N*-thioacylhydrazines **355** (X=S) with isothiocyanates, carbon disulphide or thiophosgene. Most new work in the period 1980–2020 has been focused on the preparation of bi-and tricyclic derivatives using these approaches.

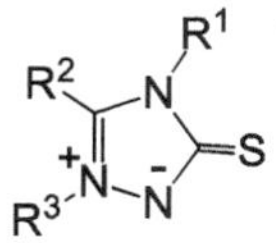
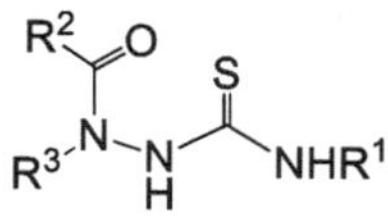
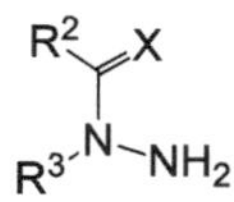

353 **354** **355**

Little new work on monocyclic derivatives has been reported. A series of 5-pyrazinyl derivatives were prepared by the acylthiosemicarbazide route.[574,575] In hot xylene 2,4-disubstituted semicarbazides rearrange and react with orthoesters to give low yields of mesoionic triazol-3-thiones **353**.[576] Reaction of thiosemicarbazides with tetracyanoethylene (TCNE) gives 5-(1-amino-2,2-dicyanovinyl) derivatives.[577] In a novel approach, nucleoside analogues have been prepared by reaction of isothiouronium reagents [R_2N^+ =C(SMe)-NMe-CS_2Me I^-] with furanosylhydazines.[578]

Careful studies by Miller and co-workers have shown that reactions of 1,4-diphenylthiosemicarbazides with aroyl chlorides lead to either mesoionic 1,2,4-triazol-3-thiones **353** (thermodynamic product) or isomeric mesoionic 1,3,4-thiadiazol-2-imines **372** (Section 5.11.11) (kinetic product) depending upon the reaction conditions used.[579–581]

A number of new bicyclic derivatives, and their tricyclic benzo equivalents, have been synthesized using established methodology. Typical products include ring systems **356**,[532,582–586] **357**,[478,587–593] and **358**,[594,595] including heterocyclic variations and benzo derivatives of the second ring. Irradiation of a quinazoline *N*-thioamidate has been reported to give a low yield of a tricyclic derivative of structure **357** (X=N).[535]

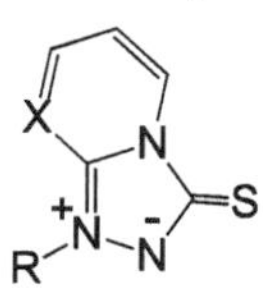
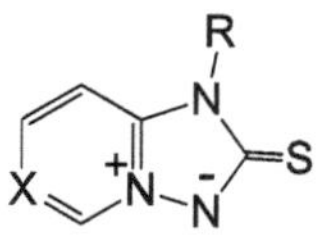
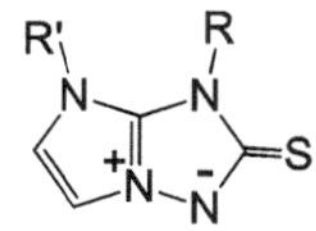

356 (X = N, CR') **357** (X = N, CR') **358** (X = N, CR')

^{1}H, ^{13}C, ^{14}N and ^{15}N NMR spectra of 1,2,4-triazol-3-thione have been discussed,[560,575] and an X-ray structure of the triphenyl derivative **353** (R^1 =R^2 =R^3 =Ph) reported.[596]The exocyclic sulfur atom can act as a nucleophile or a ligand. Complex formation with molybdenum,[597] gold,[598] or platinum[599] has been described. Michael addition of the sulfur atom to quinones has been reported,[600] and S_NAr reactions of 5-aryl substituents have been studied.[601,602]

5.10.9 1,2,4-Triazol-3-enes (1,2,4-triazolium-3-methylides) (*359*)

As far as we are aware, no new work on this ring system has been reported.

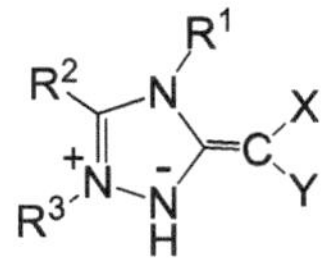

359

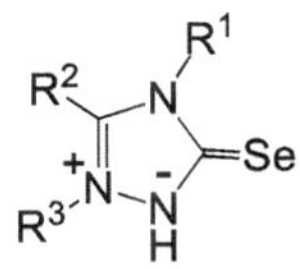

360

5.10.10 1,2,4-Triazol-3-selones (1,2,4-triazolium-3-selenolates) (*360*)

Three new mesoionic 1,2,4-triazol-3-selones **360** have been prepared by treatment of 1,4-diphenylselenosemicarbazide with acid chlorides.[603] ^{1}H, ^{13}C, ^{14}N, ^{15}N, ^{17}O, and ^{77}Se NMR studies of this ring system have been reported.[516,604]

5.11 Thiadiazoles

5.11.1 1,2,3-Thiadiazol-4-ones (1,2,3-thiadiazolium-4-olates) (*361*)

Treatment of 4-amino-1,2,3-thiadiazolium salts **362** ($Y=NH_2$) with $NaNO_2$/AcOH gives mesoionic products **361**,[605] which have also been obtained by alkaline hydrolysis of the 4-methylthio salts **362** (Y=SMe).[606] Treatment of the 3-phenyl-4,5-dichloro chloride **362** (R^1 = Ph, $Y=X=R^2$ = Cl) with $NaNO_2$ gives the 5-nitro derivative **361** (R^1 = Ph, $R^2 = NO_2$, 81%).[607] Reaction with Lawesson's reagent gave a mesoionic 1,2,3-thiadiazol-4-thione (Section 5.11.3).[608] Alkylation ($R_3O^+BF_4^-$) gives 4-alkoxy salts **362** ($X=BF_4$, Y=OR),[606] and $POCl_3$ gives 4-chloro salts **362** (Y=Cl).[30]

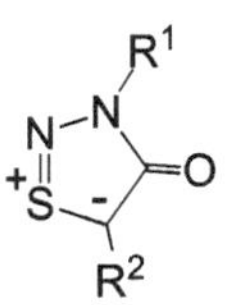

361

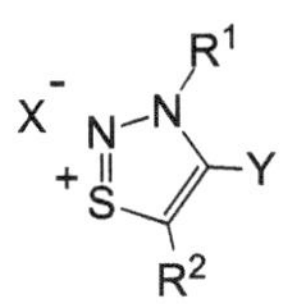

362

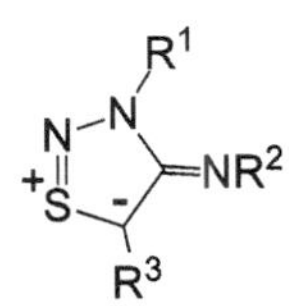

363

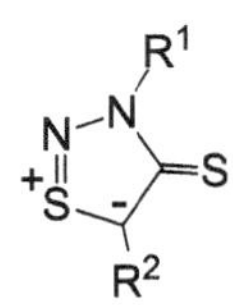

364

5.11.2 1,2,3-Thiadiazol-4-imines (1,2,3-thiadiazolium-4-aminides) (363)

N-Ethoxycarbonyl derivatives **363** ($R^2 = CO_2Et$) have been prepared by N-methylation (FSO_3Me) of the corresponding 1,2,3-thiadiazoles and subsequent treatment with $NaHCO_3$. Treatment of a 4-amino-1,2,3-thiadiazolium salt **362** ($Y = NH_2$) with Ac_2O/pyridine gave an *N*-acetyl derivative **363** ($R^2 = Ac$).[605]

5.11.3 1,2,3-Thiadiazol-4-thiones (1,2,3-thiadiazolium-4-thiolates) (364)

The *N*-4-methoxyphenyl derivative **364** ($R^1 = 4\text{-}MeOC_6H_4$, $R^2 = H$) has been prepared either by reaction of the corresponding 4-methoxy-1,2,3-thiadiazolium tetrafluoroborate **362** ($X = BF_4$, $Y = OMe$) with NaHS at −70 °C,[606] or by treatment of the *N*-4-methoxyphenyl-1,2,3-thiadiazol-4-ones **361** (Section 5.11.1) with Lawesson's reagent.[608] Six derivatives have been prepared by reaction of 1,2,3-thiadiazolium *p*-toluenesulfonates **362** ($X = TsO$, $Y = H$) with NaH/S_8 in DMF.[606] Treatment with MeI gives the iodides **262** ($X = I$, $Y = SMe$).

5.11.4 1,2,3-Thiadiazol-5-ones (1,2,3-thiadiazolium-5-olates) (365)

The *N*-phenyl derivative **365** ($R^1 = Ph$, $R^2 = H$) has been converted to the mesoionic 5-thiolate (Section 5.11.6) using Lawesson's reagent.[608] The formation of radical anions by reduction of 3-aryl derivatives **365** ($R^1 = Ar$, $R^2 = H$) using Zn or sodium dithionite has been investigated using ESR spectroscopy.[487]

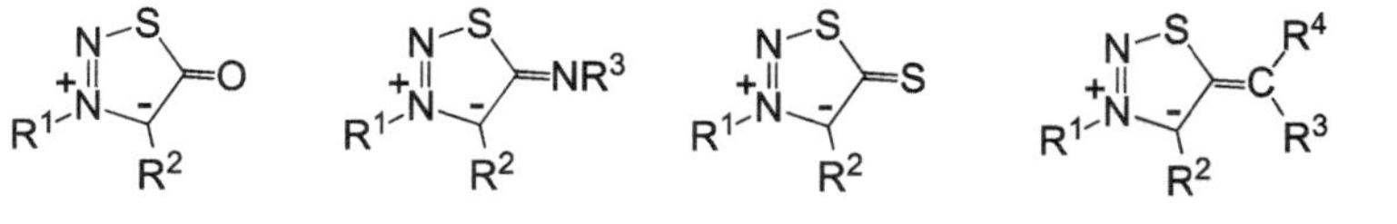

365 **366** **367** **368**

5.11.5 1,2,3-Thiadiazol-5-imines (1,2,3-thiadiazolium-5-aminides) (366)

N-Alkylation of 5-acylamino-1,2,3-thiadiazolium salts gives *N*-acyl derivatives **366** ($R^3 = COR'$). Treatment of the 5-amino salts with nitrous acid gives the *N*-nitroso derivatives **366** ($R^3 = NO$).[609] *N*-Aryl derivatives **366** ($R^3 = Ar$) have been made by N-alkylation of 5-phenylamino-1,2,3-thiadiazole or by treatment of 5-chloro-1,2,3-thiadiazolium salts with arylamines. The structure of the derivative **366** ($R^1 = Me$, $R^2 = H$, $R^3 = Ph$) has been confirmed by X-ray crystallography,[30,607] and it ^{15}N NMR spectrum reported.[610]

5.11.6 1,2,3-Thiadiazol-5-thiones (1,2,3-thiadiazolium-5-thiolates) (367)

Treatment of 3-methyl-1,2,3-thiadiazolium *p*-toluenesulfonate with NaH/S_8 in DMF gives the *N*-methyl derivative **367** (R^1 =Me, R^2 =H) (48%) together with the mesoionic isomer **364** (R^1 =Me, R^2 =H) (3%).[606] The *N*-phenyl derivative has been made in 98% yield by reaction of the corresponding 5-one **365** (R^1 =Ph, R^2 =H) (Section 5.11.4) with Lawesson's reagent.[608]

5.11.7 1,2,3-Thiadiazol-5-enes (1,2,3-thiadiazolium-5-methylides) (368)

Derivatives **368** (R^3 =CN, R^4 =CO_2R) have been made by addition of diazoalkanes to thioketenes; the structures of two products were confirmed by X-ray crystallography.[611] A phenylazo derivative **368** (R^3 =H, R^4 =NNPh) has been made by methylation (FSO_3Me or $Me_3O^+BF_4^-$) of a thiadiazole 5-hydrazone followed by base treatment of the resulting salt.[612] Treatment of a 5-chloro-1,2,3-thiadiazolium salt with cyclohexane 1,3-diones gives the products **368** (R^3R^4 =CO-$CH_2CR_2CH_2$-CO).[613]

5.11.8 1,2,4-Thiadiazol-3-ones (1,2,3-thiadiazolium-3-olates) (369)

The derivative **369** (R^1 =Me, R^2 =Ph) has been obtained in low yield (10%) by methylation of 3-hydroxy-5-phenyl-1,2,4-thiadiazole.[2,614] No other examples of this mesoionic system have been reported.

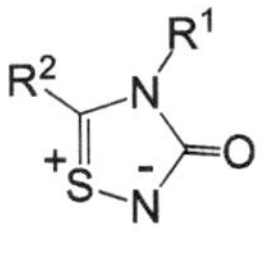

369

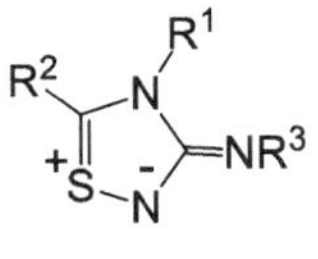

370

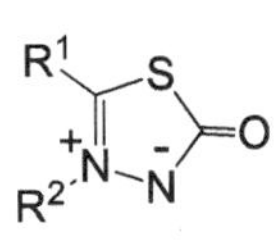

371

5.11.9 1,2,4-Thiadiazol-3-imines (1,2,3-thiadiazolium-3-aminides) (370)

Methylation of 5-phenyl-3-(toluene-*p*-sulfonamido)-1,2,4-thiadiazole ($MeFSO_3$) gives a low yield of the product **370** (R^1 =Me, R^2 =Ph, R^3 =$SO_2C_6H_4Me$-*p*) (18.5%).[2,614] No other examples of this mesoionic system have been reported.

5.11.10 1,3,4-Thiadiazol-2-ones (1,3,4-thiadiazolium-2-olates) (371)

Mesoionic 1,3,4-thiadiazol-2-ones **371** are conveniently prepared by reaction of *N*-thioacylhydrazines with phosgene or a phosgene equivalent. Using this approach, six novel derivatives have been prepared and shown to have high affinity for the insect nicotinic acetylcholine receptor.[615]

5.11.11 1,3,4-Thiadiazol-2-imines (1,3,4-thiadiazolium-2-aminides) (372)

Most approaches to the monocyclic ring system **372** involve initial formation of the stable chlorides **373**. These can be deprotonated to give the free bases which readily isomerize to 1,2,4-triazol-3-thiones **353** (Section 5.10.8). Methods of forming the chlorides **373** are: (i) reaction of *N*-thioacylhydrazines with arylisocyanide dichlorides[1]; (ii) treatment of acid chlorides with 1,4-diarylthiosemicarbazides under anhydrous conditions[579,616–618]; (iii) condensation of monoacylhydrazines with isothiocyanates[619]; and (iv) treatment of 1,4-diarylthiosemicarbazides with aldehydes in the presence of TMS-Cl or thionyl chloride.[620–623] The latter method has been employed to develop an efficient solid-phase methodology for the preparation of a library of derivatives **373**.[620] In an alternative approach, treatment of the iminophosphorane **374** with arylisocyanates gives the free bases **372** (R^1 = SMe, R^2 = Me, R^3 = Ar) (~50% yield).[624]

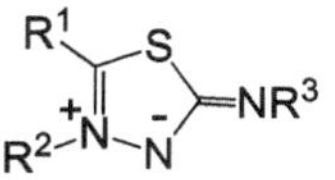

372

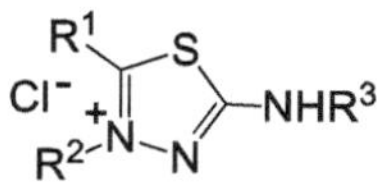

373

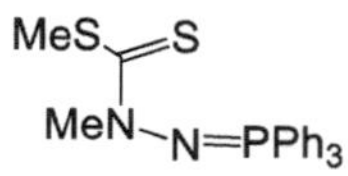

374

Further examples of the Dimroth-type rearrangement of the imines **372** to isomeric mesoionic triazol-3-thiones **353** (Section 5.10.8) have been reported,[576,625] and their ^{15}N and ^{13}C NMR spectra compared.[616]

The salts **373** have been reported to have biological activity including antitumor, antibacterial, antiprotozoal, sedative and other properties. Echevarria and co-workers have made a particular study of 5-cinnamoyl derivatives.[621–623,626–631] Montanari and co-workers have studied antibacterial properties and relevant physicochemical properties.[619,632–634]

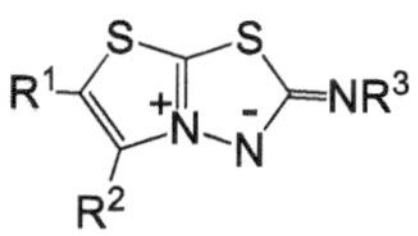

375

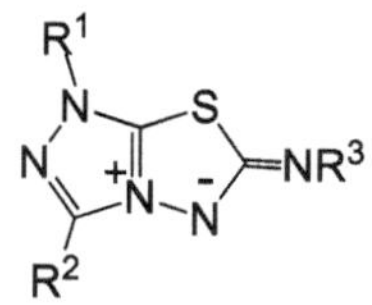

376

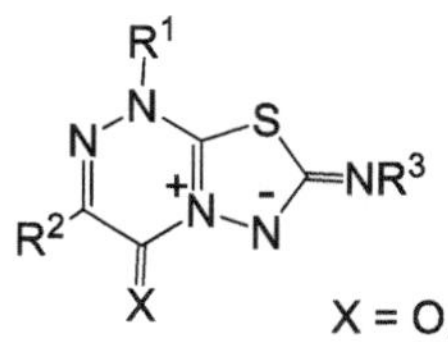

X = O,S

377

A number of bicyclic derivatives with general structures **375**,[635–637] **376**,[638] and **377**[639] have been prepared using variations of the methods used for monocyclic derivatives.

5.11.12 *1,3,4-Thiadiazol-2-thiones (1,3,4-thiadiazolium-2-thiolates)* (378)

Mesoionic 1,3,4-thiadiazol-2-thiones **378** have been known for over a century and methods of preparation are well established.[1] Most new studies are focused on potential applications as pharmaceuticals or materials and use routine synthetic procedures. Molina and co-worker have prepared the 5-SMe derivative **378** (R^1 = SMe, R^2 = Me) by reaction of the iminophosphorane **374** with carbon disulphide.[640] In attempts to form similar derivatives by reaction of hydrazinecarbodithioates (MeSCS-NR-NH_2) with carbon disulphide at 120 °C the products were the thiones **380**. A detailed mechanistic study suggests that the initially formed mesoionic products **379** undergo intermolecular S,S-*trans*-alkylation leading to the isomers **380**.[641] Treatment of 2,5-dimercapto-1,3,4-thiadiazole with 1,3-dibromopropane gives the bicyclic derivative **381**.[642] Using an established procedure, tricyclic derivatives have been prepared by the reaction 3-amino-4-thioquinazolones with carbon disulphide.[590]

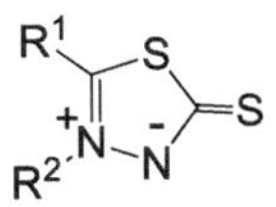

378

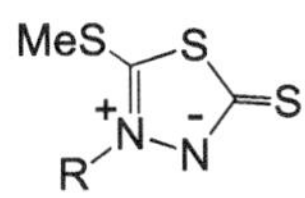

379

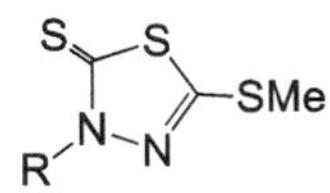

380

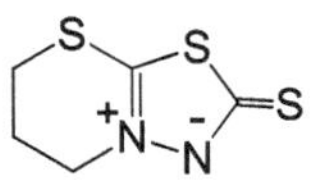

381

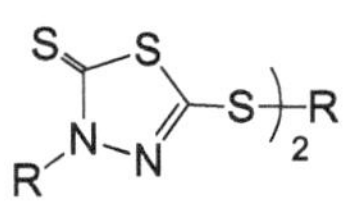

382

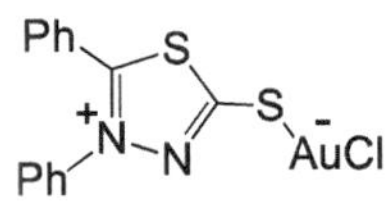

383

The exocyclic sulfur has both basic and nucleophilic character. Treatment of the SMe derivatives **379** with 1,3-dihalo reagents results in an S-alkylation/S'-demethylation sequence leading to the products **382**.[640] By a similar S-nucleophilic process, the derivative **381** on heating in nitrobenzene gives a macrocyclic dimer.[642]

For six 4,5-diaryl derivatives, pK_{BH+} values have been measured.[643] Treatment of the diphenyl derivative with sodium tetrachloroaurate(III) gives the gold complex **383**.[644] In a similar manner, the 5-*p*-chlorophenyl-4-phenyl derivative **378** (R^1 = 4-ClC_6H_4, R^2 = Ph) forms lanthanide complexes (Eu^{III}, Er^{III}).[645] Selective extraction of Hg^{II} from solution using the 4-phenyl-5-*p*-tolyl derivative **378** (R^1 = 4-MeC_6H_4, R^2 = Ph) has been investigated.[646]

Using standard procedures, a wide variety of novel derivatives have been prepared in the search for useful antibacterial activity,[647–650] and spasmolytic activity.[414]

A new polyimide material with "high chemical and thermal robustness" has been prepared by double 1,3-dipolar cycloaddition of the diphenyl derivative **378** ($R^1 = R^2 = Ph$) with a bis-maleimide.[651] Non-linear optical properties of six derivatives have been investigated.[652] X-ray photoelectron spectra have been recorded for six compounds.[653] Theoretical methods have investigated electronic spectra (AM1 and CNDO/S),[654] n-π* and p-π* electronic transitions (TD-DFT),[655] and first hyperpolarizabilities in solution (MP2).[656]

5.11.13 1,3,4-Thiadiazol-5-enes (1,3,4-thiadiazolium-5-methylides (385)

The bicyclic derivatives **385** (X=CN, CO_2Me, $CONH_2$) have been prepared by reaction of the 1-aminopyrid-2-thione **384** with ketene dithioacetals or by treatment of the methiodides **386** with acetonitrile derivatives in the presence of base (Scheme 62).[657] Tricyclic derivatives have also been prepared using the ketene dithioacetal methodology.[658]

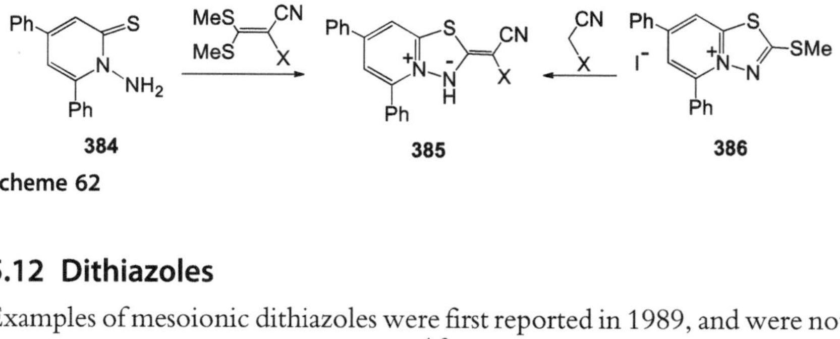

384 **385** **386**

Scheme 62

5.12 Dithiazoles

Examples of mesoionic dithiazoles were first reported in 1989, and were not, therefore, covered by earlier reviews.[1,2]

5.12.1 1,3,2-Dithiazol-4-ones (1,3,2-dithiazolium-4-olates) (388)

The 5-phenyl-1,3,2-dithiazol-4-thione **391** (R^1 =Ph) is oxidized by lead tetra-acetate to give the yellow mesoionic 1,3,2-dithiazol-4-one **388** (R^1 =Ph) (70%) (Scheme 63). This product is a stable solid but is light sensitive and in solution rapidly isomerizes to the 1,2,3-dithiazolone **387** (R^1 =Ph). 1,3-Dipolarcycloaddition occurs with DMAD in refluxing toluene to give the corresponding 5-phenylisothiazole.[659]

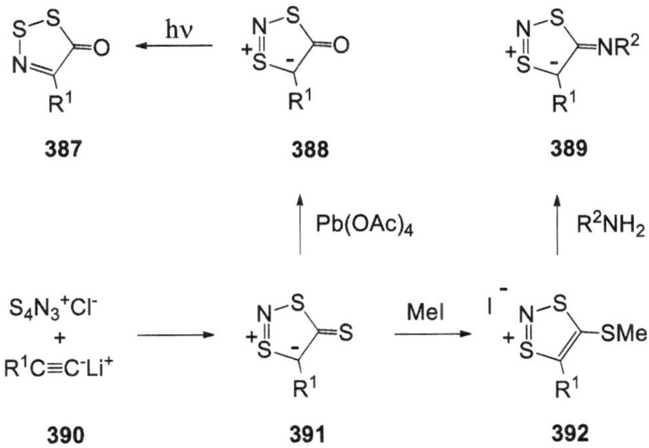

Scheme 63

5.12.2 *1,3,2-Dithiazol-4-imines (1,3,2-dithiazolium-4-aminides) (389)*

Treatment of the methiodide **392** (R^1 = Ph) with aniline in hot ethanol gives the purple 4-imine **389** ($R^1 = R^2$ = Ph) (26%).[659]

5.12.3 *1,3,2-Dithiazol-4-thiones (1,3,2-dithiazolium-4-thiolates) (391)*

Four mesoionic 1,3,2-dithiazol-4-thiones **391** (R^1 = H, Ph, ^{t}Bu, SPh) have been prepared in low yield (5–18%) by reaction of tetrathiatriazepinium chloride with lithium acetylides **390**. These thiones are green/purple solids or oils which, although stable in solution, rapidly decompose if exposed to light or oxygen. Treatment with methyl iodide gives a methiodide **392** (R^1 = Ph).[659]

5.13 Oxatriazoles

5.13.1 *1,2,3,4-Oxatriazol-5-ones (1,2,3,4-oxatriazolium-5-olates) (393)*

Common name: azasydnones

Mesoionic 1,2,3,4-oxatriazol-5-one derivatives **393** are of interest for their hypertensive effect as NO donors,[660–663] and for their potential as explosive materials. The latter properties include high density, detonation performance with thermal stability making this mesoionic ring an attractive explosophoric building block.[664–667]

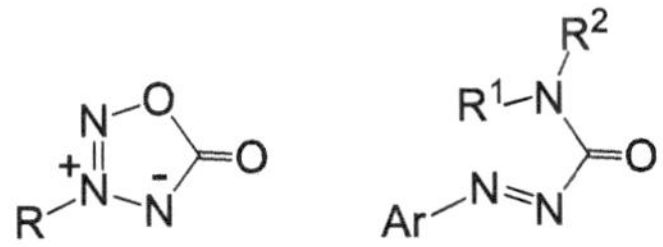

393 394

A number of methods of preparing this ring, including nitrosation of semicarbazides and coupling of diazonium salts with various reagents, have been reviewed.[1] These methods have some intrinsic disadvantages which have been summarized.[668] An improved one-pot method starting from primary amines involves diazotization ($NOBF_4$) and coupling with the potassium salt of nitroform [$KC(NO_2)_3$] in the presence of $ZnCl_2$ as catalyst. In this way a number of *N*-aryl and *N*-heteroaryl derivatives have been made in good yield. A plausible mechanism has been proposed for this cascade sequence and the X-ray structures of three *N*-furoxanyl derivatives reported.[668]

Other new preparative procedures are reaction of arylhydrazones of bromonitroformaldehyde [$ArNHN=C(Br)NO_2$] with arylazotrinitromethanes in acetic acid (65 °C)[669] or with nitrate salts in DMF or acetonitrile (20 °C).[670,671]

Studies of nitration of *N*-aryl derivatives **393** (R = Ar) have shown that the oxatriazole ring is meta-directing.[672] Reaction of chloro or nitro aryl derivatives with NH_3 or amines gives aryl azides and azocarboxamides **394**, or in some cases substitution of the aryl ring; products depend on the N-nucleophile and the substitution pattern.[673] The kinetics of the acid hydrolysis of the *N*-ethyl and four *N*-aryl derivatives have been investigated.[674,675] The influence of electronegativity on the relative reactivity in the series münchnone, sydnone, azasydnone has been discussed.[676]

In a comparative study of conjugated and cross-conjugated mesomeric betaines, the reduction potential of two mesoionic oxatriazolones were found to be in the range −0.85 to −1.22 V.[677] *Ab initio* calculations on mesoionic oxatriazolones and related species have been reported.[34,678]

5.13.2 1,2,3,4-Oxatriazol-5-imines (1,2,3,4-oxatriazolium-5-aminides) (395)

Common name: azasydnone imines

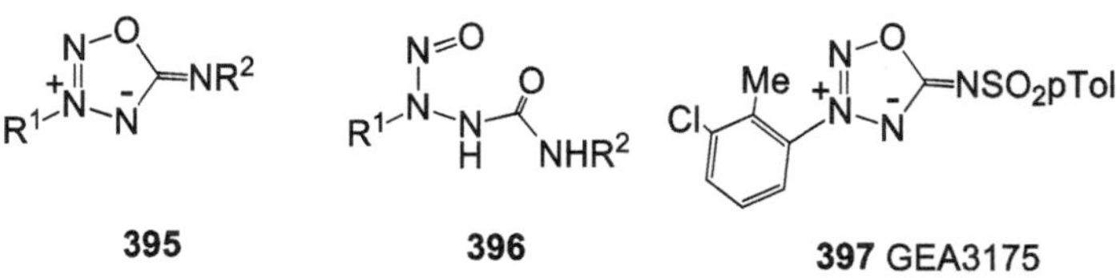

395 **396** **397** GEA3175

Much of the recent work on this ring system has been devoted to investigations of pharmacological properties. Like the sydnone imine molsidomine **293** (Section 5.8.5), mesoionic 1,2,3,4-oxatriazol-5-imines **395** are NO donors with potential applications as potent thrombolytic, antiplatelet, broncholytic and fibrinolytic agents.[679–682] The medicinal chemistry of the imines **395** has been the subject of three reviews.[6,683,684] The mode of action appears to arise from initial hydrolysis to an *N*-nitroso-semicarbazide **396**. Particular attention has focused on the derivatives GEA3175 **397**[685–696] and GEA3162.[697–699]

An interesting development has been the discovery that using mild conditions under copper catalysis 3-aryl derivatives **398** react with terminal alkynes to give 1,2,3-triazoles **401** or hydrazone derivatives **400** (Scheme 64). The outcome of the reaction depends upon the copper ligand and the alkyne. Use of 2,2'-bipyridine as copper ligand gives predominantly the 1,2,3-triazole products **401** (40–78%) whereas use of the ligand **399** favors the hydrazones **400** (44–84%). A mechanistic study suggests that both products arise from formation of a copper-coordinated [3 + 2] cycloadduct followed by extrusion of isocyanate.[700] Prior to this work only one example of a [3 + 2] addition to a mesoionic 1,2,3,4-oxatriazol-5-imine had been reported.[701]

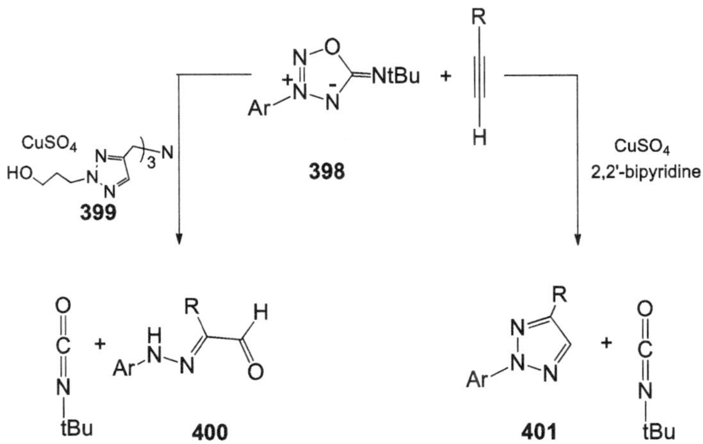

Scheme 64

Four NMR studies of this ring system (^{1}H, ^{13}C, ^{14}N, ^{15}N, ^{17}O) have been reported.[702–705] *Ab initio* calculations suggest that an anomeric interaction of the exocyclic lone pair with the ring C—Ö antibonding orbital favors the observed *trans* configuration in *N*-phenyl derivatives **395** (R^2 = Ph).[34]

5.13.3 1,2,3,4-Oxatriazol-5-thiones (1,2,3,4-oxatriazolium-5-thiolates) (402)

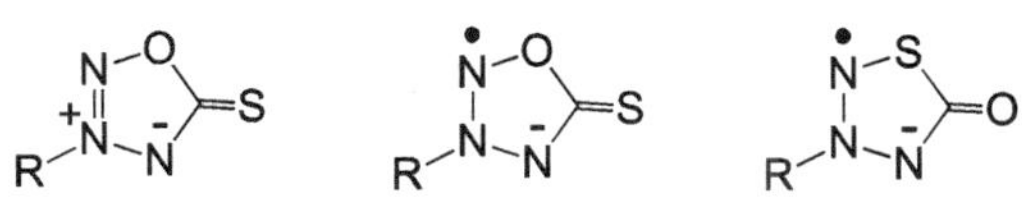

402 **403** **404**

Electrochemical studies of the 3-phenyl derivative **402** (R=Ph) have shown that it is irreversibly reduced at −1.19V to the radical anion **403** which rapidly rearranges to the isomeric radical **404**.[706] Applications in optically transparent electrodes have been investigated.[707] Detailed NMR studies of the phenyl derivative **402** (R=Ph) have been reported.[516,708,709]

5.14 Tetrazoles

An excellent comprehensive review of mesoionic tetrazoles by Moderhack covers the literature between 1980 and 2016, including "mesoionic" N-heterocyclic carbenes.[710]

5.14.1 1,2,3,4-Tetrazol-5-ones (1,2,3,4-tetrazolium-5-olates) (406)

No new general preparative route to the mesoionic 1,2,3,4-tetrazol-5-ones **406** have been reported. Treatment of the tetrazole-5-ones (or thiones) **405** (X=O,S) with tertiary alcohols (adamantan-1-ol, ^{t}BuOH) in 94% sulfuric acid gives good yields of the 3-alkyl products **406** (R^2 =adamantyl, ^{t}Bu) (Scheme 65).[711,712] Thiones **405** (X=S, R^1 =Me, Ph, CH_2CH_2OH) in acetic acid react with ethylene oxide to form the tetrazole-5-ones **406** (R^1 =Me, Ph, CH_2CH_2OH, R^2 =CH_2CH_2OH) (28–56%); this reaction is believed to occur *via* initial S-alkylation.[713] Reaction of the thioether **407** (R^2 =Me) with dimethyl sulfate followed by KOH gives the dimethyl derivative **406** (R^1 =R^2 =Me) in 60% yield.[714] Aqueous hydrolysis of the azide **408** (X=N_3, R^1 =R^2 =Ph) gives the diphenyl derivative **406** (R^1 =R^2 =Ph) (87%). The azide is prepared by nucleophilic substitution

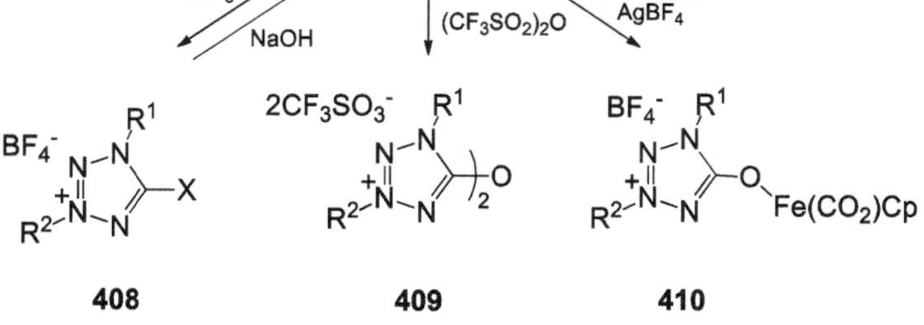

Scheme 65

(NaN_3/DMF) of the corresponding chloride **408** (X = Cl) which is conveniently made in 90% yield by $POCl_3$ treatment of **406** ($R^1 = R^2 = Ph$).[715]

Reaction of diaryl derivatives **406** ($R^1 = R^2 = Ar$) with Lawesson's reagent gives good yields of the mesoionic 1,2,3,4-tetrazol-5-thiones **421** (Section 5.14.3).[608,716]

Treatment of the diphenyl derivative with triflic anhydride gives a quantitative yield of the salt **409** ($R^1 = R^2 = Ph$) which is a convenient precursor of 1,2,3,4-tetrazol-5-enes **425** (Section 5.14.4).[489] Ultrasonication of **406** ($R^1 = R^2 = Ph$) with dicarbonyl(cyclopentadienyl)iodoiron(II) and $AgBF_4$ gives the yellow crystalline complex **410**.[490,597]

Detailed NMR studies of methyl and phenyl derivatives have been reported.[714,717,718] A DFT study supports the view that type A mesoionic tetrazoles have low classical aromaticity and modest magnetic aromaticity.[20] DFT calculations on 1-phenyltetrazol-5-one and its tautomers, including the mesoionic structure **406** ($R^1 = Ph$, $R^2 = H$) have been reported.[719]

Mesoionic 1,2,3,4-tetrazol-5-one derivatives **406** have been studied as additives for the promotion of conductivity in proton-conducting polymer gels,[720] and adamantyl derivatives have potentially useful anti-influenza activity.[721]

5.14.2 1,2,3,4-Tetrazol-5-imines (1,2,3,4-tetrazolium-5-aminides) (412)

In contrast to mesoionic heterocycles with less nitrogen atoms in the ring, mesoionic 1,2,3,4-tetrazol-5-imines with no exocyclic N-substituent **412** ($R^3 = H$) are stable. Alkylation of 5-aminotetrazole **411** (or N1 or N3 substituted derivatives) by tertiary alcohols in strong acid (H_2SO_4 or $HClO_4$) followed by base treatment gives the derivatives **412** ($R^3 = H$). In this way, for example, the 1,3-di-*tert*-butyl and 1,3-diadamantyl derivatives **412** ($R^1 = R^2 = {}^tBu$, Ad, $R^3 = H$) have been prepared,[722–724] and the di-*tert*-butyl structure confirmed by X-ray crystallography.[724] A photochemical study has shown that in solid argon under UV irradiation 2-methyl-5-aminotetrazole partially isomerizes to the mesoionic structure **412** ($R^1 = R^3 = H$, $R^2 = Me$).[725]

The 1,3-diphenyl-5-chloro salt **413** ($R^1 = R^2 = Ph$, X = Cl) with aqueous ammonia gives the derivative **412** ($R^1 = R^2 = Ph$, $R^3 = H$) but mainly the coupled product **414** ($R^1 = R^2 = Ph$); with hydroxylamine the derivative **412** ($R^1 = R^2 = Ph$, $R^3 = OH$) (62%) is formed (Scheme 66).[715] Hydrazine hydrate gives a 5,5'-bis-aminide **415** ($R^1 = R^2 = Ph$).[726] Other amino nucleophiles give analogous products **412**.[715,726] Photolysis of the diaryl azides **413** ($R^1 = R^2 = Ph$, X = N_3) gives the heteropentalenes

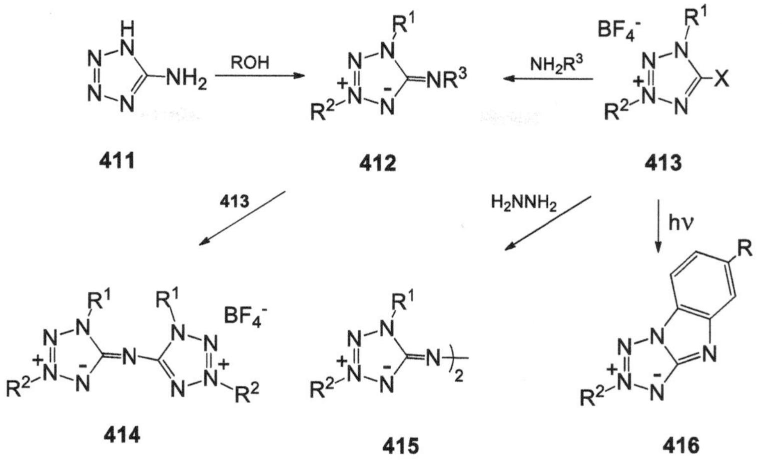

Scheme 66

416 in modest yield together with **412**. Formation of the mesomeric betaines **416** appears to occur *via* a singlet nitrene; in the presence of a triplet sensitizer the formation of **416** is completely suppressed and the yield of **412** increases to >70%.[727] Use of the 1-mesityl salt **413** (R^1 = mesityl, R^2 = Ph, X = N_3) gives alternative products in low yield.[728] Similar photolytic behavior has been observed using triazenide derivatives **412** (R^3 = N=NTs, N=NCN).[729]

Reactions with electron-deficient unsaturated reagents have been investigated. This is illustrated by reactions of the diphenyl derivatives **418** with DMAD. In the absence of an exocyclic N-substituent **418** (R^3 = H) the Michael adduct **417** is formed. If the exocyclic substituent is an aryl group **418** (R^3 = Ar), the pyrazole[3,4-*b*]pyridines **419** are formed in low yield by a mechanism that involves opening of the tetrazole ring (Scheme 67).[730]

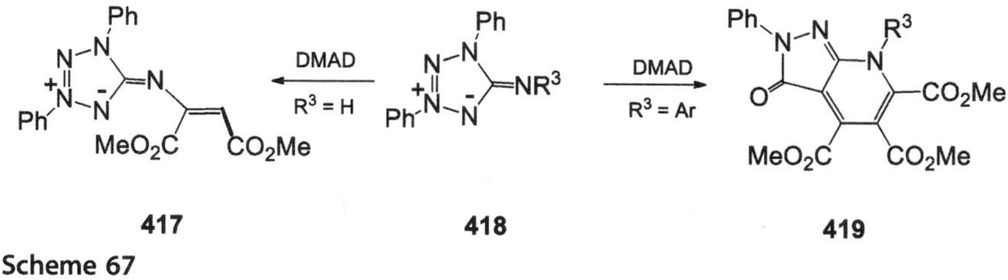

Scheme 67

The 1,3-di-*tert*-butyl derivative **412** (R^1 = R^2 = ^{t}Bu, R^3 = H) forms a $Cu_2Cl_6^{2-}$ salt when treated with $CuCl_2$,[731] and a Mn(III) complex with a macrocyclic manganese ligand.[732] The derivative **412** ($R^1 = R^2 = Ph$, $R^3 = OH$) is a precursor to the nitroso salt **413** ($R^1 = R^2 = Ph$, $X = NO$), which has potential has an oxidant,[733,734] and has been incorporated as a component of an aerobic oxidation system for benzyl alcohols.[735] Several NMR studies of the structures **412** have been reported.[703,714,718,736]

5.14.3 1,2,3,4-Tetrazol-5-thiones (1,2,3,4-tetrazolium-5-thiolates) (421)

A convenient new route to 1,3-diaryl derivatives **421** ($R^1 = R^2 = Ar$) is treatment of mesoionic 1,2,3,4-tetrazol-5-ones **406** (Section 5.14.1) with Lawesson's reagent.[608,716] Reaction of 1-substituted tetrazole-5-thiones **420** with adamantan-1-ol in sulfuric acid gives mainly the 3-adamantyl derivatives **421** ($R^2 =$ adamantyl), together with other alkylation products. Using *tert*-butyl alcohol the major products are the 3-*tert*-butyl substituted rings **421** ($R^2 = {}^tBu$) (Scheme 68).[711,712] The 1,3-dimethyl derivative **421** ($R^1 = R^2 = Me$) has been prepared in 40% yield by reaction of 3-methyl-5-methylthiotetrazole with dimethyl sulfate followed treatment with sodium sulphide.[714]

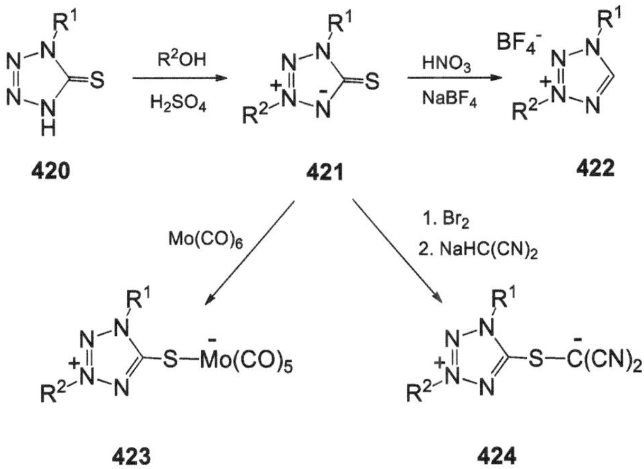

Scheme 68

Oxidation of 1,3-aryl derivatives **421** ($R^1 = R^2 = Ar$) by concentrated nitric acid and treatment with $NaBF_4$ gives the stable tetrazolium salts **422** ($R^1 = R^2 = Ar$) in good yield.[716] The nucleophilic exocyclic sulfur atom of the diphenyl derivative forms a molybdenum[0] complex **423** ($R^1 = R^2 = Ph$) (98%) with $Mo(CO)_6$,[597] and the betaine product **424** ($R^1 = R^2 = Ph$) (36%) on treatment with bromine followed by $NaHC(CN)_2$.[737]

5.14.4 1,2,3,4-Tetrazol-5-enes (1,2,3,4-tetrazolium-5-methylides) (425)

Derivatives of the type **425** ($R^3 = R^4 =$ CN) have previously been prepared by reaction of the salts **413** (X=OEt) with malononitrile and base.[2] These and other derivatives have been prepared in a similar way using the dicationic ether **409** ($R^1 = R^2 =$ Ph),[489,490] or the 5-chloro salt **413** ($R^1 = R^2 =$ Ph, X=Cl).[716] Reaction of the salt **413** ($R^1 = R^2 =$ Ph, X=OEt) with sodium cyclopentadienide gives the mesoionic ene **426** (69%) and benzo analogues have been prepared in a similar manner.[738,739] NMR studies suggest high electron density on the cyclopentadiene ring and a significant contribution of the polarized structure **427** to the ground state. The cyclopentadienyl ring is readily protonated, undergoes electrophilic substitution with a variety of reagents,[740] and forms chromium and molybdenum complexes.[597]

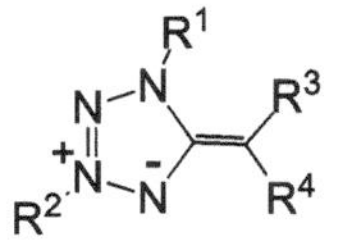

425

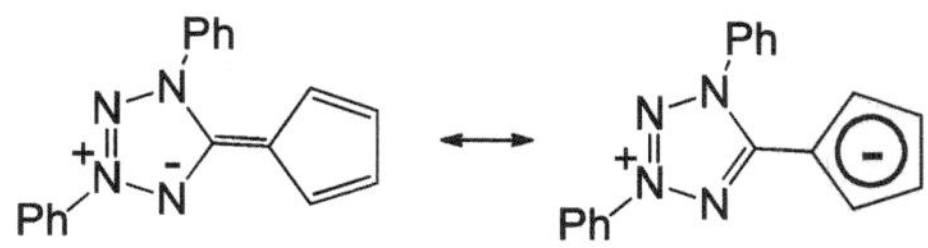

426 **427**

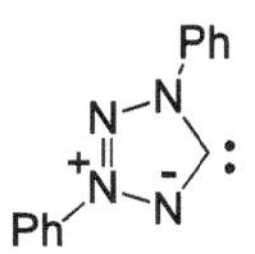

428

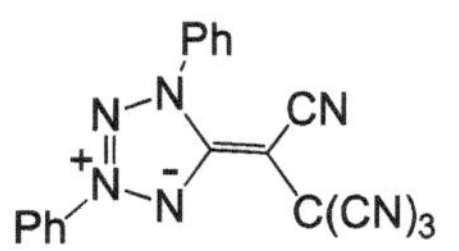

429

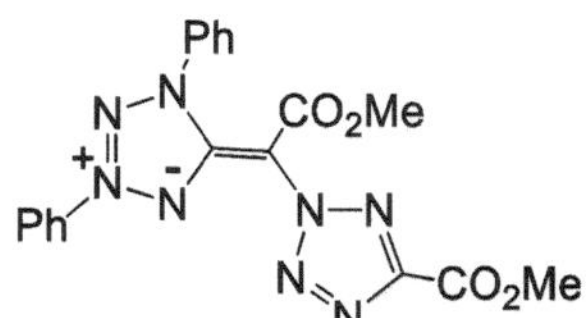

430

Treatment of the salt **413** ($R^1 = R^2 =$ Ph, X=H) with 1,8-diazabicyclo [5.4.0]undec-7-ene (DBU) generates the "mesoionic" carbene **428**. This reacts to give a variety of products in low yield, including the ene **429** (9%) (from tetracyanoethylene) and the ene **430** (15%) (from dimethyl 1,2,4,5-tetrazine-3,6-dicarboxylate). The structure **430** has been confirmed by X-ray crystallography.[741] Comparative NMR studies have been reported.[742,516]

5.14.5 1,2,3,4-Tetrazol-5-selones (1,2,3,4-tetrazolium-5-selenolates) (431)

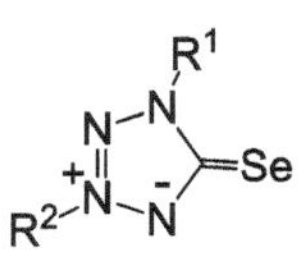

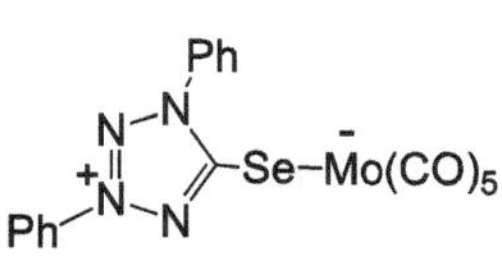

431 **432**

The selone **431** ($R^1 = R^2 = Ph$), prepared by reaction of the 5-chloro salt **413** ($R^1 = R^2 = Ph$, X=Cl) with sodium hydrogen selenide,[716] forms the molybdenum[0] complex **432** (10%) upon treatment with $Mo(CO)_6$.[597]

5.15 Thiatriazoles

5.15.1 1,2,3,4-Thiatriazol-5-ones (1,2,3,4-thiatriazolium-5-olates) (433)

New work has been restricted to the *N*-phenyl derivative **433** (R=Ph) for which an X-ray structure has been reported.[743] Electrochemical reduction of the *N*-phenyloxatriazol-5-thione **434** (R=Ph) induces rearrangement to the thiatriazole isomer **433** (R=Ph) by a mechanism that involves the respective radical anions.[706] Although thermal rearrangements of mesoionic isomers, e.g., **434** → **433**, has been reported previously (see Section 4.4.4), this is the first example of an electrochemically induced Dimroth rearrangement. The *N*-phenyl derivative **433** (R=Ph) is reversibly reduced to the radical anion with a reduction potential of −1.53 V. Treatment of compound **433** (R=Ph) with Lawesson's reagent gives the 5-thione **436** (R=Ph) (Section 5.15.3).[608]

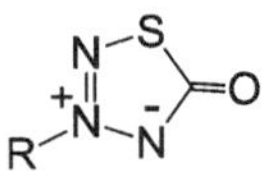

433

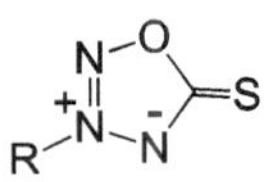

434

Several NMR studies (^1H, ^{13}C, ^{14}N, ^{15}N, ^{17}O) have been reported.[709,743,744,516] The ^{15}N NMR chemical shift of the N2 nitrogen (−39.6 ppm) clearly distinguishes the ring from isomeric oxatriazol-5-thiones **434** (12.9 ppm).[744]

The influence of this and related mesoionic ring systems on the conductivity of proton conducting polymer gels has been investigated.[720,745] The influence on the refractive index of polymer films has also been studied.[746]

5.15.2 1,2,3,4-Thiatriazol-5-imines (1,2,3,4-thiatriazolium-5-aminides) (435)

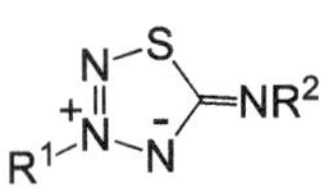

435

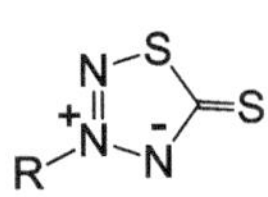

436

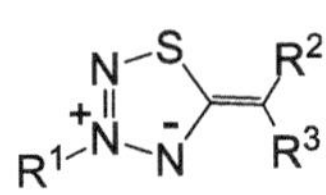

437

Several X-ray crystallographic studies of 5-arylimines **435** (R^2 = Ar) have been reported and, as with other examples in which steric interactions are minimal (see Section 4.2), the aryl substituents are *cis* with the C—Ẍ bond.[33,747,748] Comprehensive NMR studies of a range of diaryl derivatives have been published.[703,748–750] In a multiple screening study, the diphenyl derivative **435** ($R^1 = R^2 =$ Ph) was found to react with terminal alkynes in the presence of copper catalysts, but yields were low and the products did not merit further investigation.[700]

5.15.3 1,2,3,4-Thiatriazol-5-thiones (1,2,3,4-thiatriazolium-5-thiolates) (436)

The thiatriazol-5-thiones **436** are conveniently prepared by reaction of the thiatriazol-5-ones **433** with Lawesson's reagent.[608] The *N*-phenyl derivative **436** (R=Ph) is reversibly reduced to the radical anion with a reduction potential of −1.27 V. ^{1}H, ^{13}C, ^{14}N and ^{15}N NMR spectral details of the *N*-phenyl derivative **436** (R=Ph) have been reported.[749] The influence of **436** on the refractive index of polymer films has also been studied.[746]

5.15.4 1,2,3,4-Thiatriazol-5-enes (1,2,3,4-thiatriazolium-5-methylides) (437)

The structures of four *N*-phenyl derivatives have been determined by X-ray crystallography, namely **437** (a, $R^2 = R^3 =$ CN; b, $R^2 = R^3 =$ COPh; c, $R^2 = CO_2Et$, $R^3 =$ CN; d, $R^2 = CO_2Et$, $R^3 =$ COMe).[708,751] NMR details have been recorded.[708,752]

5.15.5 1,2,3,5-Thiatriazol-4-imines (1,2,3,5-thiatriazolium-4-aminides) (439)

Examples of the mesoionic ring system **439** have not previously been reported. Three 3-methyl derivatives **439** ($R^1 =$ Me, $R^2 =$ Ar) have been prepared by reaction of 1-amino-1-methyl-3-arylguanidines **438** ($R^1 =$ Me, $R^2 =$ Ar) with thionyl chloride/pyridine followed by treatment of the resulting pyridinium chloride complexes with aqueous potassium carbonate (Scheme 69). The reaction is believed to proceed *via* an intermediate sulfoxide. The phenyl derivative **439** ($R^1 =$ Me, $R^2 =$ Ph) was isolated as violet-black crystals in 40% yield.[753]

The formation of the products **439** is analogous to the formation of the bicyclic derivative **441** from the diaminoimidazole **440** (Scheme 69).[754] Structure **441** is alternatively classified as a type C heteropentalene HMB.[14]

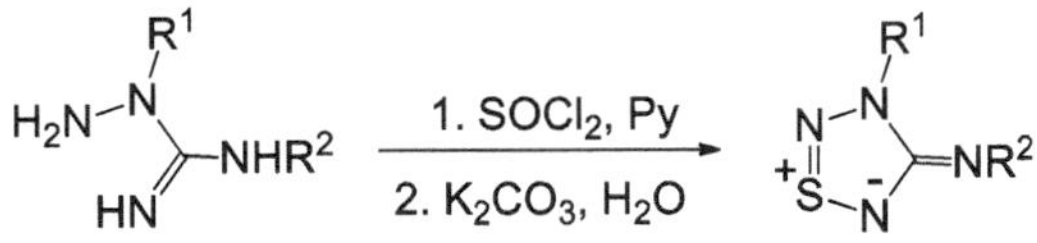

438 439

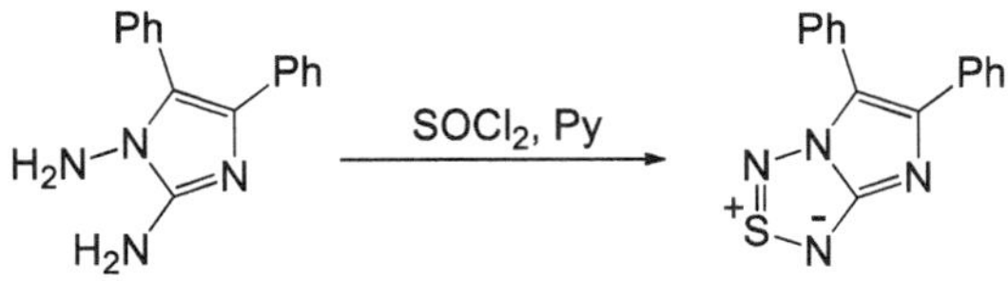

440 441

Scheme 69

5.16 Dithiadiazoles

5.16.1 1,3,2,4-Dithiadiazol-5-one (1,3,2,4-dithiadiazolium-5-olate) (442)

Common name: Roesky's ketone

1,3,2,4-Dithiadiazol-5-one **442** was first prepared in 1975 by Roesky and Wehner and has subsequently been referred to as Roesky's ketone. Pre-1982 studies have been reviewed.[2] It has been generated in the gas phase by reaction of gaseous NSCl with solid silver isocyanate (AgNCS) and characterized by He I photoelectron and mid-infrared spectroscopy.[755] Its rotational spectrum has been recorded and it was concluded that the gas phase structure closely resembles that in the solid phase.[756] The crystal structure has been re-examined using low-temperature X-ray diffraction and the bond lengths suggest a high degree of electron delocalization in the ring. The calculated NICS(0) (−11.13 ppm) and NICS(1) (−7.73 ppm) values suggest that the ring is aromatic, and these values are similar to those of other type A mesoionic rings (see Section 4.3). B3LYP/6 − 311 + G* calculations of bond lengths gave good agreement with experiment except for the C—S bond.[757] As part of a study of small five- and six-membered rings containing the NSNS fragment, the ring **442** has been investigated using other high level theoretical methods.[758,759]

442 443

5.16.2 *1,3,2,4-Dithiadiazol-5-imines (1,3,2,4-dithiadiazolium-5-aminides) (443)*

As far as we are aware, no new work on derivatives of this ring system have been reported since the subject was last reviewed.[2]

5.17 Diphosphadiazole

5.17.1 *1,3,2,4-Diphosphadiazol-4-ones (1,3,2,4-diphosphadiazolium-4-olates) (444)*

The 1,3,2,4-diphosphadiazol-4-ones **444** and their precursors **445** can be represented as mesomeric betaines (**A** and **B**) or as diradicals (**C**). The diradical representation simply implies high single electron density on the phosphorus atoms which is also inherent in the 1,3-dipole structures (associated with non-bonding molecular orbital (NBMO) character). For the purposes of this review the mesoionic representation is preferred.

The novel ring system **444** (R = 2,6-dimesitylphenyl) has been prepared by reaction of the conjugated mesomeric betaine **445** (R = 2, 6-dimesitylphenyl) with carbon monoxide. The reaction is slow under 1 bar pressure but at 60 bar 95% conversion is achieved after 9 h (Scheme 70). The red crystalline product has been fully characterized by a limited X-ray study and ^{31}P NMR (197.4 and 254.1 ppm, $^{2}J_{PP}$ 94.2 Hz).[760]

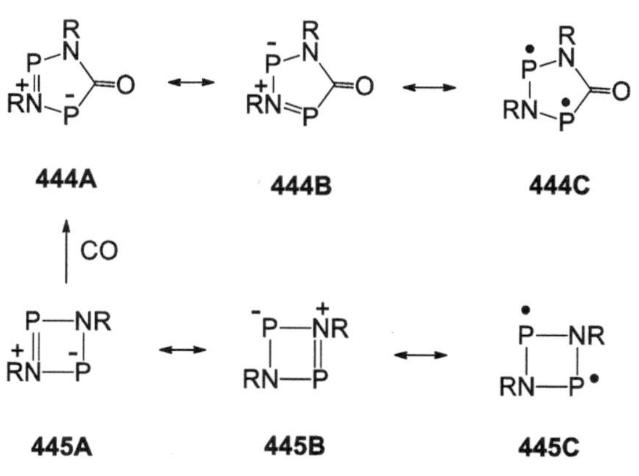

Scheme 70

During the preparation of **444** (R = 2,6-dimesitylphenyl) (Scheme 70) small amounts of the double addition product **446** (R = 2, 6-dimesitylphenyl) (<5%) (^{31}P NMR 169.9 and 246.3 ppm, $^{2}J_{PP}$ 45 Hz)

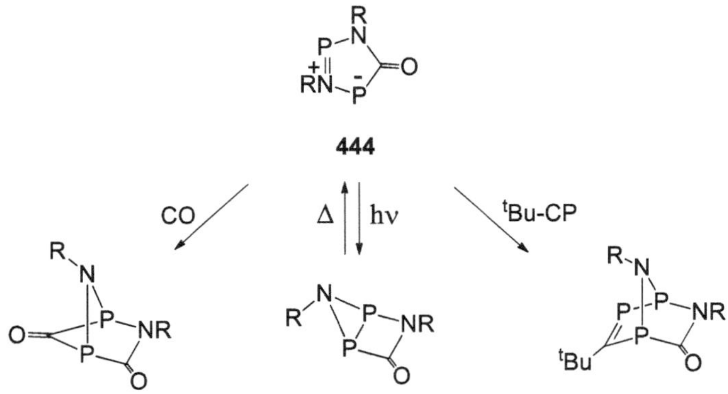

444

446 447 448

Scheme 71

were detected but could not be isolated due to fast CO release (Scheme 71). Upon UV irradiation the deep red mesoionic compound **444** (R = 2,6-dimesitylphenyl) gives the colorless bicyclic product **447** (R = 2,6-dimesitylphenyl) which rapidly reverts to **444** in solution at room temperature. Formation of bicyclic intermediates of this type have been proposed as intermediates in many photochemical studies of mesoionic rings. The phosphaalkyne ^{t}Bu-C≡P reacts selectively in high yield to give the 1,3-dipolar cycloadduct **448** (R = 2,6-dimesitylphenyl).[760]

5.17.2 1,3,2,4-Diphosphadiazol-4-imines (1,3,2,4-Diphosphadiazolium-4-aminides) (449)

Reaction of the reagent **445** (R^1 = 2,6-dimesitylphenyl) with one equivalent of isonitrile (e.g., R^2 = ^{t}Bu or 2,6-dimethylphenyl) gives the corresponding diphosphadiazol-4-imines **449**. The X-ray crystal structures of the derivatives **449** and **451** (R^1 = 2,6-dimesitylphenyl, R^2 = 2,6-dimethylphenyl) have been reported. The chemistry of these species is similar to that of the diphosphadiazolones **444**. Use of excess isonitrile results in formation of the adducts **451**, UV irradiation leads to reversible valence tautomerism to the colorless "housane" products **452**, and tBu-C≡P adds regioselectively to give cycloadducts **453** (Scheme 72). Detailed NMR, computational and substituent effect studies have been reported including the use of bis-isonitriles to give polyfunctional products.[761]

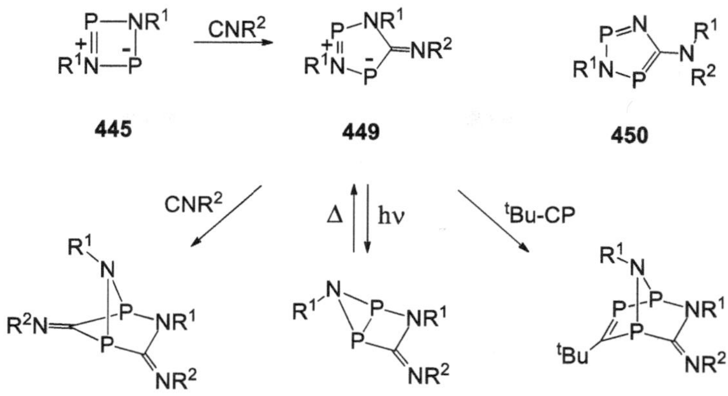

Scheme 72

In reactions of the hypersilyl reagent **445** ($R^1 = (Me_3Si)_3Si$) with isonitriles ($R^2 = {}^tBu$, 2,6-dimethylphenyl, mesityl) the products are the yellow heterocyclopentadienes **450** in which a silyl substituent has migrated to the exocyclic nitrogen atom. Only in the case of the isonitrile (R^2 = 2,6-dimesitylphenyl) was the anticipated blue mesoionic product **449** ($R^1 = (Me_3Si)_3Si$, R^2 = 2,6-dimesitylphenyl) formed. Migration of the silyl substituent appears to be controlled by steric hindrance of the isonitrile substituent.[762]

5.18 Arsenic derivatives

The mesoionic 1,2,3,4-arsatriazol-5-imine **454** has been prepared (83%, red needles) by isocyanide insertion into the reagent **455** (X=N, Y=As) and the structure confirmed by X-ray crystallography (Scheme 73). The phosphorus analogue **456** was generated in a similar way (λ_{max} 490 nm) but could not be isolated.[763]

In a similar manner, the mesoionic 1,3,2,4-arsaphosphadiazol-5-imine **458** has been prepared (68%) and its structure (P-As 3.05Å) confirmed by X-ray diffraction. During X-ray irradiation isomerization to the "housane" structure **457** was observed and its structure determined (P-As 2.29 Å). In contrast to products **454** and **456** which did not react, the phosphaalkyne ^{t}Bu-CP and compound **458** gave the cycloadduct **459** (78%) with selective P—P bond formation (Scheme 73).[763]

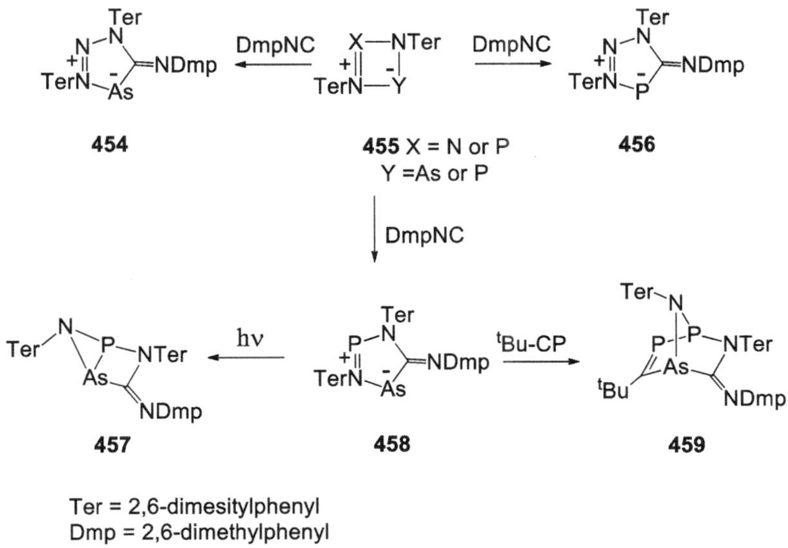

Scheme 73

6. Conclusions

The chemistry of Type A mesoionic heterocycles has continued to be an active area during the period under review. Fig. 10 shows the annual publication rates, which do not include the sydnone and sydnone imine literature. In the same period about 400 papers were published on mesoionic sydnones and sydnone imines. Clearly there has been a sustained level of interest over the last four decades following on from the revelation of 1,3-dipolar reactivity by Huisgen and others in the 1960s and 1970s.

Cycloadditions and subsequent fragmentations of mesoionic rings continue to be of synthetic interest. Systems of particular interest are the isomünchnones (Section 5.2.1), the münchnones (Section 5.2.4), the thioisomünchnones (Section 5.5.1) and the sydnones.[3] Attributes that contribute to their synthetic value are their relatively easy accessibility combined with high reactivity. As the number of nitrogen atoms in the ring increases so does the thermodynamic stability. This provides a pool of heterocycles that are of potential interest as pharmaceutical, agrochemicals, electronic components, additives and new materials. We hope that this review of the recent chemistry of type A mesoionic rings will foster further interest in these areas.

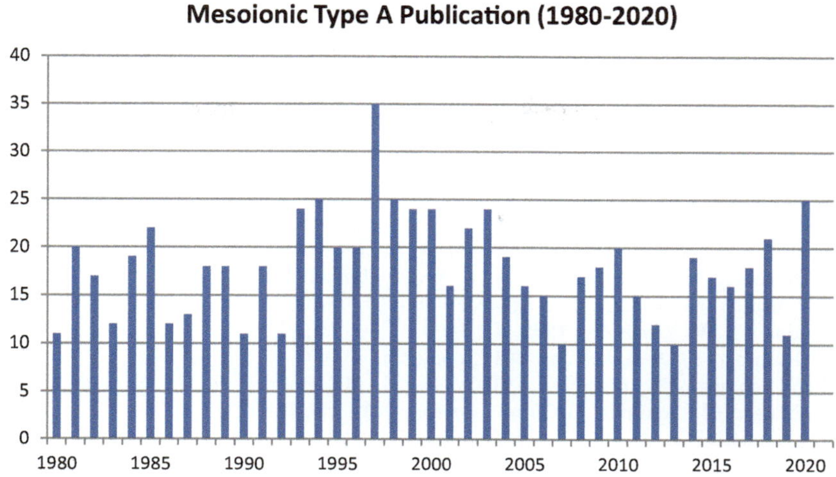

Fig. 10 Annual mesoionic type A publications during the period 1980–2020 (not including sydnones and sydnone imines).

References

1. Ollis, W. D.; Ramsden, C. A. *Adv. Heterocycl. Chem.* **1976**, *19*, 1.
2. Newton, C. G.; Ramsden, C. A. *Tetrahedron* **1982**, *38*, 2965.
3. Cherepanov, I. A.; Moiseev, S. K. *Adv. Heterocycl. Chem.* **2020**, *131*, 49.
4. Browne, D. L.; Harrity, J. P. A. *Tetrahedron* **2010**, *66*, 553.
5. Gupta, R. R.; Kumar, M.; Gupta, V. *Heterocyclic Chemistry*, Vol. II; Springer: Berlin, Heidelberg, 1999.
6. Kawase, M.; Sakagami, H.; Motohashi, N. *Top. Heterocycl. Chem.* **2009**, *16*, 135.
7. Vivancos, A.; Segarra, C.; Albrecht, M. *Chem. Rev.* **2018**, *19*, 9493.
8. Guisado-Barrios, G.; Soleilhavoup, M.; Bertrand, G. *Acc. Chem. Res.* **2018**, *51*, 3236.
9. Dewar, M. J. S.; Dougherty, R. C. *The PMO Theory of Organic Chemistry*; Plenum Press: New York, 1975.
10. Ramsden, C. A. *Chem. Soc. Rev.* **1994**, *23*, 111.
11. Ramsden, C. A. *Prog. Heterocycl. Chem.* **2016**, *28*, 1.
12. Ramsden, C. A. *Tetrahedron* **2013**, *69*, 4146.
13. Ramsden, C. A. *Adv. Heterocycl. Chem.* **1980**, *26*, 1.
14. Ramsden, C. A. *Tetrahedron* **1977**, *33*, 3203.
15. Ramsden, C. A. *J. Chem. Soc., Chem. Commun.* **1977**, 109.
16. Kratky, C.; Kappe, T. *J. Heterocycl. Chem.* **1981**, *18*, 881.
17. Ollis, W. D.; Stanforth, S. P.; Ramsden, C. A. *Tetrahedron* **1985**, *41*, 2239.
18. Potts, K. T.; Dery, M. O. *J. Organomet. Chem.* **1990**, *55*, 2884.
19. Padwa, A.; Harring, S. R.; Semones, M. A. *J. Organomet. Chem.* **1998**, *63*, 44.
20. Oziminski, W. P.; Ramsden, C. A. *Tetrahedron* **2015**, *71*, 7191.
21. Neugebauer, F. A.; Fischer, H.; Krieger, C. *Tetrahedron Lett.* **1984**, *25*, 629.
22. Schmidt, A.; Guan, Z. *Synthesis* **2012**, 3251.
23. Schmidt, A.; Weichmann, S.; Freese, T. *ARKIVOC* **2013**, *i*, 424.
24. Dewar, M. J. S.; Turchi, I. J. *J. Chem. Soc., Perkin Trans. 2* **1976**, 548.
25. Champagne, P. A.; Houk, K. N. *J. Organomet. Chem.* **2017**, *82*, 10980.
26. Barnighausen, H.; Jellinek, F.; Munnik, J.; Vos, A. *Acta Crystallogr.* **1963**, *16*, 471.

27. Ottersen, T. *Acta Chem. Scand. A* **1975**, *29*, 799.
28. King, T. J.; Preston, P. N.; Suffolk, J. S.; Turnbull, K. *J. Chem. Soc., Perkin Trans. 2* **1979**, 1751.
29. Ottersen, T.; Christophersen, C.; Treppendahl, S. *Acta Chem. Scand. A* **1975**, *29*, 45.
30. Kozinskij, V. A.; Zelenskaja, O. V.; Brückner, S.; Malpezzi, L. *J. Heterocycl. Chem.* **1984**, *21*, 1889.
31. Cannon, J. R.; Raston, C. L.; White, A. H. *Aust. J. Chem.* **1980**, *33*, 2237.
32. Danopoulos, A. A.; Monakhov, K. Y.; Braunstein, P. *Chem. – Eur. J.* **2013**, *19*, 450.
33. Staszewski, P.; Lipkowski, J.; Jaźwiński, J.; Staszewska, O.; Stefaniak, L.; Urbańczyk-Lipkowska, Z.; Webb, G. A. *J. Chem. Crystallogr.* **1998**, *28*, 227.
34. Ramsden, C. A.; Oziminsky, W. P. *Struct. Chem.* **2021**, *32*, 2075.
35. Simas, A. M.; Miller, J.; de Athayde Filho, P. F. *Can. J. Chem.* **1998**, *76*, 869.
36. Bird, C. W. *Tetrahedron* **1985**, *41*, 1409.
37. Ramsden, C. A. *Tetrahedron* **2010**, *66*, 2695.
38. Wiechmann, S.; Freese, T.; Drafz, M. H. H.; Hübner, E. G.; Namyslo, J. C.; Nieger, M.; Schmidt, A. *Chem. Commun.* **2014**, *50*, 11822.
39. Gingrich, H.L.; Baum, J.S. *The Chemistry of Heterocyclic Compounds, Chapter 4*, Vol. 45, Ed: I.J. Turchi, John Wiley and Sons, Ltd, pp 731-922.
40. Gribble, G. W. Mesoionic Oxazoles. In *The Chemistry of Heterocyclic Compounds, Vol. 60. Oxazoles: Synthesis, Reactions, and Spectroscopy, Part A*; Palmer, D. A., Ed.; Wiley: New York, 2003. Chapter 4.
41. Padwa, A.; Bur, S. *Adv. Heterocycl. Chem.* **2016**, *119*, 241.
42. Padwa, A. *ARKIVOC* **2018**, *iv*, 23.
43. Gribble, G. W. Mesoionic Ring Systems. In *Synthetic Applications of 1,3-Dipolar Cycloaddition Chemistry Toward Heterocycles and Natural Products*; Padwa, A., Pearson, W. H., Eds.; Wiley: New York, 2003; pp. 681–753.
44. Babaev, E. V. *Chem. Heterocycl. Compd.* **2012**, *48*, 59.
45. Osterhout, M. H.; Nadler, W. R.; Padwa, A. *Synthesis* **1994**, 123.
46. Ávalos, M.; Babiano, R.; Cintas, P.; Jiménez, J. L.; Palacios, J. C. *Acc. Chem. Res.* **2005**, *38*, 460.
47. Badami, B. V. *Resonance* **2006**, *11*, 40.
48. de la Concepción, G. J.; Martínez, R. F.; Cintas, P.; Jiménez, J. L. Cycloadditions with Mesoionic Dipoles: Strategy and Control. In *Targets in Heterocyclic Systems*; Attanasi, O., Merino, P., Spinelli, D., Eds.; Vol. 21; Italian Society of Chemistry: Rome, Italy, 2018; pp. 228–253. ch. 11.
49. Porte, K.; Riomet, M.; Figliola, C.; Audisio, D.; Taran, F. *Chem. Rev.* **2021**, *121*, 6718.
50. Avalos, M.; Babiano, R.; Diánez, M. J.; Espinosa, J.; Estrada, M. D.; Jiménez, J. L.; López-Castro, A.; Méndez, M. M.; Palacios, J. C. *Tetrahedron* **1992**, *48*, 4193.
51. de la Concepción, J. G.; Ávalos, M.; Babiano, R.; Cintas, P.; Jiménez, J. L.; Light, M. E.; Palacios, J. C. *Tetrahedron* **2016**, *72*, 4665.
52. Potts, K. T.; Huang, F.; Khattak, R. K. *J. Organomet. Chem.* **1977**, *42*, 1644.
53. Morin, M. S. T.; St-Cyr, D. J.; Arndtsen, B. A.; Krenske, E. H.; Houk, K. N. *J. Am. Chem. Soc.* **2018**, *135*, 17349.
54. César, V.; Tourneux, J.-C.; Vujkovic, N.; Lavigne, G. *Chem. Commun.* **2012**, *48*, 2349.
55. Danopoulos, A. A.; Braunstein, P. *Chem. Commun.* **2014**, *50*, 3055.
56. Ramsden, C. A.; Oziminski, W. P. *Tetrahedron* **2015**, *71*, 6846.
57. Biju, A.; Hirano, K.; Frohlich, R.; Glorius, F. *Chem.-Asian J.* **2009**, *4*, 1786.
58. Ramsden, C. A.; Oziminski, W. P. *J. Organomet. Chem.* **2016**, *81*, 10295.
59. Ramsden, C. A.; Oziminski, W. P. *J. Organomet. Chem.* **2017**, *82*, 12485.
60. Hamaguchi, M.; Tomida, N.; Iyama, Y.; Oshima, T. *J. Organomet. Chem.* **2006**, *71*, 5162.

61. Hamaguchi, M.; Tomida, N.; Iyama, Y. *J. Organomet. Chem.* **2007**, *72*, 1326.
62. Hanley, R. N.; Ollis, W. D.; Ramsden, C. A.; Rowlands, G.; Sutton, L. E. *J. Chem. Soc., Perkin Trans. 1* **1978**, 600.
63. Ollis, W. D.; Ramsden, C. A. *Adv. Heterocycl. Chem.* **1976**, *19*, 65.
64. Hamaguchi, M.; Tomida, N.; Mochizuki, E.; Oshima, T. *Tetrahedron Lett.* **2003**, *44*, 7945.
65. Berk, H. C.; Zwikelmaier, K. E.; Franz, J. E. *Synth. Commun.* **1980**, *10*, 707.
66. Kawase, M.; Kurihara, T. *Tetrahedron Lett.* **1994**, *35*, 8209.
67. Kawase, M.; Saito, S.; Kurihara, T. *Chem. Pharm. Bull.* **2000**, *48*, 1338.
68. Hamaguchi, M.; Nagai, T. *J. Chem. Soc., Chem. Commun.* **1985**, 1319.
69. Hamaguchi, M.; Nagai, T. *J. Chem. Soc., Chem. Commun.* **1985**, 190.
70. Hamaguchi, M.; Tomida, N.; Mochizuki, E.; Oshima, T. *Tetrahedron Lett.* **2005**, *46*, 1259.
71. Maier, M. E.; Evertz, K. *Tetrahedron Lett.* **1988**, *29*, 1677.
72. Maier, M. E.; Schöffing, B. *Chem. Ber.* **1989**, *122*, 1081.
73. Padwa, A.; Hertzog, D. L.; Chinn, R. L. *Tetrahedron Lett.* **1989**, *30*, 4077.
74. Doyle, M. P.; Pieters, R. J.; Taunton, J.; Pho, H. Q.; Padwa, A.; Hertzog, D. L.; Precedo, L. *J. Organomet. Chem.* **1991**, *56*, 820.
75. Ibata, T.; Hamaguchi, M. *Tetrahedron Lett.* **1974**, *15*, 4475.
76. Liu, F.; Austin, D. J. *Org. Lett.* **2001**, *3*, 2273.
77. Savinov, S. N.; Austin, D. J. *Org. Lett.* **2002**, *4*, 1415.
78. Bonderoff, S. A.; Padwa, A. *J. Organomet. Chem.* **2017**, *82*, 642.
79. Xu, C.; Wang, K.; Li, D.; Lin, L.; Feng, X. *Angew. Chem., Int. Ed.* **2019**, *58*, 18438.
80. Oguri, H.; Schreiber, S. L. *Org. Lett.* **2005**, *7*, 47.
81. Padwa, A.; Prein, M. *Tetrahedron Lett.* **1996**, *37*, 6981.
82. Prein, M.; Manley, P. J.; Padwa, A. *Tetrahedron* **1997**, *53*, 7777.
83. Whitehouse, D. L.; Nelson, K. H., Jr.; Savinov, S. N.; Lowe, R. S.; Austin, D. J. *Bioorg. Med. Chem.* **1998**, *6*, 1273.
84. Lam, H.; Qureshi, Z.; Wegmann, M.; Lautens, M. *Angew. Chem., Int. Ed.* **2018**, *57*, 16185.
85. Kuethe, J.; Padwa, A. *Tetrahedron Lett.* **1997**, *38*, 1505.
86. Padwa, A.; Heidelbaugh, T. M.; Kuethe, J. T. *J. Organomet. Chem.* **1999**, *64*, 2038.
87. Padwa, A.; Heidelbaugh, T. M.; Kuethe, J. T. *J. Organomet. Chem.* **2000**, *65*, 2368.
88. Haddadin, M. J.; Kattan, A. M.; Freeman, J. P. *J. Organomet. Chem.* **1982**, *47*, 723.
89. Haddadin, M. J.; Tanus, H. T. *Heterocycles* **1984**, *22*, 773.
90. Padwa, A.; Hertzog, D. L. *Tetrahedron* **1993**, *49*, 2589.
91. Sheehan, S. M.; Padwa, A. *J. Organomet. Chem.* **1997**, *62*, 438.
92. Kappe, C. O.; Peters, K.; Peters, E.-M. *J. Organomet. Chem.* **1997**, *62*, 3109.
93. Padwa, A.; Prein, M. *J. Organomet. Chem.* **1997**, *62*, 6842.
94. Angell, R.; Drew, M. G. B.; Fengler-Veith, M.; Finch, H.; Harwood, L. M.; Jahans, A. W.; Tucker, T. T. *Tetrahedron Lett.* **1997**, *38*, 3107.
95. Angell, R.; Fengler-Veith, M.; Finch, H.; Harwood, L. M.; Tucker, T. T. *Tetrahedron Lett.* **1997**, *38*, 4517.
96. Padwa, A.; Prein, M. *Tetrahedron* **1998**, *54*, 6957.
97. Lou, S.; Dai, P.; Schaus, S. E. *J. Organomet. Chem.* **2007**, *72*, 9998.
98. Zhang, X.; Schmitt, A. C.; Decicco, C. P. *Tetrahedron Lett.* **2002**, *43*, 9663.
99. Savinov, S. N.; Austin, D. J. *Chem. Commun.* **1999**, 1813.
100. Savinov, S. N.; Austin, D. J. *Org. Lett.* **2002**, *4*, 1419.
101. Kanazawa, A.; Muniz, M. N.; Baumlová, B.; Ljungdahl, N.; Greene, A. E. *Synlett* **2008**, 2275.
102. Suga, H.; Hashimoto, Y.; Yasumura, S.; Takezawa, R.; Itoh, K.; Kakehi, A. *J. Organomet. Chem.* **2013**, *78*, 10840.

103. Raolji, G. B.; Garcon, S.; Greene, A. E.; Kanazawa, A. *Angew. Chem., Int. Ed.* **2003**, *42*, 5059.
104. Mmutlane, E. M.; Harris, J. M.; Padwa, A. *J. Organomet. Chem.* **2005**, *70*, 8055.
105. Nair, V.; Santhi, V.; Sheela, K. C.; Radhakrishnan, K. V.; Rath, N. P. *Synthesis* **2003**, 1559.
106. Gonzdlez, R.; Knight, B. W.; Wudl, F.; Semones, M. A.; Padwa, A. *J. Organomet. Chem.* **1994**, *59*, 7949.
107. Padwa, A.; Sheehan, S. M.; Straub, C. S. *J. Organomet. Chem.* **1999**, *64*, 8648.
108. Harris, J. M.; Padwa, A. *Org. Lett.* **2003**, *5*, 4195.
109. De, S.; Chen, L.; Zhang, S.; Gilbertson, S. R. *ACS Comb. Sci.* **2013**, *15*, 340.
110. Hertzog, D. L.; Austin, D. J.; Nadler, W. R.; Padwa, A. *Tetrahedron Lett.* **1992**, *33*, 4731.
111. Padwa, A.; Hertzog, D. L.; Nadler, W. R.; Osterhout, M. H.; Price, A. T. *J. Organomet. Chem.* **1994**, *59*, 1418.
112. Marino, J. P., Jr.; Osterhout, M. H.; Price, A. T.; Semones, M. A.; Padwa, A. *J. Organomet. Chem.* **1994**, *59*, 5518.
113. Padwa, A.; Hertzog, D. L.; Nadler, W. R. *J. Organomet. Chem.* **1994**, *59*, 7072.
114. Padwa, A.; Austin, D. J.; Price, A. T. *Tetrahedron Lett.* **1994**, *35*, 7159.
115. Marino, J. P., Jr.; Osterhout, M. H.; Padwa, A. *J. Organomet. Chem.* **1995**, *60*, 2704.
116. Padwa, A.; Austin, D. J.; Price, A. T.; Weingarten, M. D. *Tetrahedron* **1996**, *52*, 3247.
117. Padwa, A.; Brodney, M. A.; Marino, J. P., Jr.; Osterhout, M. H.; Price, A. T. *J. Organomet. Chem.* **1997**, *62*, 67.
118. Padwa, A.; Brodney, M. A.; Marino, J. P., Jr.; Sheehan, S. M. *J. Organomet. Chem.* **1997**, *62*, 78.
119. Jauk, B.; Belaj, F.; Kappe, C. O. *J. Chem. Soc., Perkin Trans. 1* **1999**, 307.
120. Brodney, M. A.; Padwa, A. *J. Organomet. Chem.* **1999**, *64*, 556.
121. Jauk, B.; Pernat, T.; Kappe, C. O. *Molecules* **2000**, *5*, 227.
122. England, D. B.; Eagan, J. M.; Merey, G.; Anac, O.; Padwa, A. *Tetrahedron* **2008**, *64*, 988.
123. Muthusamy, S.; Srinivasan, P. *Tetrahedron* **2009**, *65*, 1567.
124. Grillet, F.; Sabot, C.; Anderson, R.; Babjak, M.; Greene, A. E.; Kanazawa, A. *Tetrahedron* **2011**, *67*, 2579.
125. Muthusamy, S.; Gangadurai, C. *Synthesis* **2016**, 2213.
126. Straub, C. S.; Padwa, A. *Org. Lett.* **1999**, *1*, 83.
127. Gowravaram, M. R.; Gallop, M. A. *Tetrahedron Lett.* **1997**, *38*, 6973.
128. Whitehouse, D. L.; Nelson, K. H., Jr.; Savinov, S. N.; Austin, D. J. *Tetrahedron Lett.* **1997**, *38*, 7139.
129. Mathias, L. J.; Moore, D. R. *J. Am. Chem. Soc.* **1985**, *107*, 5817.
130. Moore, D. R.; Mathias, L. J. *Macromolecules* **1986**, *19*, 1530.
131. Ruf, S. G.; Bergstrasser, U.; Regitz, M. *Tetrahedron* **2000**, *56*, 63.
132. Drew, M. G. B.; Fengler-Veith, M.; Harwood, L. M.; Jahans, A. W. *Tetrahedron Lett.* **1997**, *38*, 4521.
133. Gan, Y.; Harwood, L. M.; Richards, S. C.; Smith, I. E. D.; Vinader, V. *Tetrahedron: Asymmetry* **2009**, *20*, 723.
134. Bach, P.; Bergstaesser, U.; Leininger, S.; Regitz, M. *Bull. Soc. Chim. Fr.* **1997**, *134*, 927.
135. Hamaguchi, M.; Ibata, T. *J. Chem. Soc., Chem. Commun.* **1980**, 230.
136. Muthusamy, S.; Gunanathan, C. *Chem. Commun.* **2003**, 440.
137. Freeman, J. P.; Haddadin, M. J.; Rastetter, W. H.; Chancellor, T. *Tetrahedron Lett.* **1982**, *23*, 815.
138. Gotthardt, H.; Oppermann, M. *J. Chem. Soc., Chem. Commun.* **1985**, 1154.
139. Gribble, G. W. In *Synthetic Applications of 1,3-Dipolar Cycloaddition Chemistry Towards Heterocycles and Natural Products*; Padwa, A., Pearson, W. H., Eds.; John Wiley & Sons: Hoboken, NJ, 2002; pp. 681–755.

140. Gribble, G. W. Chapter 4: Mesoionic Oxazoles. In *The Chemistry of Heterocyclic Compounds*; Palmer, D. C., Ed.; Vol. 60; Wiley: New York, 2003.
141. Reissig, H.-U.; Zimmer, R. *Angew. Chem., Int. Ed.* **2014**, *53*, 9708.
142. Qun, M.; Huailin, S. *Chin. J. Org. Chem.* **2016**, *36*, 913.
143. Mjalli, A. M. M.; Sarshar, S.; Baiga, T. J. *Tetrahedron Lett.* **1996**, *37*, 2943.
144. Samanta, S. K.; Yli-Kauhaluoma, J. *J. Comb. Chem.* **2005**, 7, 142.
145. Hamper, B. C.; Jerome, K. D.; Yalamanchili, G.; Walker, D. M.; Chott, R. C.; Mischke, D. A. *Biotechnol. Bioeng.* **2000**, *71*, 28.
146. Kakaawla, T. K. K.; Hartley, W. C.; Harrity, J. P. A. *Eur. J. Org. Chem.* **2016**, 2789.
147. Mazgarova, G. G.; Fatykhov, A. A.; Gataullin, R. R. *Russ. J. Org. Chem.* **2014**, *50*, 1155.
148. Bildeau, M. T.; Cunningham, A. M. *J. Organomet. Chem.* **1998**, *63*, 2800.
149. Pandey, P. S.; Rao, T. S. *Bioorg. Med. Chem. Lett.* **2004**, *14*, 129.
150. Gerster, H.; Maas, G. Z. *Naturforsch* **2008**, *63b*, 384.
151. Lopchuk, J. M.; Gribble, G. W. *Tetrahedron Lett.* **2015**, *56*, 3208.
152. Okano, T.; Uekawa, T.; Morishima, N.; Eguchi, S. *J. Organomet. Chem.* **1991**, *56*, 5259.
153. Funabiki, K.; Ishihara, T.; Yamanaka, H. *J. Fluor. Chem.* **1995**, *71*, 5.
154. Rybakov, V. B.; Zhukov, S. G.; Pasichnichenko, K. Y.; Babaev, E. V. *Russ. J. Coord. Chem.* **2000**, *26*, 670.
155. Hershenson, F. M.; Pavia, M. R. *Synthesis* **1988**, 999.
156. Peddibhotla, S.; Jayakumar, S.; Tepe, J. J. *Org. Lett.* **2002**, *4*, 3533.
157. Peddibhotla, S.; Tepe, J. J. *Synthesis* **2003**, 1433.
158. Peddibhotla, S.; Tepe, J. J. *J. Am. Chem. Soc.* **2004**, *126*, 12776.
159. Melhado, A. D.; Luparia, M.; Toste, F. D. *J. Am. Chem. Soc.* **2007**, *129*, 12638.
160. Melhado, A. D.; Amarante, G. W.; Wang, Z. J.; Luparia, M.; Toste, F. D. *J. Am. Chem. Soc.* **2011**, *133*, 3517.
161. Marco-Martinez, J.; Reboredo, S.; Izquierdo, M.; Marcos, V.; López, J. L.; Filippone, S.; Martin, N. *J. Am. Chem. Soc.* **2014**, *136*, 2897.
162. Merlic, C. A.; Baur, A.; Aldrich, C. C. *J. Am. Chem. Soc.* **2000**, *122*, 7398.
163. Dghaym, R. D.; Dhawan, R.; Arndtsen, B. A. *Angew. Chem., Int. Ed.* **2001**, *40*, 3228.
164. Dhawan, R.; Dghaym, R. D.; Arndtsen, B. A. *J. Am. Chem. Soc.* **2003**, *125*, 1474.
165. Worrall, K.; Xu, B.; Bontemps, S.; Arndtsen, B. A. *J. Organomet. Chem.* **2011**, *76*, 170.
166. Siamaki, A. R.; Sakalauskas, M.; Arndtsen, B. A. *Angew. Chem., Int. Ed.* **2011**, *50*, 6552.
167. Dhawan, R.; Arndtsen, B. A. *J. Am. Chem. Soc.* **2004**, *126*, 468.
168. Lu, Y.; Arndtsen, B. A. *Angew. Chem., Int. Ed.* **2008**, *47*, 5430.
169. Leitch, D. C.; Kayser, L. V.; Han, Z.-H.; Siamaki, A. R.; Keyzer, E. N.; Gefen, A.; Arndtsen, B. A. *Nat. Commun.* **2015**, *6*, 7441.
170. Nan'ya, S.; Goto, S.; Butsugan, Y. *J. Heterocycl. Chem.* **1990**, *27*, 1519.
171. Nan'ya, S.; Kurachi, Y.; Butsugan, Y. *J. Heterocyl. Chem.* **1991**, *28*, 1853.
172. Lopchuk, J. M.; Hughes, R. P.; Gribble, G. W. *Org. Lett.* **2013**, *15*, 5218.
173. Avalos, M.; Babiano, R.; Cabanillas, A.; Cintas, P.; Jimenez, J. L.; Palacios, J. C.; Aguilar, M. A.; Corchado, J. C.; Espinosa-Garcia, J. *J. Organomet. Chem.* **1996**, *61*, 7291.
174. González, M. M.; Hernandez-Castillo, D.; Montero-Cabrera, L. A.; Miranda-Quintana, R. A. *Int. J. Quantum Chem.* **2017**, *117*, e25444.
175. Kakaawla, T. K. K.; Harrity, J. P. A. *Org. Lett.* **2018**, *20*, 201.
176. Saijo, R.; Kawase, M. *Eur. J. Org. Chem.* **2019**, 1535.
177. Avalos, M.; Babiano, R.; Bautista, I.; Fernandez, J. I.; Jiménez, J. L.; Palacios, J. C.; Plumet, J.; Rebolledo, F. *Carbohydr. Res.* **1989**, *186*, c7.
178. Lopchuk, J. M.; Gribble, G. W. *Heterocycles* **2011**, *82*, 1617.

179. Lopchuk, J. M.; Gribble, G. W. *Tetrahedron Lett.* **2011**, *52*, 4106.
180. Lopchuk, J. M.; Song, M.; Butler, B.; Gribble, G. W. *Synthesis* **2015**, *47*, 2776.
181. Gribble, G. W.; Pelkey, E. T.; Switzer, F. L. *Synlett* **1998**, 1061.
182. Gribble, G. W.; Pelkey, E. T.; Simon, W. M.; Trujillo, H. A. *Tetrahedron* **2000**, *56*, 10133.
183. Pechenkin, S. Y.; Starosotnikov, A. M.; Bastrakov, M. A.; Kachala, V. V.; Shevelev, S. A. *Mendeleev Commun.* **2012**, *22*, 35.
184. Texier, F.; Mazari, M.; Yebdri, O.; Tonnard, F.; Carrié, R. *Tetrahedron* **1990**, *46*, 3515.
185. Grassi, G.; Foti, F.; Risitano, F.; Zona, D. *Tetrahedron Lett.* **2005**, *46*, 1061.
186. Clerici, F.; Gelmi, M. L.; Trimarco, P. *Tetrahedron* **1998**, *54*, 5763.
187. Debaerdmaeker, T.; Schröer, W.-D.; Friedrichsen, W. *Liebigs Ann. Chem.* **1981**, *502*.
188. Deem, M. L. *Org. Prep. Proced. Int. New J. Org. Synth.* **1981**, *13*, 414.
189. Maryanoff, C. A.; Karash, C. B.; Turchi, I. J.; Corey, E. R.; Maryanoff, B. E. *J. Organomet. Chem.* **1989**, *54*, 3790.
190. Maryanoff, C. A.; Turchi, I. J. *Heterocycles* **1993**, *35*, 649.
191. Martin, H.-D.; Mais, F.-J.; Mayer, B.; Hecht, H.-J.; Hekman, M.; Steigel, A. *Monatsh. Chem.* **1983**, *114*, 1145.
192. Gribble, G. W.; Sponholtz, W. R., III; Switzer, F. L.; D'Amato, F. J.; Byrn, M. P. *Chem. Commun.* **1997**, 993.
193. Lopes, S. M. M.; Laranjo, M.; Serra, A. C.; Abrantes, A. M.; Rocha Gonsalves, A. M. d.'A.; Botelho, M. F.; Beja, A. M.; Silva, M. R.; Pinho e Melo, T. M. V. D. *J. Heterocycl. Chem.* **2010**, *47*, 960.
194. Brancatelli, G.; Bruno, G.; Nicolo, F.; Cordaro, M.; Grassi, G.; Risitano, F.; Scala, A. *J. Mol. Struct.* **2011**, *990*, 132.
195. Cordaro, M.; Grassi, G.; Rescifina, A.; Chiacchio, U.; Risitano, F.; Scala, A. *J. Mol. Struct.* **2011**, *991*, 143.
196. Cordaro, M.; Grassi, G.; Rescifina, A.; Chiacchio, U.; Risitano, F.; Scala, A. *Tetrahedron* **2011**, *67*, 608.
197. Cordaro, M.; Grassi, G.; Risitano, F.; Scala, A. *Tetrahedron* **2010**, *66*, 2713.
198. Baggi, P.; Clerici, F.; Gelmi, M. L.; Mottadelli, S. *Tetrahedron* **1995**, *51*, 2455.
199. Clerici, F.; Gelmi, M. L.; Soave, R.; Valle, M. *Tetrahedron* **1998**, *54*, 11285.
200. Padwa, A.; Gingrich, H. L.; Lim, R. *Tetrahedron Lett.* **1980**, *21*, 3419.
201. Padwa, A.; Gingrich, H. L.; Lim, R. *J. Organomet. Chem.* **1982**, *47*, 2447.
202. Padwa, A.; Lim, R.; MacDonald, J. G.; Gingrich, H. L.; Kellar, S. M. *J. Organomet. Chem.* **1985**, *50*, 3816.
203. Mazgarova, G. G.; Fatykhov, A. A.; Gataullin, R. R. *Russ. J. Org. Chem.* **2014**, *50*, 1346.
204. Bizhanova, G. G.; Suponitsky, K. Y.; Vakhitova, Y. V.; Gataullin, R. R. *Tetrahedron* **2017**, *73*, 109.
205. Gataullin, R. R.; Ibatullina, Z. A.; Meshcheryakova, E. S.; Fatykhov, A. A.; Khalilov, L. M. *Russ. J. Org. Chem.* **2017**, *53*, 697.
206. Aleksandrova, L. A.; Fatykhov, A. A.; Gataullina, A. R.; Kantor, E. A.; Gataullin, R. R. *Russ. J. Org. Chem.* **2018**, *54*, 1519.
207. Bélanger, G.; April, M.; Dauphin, É.; Roy, S. *J. Organomet. Chem.* **2007**, *72*, 1104.
208. Sugimoto, K.; Miyakawa, Y.; Tokuyama, H. *Tetrahedron* **2015**, *71*, 3619.
209. Berk, H. C.; Franz, J. E. *Synth. Commun.* **1981**, *11*, 267.
210. Padwa, A.; Burgess, E. M.; Gingrich, H. L.; Roush, D. M. *J. Organomet. Chem.* **1982**, *47*, 786.
211. Anderson, W. K.; Chang, C.-P.; McPherson, H. L. *J. Med. Chem.* **1983**, *26*, 1333.
212. Croce, P. D.; La Rosa, C. *Heterocycles* **1988**, *27*, 2825.
213. Croce, P. D.; Gariboldi, P.; La Rosa, C. *J. Heterocycl. Chem.* **1987**, *24*, 1793.

214. Roth, B. D.; Blankley, C. J.; Chucholowski, A. W.; Ferguson, E.; Hoefle, M. L.; Ortwine, D. F.; Newton, R. S.; Sekerke, C. S.; Sliskovic, D. R.; Stratton, C. D.; Wilson, M. W. *J. Med. Chem.* **1991**, *34*, 357.
215. Györgydeák, Z.; Szilágyi, L.; Kajtár, J.; Argay, G.; Kálmán, A. *Monatsh. Chem.* **1994**, *125*, 189.
216. Santiago, B.; Dalton, C. R.; Huber, E. W.; Kane, J. M. *J. Organomet. Chem.* **1995**, *60*, 4947.
217. Hutchison, D. R.; Nayyar, N. K.; Martinelli, M. J. *Tetrahedron Lett.* **1996**, *37*, 2887.
218. Granier, T.; Gaiser, F.; Hintermann, L.; Vasella, A. *Helv. Chim. Acta* **1997**, *80*, 1443.
219. Nayyar, N. K.; Hutchison, D. R.; Martinelli, M. J. *J. Organomet. Chem.* **1997**, *62*, 982.
220. Coppola, B. P.; Noe, M. C.; Hong, S. S.-K. *Tetrahedron Lett.* **1997**, *38*, 7159.
221. Pinho e Melo, T. M. V. D.; Gomes, C. S. B.; Gonsalves, A. M.d.'A. R.; Paixão, J. A.; Beja, A. M.; Silva, M. R.; da Veiga, L. A. *Tetrahedron* **2002**, *58*, 5093.
222. Pinho e Melo, T. M. V. D.; Gomes, C. S. B.; Soares, M. I. L.; Gonsalves, A. M.d.'A. R.; Paixão, J. A.; Beja, A. M.; Silva, M. R. *J. Heterocycl. Chem.* **2004**, *41*, 493.
223. Angle, S. R.; Qian, X. L.; Pletnev, A. A.; Chinn, J. *J. Organomet. Chem.* **2007**, *72*, 2015.
224. Park, W. K. C.; Kennedy, R. M.; Larsen, S. D.; Miller, S.; Roth, B. D.; Song, Y.; Steinbaugh, B. A.; Sun, K.; Tait, B. D.; Kowala, M. C.; Trivedi, B. K.; Auerbach, B.; Askew, V.; Dillon, L.; Hanselman, J. C.; Lin, Z.; Lu, G. H.; Robertson, A.; Sekerke, C. *Bioorg. Med. Chem. Lett.* **2008**, *18*, 1151.
225. Harju, K.; Manevski, N.; Yli-Kauhaluoma, J. *Tetrahedron* **2009**, *65*, 9702.
226. Shahbazi, S.; Kaur, J.; Kuanar, A.; Kar, D.; Singh, S.; Sobti, R. C. *Assay Drug Dev. Technol.* **2017**, *15*, 342.
227. Coppola, B. P.; Noe, M. C.; Schwartz, D. J.; Abdon, R. L., II; Trost, M. B. *Tetrahedron* **1994**, *50*, 9.
228. Antonucci, T.; Warmus, J. S.; Hodges, J. C.; Nickell, D. G. *Aniviral Chem. Chemotherapy* **1995**, *6*, 98.
229. Firoozi, N.; Torres, G. M.; Arndtsen, B. A. *J. Organomet. Chem.* **2016**, *81*, 11145.
230. Sainsbury, M.; Strange, R. H.; Woodward, P. R.; Barsanti, P. A. *Tetrahedron* **1993**, *49*, 2065.
231. Pinho e Melo, T. M. V. D.; Soares, M. I. L.; Barbosa, D. M.; Rocha Gonsalves, A. M. d.'A.; Beja, A. M.; Paixao, J. A.; Silva, M. R.; da Veiga, L. A. *Tetrahedron* **2000**, *56*, 3419.
232. Vasilescu, M.; Dumitrascu, F.; Lemmetyinen, H.; Tkachenko, N. *J. Fluoresc.* **2004**, *14*, 443.
233. Vasilescu, M.; Bandula, R.; Dumitrascu, F.; Draghici, C.; Lemmetyinen, H. *Rev. Roum. Chim.* **2004**, *49*, 905.
234. Caproiu, M. T.; Dumitrascu, F.; Caira, M. R. *Rev. Chim. (Bucharest)* **2008**, *59*, 1242.
235. Dumitrascu, F.; Caira, M. R.; Draghici, B.; Caproiu, M. T.; Dumitrescu, D. G. *Synlett* **2008**, 813.
236. Maffrand, J.-P. *Heterocycles* **1981**, *16*, 35.
237. Ladurée, D.; Lancelot, J.-C.; Robba, M. *Tetrahedron Lett.* **1985**, *26*, 1295.
238. Anderson, W. K.; Mach, R. H. *J. Med. Chem.* **1987**, *30*, 2109.
239. Lalezari, I.; Schwartz, E. L. *J. Med. Chem.* **1988**, *31*, 1427.
240. Ladurée, D.; Lancelot, J.-C.; Robba, M.; Chenu, E.; Mathé, G. *J. Med. Chem.* **1989**, *32*, 456.
241. Pinho e Melo, T. M. V. D.; Barbosa, D. M.; Ramos, P. J. R. S.; Rocha Gonsalves, A. M.d.'A.; Gilchrist, T. L.; Beja, A. M.; Paixão, J. A.; Silva, M. R.; da Veiga, L. A. *J. Chem. Soc., Perkin Trans. 1* **1999**, 1219.
242. Croce, P. D.; La Rosa, C. *Heterocycles* **2001**, *55*, 1843.
243. Pinho e Melo, T. M. V. D.; Soares, M. I. L.; Rocha Gonsalves, A. M.d.'A.; Paixão, J. A.; Beja, A. M.; Silva, M. R.; da Veiga, L. A.; Pessoa, J. C. *J. Organomet. Chem.* **2002**, *67*, 4045.

244. Castro, J.; Coteron, J. M.; Fraile, M. T.; Silvestre, G.-O.; Gomez de las Heras, F.; Martin-Cuesta, A. *Tetrahedron Lett.* **2002**, *43*, 1851.
245. Gao, D.; Zhai, H.; Parvez, M.; Back, T. G. *J. Organomet. Chem.* **2008**, *73*, 8057.
246. Fang, Y.; Larock, R. C.; Shi, F. *Asian J. Org. Chem.* **2014**, *3*, 55.
247. Lopchuk, J. M.; Gribble, G. W. *Tetrahedron Lett.* **2014**, *55*, 2809.
248. Consonni, R.; Dalla Croce, P.; Ferraccioli, R.; La Rosa, C. *J. Chem. Res.* **1991**, 7, 188.
249. Bonati, L.; Ferraccioli, R.; Moro, G. *J. Phys. Org. Chem.* **1995**, 7, 452.
250. Dalla Croce, P.; Ferraccioli, R.; La Rosa, C. *Tetrahedron* **1995**, *51*, 9385.
251. Saijo, R.; Kurihara, K.; Akira, K.; Uno, H.; Kawase, M. *Tetrahedron Lett.* **2013**, *54*, 4418.
252. Saijo, R.; Uno, H.; Kawase, M. *Heterocycles* **2016**, *92*, 2047.
253. Saijo, R.; Sekiya, H.; Tamai, E.; Kurihara, K.; Maki, J.; Sakagami, H.; Kawase, M. *Chem. Pharm. Bull.* **2017**, *65*, 365.
254. Wang, J.-K.; Zong, Y.-X.; Yue, G.-R. *Synlett* **2005**, 1135.
255. Lira, B. F.; de Filho, P. F. A.; Miller, J.; Simas, A. M.; de Dias, A. F.; Viera, M. S. *Molecules* **2002**, 7, 791.
256. Cantillo, D.; Ávalos, M.; Babiano, R.; Cintas, P.; Jiménez, J. L.; Light, M. E.; Palacios, J. C.; Rodrigues, V. *Org. Biomol. Chem.* **2010**, *8*, 5367.
257. Cantillo, D.; Ávalos, M.; Babiano, R.; Cintas, P.; Jiménez, J. L.; Light, M. E.; Palacios, J. C.; Porro, R. *J. Organomet. Chem.* **2014**, *79*, 4201.
258. Friedrichsen, W.; Schwarz, I.; Debaerdemacker, T. Z. *Naturforsch* **1982**, *37b*, 663.
259. Dalla Croce, P.; Ferraccioli, R.; La Rosa, C. *Tetrahedron* **1999**, *55*, 201.
260. Saijo, R.; Hagimoto, Y.; Kawase, M. *Org. Lett.* **2010**, *12*, 4776.
261. Saijo, R.; Kawase, M. *Tetrahedron Lett.* **2012**, *53*, 2782.
262. Kawase, M. *J. Chem. Soc., Chem. Commun.* **1994**, 2101.
263. Kawase, M.; Saito, S. *Chem. Pharm. Bull.* **2000**, *48*, 410.
264. Kawase, M.; Saito, S.; Kurihara, T. *Heterocycles* **1995**, *41*, 1617.
265. Kawase, M.; Koiwai, H.; Saito, S.; Kurihara, T. *Tetrahedron Lett.* **1998**, *39*, 6189.
266. Kawase, M.; Saito, S.; Kurihara, T. *Chem. Pharm. Bull.* **2001**, *49*, 461.
267. Saijo, R.; Uno, H.; Mori, S.; Kawase, M. *Chem. Commun.* **2016**, *52*, 8006.
268. Kawase, M.; Koiwai, H.; Yamano, A.; Miyamae, H. *Tetrahedron Lett.* **1998**, *39*, 663.
269. Kawase, M.; Koiwai, H. *Chem. Pharm. Bull.* **2008**, *56*, 433.
270. Saijo, R.; Kurihara, K.; Akira, K.; Kawase, M. *Heterocycles* **2013**, *87*, 115.
271. Lebedev, S. N.; Cherepanov, I. A.; Kalinin, V. N. *Mendeleev Commun.* **2003**, *13*, 215.
272. St. Cyr, D. J.; Martin, N.; Arndtsen, B. A. *Org. Lett.* **2007**, *9*, 449.
273. Arevalo, M. J.; Kielland, N.; Masdeu, C.; Miguel, M.; Isambert, N.; Lavilla, R. *Eur. J. Org. Chem.* **2009**, 617.
274. Perrin, S.; Monnier, K.; Laude, B.; Kubicki, M.; Blacque, O. *Eur. J. Org. Chem.* **1999**, 297.
275. Gotthardt, H.; Schoy-Tribbensee, S.; Feist, U. *Angew. Chem., Int. Ed. Engl.* **1982**, *21*, 779.
276. Gotthardt, H.; Feist, U.; Schoy-Tribbensee, S. *Chem. Ber.* **1985**, *118*, 774.
277. Gotthardt, H.; Kinzelmann, H.-G. *Chem. Ber.* **1986**, *119*, 2308.
278. Gotthardt, H.; Feist, U.; Schoy-Tribbensee, S. *Chem. Ber.* **1985**, *118*, 785.
279. Gotthardt, H.; Kinzelmann, H.-G.; Feist, U.; Buddrus, J. *Chem. Ber.* **1986**, *119*, 2317.
280. Haas, A.; Krachter, H.-U. *Chem. Ber.* **1988**, *121*, 1833.
281. Silva, A. M. S.; da Rocha, G. B.; Menezes, P. H.; Miller, J.; Simas, A. M. *J. Braz. Chem. Soc.* **2005**, *16*, 583.
282. Sponholtz, W. R., III; Trujillo, H. A.; Gribble, G. W. *Tetrahedron Lett.* **2000**, *41*, 1687.
283. Benhamou, L.; César, V.; Gornitzka, H.; Lugan, N.; Lavigne, G. *Chem. Commun.* **2009**, *45*, 4720.

284. Benhamou, L.; Vujkovic, N.; César, V.; Gornitzka, H.; Lugan, N.; Lavigne, G. *Organometallics* **2010**, *29*, 2616.
285. Hirota, T.; Katsuta, K.; Namba, T.; Sasaki, K.; Hayakawa, S. *J. Heterocycl. Chem.* **1985**, *22*, 187.
286. Nagamatsu, T.; Hantani, Y.; Sasaki, K.; Ohtomo, H.; Nakayama, T.; Hirota, T. *J. Heterocycl. Chem.* **1993**, *30*, 233.
287. Edstrom, E. D.; Wei, Y.; Gordon, M. *J. Organomet. Chem.* **1994**, *59*, 2473.
288. Abe, N.; Fukumoto, Y.; Fujii, H.; Kakehi, A. *Heterocycles* **2002**, *56*, 171.
289. Bossio, R.; Marcaccini, S.; Pepino, R.; Polo, C.; Valle, G. *Synthesis* **1989**, 641.
290. Erguven, H.; Leitch, D. C.; Keyzer, E. N.; Arndtsen, B. A. *Angew. Chem., Int. Ed.* **2017**, *56*, 6078.
291. Roy, S. A.; Zghei, J.; Zhou, C.; Arndtsen, B. A. *Chem. Sci.* **2021**, *12*, 2251.
292. Lindner, A. S.; Nieger, M.; Schmidt, A. *Tetrahedron* **2009**, *65*, 7591.
293. Pérez-Rentero, S.; Kielland, N.; Terrazas, M.; Lavilla, R.; Eritja, R. *Bioconjugate Chem.* **2010**, *21*, 1622.
294. Kielland, N.; Vendrell, M.; Lavilla, R.; Chang, Y.-T. *Chem. Commun.* **2012**, *48*, 7401.
295. Sintes, M.; De Moliner, F.; Caballero-Lima, D.; Denning, D. W.; Read, N. D.; Kielland, N.; Vendrell, M.; Lavilla, R. *Bioconjugate Chem.* **2016**, *27*, 1430.
296. Tarumi, Y.; Moriguchi, K.; Atsumi, T. *J. Heterocycl. Chem.* **1984**, *21*, 529.
297. Newton, C. G.; Ollis, W. D.; Podmore, M. L.; Wright, D. E. *J. Chem. Soc., Perkin Trans. 1* **1984**, 63.
298. Newton, C. G.; Ollis, W. D.; Wright, D. E. *J. Chem. Soc., Perkin Trans. 1* **1984**, 69.
299. Friedrichsen, W.; Schroer, W. D.; Debaerdemaeker, T. *Liebigs Ann. Chem.* **1980**, 1850.
300. Ramsden, C. A. *Adv. Heterocycl. Chem.* **2010**, *100*, 1.
301. César, V.; Mallardo, V.; Nano, A.; Dahm, G.; Lugan, N.; Lavigne, G.; Bellemin-Laponnaz, S. *Chem. Commun.* **2015**, *51*, 5271.
302. Potts, K. T.; Chen, S. J. *J. Organomet. Chem.* **1977**, *42*, 1639.
303. Oziminski, W. P.; Ramsden, C. A. *RSC Adv.* **2018**, *8*, 14833.
304. Benjelloun, A. T.; Morel, G.; Marchand, E. *Heteroat. Chem.* **2000**, *11*, 16.
305. Benjelloun, A. T.; Morel, G.; Marchand, E. *Heteroat. Chem.* **2001**, *12*, 617.
306. Morel, G. *Synlett* **2003**, 2167.
307. Edward, J. T.; Sheffler, R. H. *J. Organomet. Chem.* **1985**, *50*, 4855.
308. Jaźwiński, J.; Kamieński, B.; Staszewska-Krajewska, O.; Webb, G. *J. Mol. Struct.* **2003**, *646*, 1.
309. Hansmann, M. M.; Antoni, P. W.; Pescht, H. *Angew. Chem., Int. Ed.* **2020**, *59*, 5782.
310. Merschel, A.; Rottschafer, D.; Neumann, B.; Stammler, H.-G.; Ghadwal, R. S. *Organometallics* **2020**, *39*, 1719.
311. Merschel, A.; Glodde, T.; Neumann, B.; Stammler, H.-G.; Ghadwal, R. S. *Angew. Chem., Int. Ed.* **2021**, *60*, 2969.
312. Padwa, A.; Harring, S. R.; Hertzog, D. L.; Nadler, W. R. *Synthesis* **1994**, 993.
313. Potts, K. T.; Houghton, E.; Singh, U. P. *J. Organomet. Chem.* **1974**, *39*, 3627.
314. Avalos, M.; Babiano, R.; Cabanillas, A.; Cintas, P.; Higes, F. J.; Jimenez, J. L.; Palacios, J. C. *J. Organomet. Chem.* **1996**, *61*, 3738.
315. Potts, K. T.; Chen, S. J.; Kane, J.; Marshall, J. L. *J. Organomet. Chem.* **1977**, *42*, 1633.
316. Hertzog, D. L.; Nadler, W. R.; Zhang, Z. J.; Padwa, A. *Tetrahedron Lett.* **1992**, *33*, 5877.
317. Michael, J. P.; de Koning, C. B.; van der Westhuyzen, C. W.; Fernandes, M. A. *J. Chem. Soc., Perkin Trans. 1* **2001**, 2055.
318. Wamhoff, H.; Hohmann, C.; Sohar, P. *Org. Lett.* **2000**, *2*, 581.
319. Gardner, J. O.; Beard, C. C.; Rotstein, D. M. *J. Organomet. Chem.* **1982**, *47*, 1136.
320. Fedotov, K. V.; Romanov, N. N.; Tolmachev, A. I. *Chem. Heterocycl. Compd.* **1983**, *19*, 492.

321. Walker, K. A. M.; Sjogren, E. B.; Matthews, T. R. *J. Med. Chem.* **1985**, *28*, 1673.
322. Fedotov, K. V.; Romanov, N. N. *Chem. Heterocycl. Compd.* **1989**, *25*, 1399.
323. Santagati, A.; Modica, M.; Scolaro, L. M.; Santagatia, M. *J. Chem. Res.* **1999**, 86.
324. Wasfy, A. A. F. *Heteroat. Chem.* **2003**, *14*, 576.
325. Fedotov, K. V.; Shumeiko, T. P.; Romanov, N. N.; Tolmachev, A. I.; Ilchenko, A. Y. *Chem. Heterocycl. Compd.* **2011**, *47*, 622.
326. Matsumoto, S.; Kimura, K.; Sano, T.; Akazome, M.; Ogura, K. *J. Photopolym. Sci. Technol.* **2009**, *22*, 677.
327. Padwa, A.; Kinder, F. R.; Zhi, L. *Synlett* **1991**, 287.
328. Padwa, A.; Kinder, F. R.; Nadler, W. R.; Zhi, L. *Heterocycles* **1993**, *35*, 367.
329. Moody, C. J.; Slawin, A. M. Z.; Willows, D. *Org. Biomol. Chem.* **2003**, *1*, 2716.
330. Potts, K. T.; Murphy, P. *J. Chem. Soc., Chem. Commun.* **1984**, 1348.
331. Gotthardt, H.; Pflaumbaum, W. *Chem. Ber.* **1987**, *120*, 1017.
332. Heidelbaugh, T. M.; Liu, B.; Padwa, A. *Tetrahedron Lett.* **1998**, *39*, 4757.
333. Cavalleri, P.; Clerici, F.; Erba, E.; Trimarco, P. *Chem. Ber.* **1992**, *125*, 883.
334. Arevalo, M. J.; Avalos, M.; Babiano, R.; Cintas, P.; Hursthouse, M. B.; Jimenez, J. L.; Light, M. E.; Palacios, J. C. *Tetrahedron: Asymmetry* **2002**, *13*, 223.
335. Arevalo, M. J.; Avalos, M.; Babiano, R.; Cintas, P.; Jimenez, J. L.; Light, M. E.; Palacios, J. C. *Tetrahedron* **2006**, *62*, 6909.
336. Padwa, A.; Beall, L. S.; Heidelbaugh, T. M.; Liu, B.; Sheehan, S. M. *J. Organomet. Chem.* **2000**, *65*, 2684.
337. Hamid, A.; Oulyadi, H.; Daich, A. *Tetrahedron* **2006**, *62*, 6398.
338. Padwa, A.; Zhang, Z. J. *Heterocycles* **1994**, *37*, 441.
339. De la Concepción, J. G.; Avalos, M.; Cintas, P.; Jimenez, J. L.; Light, M. E. *Org. Biomol. Chem.* **2018**, *16*, 4778.
340. Areces, P.; Avalos, M.; Babiano, R.; Cintas, P.; Gonzalez, L.; Hursthouse, M. B.; Jimenez, J. L.; Light, M. E.; Lopez, I.; Palacios, J. C.; Silvero, G. *Eur. J. Org. Chem.* **2001**, 2135.
341. Baudy, M.; Robert, A.; Guimon, C. *Tetrahedron* **1982**, *38*, 1241.
342. Avalos, M.; Babiano, R.; Cintas, P.; Diaz, J.; Hursthouse, M. B.; Jimenez, J. L.; Light, M. E.; Lopez, I.; Palacios, J. C. *Tetrahedron Lett.* **2003**, *44*, 4657.
343. Avalos, M.; Babiano, R.; Cabanillas, A.; Cintas, P.; Dianez, M. J.; Estrada, M. D.; Jimenez, J. L.; Lopez-Castro, A.; Palacios, J. C.; Garrido, S. P. *J. Chem. Soc., Chem. Commun.* **1995**, 2213.
344. Kato, H.; Kobayashi, T.; Ciobanu, M.; Kakehi, A. *Tetrahedron* **1997**, *53*, 9921.
345. Avalos, M.; Babiano, R.; Cintas, P.; Clemente, F. R.; Gordillo, R.; Jimenez, J. L.; Palacios, J. C.; Raithby, P. R. *J. Organomet. Chem.* **2000**, *65*, 5089.
346. Avalos, M.; Babiano, R.; Cintas, P.; Clemente, F. R.; Gordillo, R.; Jimenez, J. L.; Palacios, J. C. *J. Organomet. Chem.* **2001**, *66*, 5139.
347. Potts, K. T.; Dery, M. O.; Juzukonis, W. A. *J. Organomet. Chem.* **1989**, *54*, 1077.
348. Potts, K. T.; Dery, M. O. *J. Chem. Soc., Chem. Commun.* **1986**, 561.
349. De la Concepcion, J. G.; Avalos, M.; Cintas, P.; Jimenez, J. L.; Light, M. E. *Org. Biomol. Chem.* **2018**, *16*, 3438.
350. Baudy, M.; Robert, A.; Guimon, C. *Tetrahedron* **1982**, *38*, 2129.
351. Arevalo, M. J.; Avalos, M.; Babiano, R.; Cintas, P.; Hursthouse, M. B.; Jimenez, J. L.; Light, M. E.; Lopez, I.; Palacios, J. C. *Tetrahedron* **2000**, *56*, 1247.
352. Ill-Yun, J.; Yong-Jin, Y.; Seo, R. K.; Cheol, S. G.; Chul, S. S. *Chem. Lett.* **1994**, 859.
353. Potts, K. T.; Bordeaux, K. G.; Kuehnling, W. R.; Salsbury, R. L. *J. Organomet. Chem.* **1985**, *50*, 1666.
354. Fischer, U.; Mohler, H.; Schneider, F.; Widmer, U. *Helv. Chim. Acta* **1990**, *73*, 763.
355. Areces, P.; Avalos, M.; Babiano, R.; Gonzalez, L.; Jimenez, J. L.; Palacios, J. C.; Pilo, M. D. *Carbohydr. Res.* **1991**, *222*, 99.

356. Santagati, A.; Granata, G.; Santagati, M. *J. Chem. Res.* **2001**, *8*, 304.
357. De la Concepcion, J. G.; Avalos, M.; Babiano, R.; Cintas, P.; Jimenez, J. L.; Light, M. E.; Palacios, J. C. *Tetrahedron* **2017**, *73*, 1551.
358. Areces, P.; Avalos, M.; Babiano, R.; Gonzalez, L.; Jimenez, J. L.; Mendez, M. M.; Palacios, J. C. *Tetrahedron Lett.* **1993**, *34*, 2999.
359. Ávalos, M.; Babiano, R.; Cintas, P.; Hursthouse, M. B.; Jiménez, J. L.; Light, M. E.; Lopez, I.; Palacios, J. C. *Chem. Commun.* **1999**, 1589.
360. Avalos, M.; Babiano, R.; Cintas, P.; Hursthouse, M. B.; Jimenez, J. L.; Light, M. E.; Lopez, I.; Palacios, J. C.; Silvero, G. *Chem. – Eur. J.* **2001**, 7, 3033.
361. Avalos, M.; Babiano, R.; Cintas, P.; Clemente, F. R.; Gordillo, R.; Hursthouse, M. B.; Jimenez, J. L.; Light, M. E.; Palacios, J. C. *Tetrahedron: Asymmetry* **2001**, *12*, 2265.
362. Avalos, M.; Babiano, R.; Cintas, P.; Clemente, F. R.; Gordillo, R.; Jimenez, J. L.; Palacios, J. C. *J. Organomet. Chem.* **2003**, *68*, 6338.
363. Sheradsky, T.; ltzhak, N. *J. Chem. Soc., Perkin Trans. 1* **1987**, 1979.
364. Sheradsky, T.; ltzhak, N. *J. Chem. Soc., Perkin Trans. 1* **1989**, 33.
365. Sanchez, B.; Bravo, J. L.; Arévalo, M. J.; López, I.; Light, M. E.; Silvero, G. *Aust. J. Chem.* **2009**, *62*, 356.
366. Arevalo, M. J.; Avalos, M.; Babiano, R.; Cintas, P.; Hursthouse, M. B.; Jimenez, J. L.; Light, M. E.; Lopez, I.; Palacios, J. C. *Tetrahedron Lett.* **1999**, *40*, 8675.
367. Sanchez, B.; Lopez, I.; Light, M. E.; Silvero, G.; Bravo, J. L. *Eur. J. Org. Chem.* **2010**, 1648.
368. De la Concepción, J. G.; Avalos, M.; Jimenez, J. L.; Cintas, P. *Org. Biomol. Chem.* **2020**, *18*, 6328.
369. Sheradsky, T.; ltzhak, N. *J. Chem. Soc., Perkin Trans. 1* **1992**, 2389.
370. Cantillo, D.; Avalos, M.; Babiano, R.; Cintas, P.; Jimnez, J. L.; Light, M. E.; Palacios, J. C. *Org. Lett.* **2008**, *10*, 1079.
371. Cantillo, D.; Avalos, M.; Babiano, R.; Cintas, P.; Jimenez, J. L.; Light, M. E.; Palacios, J. C. *J. Organomet. Chem.* **2009**, *74*, 3698.
372. Cantillo, D.; Avalos, M.; Babiano, R.; Cintas, P.; Jimenez, J. L.; Light, M. E.; Palacios, J. C. *J. Organomet. Chem.* **2009**, *74*, 7644.
373. Avalos, M.; Babiano, R.; Cintas, P.; Diaz, J.; Palacios, J. C.; Jimenez, J. L.; Lopez, I.; Palacios, J. C. *Eur. J. Org. Chem.* **2004**, 2805.
374. Avalos, M.; Babiano, R.; Cintas, P.; Diaz, J.; Jimenez, J. L.; Lopez, I.; Palacios, J. C. *Tetrahedron* **2006**, *62*, 11979.
375. Zaleska, B.; Ciez, D.; Falk, H. *Monatsh. Chem.* **1996**, *127*, 1251.
376. Kato, H.; Wakao, K.; Yamada, A.; Kojima, M. *J. Chem. Soc., Chem. Commun.* **1984**, 1558.
377. Kato, H.; Wakao, K.; Yamada, A.; Mutoh, Y. *J. Chem. Soc., Perkin Trans. 1* **1988**, 189.
378. Sheradsky, T.; Zbaida, D. *J. Heterocycl. Chem.* **1983**, *28*, 245.
379. Benjelloun, A. T.; Morel, G.; Marchand, E. Y. *Phosphorus, Sulfur Silicon Relat. Elem.* **1998**, *133*, 195.
380. Barton, D. H. R.; Liu, W. *Tetrahedron Lett.* **1997**, *38*, 2435.
381. Souizi, A.; Robert, A. *Synthesis* **1982**, 1059.
382. Bssaibis, M.; Robert, A.; Lemaguerès, P.; Ouahab, L.; Carlier, R.; Tallec, A. *J. Chem. Soc., Chem. Commun.* **1993**, 601.
383. Bssaibis, M.; Robert, A.; Souizi, A. A. *J. Chem. Soc., Perkin Trans. 1* **1994**, 1469.
384. Takezawa, K.; Nomura, K.; Yoshii, E.; Masuda, K. *Chem. Pharm. Bull.* **1984**, *32*, 4637.
385. Hagimoto, Y.; Saijo, R.; Kawase, M. *Heterocycles* **2014**, *89*, 709.
386. Kato, H.; Kobayashi, T.; Horie, K.; Oguri, K.; Moriwaki, M. *J. Chem. Soc., Perkin Trans. 1* **1993**, 1055.
387. Kato, H.; Kobayashi, T.; Tokue, K.; Shirasawa, S. *J. Chem. Soc., Perkin Trans. 1* **1993**, 1617.

388. Friedrichsen, W.; Schwarz, I.; Epe, B.; Hesse, K.-F. Z. *Naturforsch* **1981**, *36b*, 622.
389. Moura, G. L. C.; Simas, A. M. *Chem. Mater.* **2008**, *20*, 4142.
390. Berrée, F.; Malvaut, Y.; Marchand, E.; Morel, G. *J. Organomet. Chem.* **1993**, *58*, 6022.
391. Berrée, F.; Morel, G. *Tetrahedron* **1995**, *51*, 7019.
392. Morel, G.; Marchand, E.; Benjelloun, A. T.; Sinbandhit, S.; Guillou, O.; Gall, P. *Eur. J. Org. Chem.* **1998**, 2631.
393. Babaev, E. V.; Bush, A. A.; Orlova, I. A.; Rybakov, V. B.; Iwataki, I. *Russ. Chem. Bull.* **2005**, *54*, 231.
394. Shahrekipour, F.; Heydari, R.; Tahamipour, B.; Saravani, H.; Graiff, C. *Phosphorus, Sulfur Silicon Relat. Elem.* **2014**, *189*, 263.
395. Maity, A.; Chakraborty, D.; Hazra, A.; Bharitkar, Y. P.; Kundu, S.; Maulik, P. R.; Mondal, N. B. *Tetrahedron Lett.* **2014**, *55*, 3059.
396. De Athayde-Filho, P. F.; Miller, J.; Simas, A. M.; Lira, B. F.; De Souza Luis, J. A.; Zuckerman-Schpector, J. *Synthesis* **2003**, 685.
397. Lira, B. F.; Miller, J.; Simas, A. M.; de Athayde-Filho, P. F.; Dias, A. F.; Silva, R. O.; de Oliveira, V. C. *ARKIVOC* **2004**, *vi*, 12.
398. Lyra, B. F.; de Morais, S. A.; Rocha, G. B.; Miller, J.; Moura, G. L.; Simas, A. M.; Peppe, C.; de Athayde-Filho, P. F. *J. Braz. Chem. Soc.* **2010**, *21*, 934.
399. Babaev, E. V.; Rybakov, V. B.; Orlova, I. A.; Bush, A. A.; Maerle, K. V.; Nasonov, A. F. *Russ. Chem. Bull.* **2004**, *53*, 176.
400. Zama, S.; Bouraiou, A.; Bouacida, S.; Roisnel, T.; Belfaitah, A. *Tetrahedron Lett.* **2013**, *54*, 5605.
401. Moura, G. L. C.; Simas, A. M.; Miller, J. *Chem. Phys. Lett.* **1996**, *257*, 639.
402. Rakov, N.; Araujo, C. B.; Athayde-Filho, P. F.; da Rocha, G. B.; Miller, J.; Simas, A. M. *Chem. Phys. Lett.* **2000**, *332*, 13.
403. Rakov, N.; de Araujo, C. B.; Rocha, G. B.; Simas, A. M.; de Athayde-Filho, P. A. F.; Miller, J. *Appl. Opt.* **2001**, *40*, 1389.
404. De Araújo, C. B.; Gomes, A. S. L.; Borissevitch, I. E. *Mol. Cryst. Liq. Cryst.* **2002**, *378*, 113.
405. De Araujo, C. B.; Gomes, A. S. L.; Borissevitch, I. E. *Mol. Cryst. Liq. Cryst.* **2002**, *374*, 357.
406. Pilla, V.; de Araújo, C. B.; Lira, B. F.; Simas, A. M.; Miller, J.; Athayde-Filho, P. F. *Opt. Commun.* **2006**, *264*, 225.
407. Bosco, C. A. C.; Maciel, G. S.; Rakov, N.; de Araújo, C. B.; Acioli, L. H.; Simas, A. M.; de Athayde-Filho, P. F.; Miller, J. *Chem. Phys. Lett.* **2007**, *449*, 101.
408. de Morais, S. A.; da Silva Morais, C. R.; de Athayde Filho, P. F.; Lira, B. F.; do Nascimento, R. S. T. R. *J. Therm. Anal. Calorim.* **2009**, *97*, 437.
409. de Morais, S. A.; da Morais, S. C. R.; de Filho, P. F. A.; Lira, B. F.; de Souza, M. A. F. *J. Alloys Compd.* **2010**, *495*, 598.
410. Barbosa-Silva, R.; Nogueira, M.A.M.; Souza, H.D.S.; Lira, B. F; de Athayde-Filho, P.F.; de Araújo, C.B. *J. Phys. Chem. C* **2019**, *123*, 677.
411. Liu, Z.; Tian, Z.; Lu, T.; Hua, S. *J. Phys. Chem. A* **2020**, *124*, 5563.
412. Hua, S.; Liu, Z.; Lu T. ChemRxiv. n.d., 10.26434/chemrxiv.13525505.
413. Cavalcante, K. V. M.; de Azevedo Correia, N.; Dias, K. L. G.; Silva, D. F.; Silva-Filho, J. C.; Araújo, I. G. A.; Lira, B. F.; de Athayde-Filho, P. F.; Medeiros, I. A. *J. Pharmacol. Sci.* **2009**, *110*, 29.
414. Luis, J. A. S.; De Aquino, T. M.; Lira, B. F.; Filho, P. F. A.; Scotti, M. T.; Scotti, L.; De Moura, R. O.; Junior, F. J. B. M. *Acta Pharm.* **2014**, *64*, 233.
415. Peixoto, I. N.; Souza, H. D. S.; Lira, B. F.; Silva, D. F.; Lima, E. O.; Barbosa-Filho, J. M.; Athayde-Filho, P. F. *J. Braz. Chem. Soc.* **2016**, *27*, 1807.

416. de Mascena Costa, L. A.; Harmon, A. C.; Teixeira, A. A. C.; de Lima, F. C. S.; de Sousa Araújo, S.; Del Piero, F.; da Silva Souza, H. D.; de Athayde Filho, P. F.; Junior, S. A.; de Mascena Diniz Maia, M.; Wischral, A.; Filho, M. A. G.; Mathis, J. M. *Breast Cancer Basic Clin. Res.* **2020**, *14*, 1.
417. de Mascena Costa, L. A.; Debnath, D.; Harmon, A. C.; de Sousa Araújo, S.; da Silva Souza, H. D.; de Athayde Filho, P. F.; Wischral, A.; Filho, M. A. G.; Mathis, J. M. *Oncol. Lett.* **2020**, *20*, 2291.
418. Gotthardt, H.; Oppermann, M. *Chem. Ber.* **1986**, *119*, 2094.
419. Rybakov, V. B.; Bush, A. A.; Troyanov, S. I.; Babaev, E. V.; Kemnitz, E. *Acta Crystallogr.* **2006**, *E62*. o1673.
420. Gotthardt, H.; Pflaumbaum, W. *Chem. Ber.* **1987**, *120*, 61.
421. Gotthardt, H.; Pflaumbaum, W. *Chem. Ber.* **1987**, *120*, 411.
422. Kumar, R. A.; Pattanayak, M. R.; Yen-Pon, E.; Eliyan, J.; Porte, K.; Bernard, S.; Riomet, M.; Thuery, P.; Audisio, D.; Taran, F. *Angew. Chem., Int. Ed.* **2019**, *58*, 14544.
423. Tono, M.; Aoto, H.; Matsudaira, Y.; Sugiyama, T.; Kajitani, M.; Akiyama, T.; Sugimori, A. *Tetrahedron Lett.* **1991**, *32*, 4023.
424. Gotthardt, H.; Huss, O. M. *Liebigs Ann. Chem.* **1981**, 347.
425. Anderson, W. K.; Jones, A. N. *J. Med. Chem.* **1984**, *27*, 1559.
426. Gotthardt, H.; Böhm, F.-R.; Brauer, D. J.; Weisshuhn, C. M.; Wilke, C. J. *Chem. Ber.* **1988**, *121*, 95.
427. Gotthardt, H.; Pflaumbaum, W.; Gutowski, P. *Chem. Ber.* **1988**, *121*, 313.
428. Meazza, G.; Zanardi, G.; Guglielmetti, G.; Piccardi, P. *J. Fluor. Chem.* **1997**, *82*, 175.
429. Rosch, W.; Richter, H.; Regitz, M. *Chem. Ber.* **1987**, *120*, 1809.
430. Markl, G.; Eckl, E.; Jakobs, U.; Ziegler, M. L.; Nuber, B. *Tetrahedron Lett.* **1987**, *28*, 2119.
431. Kato, H.; Aoki, N.; Kawamura, Y.; Yoshino, K. *J. Chem. Soc., Perkin Trans. 1* **1985**, 1245.
432. Kato, H.; Kobayashi, T.; Ciobanu, M.; Iga, H.; Akutsu, A.; Kakehi, A. *Chem. Commun.* **1996**, 1011.
433. Kato, H.; Toda, S. *Chem. Commun.* **1982**, 510.
434. Kato, H.; Arikawa, Y.; Hashimoto, M.; Masuzawa, M. *J. Chem. Soc., Chem. Commun.* **1983**, 938.
435. Kobayashi, T.; Minemura, H.; Kato, H. *Heterocycles* **1995**, *40*, 311.
436. Kawamura, Y.; Akaishi, K.; Nishiuchi, M.; Tsukayama, M. *Mol. Cryst. Liq. Cryst.* **2002**, *376*, 219.
437. Friedrichsen, W.; Schröer, W.-D.; Schwarz, I.; Böttcher, A. Z. *Naturforsch* **1981**, *36b*, 609.
438. Gotthardt, H.; Oppermann, M. *Tetrahedron Lett.* **1985**, *26*, 1627.
439. Bertho, F.; Robert, A.; Batail, P.; Robin, P. *Tetrahedron* **1990**, *46*, 433.
440. Bssaibis, M.; Robert, A.; Souizi, A. *J. Chem. Soc., Chem. Commun.* **1993**, 998.
441. Souizi, A.; Robert, A. *Tetrahedron* **1984**, *40*, 1817.
442. Sugimori, A.; Matsudaira, Y.; Tono, M.; Kajitani, M.; Akiyama, T.; Sugiyama, T. *Chem. Lett.* **1991**, *20*, 269.
443. Hiroshi, K.; Kazue, T.; Haruki, E.; Yoshiko, T. *Chem. Lett.* **1980**, *9*, 717.
444. Gotthardt, H.; Huss, O. M.; Schoy-Tribbensee, S. *Chem. Ber.* **1981**, *114*, 285.
445. Kato, H.; Tani, K.; Kurumisawa, H.; Tamura, Y. *Heterocycles* **1987**, *26*, 1313.
446. Kato, H.; Shiba, T.; Aoki, N.; Iijima, H. *J. Chem. Soc., Perkin Trans. 1* **1982**, 1885.
447. Jorgensen, M.; Lerstrup, K.; Bechgaard, K. *Synth. Met.* **1991**, *41–43*, 2561.
448. Jorgensen, M.; Lerstrup, K. A.; Bechgaard, K. *J. Organomet. Chem.* **1991**, *56*, 5684.

449. Oliveira, M. B.; Miller, J.; Pereira, A. B.; Galembeck, S. E.; Moura, G. L. C.; Simas, A. M. *Phosphorus, Sulfur Silicon Relat. Elem.* **1996**, *108*, 75.
450. de Almeida, P. A.; Sarmento da Silva, T. M.; Echevarria, A. *Heterocycl. Commun.* **2002**, *8*, 593.
451. Cheung, K.-K.; Galembeck, S.; Miller, J.; de Oliveira, M. B.; Pereira, A. B.; Simas, A. M. *Acta Crystallogr.* **1992**, *C48*, 523.
452. Cheung, K.-K.; Galembeck, S.; Miller, J.; de Oliveira, M. B.; Pereira, A. B.; Simas, A. M. *Acta Crystallogr.* **1991**, *C47*, 2630.
453. Hjorth, M.; Thorup, N. *Acta Crystallogr. C* **1991**, *C47*, 1548.
454. Bellec, N.; Lorcy, D.; Robert, A. *Synthesis* **1998**, 1442.
455. Bertho, F.; Talham, D.; Robert, A.; Batail, P.; Megtert, S.; Robin, P. *Mol. Cryst. Liq. Cryst.* **1988**, *156*, 339.
456. Bertho-Thoraval, F.; Robert, A.; Souizi, A.; Boubekeur, K.; Batail, P. *J. Chem. Soc., Chem. Commun.* **1991**, 843.
457. Lorey, D.; Robert, A.; Triki, S.; Ouahab, L.; Robin, P.; Chastaing, E. *Tetrahedron Lett.* **1992**, *33*, 7341.
458. Le Paillard, M. P.; Robert, A. *Bull. Soc. Chim. Fr.* **1992**, *129*, 205.
459. Fourmigue, M.; Johannsen, I.; Boubekeur, K.; Nelson, C.; Batail, P. *J. Am. Chem. Soc.* **1993**, *115*, 3752.
460. Le Paillard, M. P.; Robert, A.; Garrigou-Lagrange, C.; Delhaes, P.; Le Magueres, P.; Ouahab, L.; Toupet, L. *Synth. Met.* **1993**, *58*, 223.
461. Lorcy, D.; Le Paillard, M.-P.; Robert, A. *Tetrahedron* **1993**, *34*, 5289.
462. Boubekeur, K.; Lenoir, C.; Batail, P.; Carlier, R.; Tallec, A.; Le Paillard, M.-P.; Lorcy, D.; Robert, A. *Angew. Chem., Int. Ed. Engl.* **1994**, *33*, 1379.
463. Jørgensen, M.; Bechgaard, K.; Bjoernholm, T.; Sommer-Larsen, P.; Hansen, L. G.; Schaumburg, K. *J. Organomet. Chem.* **1994**, *59*, 5877.
464. Abashev, G. G.; Shklyaeva, E. V.; Russkikh, V. S. *Russ. J. Org. Chem.* **1997**, *33*, 1652.
465. Dahlstedt, E.; Hellberg, J. *Synth. Met.* **2001**, *119*, 181.
466. Abashev, G. G.; Shklyaeva, E. V.; Tenishev, A. G.; Kazheva, O. N.; Shilov, G. V.; Dyachenko, O. A. *Synth. Met.* **2003**, *133–134*, 329.
467. Hellberg, J.; Dahlstedt, E.; Woldegiorgis, A. *Eur. J. Org. Chem.* **2004**, 1455.
468. Dahlstedt, E.; Hellberg, J.; Petoral, R. M., Jr.; Uvdal, K. *J. Mater. Chem.* **2004**, *14*, 81.
469. Abashev, G. G.; Bushueva, A. Y.; Lebedev, K. Y.; Shklyaeva, E. V. *Russ. J. Org. Chem.* **2007**, *43*, 135.
470. Guerro, M.; Pham, N. H.; Massue, J.; Bellec, N.; Lorcy, D. *Tetrahedron* **2008**, *64*, 5285.
471. Montanari, C. A.; Giesbrecht, A. M.; Sandall, J. P. B.; Miyata, Y.; Miller, J. *Heterocycl. Commun.* **1996**, *2*, 71.
472. Gibson, H. H., Jr.; Weissinger, K.; Abashawl, A.; Hall, G.; Lawshae, T.; LeBlanc, K.; Moody, J.; Lwowski, W. *J. Organomet. Chem.* **1986**, *51*, 3858.
473. Alajarin, M.; Molina, P. *Tetrahedron Lett.* **1980**, *21*, 4025.
474. Molina, P.; Alajarin, M.; Arques, A.; Benzal, R. *J. Chem. Soc., Perkin Trans. 1* **1982**, 351.
475. Andronati, S. A.; Yavorskii, A. S.; Simonov, Y. A.; Pavlovskii, V. I.; Dvorkin, A. A.; Chumakov, Y. M. *Zh. Org. Khim.* **1993**, *29*, 1582.
476. Dessolin, M.; Golfier, M.; Prange, T. *J. Organomet. Chem.* **1985**, *50*, 4461.
477. Molina, P.; Alajarin, M.; Arques, A.; Benzal, R.; Hernandez, H. *Tetrahedron Lett.* **1983**, *24*, 3523.
478. Molina, P.; Alajarin, M.; Arques, A.; Benzal, R.; Hernandez, H. *J. Chem. Soc., Perkin Trans. 1* **1984**, 1891.
479. Aranda, G.; Dessolin, M.; Golfier, M.; Guillerez, M.-G. *Org. Magn. Reson.* **1982**, *18*, 159.
480. Dahn, H.; Ung-Truong, M. N. *Helv. Chim. Acta* **1988**, *71*, 241.

481. Bocian, W.; Jaźwiński, J.; Staszewska, O.; Wiench, J. W.; Stefaniak, L.; Webb, G. A. *Chem. Heterocycl. Compd.* **1996**, *32*, 1358.
482. De la Concepción, J. G.; Ávalos, M.; Cintas, P.; Jiménez, J. L. *Chem. – Eur. J.* **2018**, *24*, 7507.
483. Molina, P.; Alajarin, M.; de Vega, M. J. P.; de la Foces-Foces, C. M.; Cano, F. *Chem. Ber.* **1988**, *121*, 1495.
484. Jeon, D. J.; Shin, S. C.; Lee, Y. Y. *Heterocycles* **1997**, *45*, 2101.
485. Abdualkader, A. M.; Taher, M.; Yusoff, N. I. N. *Int. J. Pharm. Pharm. Sci.* **2017**, *9*, 1.
486. Albota, F.; Stanescu, M. D. *Rev. Roum. Chim.* **2017**, *62*, 711.
487. Stuhlmann, F.; Domschke, G.; Bartl, A.; Omelka, L. *Magn. Reson. Chem.* **1998**, *36*, 8.
488. Araki, S.; Mizuya, J.; Butsugan, Y. *Chem. Lett.* **1984**, *13*, 1045.
489. Araki, S.; Mizuya, J.; Butsugan, Y. *J. Chem. Soc., Perkin Trans. 1* **1985**, 2439.
490. Araki, S.; Mizuya, J.; Aoyama, N.; Butsugan, Y. *J. Chem. Soc., Perkin Trans. 1* **1995**, 1989.
491. Gotthardt, H.; Reiter, F.; Kromer, C. *Liebigs Ann. Chem.* **1981**, 1025.
492. Harrit, N.; Holm, A.; Dunkin, I. R.; Poliakoff, M.; Turner, J. J. *J. Chem. Soc., Perkin Trans. 2* **1987**, 1227.
493. Lu, D.; Nadas, J.; Zhang, G.; Johnson, W.; Zweier, J. L.; Cardounel, A. J.; Villamena, F. A.; Wang, P. G. *J. Am. Chem. Soc.* **2007**, *129*, 5503.
494. Eilertsen, M.; Allin, S. M.; Pearson, R. J. *Bioorg. Med. Chem. Lett.* **2018**, *28*, 1106.
495. Kato, H.; Toda, S.; Arikawa, Y.; Masuzawa, M.; Hashimoto, M.; Ikoma, K.; Wang, S.-Z.; Miyasaka, A. *J. Chem. Soc., Perkin Trans. 1* **1990**, 2035.
496. Kato, A.; Tono, M.; Hisamatsu, N.; Nozawa, S.; Sugiyama, T.; Kajitani, M.; Takeo, A.; Sugimori, A. *Chem. Lett.* **1992**, *21*, 243.
497. Cui, H.; Shen, J.; Lu, D.; Zhang, T.; Zhang, W.; Sun, D.; Wang, P. G. *Cancer Chemother. Pharmacol.* **2008**, *62*, 509.
498. Shafran, E. A.; Bakulev, V. A.; Rozin, Y. A.; Shafran, Y. M. *Chem. Heterocycl. Compd.* **2008**, *44*, 1040.
499. Abbott, P. A.; Bonnert, R. V.; Caffrey, M. V.; Cage, P. A.; Cooke, A. J.; Donald, D. K.; Furber, M.; Hill, S.; Withnall, J. *Tetrahedron* **2002**, *58*, 3185.
500. Nein, Y. I.; Pospelova, T. A.; Bakulev, V. A.; Morzherin, Y. Y. *Chem. Heterocycl. Compd.* **2005**, *41*, 940.
501. Nein, Y. I.; Gladkova, S. V.; Pospelova, T. A.; Morzherin, Y. Y. *Chem. Heterocycl. Compd.* **2006**, *42*, 1472.
502. Nein, I. Y.; Morzherin, Y. Y.; Rozin, Y. A.; Bakulev, V. A. *Chem. Heterocycl. Compd.* **2002**, *38*, 1144.
503. Rozin, Y. A.; Savel'eva, E. A.; Morzherin, Y. Y.; Dehaen, W.; Toppet, S.; Van Meervelt, L.; Bakulev, V. A. *J. Chem. Soc., Perkin Trans. 1* **2002**, 211.
504. Nein, Y. I.; Polyakova, A. Y.; Morzherin, Y. Y.; Savel'eva, E. A.; Rozin, Y. A.; Bakulev, V. A. *Russ. J. Org. Chem.* **2004**, *40*, 879.
505. Savel'eva, E. A.; Rozin, Y. A.; Kodess, M. I.; Van Meervelt, L.; Dehaen, W.; Morzherin, Y. Y.; Bakulev, V. A. *Tetrahedron* **2004**, *60*, 5367.
506. Nein, Y. I.; Savel'eva, E. A.; Rozin, Y. A.; Bakulev, V. A.; Morzherin, Y. Y. *Chem. Heterocycl. Compd.* **2006**, *42*, 412.
507. Nein, Y. I.; Morzherin, Y. Y. *Chem. Heterocycl. Compd.* **2014**, *50*, 1021.
508. Khazhieva, I. S.; Demkin, P. M.; Nein, J. I.; Glukhareva, T. V.; Morzherin, Y. Y. *Chem. Heterocycl. Compd.* **2015**, *51*, 199.
509. Nein, Y. I.; Glukhareva, T. V.; Sadovskova, K. A.; Morzherin, Y. Y. *Chem. Heterocycl. Compd.* **2016**, *52*, 716.
510. Khazhieva, I. S.; Demkin, P. M.; Eltsov, O. S.; Nein, Y. I.; Kalinina, T. A.; Fan, Z.-J.; Chen, L.; Zhao, B.; Glukhareva, T. V.; Morzherin, Y. Y. *Chem. Heterocycl. Compd.* **2018**, *54*, 956.

511. Petronilho, A.; Müller-Bunz, H.; Albrecht, M. *Chem. Commun.* **2012**, *48*, 6499.
512. Xu, X.; Li, L.; Zhang, Z.; Yan, X. *Tetrahedron* **2018**, *74*, 6846.
513. Zhao, S.; Yu, R.; Chen, W.; Liu, M.; Wu, H. *Org. Lett.* **2015**, *17*, 2828.
514. Tsoleridis, C. A.; Alexandrou, N. E. *J. Chem. Soc., Perkin Trans. 2* **1985**, 1123.
515. Tsoleridis, C. A.; Alexandrou, N. E. *Magn. Reson. Chem.* **1985**, *23*, 842.
516. Stefaniak, L.; Jaźwiński, J. *Chem. Heterocycl. Compd.* **1995**, *31*, 1027.
517. Bocian, W.; Wiench, J. W.; Stefaniak, L.; Webb, G. A. *Magn. Reson. Chem.* **1996**, *34*, 453.
518. Wall, R. J.; Bell, D. R.; Bazzi, R.; Fernandes, A.; Rose, M.; Rowlands, J. C.; Mellor, I. R. *Toxicology* **2012**, *302*, 140.
519. Bouchemella, K.; Fauché, K.; Anak, B.; Jouffret, L.; Bencharif, M.; Cisnetti, F. *New J. Chem.* **2018**, *42*, 18969.
520. Papageorgiou, M.; Kokkou, S. C.; Rentzeperis, P. J.; Tsoleridis, C. *Acta Crystallogr.* **1983**, *C39*, 1581.
521. Ohmatsu, K.; Suzuki, R.; Furukawa, Y.; Sato, M.; Ooi, T. *ACS Catal.* **2020**, *10*, 2627.
522. Vaddamanu, M.; Velappan, K.; Prabusankar, G. *New J. Chem.* **2020**, *44*, 129.
523. Wang, H.; Zhang, B.; Yan, X.; Guo, S. *Dalton Trans.* **2018**, *47*, 528.
524. Yan, X.; Zhang, B.; Zhang, X.; Wang, H.; Duan, Y.; Guo, S. *Appl. Organomet. Chem.* **2020**, e5885.
525. Zhang, Z.; Huang, S.; Huang, L.; Xu, X.; Zhao, H.; Yan, X. *J. Organomet. Chem.* **2020**, *85*, 12036.
526. Liang, Q.; Hayashi, K.; Song, D. *Organometallics* **2020**, *39*, 4115.
527. Beerhues, J.; Aberhan, H.; Streit, T.-N.; Sarkar, B. *Organometallics* **2020**, *39*, 4557.
528. Vaddamanu, M.; Prabusankar, G. *Eur. J. Inorg. Chem.* **2020**, *25*, 2403.
529. Ziman, S. D. *J. Heterocycl. Chem.* **1980**, *17*, 1319.
530. Lwowski, W.; Kanemasa, S.; Murray, R. A.; Ramakrishnan, V. T.; Thiruvengadam, T. K.; Yoshida, K.; Subbaraj, A. *J. Organomet. Chem.* **1986**, *51*, 1719.
531. Huber, E. W.; Kane, J. M. *J. Heterocycl. Chem.* **1990**, *27*, 1957.
532. Marley, H.; McCullough, K. J.; Preston, P. N.; Wright, S. H. B. *J. Chem. Soc., Perkin Trans. 1* **1988**, 351.
533. Akio, S.; Shimizu, B. *Bull. Chem. Soc. Jpn.* **1983**, *56*, 2969.
534. Akio, S.; Bunji, S. *Bull. Chem. Soc. Jpn.* **1983**, *56*, 2974.
535. Lempert-Sréter, M.; Lempert, K.; Møller, J. *J. Chem. Soc., Perkin Trans. 1* **1984**, 1143.
536. Gineinah, M. M.; Ismaiel, A. E.-K. M.; El-Kerdawy, M. M.; Glennon, R. A. *J. Heterocycl. Chem.* **1990**, *27*, 723.
537. Marley, H.; McCullough, K. J.; Preston, P. N.; Wright, S. H. B. *J. Chem. Soc., Chem. Commun.* **1987**, 112.
538. Marley, H.; Wright, S. H. B.; Preston, P. N. *J. Chem. Soc., Perkin Trans. 1* **1989**, 1727.
539. Kim, D. K.; O'Shea, K. E. *J. Am. Chem. Soc.* **2004**, *126*, 700.
540. Busch, M. *Ber. Dtsch. Chem. Ges.* **1905**, *38*, 861.
541. Färber, C.; Leibold, M.; Bruhn, C.; Maurer, M.; Siemeling, U. *Chem. Commun.* **2012**, *48*, 227.
542. Kriven'ko, A. P.; Morozova, N. A. *Russ. J. Appl. Chem.* **2006**, *79*, 506.
543. Grant, J.; Lu, Z.; Tucker, D. E.; Hockin, B. M.; Yufit, D. S.; Fox, M. A.; Kataky, R.; Chechik, V.; O'Donoghue, A. C. *Nat. Commun.* **2017**, *8*, 15088.
544. Thie, C.; Bruhn, C.; Leibold, M.; Siemeling, U. *Molecules* **2017**, *22*, 1133.
545. Ersan, S.; Jelonek, S.; Sieler, J.; Schulze, K. *Tetrahedron* **1996**, *52*, 791.
546. Chernyshev, V. M.; Mazharova, A. G.; Rybakov, V. B. *Acta Crystallogr.* **2011**, *E67*, o870.
547. Chernyshev, V. M.; Vlasova, A. G.; Astakhov, A. V.; Shishkina, S. V.; Shishkin, O. V. *J. Organomet. Chem.* **2015**, *80*, 375.
548. Szabo, J.; Greiner, J.; Maas, G. *Beilstein J. Org. Chem.* **2017**, *13*, 579.

549. Marley, H.; McCullough, K. J.; Preston, P. N.; Wright, S. H. B. *J. Chem. Soc., Chem. Commun.* **1988**, 506.
550. Bishop, B. C.; Marley, H.; McCullough, K. J.; Preston, P. N.; Wright, S. H. B. *J. Chem. Soc., Perkin Trans. 1* **1993**, 705.
551. Crabb, D. L.; Main, D. A.; Morley, J. O.; Preston, P. N.; Wright, S. H. B. *J. Chem. Soc., Perkin Trans. 2* **1997**, 49.
552. Bishop, B. C.; Marley, H.; Preston, P. N.; Wright, S. H. B. *J. Chem. Soc., Perkin Trans. 1* **1999**, 1527.
553. Astakhov, A. V.; Chernyshev, V. M. *Chem. Heterocycl. Compd.* **2010**, *46*, 1144.
554. Chernyshev, V. M.; Astakhov, A. V.; Starikova, Z. A. *Tetrahedron* **2010**, *66*, 3301.
555. Pyatakov, D. A.; Sokolov, A. N.; Astakhov, A. V.; Chernenko, A. Y.; Fakhrutdinov, A. N.; Rybakov, V. B.; Chernyshev, V. V.; Chernyshev, V. M. *J. Organomet. Chem.* **2015**, *80*, 10694.
556. Pyatakov, D. A.; Astakhov, A. V.; Sokolov, A. N.; Fakhrutdinov, A. N.; Fitch, A. N.; Rybakov, V. B.; Chernyshev, V. V.; Chernyshev, V. M. *Tetrahedron Lett.* **2017**, *58*, 748.
557. Chernyshev, V. M.; Sokolov, A. N.; Khoroshkin, D. A.; Taranushich, V. A. *Russ. J. Org. Chem.* **2008**, *44*, 715.
558. Chernyshev, V. M.; Pyatakov, D. A.; Sokolov, A. N.; Astakhov, A. V.; Gladkov, E. S.; Shishkina, S. V.; Shishkin, O. V. *Tetrahedron* **2014**, *70*, 684.
559. Chernyshev, V. M.; Pyatakov, D. A.; Astakhov, A. V.; Sokolov, A. N.; Fakhrutdinov, A. N.; Rybakov, V. B.; Chernyshev, V. V. *Tetrahedron* **2015**, *71*, 6259.
560. Bocian, W.; Jaźwiński, J.; Stefaniak, L.; Webb, G. A. *Magn. Reson. Chem.* **1995**, *33*, 134.
561. Morley, J. O. *J. Mol. Struct.: THEOCHEM* **1996**, *365*, 1.
562. Ridha, S. M. A.; Saleh, Z. A.; Askar, F. W. *Physical Chemistry* **2015**, *5*, 6.
563. Sambade, D.; Collins, C.; Parkin, G. *J. Mol. Struct.* **2021**, *1231*, 129682.
564. Hitzel, S.; Färber, C.; Bruhn, C.; Siemeling, U. *Organometallics* **2014**, *33*, 425.
565. Thie, C.; Hitzel, S.; Wallbaum, L.; Bruhn, C.; Siemeling, U. *J. Organomet. Chem.* **2016**, *821*, 112.
566. Sevim, M.; Kavukcu, S. B.; Kınal, A.; Şahinb, O.; Türkmen, H. *Dalton Trans.* **2020**, *49*, 16903.
567. Quinlivan, P. J.; Loo, A.; Shlian, D. G.; Martinez, J.; Parkin, G. *Organometallics* **2021**, *40*, 166.
568. Amit, E.; Berg, I.; Gross, E. *Chem. – Eur. J.* **2020**, *26*, 13046.
569. Gerhard, M.; Raphael, K. Z. *Naturforsch* **2020**, *75b*, 1065.
570. Halanicki, A.; Maj, M. *Talanta* **1975**, *22*, 767.
571. Chiou, S. J.; Gran, T.; Meloan, C. E.; Danen, W. C. *Anal. Lett.* **1981**, *14*, 865.
572. Hioki, A.; Watanabe, T.; Terajima, K.; Fudagawa, N.; Kubota, M.; Kawase, A. *Anal. Sci.* **1990**, *6*, 757.
573. Racheva, P. V.; Gavazov, K. B.; Lekova, V. D.; Dimitrov, A. N. *J. Mater.* **2013**, *7*, 897343.
574. Shutske, G. M. *J. Heterocycl. Chem.* **1981**, *18*, 1017.
575. Shutske, G. M.; Agnew, M. N. *J. Heterocycl. Chem.* **1981**, *18*, 1025.
576. Buccheri, F.; Cusmano, G.; Gruttadauria, M.; Noto, R.; Werber, G. *J. Heterocycl. Chem.* **1997**, *34*, 1447.
577. Hassan, A. A.; El-Shaieb, K. M. A.; Mohamed, N. K.; Tawfeek, H. N.; Bräse, S.; Nieger, M. *Tetrahedron Lett.* **2014**, *55*, 2385.
578. Yokoyama, M.; Ikuma, T.; Obara, N.; Togo, H. *J. Chem. Soc., Perkin Trans. 1* **1990**, 3243.
579. Echevarria, A.; Galembeck, S. E.; Maciel, M. A. M.; Miller, J.; Montanari, C. A.; Rumjanek, V. M.; Simas, A. M.; Sandall, J. B. P. *Heterocycl. Commun.* **1995**, *1*, 129.

580. De Athayde Filho, P. F.; Miller, J.; Thomas, G.; de Araujo, C. C. *Heterocycl. Commun.* **1996**, *2*, 573.
581. Soares de Oliveira, C.; Dos Santos Falcão-Silva, V.; Siqueira-Júnior, J. P.; Harding, D. P.; Lira, B. F.; Lorenzo, J. G. F.; Barbosa-Filho, J. M.; Filgueiras de Athayde-Filho, P. *Molecules* **2011**, *16*, 2023.
582. Postovskii, I. Y.; Kotovskaya, S. K.; Mokrushina, G. A.; Il'enko, V. I.; Platonov, V. G.; Polonyankan, A. V. *Pharm. Chem. J.* **1980**, *14*, 123.
583. Kurasawa, Y.; Kawano, T.; Katoh, R.; Takada, A.; Kim, H. S.; Okamoto, Y. *J. Heterocycl. Chem.* **1992**, *29*, 1337.
584. Shin, S. C.; Jeon, D. J.; Jang, K. A.; Lee, Y. Y. *Bull. Korean Soc. Chem.* **1994**, *15*, 517.
585. Lee, K.-J.; Lee, Y.-S.; Choi, D. O. *Bull. Korean Chem. Soc.* **1997**, *18*, 1226.
586. Kim, H. S.; Kim, T. E.; Kwag, S. T.; Park, Y. T.; Hong, Y. S.; Okamoto, Y.; Kurasawa, Y. *J. Heterocycl. Chem.* **1997**, *34*, 1539.
587. Molina, P.; Arques, A.; Cartagena, I.; Velcarcel, M. V. *Synthesis* **1984**, 881.
588. Molina, P.; Arques, A.; Cartagena, I.; Valcarcel, M. V. *J. Heterocycl. Chem.* **1986**, *23*, 43.
589. Molina, P.; Arques, A.; Alias, M. A.; Saiz, A. L. L. M.; Foces-Foces, M. C. *Liebigs Ann. Chem.* **1989**, 1055.
590. Wasfy, A. A. F. *Heteroat. Chem.* **2003**, *14*, 581.
591. Basyouni, W. M.; El-Bayouki, K. A. M.; Hosni, H. M. *J. Chem. Res.* **2003**, 755.
592. Girgis, A. S.; Barsoum, F. F. *Eur. J. Med. Chem.* **2009**, *44*, 1972.
593. Kovalenko, S. M.; Drushlyak, O. G.; Mariutsa, I. O. *J. Sulfur Chem.* **2020**, *41*, 388.
594. Molina, P.; Lorenzo, A.; Claramunt, R. M.; Elguero, J. *Tetrahedron Lett.* **1984**, *25*, 5427.
595. Molina, P.; Lorenzo, A.; Claramunt, R. M.; Elguero, J. *Tetrahedron* **1986**, *42*, 2121.
596. Cheung, K. K.; Echevarria, A.; Galembeck, S.; Maciel, M. A. M.; Miller, J.; Rumjanek, V. M.; Simas, A. M. *Acta Crystallogr.* **1993**, *C49*, 1092.
597. Araki, S.; Butsugan, Y. *Chem. Ber.* **1993**, *126*, 1157.
598. Deaton, J. C.; Luss, H. R. *J. Chem. Soc., Dalton Trans.* **1999**, 3163.
599. Teles, W. M.; Machado, F. C.; Montanari, C. A.; Filgueiras, C. A. L.; Jenkins, T. C. *Transition Met. Chem.* **2005**, *30*, 733.
600. Altland, H. W.; Briffa, B. F., Jr. *J. Organomet. Chem.* **1985**, *50*, 433.
601. Echevarria, A.; Miller, J. *J. Chem. Soc., Perkin Trans. 2* **1989**, 1425.
602. de Athayde-Filho, P. F.; Miller, J.; Simas, A. M. *Synthesis* **2000**, 1565.
603. de Athayde-Filho, P. F.; Simas, A. M.; Goncalves, S. M. C.; Miller, J. *Phosphorus, Sulfur Silicon Relat. Elem.* **2000**, *161*, 115.
604. Bocian, W.; Jaźwiński, J.; Stefaniak, L. *Pol. J. Chem.* **1995**, *69*, 85.
605. Masuda, K.; Adachi, J.; Morita, H.; Nomura, K. *Chem. Pharm. Bull.* **1981**, *29*, 1743.
606. Adachi, J.; Takahata, H.; Nomura, K.; Masuda, K. *Chem. Pharm. Bull.* **1983**, *31*, 1746.
607. Alekseenko, T. A.; Bazhbeuk-Melikova, T. S.; Zelenskaya, O. V.; Kozinskii, V. A. *Chem. Heterocycl. Compd.* **1989**, *25*, 1301.
608. Araki, S.; Goto, T.; Butsugan, Y. *Bull. Chem. Soc. Jpn.* **1988**, *61*, 2977.
609. Masuda, K.; Adachi, J.; Nate, H.; Takahata, H.; Nomura, K. *J. Chem. Soc. Perkin Trans.* **1981**, *1*, 1591.
610. L'abbé, G.; Delbeke, P.; Bastin, L.; Dehaen, W.; Toppet, S. *J. Heterocycl. Chem.* **1993**, *20*, 301.
611. Capuano, L.; Boschat, P.; Müller, I.; Zander, R.; Schramm, V.; Hädicke, E. *Chem. Ber.* **1983**, *116*, 2058.
612. L'abbé, G.; Frederix, A.; Toppet, S.; Declercq, J.-P. *J. Heterocycl. Chem.* **1991**, *28*, 477.
613. L'abbé, G.; Bastin, L.; Vlieghe, D.; Van Meervelt, L. *J. Chem. Soc., Perkin Trans. 1* **1993**, 3051.
614. Newton, G.; Ollis, W. D.; Wright, D. E. *J. Chem. Soc., Perkin Trans. 1* **1984**, 75.
615. Kuzmina, O. M.; Weisel, M.; Narine, A. A. *Eur. J. Org. Chem.* **2019**, 5527.

616. Montanari, C. A.; Sandall, J. P. B.; Miyata, Y.; Miller, J. *J. Chem. Soc., Perkin Trans. 2* **1994**, 2571.
617. Maciel, M. A. M.; Echevarria, A.; Rumjanek, V. M. *Quim. Nova* **1998**, *21*, 569.
618. Britto, M. M.; Almeida, T. M. G.; Leitao, A.; Donnici, C. L.; Lopes, M. T. P.; Montanari, C. A. *Synth. Commun.* **2006**, *36*, 3359.
619. Britto, M. M.; Montanari, C. A.; Donnici, C. L.; Cass, Q. B. *Heterocycl. Commun.* **1998**, *4*, 209.
620. Kappel, J. C.; Yokum, T. S.; Barany, G. *J. Comb. Chem.* **2004**, *6*, 746.
621. dos Reis, C. M.; Echevarria-Lima, J.; Miranda, A. F.; Echevarria, A. *J. Braz. Chem. Soc.* **2011**, *22*, 1505.
622. Lopes, N. D.; Chaves, O. A.; de Oliveira, M. C. C.; Sant'Anna, C. M. R.; Sousa-Pereira, D.; Netto-Ferreira, J. C.; Echevarria, A. *Int. J. Biol. Macromol.* **2018**, *112*, 1062.
623. Sousa-Pereira, D.; Silva de Oliveira, T.; Paiva, R. O.; Chaves, O. A.; Netto-Ferreira, J. C.; Echevarria-Lima, J.; Echevarria, A. *Molecules* **2020**, *25*, 2537.
624. Molina, P.; Tárraga, A.; Espinosa, A. *Synthesis* **1989**, 923.
625. Buccheri, F.; Cusmano, G.; Noto, R.; Rainieri, R. G. *J. Heterocycl. Chem.* **1987**, *24*, 521.
626. Grynberg, N.; Santos, A. C.; Echevarria, A. *Anti-Cancer Drugs* **1997**, *8*, 88.
627. Cadena, S. M. S. C.; Carnieri, E. G. S.; Echevarria, A.; de Oliveira, M. B. M. *Cell Biochem. Funct.* **2002**, *20*, 31.
628. Mendez-Sanchez, S. C.; Martinez, G. R.; Romao, S.; Echevarria, A.; Silva, E. F.; Rocha, M. E. M.; Noleto, G. R.; Carnieri, E. G. S.; Cadena, S. M. S. C.; de Oliveira, M. B. M. *Chem.-Biol. Interact.* **2009**, *179*, 125.
629. Pires, A. R. A.; Noleto, G. R.; Echevarria, A.; Reis, C. M.; Rocha, M. E. M.; Carnieri, E. G. S.; Martinez, G. R.; Noleto, G. R.; Cadena, S. M. S. C. *Chem.-Biol. Interact.* **2011**, *189*, 17.
630. Rodrigues, R. F.; Castro-Pinto, D.; Echevarria, A.; dos Reis, C. M.; Del Cistia, C. N.; Sant'Anna, C. M. R.; Teixeira, F.; Castro, H.; Canto-Cavalheiro, M.; Leon, L. L.; Tomas, A. *Bioorg. Med. Chem.* **2012**, *20*, 1760.
631. Gozzi, G. J.; Pires, A.d. R. A.; Valdameri, G.; Rocha, M. E. M.; Martinez, G. R.; Noleto, G. R.; Acco de Souza, A. C. E. A.; Echevarria, A.; dos Reis, C. M.; Di Pietro, A.; Cadena, S. M. S. C. *PLoS One* **2015**, *10*, e0130046.
632. Montanari, C. A.; Amaral, A. T.; Giesbrecht, A. *Pharm. Sci.* **1997**, *3*, 565.
633. Britto, M. M.; Montanari, C. A.; Donnici, C. L.; Cass, Q. B. *J. Liq. Chromatogr. Relat. Technol.* **1999**, *22*, 357.
634. Almeida, T. M. G.; Leitao, A.; Montanari, M. L. C.; Montanari, C. A. *Chem. Biodiversity* **2005**, *2*, 1691.
635. Jäger, G.; Heitzer, H. *Synthesis* **1981**, 704.
636. Molina, P.; Arques, A.; Velasco, M. D.; Villalgordo, J. M. *Synthesis* **1988**, 729.
637. Molina, P.; Arques, A.; Alias, A. *Tetrahedron* **1992**, *48*, 1285.
638. Molina, P.; Alajarín, M.; de Vega, M. J. P. *Liebigs Ann. Chem.* **1986**, 1540.
639. Molina, P.; Alajarín, M.; López-Lázaro, A. *Tetrahedron* **1991**, *47*, 6747.
640. Molina, P.; Espinosa, A.; Tarraga, A.; Cano, F. H.; Foces-Foces, M. C. *J. Chem. Soc., Perkin Trans. 1* **1991**, 1159.
641. Espinosa, A.; Garcia, R.; Molina, P.; Tarraga, A. *Org. Biomol. Chem.* **2010**, *8*, 1623.
642. Bottino, F.; Chiacchio, U.; Fronczek, F. R.; Pappalardo, S. *J. Organomet. Chem.* **1989**, *54*, 2024.
643. Kiwan, A. M.; Wanas, G. A. *J. Chem. Soc., Perkin Trans. 2* **1981**, 1534.
644. de Athayde Filho, P. F.; Howe, B. P.; Miller, J. *Met.-Based Drugs* **1997**, *4*, 273.

645. do Nascimento, R. S. T. R.; da Silva Morais, C. R.; Lira, H. L.; de Morais, S. A.; de Athayde Filho, P. F.; Lucena, L. F. L.; de Souza, A. G.; Campos, G. B. *J. Alloys Compd.* **2010**, *495*, 603.
646. Aly, H. F.; Kiwan, A. M. *J. Radioanalytical Chem.* **1981**, *62*, 15.
647. Asundaria, S. T.; Patel, S. A.; Mehta, K. M.; Patel, K. C. S. *Afr. J. Chem.* **2010**, *63*, 141.
648. Asundaria, S. T.; Patel, K. C.; Mehta, K. M. *Phosphorus, Sulfur Silicon Relat. Elem.* **2011**, *186*, 1554.
649. Asundaria, S. T.; Patel, N. S.; Patel, K. C. *Med. Chem. Res.* **2012**, *21*, 1199.
650. Akbari, V. K.; Savaliya, P. P.; Patel, K. C. *Phosphorus, Sulfur Silicon Relat. Elem.* **2014**, *189*, 711.
651. Kalluraya, B.; Kaushik, B. R.; Somashekarappa, H. M. *J. Modern Mat.* **2020**, 7, 17.
652. Bezerra, A.; Gomes, A. S.; Athayde-Filho, P.; da Rocha, G.; Miller, J.; Simas, A. *Chem. Phys. Lett.* **1999**, *309*, 421.
653. Chu, C.-H.; Hui, X.-P.; Xu, P.-F.; Zhang, Z.-Y.; Li, Z.-C.; Liao, R.-A. *J. Chin. Chem. Soc.* **2000**, *47*, 943.
654. Giuseppe, B. *Bull. Chem. Soc. Jpn.* **1990**, *63*, 2991.
655. Fonseca, T. L.; de Oliveira, H. C. B.; Castro, M. A. *Chem. Phys. Lett.* **2008**, *457*, 119.
656. Torres, E. M.; Junior, L. A.; Georg, H. C.; Castro, M. A.; Fonseca, T. L. *Chem. Phys. Lett.* **2019**, *736*, 136798.
657. Molina, P.; Arques, A.; Cartagena, I.; Fresneda, P. M. *Synth. Commun.* **1987**, *17*, 1929.
658. Molina, P.; Arques, A.; Cartagena, I.; Alias, M. A.; de la Concepción Foces-Foces, M.; Cano, F. H. *Liebigs Ann. Chem.* **1988**, 133.
659. Dunn, P. J.; Rees, C. W. *J. Chem. Soc., Perkin Trans. 1* **1989**, 2489.
660. Lund, M. Q.; Keir, L. B.; Glennon, R. A.; Egle, J. L. *J. Med. Chem.* **1982**, *25*, 1503.
661. Henning, R. *Nachr. Chem., Tech. Lab.* **1993**, *41*, 412.
662. Rehse, K.; König, P. *Arch. Pharm.* **1995**, *328*, 137.
663. Schonafinger, K. *Farmaco* **1999**, *54*, 316.
664. Dalinger, I. L.; Serushkina, O. V.; Muravyev, N. V.; Meerov, D. B.; Miroshnichenko, E. A.; Kon'kova, T. S.; Suponitsky, K. Y.; Vener, M. V.; Sheremetev, A. B. *J. Mater. Chem. A* **2018**, *6*, 18669.
665. Serushkin, V. V.; Sinditskii, V. P.; Filatov, S. A.; Kulagina, P. D.; Nguyen, V. T.; Vatsadze, I. A.; Dalinger, I. L.; Sheremetev, A. B. *Int. J. Energ. Mater. Chem. Propulsion* **2018**, *17*, 147.
666. Gettings, M. L.; Thoenen, M. T.; Byrd, E. F. C.; Sabatini, J. J.; Zeller, M.; Piercey, D. G. *Chem. – Eur. J.* **2020**, *26*, 14530.
667. Gettings, M.; Piercey, D. *Energ. Mater. Front.* **2020**, *1*, 136.
668. Zhilin, E.; Bystrov, D.; Ananyev, I.; Fershtat, L.; Makhova, N. *Chem. – Eur. J.* **2019**, *25*, 14284.
669. Martynova, M. N.; Pevzner, M. S.; Smorygo, N. A.; Serebryakova, N. M. *Chem. Heterocycl. Compd.* **1981**, *17*, 1238.
670. Shevelev, S. A.; Dalinger, I. L.; Gulevskaya, V. I.; Cherkasova, T. I. *Chem. Heterocycl. Compd.* **1997**, *33*, 880.
671. Shevelev, S. A.; Dalinger, I. L.; Gulevskaya, V. I.; Cherkasova, T. I.; Vinogradov, V. M.; Ugrak, B. I.; Starosotnikov, A. M. *Chem. Heterocycl. Compd.* **1999**, *35*, 363.
672. Dalinger, I. L.; Serushkina, O. V.; Lipilin, D. L.; Anisimov, A. A.; Suponitsky, K. Y.; Sheremetev, A. B. *ChemPlusChem* **2019**, *84*, 802.
673. Dalinger, I. L.; Kormanov, A. V.; Shkineva, T. K.; Sheremetev, A. B.; Asian, J. *Organomet. Chem.* **2020**, *9*, 811.
674. Said, Z.; Tillett, J. G. *J. Chem. Soc., Perkin Trans. 2* **1980**, 1437.
675. Said, Z.; Tillett, J. G. *J. Chem. Soc., Perkin Trans. 2* **1982**, 701.
676. Ramsden, C. A. *ARKIVOC* **2020**, *iv*, 12.
677. Ames, J. R.; Potts, K. T.; Ryan, M. D.; Kovacic, P. *Life Sci.* **1986**, *39*, 1085.

678. Jones, L. F.; Shillady, D. D.; Duong, A.; Kier, L. B. *Int. J. Quantum Chem.* **1989**, *36*, 465.
679. Karup, G.; Preikschat, H.; Wilhelmsen, E. S.; Pedersen, S. B.; Marcinkiewicz, E.; Cieślik, K.; Gryglewski, R. J. *Pol. J. Pharmacol.* **1994**, *46*, 541.
680. Kankaanranta, H; Rydell, E.; Petersson, A.S.; Holm, P.; Moilanen, E.; Corell, T.; Karup, G.; Vuorinen, P.; Pedersen, S.B.; Wennmalm, A.; Metsä-Ketelä, T. *Br. J. Pharmacol.* **1996**, *117*, 401.
681. Vaali, K.; Li, L.; Redemann, B.; Paakkari, I.; Vapaatalo, H. *J. Pharm. Pharmacol.* **1996**, *48*, 1309.
682. Szekeres, M.; Dezsi, L.; Monos, E.; Metsa-Ketela, T. *Acta Physiol. Scand.* **1997**, *161*, 55.
683. Gryglewski, R.; Marcinkiewicz, E.; Robak, J.; Michalska, Z.; Madej, J. *Curr. Pharm. Des.* **2002**, *8*, 167.
684. Wang, P. G.; Xian, M.; Tang, X.; Wu, X.; Wen, Z.; Cai, T.; Janczuk, A. J. *Chem. Rev.* **2002**, *102*, 1091.
685. Grenegard, M.; Gustafsson, M. C.; Andersson, R. G. G.; Bengtsson, T. *Br. J. Pharmacol.* **1996**, *118*, 2140.
686. Kosonen, O.; Kankaanranta, H.; Vuorinen, P.; Moilanen, E. *Eur. J. Pharmacol.* **1997**, *337*, 55.
687. Rydberg, I. G. M. J.; Andersson, R. G. G.; Grenegard, M. *Eur. J. Pharmacol.* **1997**, *329*, 175.
688. Kosonen, O.; Kankaanranta, H.; Malo-Ranta, U.; Ristimäki, A.; Moilanen, E. *Br. J. Pharmacol.* **1998**, *125*, 247.
689. Hernandez, M.; Elmedal, B.; Mulvany, M. J.; Simonsen, U. *Br. J. Pharmacol.* **1998**, *123*, 895.
690. Kosonen, O.; Kankaanranta, H.; Malo-Ranta, U.; Moilanen, E. *Eur. J. Pharmacol.* **1999**, *382*, 111.
691. Kosonen, O.; Kankaanranta, H.; Uotila, J.; Moilanen, E. *Eur. J. Pharmacol.* **2000**, *394*, 149.
692. Persson, A. K. A.; Palmér, L.; Gunnarsson, P.; Grenegard, M. *Eur. J. Pharmacol.* **2004**, *496*, 1.
693. Elmedal, B.; Mulvany, M.; Simonsen, U. *Eur. J. Pharmacol.* **2005**, *516*, 78.
694. Persson, A. A.; Zalavary, S.; Lindström, E.; Whiss, P. A.; Bengtsson, T.; Grenegard, M. *Eur. J. Pharmacol.* **2005**, *517*, 149.
695. Laursen, B. E.; Stankevicius, E.; Pilegaard, H.; Mulvany, M.; Simonsen, U. *Cardiovasc. Drug. Rev.* **2006**, *24*, 247.
696. Laursen, E. M.; Mulvany, J.; Simonsen, U. *Pulm. Pharmacol. Ther.* **2006**, *19*, 179.
697. Lahteenmaki, T.; Sievi, E.; Vapaatalo, H. *Br. J. Pharmacol.* **1998**, *125*, 402.
698. Holm, P.; Kankaanranta, H.; Metsä-Ketelä, T.; Moilanen, E. *Eur. J. Pharmacol.* **1998**, *346*, 97.
699. Hsu, M. F.; Chen, Y. S.; Huang, L. J.; Tsao, L. T.; Kuo, S. C.; Wang, J. P. *Eur. J. Pharmacol.* **2006**, *535*, 43.
700. Decuypere, E.; Bernard, S.; Feng, M.; Porte, K.; Riomet, M.; Thuery, P.; Audisio, D.; Taran, F. *ACS Catal.* **2018**, *8*, 11882.
701. Hanley, R. N.; Ollis, W. D.; Ramsden, C. A. *J. Chem. Soc., Perkin Trans. 1* **1979**, 736.
702. Bocian, W.; Jaźwínski, J.; Stefaniak, L.; Webb, G. A. *Chem. Heterocycl. Compd.* **1995**, *31*, 1103.
703. Jazwinski, J.; Rozwadowski, Z.; Magiera, D.; Duddeck, H. *Magn. Reson. Chem.* **2003**, *41*, 315.
704. Jawinski, J.; Duddeck, H. *Magn. Reson. Chem.* **2003**, *41*, 921.
705. Jaźwiński, J.; Staszewska-Krajewska, O. *J. Mol. Struct.* **2004**, *687*, 23.
706. Stuhlmann, F.; Domschke, G.; Bartl, A.; Neudeck, A.; Petr, A.; Omelka, L. *Magn. Reson. Chem.* **1997**, *35*, 124.
707. Neudeck, A.; Kress, L. *J. Electroanal. Chem.* **1997**, *437*, 141.

708. Jaźwiński, J.; Stefaniak, L.; Ishikawa, S.; Yamaguchi, M.; Lipkowski, J.; Słowikowska, J.; Webb, G. *J. Mol. Struct.* **1995**, *344*, 227.
709. Jazwinski, J.; Staszewska, O.; Stefaniak, L.; Webb, G. A. *J. Mol. Struct.* **1996**, *377*, 167.
710. Moderhack, D. *Heterocycles* **2016**, *92*, 185.
711. Logvinov, A. V.; Saraev, V. V.; Polyakova, I. N.; Strelenko, Y. A.; Golod, E. L. *Russ. J. Gen. Chem.* **2007**, *77*, 2186.
712. Logvinov, A. V.; Polyakova, I. N.; Golod, E. L. *Russ. J. Gen. Chem.* **2009**, *79*, 2220.
713. Awadallah, A.; Rademacher, P.; Boese, R. *J. Prakt. Chem.* **1995**, *337*, 636.
714. Bocian, W.; Jazwinski, J.; Kozminski, W.; Stefaniak, L.; Webb, G. A. *J. Chem. Soc., Perkin Trans. 2* **1994**, 1327.
715. Araki, S.; Yamamoto, K.; Yagi, M.; Inoue, T.; Fukagawa, H.; Hattori, H.; Yamamura, H.; Kawai, M.; Butsugan, Y. *Eur. J. Org. Chem.* **1998**, 121.
716. Araki, S.; Wanibe, Y.; Uno, F.; Morikawa, A.; Yamamoto, K.; Chiba, K.; Butsugan, Y. *Chem. Ber.* **1993**, *126*, 1149.
717. Bocian, W.; Jaźwiński, J.; Koźmiński, W.; Stefaniak, L.; Webb, G. A. *Magn. Reson. Chem.* **1994**, *32*, 284.
718. Jaźwiński, J.; Staszewska, O.; Stefaniak, L.; Araki, S.; Webb, G. A. *J. Mol. Struct.* **2000**, *523*, 103.
719. Gómez-Zavaglia, A.; Reva, I. D.; Frija, L.; Cristiano, M. L.; Fausto, R. *J. Photochem. Photobiol. A: Chemistry* **2006**, *179*, 243.
720. Wieczorek, W.; Zukowska, G.; Ostrovskii, D.; Florjanczyk, Z. *J. Phys. Chem. B* **2001**, *105*, 9686.
721. Zarubaev, V. V.; Golod, E. L.; Anfimov, P. M.; Shtro, A. A.; Saraev, V. V.; Gavrilov, A. S.; Logvinov, A. V.; Kiselev, O. I. *Bioorg. Med. Chem.* **2010**, *18*, 839.
722. Voitekhovich, S. V.; Gaponik, P. N.; Lyakhov, A. S.; Ivashkevich, O. A. *Tetrahedron* **2008**, *64*, 8721.
723. Logvinov, A. V.; Polyakova, I. N.; Golod, E. L. *Russ. J. Gen. Chem.* **2010**, *80*, 2366.
724. Budevich, V. A.; Voitekhovich, S. V.; Zuraev, A. V.; Matulis, V. E.; Matulis, V. E.; Lyakhov, A. S.; Ivashkevich, L. S.; Beilstein, I. O. A. *J. Organomet. Chem.* **2021**, *17*, 385.
725. Gómez-Zavaglia, A.; Reva, I. D.; Frija, L.; Cristiano, M. L.; Fausto, R. *J. Phys. Chem. A* **2005**, *109*, 7967.
726. Araki, S.; Yamamoto, K.; Inoue, T.; Fujimoto, K.; Yamamura, H.; Kawai, M.; Butsugan, Y.; Zhou, J.; Eichhorn, E.; Rieker, A.; Huber, M. *J. Chem. Soc., Perkin Trans. 2* **1999**, 985.
727. Araki, S.; Hattori, H.; Shimizu, N.; Ogawa, K.; Yamamura, H.; Kawai, M. *J. Heterocycl. Chem.* **1999**, *36*, 863.
728. Araki, S.; Hattori, H.; Yamamura, H.; Kawai, M. *J. Heterocycl. Chem.* **2000**, *37*, 1129.
729. Araki, S.; Hattori, H.; Ogawa, K.; Kuzuya, M.; Inoue, T.; Yamamura, H.; Kawai, M. *J. Chem. Soc., Perkin Trans. 1* **2001**, 2476.
730. Araki, S.; Kuzuya, M.; Hamada; Nogura, K. M.; Ohata, N. *Org. Biomol. Chem.* **2003**, *1*, 978.
731. Voitekhovich, S. V.; Lyakhov, A. S.; Ivashkevich, L. S.; Schleife, F.; Schnorr, R.; Kersting, B.; Gaponik, P. N. *Inorg. Chim. Acta* **2014**, *419*, 124.
732. Vaddypally, S.; Tomlinson, W.; O'Sullivan, O. T.; Ding, R.; Van Vliet, M.; Wayland, B. B.; Hooper, J. P.; Zdilla, M. J. *J. Am. Chem. Soc.* **2019**, *141*, 5699.
733. Matsukawa, Y.; Hirashita, T.; Araki, S. *Tetrahedron* **2017**, *73*, 6052.
734. Matsukawa, Y.; Hirashita, T.; Araki, S. *Bull. Chem. Soc. Jpn.* **2019**, *92*, 540.
735. Hirashita, T.; Matsukawa, Y. *Synlett* **2019**, 315.
736. Deev, S. L.; Paramonov, A. S.; Shestakova, T. S.; Khalymbadzha, I. A.; Chupakhin, O. N.; Subbotina, J. O.; Eltsov, O. S.; Slepukhin, P. A.; Rusinov, V. L.; Arseniev, A. S.; Shenkarev, Z. O.; Beilstein, J. *Organomet. Chem.* **2017**, *13*, 2535.

737. Araki, S.; Butsugan, Y. *Chem. Lett.* **1985**, *14*, 1639.
738. Araki, S.; Butsugan, Y. *J. Chem. Soc., Chem. Commun.* **1983**, 789.
739. Araki, S.; Butsugan, Y. *J. Chem. Soc., Perkin Trans. 1* **1984**, 2545.
740. Araki, S.; Butsugan, Y. *Tetrahedron Lett.* **1984**, *25*, 441.
741. Araki, S.; Yokoi, K.; Sato, R.; Hirashita, T.; Setsune, J.-I. *J. Heterocyclic Chem.* **2009**, *46*, 164.
742. Koźmiński, W.; Jaźwiński, J.; Stefaniak, L.; Webb, G. A. *Magn. Reson. Chem.* **1994**, *32*, 746.
743. Jaźiwińsk, J.; Staszewska, O.; Staszewski, P.; Stefaniak, L.; Wiench, J. W.; Webb, G. A. *J. Mol. Struct.* **1999**, *475*, 181.
744. Jaźwiński, J.; Stefaniak, L.; Webb, G. A. *Magn. Reson. Chem.* **1988**, *26*, 1012.
745. Wieczorek, W.; Zukowska, G.; Borkowska, R.; Chung, S.; Greenbaum, S. *Electrochim. Acta* **2001**, *46*, 1427.
746. Kato, Y.; Horie, K. *Macromol. Chem. Phys.* **2002**, *203*, 2290.
747. Jaźwiński, J.; Staszewska, O.; Stefaniak, L.; Staszewski, P.; Webb, G. A. *J. Mol. Struct.* **1999**, *477*, 143.
748. Jaźwiński, J. *J. Mol. Struct.* **2000**, *524*, 105.
749. Jaźwiński, J.; Stefaniak, L. *Magn. Reson. Chem.* **1993**, *31*, 447.
750. Jaźwiński, J. *J. Mol. Struct.* **1998**, *443*, 27.
751. Slowikowska, J.; Jaźwiński, J.; Stefaniak, L.; Webb, G. A. *J. Mol. Struct.* **1998**, *442*, 175.
752. Bocian, W.; Jaźwiński, J.; Stefaniak, L.; Webb, G. A. *J. Chem. Soc., Perkin Trans. 2* **1994**, 1467.
753. Felding, G.; Holm, A. *Acta Chem. Scand. Ser. B* **1988**, *42*, 63.
754. Potts, K. T.; Cody, R. D.; Dennis, R. J. *J. Organomet. Chem.* **1981**, *46*, 4065.
755. deLaat, R. H.; Durham, L.; Livingstone, E. G.; Westwood, N. P. C. *J. Phys. Chem.* **1993**, *97*, 11216.
756. Blockhuys, F.; Tersago, K.; Shlykov, S. A.; Konrad, A.; Christen, D. *J. Mol. Struct.* **2010**, *978*, 147.
757. Van Droogenbroeck, J.; Tersago, K.; Van Alsenoy, C.; Aucott, S. M.; Milton, H. L.; Woollins, J. D.; Blockhuys, F. *Eur. J. Inorg. Chem.* **2004**, 3798.
758. Van Droogenbroeck, J.; Tersago, K.; Alsenoy, C. V.; Blockhuys, F. *Chem. Phys. Lett.* **2004**, *399*, 516.
759. Tersago, K.; Oláh, J.; Martin, J. M. L.; Veszprémi, T.; Van Alsenoy, C.; Blockhuys, F. *Chem. Phys. Lett.* **2005**, *413*, 440.
760. Hinz, A.; Schultz, A.; Villinger, A. *Angew. Chem., Int. Ed.* **2015**, *54*, 2776.
761. Hinz, A.; Schultz, A.; Villinger, A. *J. Am. Chem. Soc.* **2015**, *137*, 9953.
762. Beer, H.; Bresien, J.; Michalik, D.; Rölke, A.-K.; Schultz, A.; Villinger, A.; Wustrack, R. *J. Organomet. Chem.* **2020**, *85*, 14435.
763. Hinz, A.; Schulz, A.; Villinger, A. *Chem. Sci.* **2016**, 7, 745.

CHAPTER FOUR

Type B mesoionic compounds (1980–2020)

Florea Dumitrascu[a] and Christopher A. Ramsden[b,*]
[a]Center for Organic Chemistry, "C.D. Nenitzescu" Romanian Academy, Bucharest, Romania
[b]Lennard-Jones Laboratories, School of Physical and Geographical Sciences, Keele University, Staffordshire, United Kingdom
*Corresponding author: e-mail address: c.a.ramsden@keele.ac.uk

Contents

Abstract

This review covers the chemistry of type B mesoionic rings published in the period 1980–2020 together with some work published in the first half of 2021. Factors contributing to the stability of type B mesoionic rings are identified and known and unknown systems are tabulated. The difference between type A and type B mesoionic rings and their classification as conjugated and semi-conjugated heterocyclic mesomeric betaines, respectively, is discussed. Important aspects of the chemistry of type B mesoionic rings are (i) valence tautomerism to an acyclic isomer and subsequent reaction and (ii) cycloadditions involving an exocyclic atom or group.

Keywords: Mesoionic, Type B, Mesomeric betaines, Cycloadditions, Valence tautomerism, Aromaticity

Advances in Heterocyclic Chemistry, Volume 137
ISSN 0065-2725
https://doi.org/10.1016/bs.aihch.2021.10.003

1. Introduction

Contrary to earlier definitions, mesoionic rings in general have low classical aromaticity and have ring π electrons in excess of a sextet. They can be broadly defined as five-membered heterocycles that cannot be represented satisfactorily by any one covalent or polar structure. Within this definition two distinct types of mesoionic heterocycle, which are structural isomers, are recognized and these are described as type A and type B. Type A systems have been reviewed in the preceding chapter. This chapter reviews the chemistry of type B systems for the period 1980–2020 and updates the coverage of two previous reviews,[1, 2] which are reproduced in this volume as Appendices 1 and 2: "Meso-ionic compounds" by Ollis and Ramsden and "Meso-ionic heterocycles (1976–1980)" by Newton and Ramsden.

The earliest example of a type B mesoionic heterocycle was dehydrodithizone **1**, prepared by Fischer and Besthorn in 1882. The structural relationship between type A and type B rings is illustrated by comparing dehydrodithizone **1** with its type A isomer **2**.

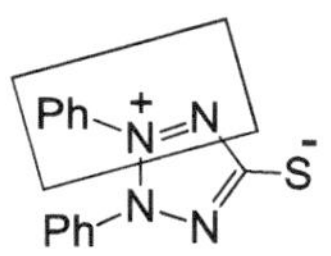

1 Type B

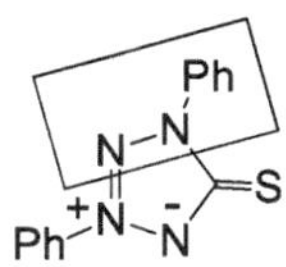

2 Type A

Representing mesoionic rings by formulae using a full circle with positive charge and an exocyclic negative charge (e.g., structure **29** in Section 3.3) is no longer regarded as useful. We prefer the use of resonance hybrids of the type **4a–c** which satisfactorily portray resonance stabilized heterocycles.

In generalized structures, Fig. 1 shows the origins of the π electrons in type A and type B mesoionic rings. In particular, in type B rings **3** the two heteroatoms donating a lone pair (b and c) are adjacent whereas in type A rings the lone pair donors (a and c) have a 1,3 relationship. Structures **4** and **6** show generalized examples of dipolar representations of these mesoionic rings. It is significant that the type B rings **4** are not associated with a 1,3-dipolar fragment, unlike the type A rings **6** (**6a**: $b{=}c^{+}{-}d^{-}$).

Although the choice of a single resonance hybrid does not adequately represent mesoionic rings, in the preceding chapter we have used 1,3-dipolar structures, e.g., **2** and **6a**, as a general representation of type A mesoionic

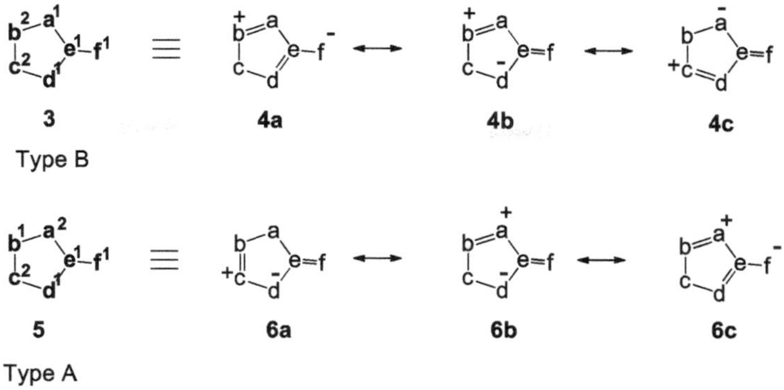

Fig. 1 The general structures of type A and type B mesoionic heterocycles **3** and **5** showing the lone-pair donating heteroatoms and contributing resonance hybrids **4a–c** and **6a–c**.

compounds since this is conveniently consistent with their participation in 1,3-dipolar cycloadditions. The structure and chemistry of type B mesoionic rings are quite different and in this chapter we represent type B rings by structures based on the general representation **4a**, which is consistent with their greater polarity and general modes of reaction. In accord with the general structure **4a**, type B mesoionic derivatives are named in this review using the commonly used olate, thiolate, aminide nomenclature. Thus dehydrodithizone **1** is named as 2,3-diphenyl-1,2,3,4-tetrazolium-5-thiolate.

Type A and B mesoionic compounds are both examples of a much larger class of heterocycles described as heterocyclic mesomeric betaines (HMBs). As described previously,[3, 4] HMBs are characterized by conjugated fragments containing an odd number of atoms that are either (i) separated by lone-pair donating heteroatoms or (ii) form a cross-conjugated fragment of the ring. Since odd conjugated fragments can only be represented as anions or cations (or radicals), this gives rise to the unusual dipolar representation of these species. Type A and type B mesoionic rings illustrate both these situations. Structure **7** shows a generalized type A ring in which the two odd fragments (b and d—e—f) are separated by the heteroatoms a and c. A typical example is *N*-phenylsydnone **8**. In contrast, the type B rings contain a cross-conjugated fragment as shown in structure **9** and illustrated by the 1,2-diphenyltetrazolium-4-olate **10**.

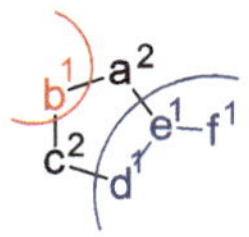

7 Type A

8

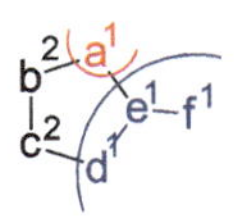

9 Type B

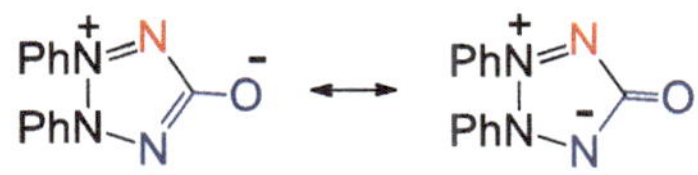

10

2. The classification of type B mesoionic heterocycles as semi-conjugated HMBs

Odd conjugated fragments, including those associated with heterocyclic mesomeric betaines, have special properties. The conjugated atoms can be divided into two sets, starred (*) and unstarred (°) as shown in Fig. 2.[5] Associated with these odd fragments are non-bonding molecular orbitals that are restricted to the starred atoms and vanish on the unstarred atoms (Fig. 2). For this reason, it is highly significant whether in HMB ring structures the heteroatoms are adjacent to a starred atom (bonding interaction) or unstarred atom (no bonding interaction). This gives rise to three distinct classes of heterocyclic mesomeric betaine: conjugated, cross-conjugated and semi-conjugated (Fig. 3).

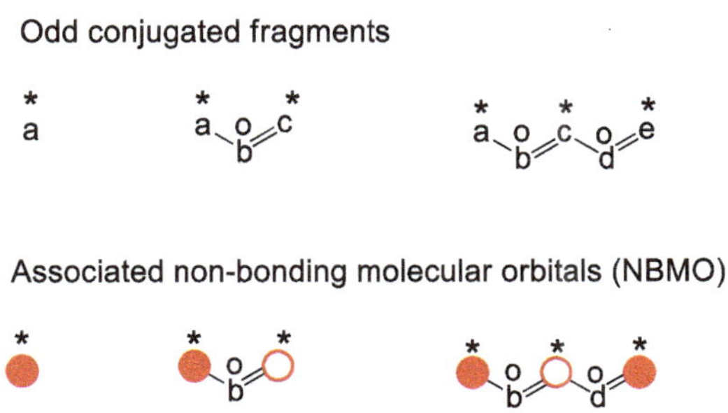

Fig. 2 Bonding characteristics of odd π-conjugated molecular fragments.

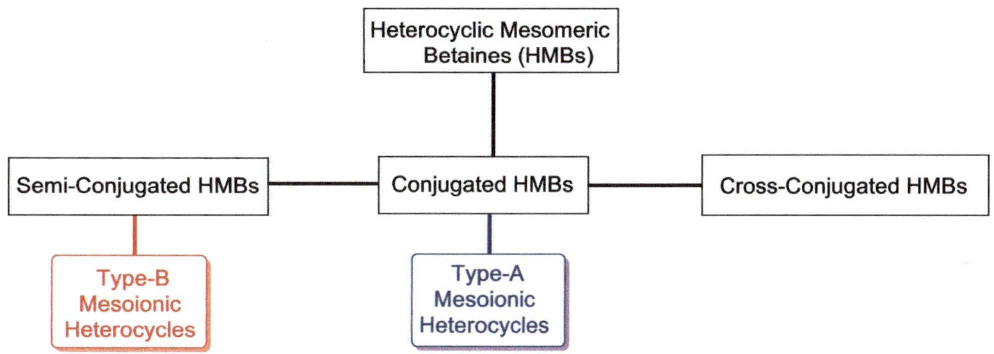

Fig. 3 The classification of type A and B mesoionic heterocycles.

In conjugated HMBs at least one of the 2π heteroatoms is attached to starred positions of both odd fragments, e.g., structure **11** in which the bottom nitrogen is fully conjugated with both odd fragments via starred positions. This situation also includes type A mesoionic rings, e.g., nitron **12**, which form a subgroups of conjugated HMBs (Fig. 3).

11

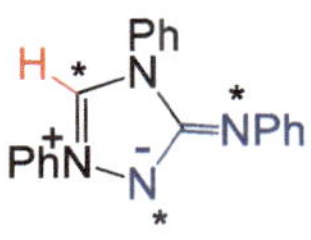

12 nitron

In cross-conjugated HMBs neither of the 2π heteroatoms is attached to a starred position of one of the odd fragments, e.g., **13**. In this case the negative and positive fragments are restricted to different parts of the ring and they are a HMB subgroup (Fig. 3) that is often associated with 1,4-dipolar cycloadditions.

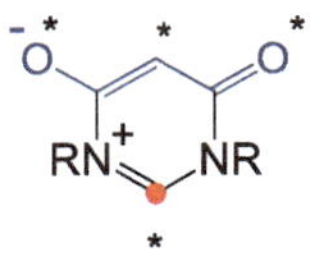

13

In semi-conjugated HMBs the two 2π heteroatoms are each attached to different odd fragments by a starred position. Semi-conjugated HMBs are rare and are illustrated by the tetrazene **14**. Note in particular that the top nitrogen interacts with a starred position of the red fragment but with an unstarred nodal position (no interaction) with the blue fragment **15**. The opposite situation occurs for the bottom nitrogen. This defines a very small subgroup of HMBs (Fig. 3) to which type B mesoionic rings belong.

Consider the type B system **16**. This contains a fragment in which the odd components (red and blue) are cross-conjugated. The top Ph-nitrogen has a bonding interaction with the red fragment but not with the blue fragment as shown in structure **17**. The bottom Ph-nitrogen has the opposite interactions. The situation is analogous to the structure **14** and for the same reasons type B systems are classified as semi-conjugated HMBs. In an earlier review type B systems were classified as conjugated HMBs,[6] but this is incorrect; their bonding is different and they are not associated with a 1,3-dipole fragment ($a{=}b^{+}{-}c^{-}$). Also, they are not cross-conjugated HMBs. An examination of the possible resonance hybrids reveals that neither the positive nor negative charges are restricted to one fragment. (In a rigorous analysis they are classified as pseudo-semi-conjugated HMBs but for most purposes this distinction is unnecessary. For an explanation of this sub-classification see Chapter "Heterocyclic mesomeric betaines: An overview" by Ramsden in this volume.[4])

14

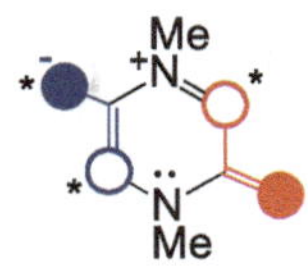

15

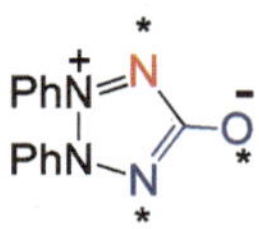

16

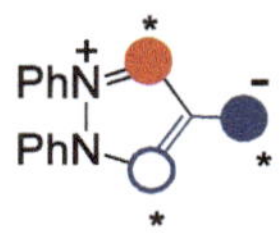

17

3. Type B mesoionic heterocycles—An overview

3.1 Known mesoionic systems of type B

If the atoms or groups a, b, c, d, e and f are selected from suitably substituted carbon, nitrogen, oxygen and sulfur atoms, 84 discrete mesoionic systems are described by the general formula **4** (Table 1).[1] However, some of these are more stable as valence tautomers (see Section 3.4.2). In 1976 eight type B ring systems were known. In 1980 this had increased to 13 and since then no new ring systems have been characterized. The known systems and the sections where they are discussed are listed in Table 1. Where unknown systems are included in Table 1 this is because attempted preparation or computational studies have been reported.

In surveying a subject it is important to assess what is not known as well as what is known. Table 1 provides insight into known and unknown type B mesoionic rings.

Type B mesoionic rings contain a heteroatom–heteroatom fragment (b—c), which for symmetrical rings can be O—O, RN—NR or S—S. These rings can be regarded as being derived by union of the heteroatom pair (b—c) **18** with a cross-conjugate fragment (a—e(d)=f) (Fig. 4). Examination of columns b, and c of Table 1 reveals how many type B mesoionic rings with each heteroatom pair (a—b) are known, and these are shown in brackets for each fragment in Fig. 4.

Table 1 Known and unknown mesoionic heterocycles of type B.

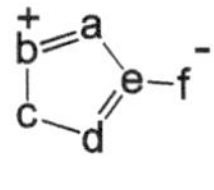

4

		Atom or group						
Parent system	**Section**	**a**	**b**	**c**	**d**	**e**	**f**	
A. Dioxoles								
1,2-Dioxolium-4-olates	4.1	CR	O	O	CR	C	O	Unknown
B. Oxazoles								
1,2-Oxazolium-4-olates	4.2.1	CR	O	NR	CR	C	O	Known
1,2-Oxazolium-4-aminides	4.2.2	CR	O	NR	CR	C	NR	Known
C. Diazoles								
1,2-Diazolium-4-olates	4.3.1	CR	NR	NR	CR	C	O	Known
1,2-Diazolium-4-aminides	4.3.2	CR	NR	NR	CR	C	NR	Known
D. Thiazoles								
1,2-Thiazolium-4-olates	4.4.1	CR	S	NR	CR	C	O	Known
1,2-Thiazolium-4-aminides	4.4.2	CR	S	NR	CR	C	NR	Known
1,2-Thiazolium-4-methylides	4.4.3	CR	S	NR	CR	C	CR_2	Unknown
E. Dithioles								
1,2-Dithiolium-4-olates	4.5.1	CR	S	S	CR	C	O	Known
1,2-Dithiolium-4-thiolates	4.5.2	CR	S	S	CR	C	S	Known
F. Thiadiazoles								
1,2,5-Thiadiazolium-3-olates	4.6.1	N	S	NR	CR	C	O	Known
1,2,5-Thiadiazolium-3-aminides	4.6.2	N	NR	NR	N	C	NR	Unknown
G. Dithiadiazoles								
1,2,3,5-Dithiadiazolium-4-thiolates	4.7.1	N	S	S	N	C	S	Unknown
H. Tetrazoles								
1,2,3,4-Tetrazolium-5-olates	4.8.1	N	NR	NR	N	C	O	Known
1,2,3,4-Tetrazolium-5-aminides	4.8.2	N	NR	NR	N	C	NR	Known
1,2,3,4-Tetrazolium-5-thiolates	4.8.3	N	NR	NR	N	C	S	Known
1,2,3,4-Tetrazolium-5-methylides	4.8.4	N	NR	NR	N	C	CR_2	Known

Inspection of Fig. 4 reveals that rings with oxygen are little known, which may be attributable to weak O—O, O—NR and O—S bonds and associated ring opening to a more stable valence tautomer. Although oxazole and thiazole derivatives are known (Sections 4.2 and 4.4), their known chemistry is meagre. Ring nitrogens at positions 3 and 4 provide some stabilization and the chemistry of thiadiazolium-3-olates

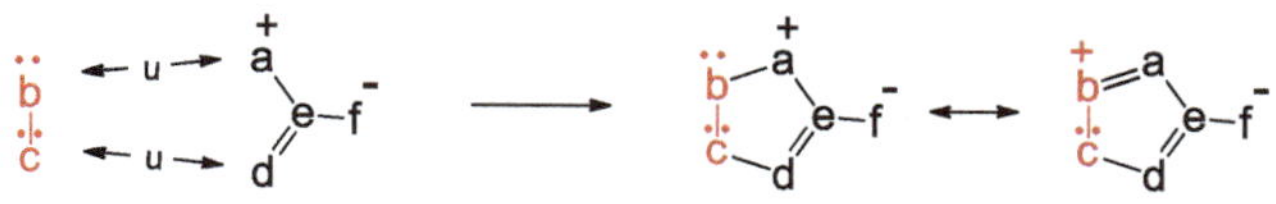

18 4

(0) (6) (2) (2) (0) (3)

Fig. 4 The numbers in brackets show the number of known type B ring systems containing the six possible ring fragments b–c containing O, N or S.

(Section 4.6.1) is a little more robust. The chemistry of type B systems is dominated by that of tetrazoles where the strong RN—NR bond and additional ring nitrogens provide stability. Surprisingly, triazole and dithiazole derivatives have not been reported, as far as we are aware, and there are some other obvious gaps in Table 1.

3.2 Structural aspects of type B mesoionic rings

Most studies have been focused on symmetrical derivatives and these molecules are best represented as resonance hybrids of structures **19** and **20**. The exocyclic bonds e—f are neither single nor double. When the ring atoms a are electronegative nitrogens the form **20** appears to make a greater contribution and is summarized by structure **21**. If atoms a are less electronegative carbon atoms, the contribution of form **19** increases, as summarized in structure **22**.

19 20

21 22

This view is supported by X-ray crystallography and theoretical studies. Most studies have been on tetrazoles. Two structures that are not tetrazoles

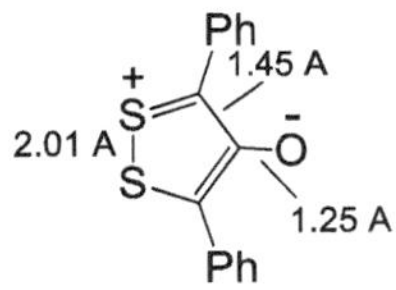

23

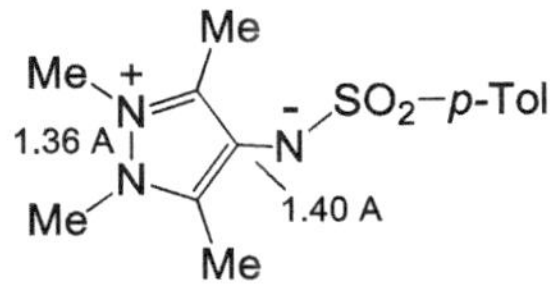

24

Fig. 5 Bond lengths for two non-tetrazole type B rings determined by X-ray crystallography.

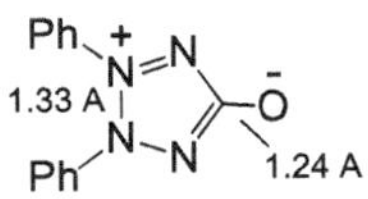

25

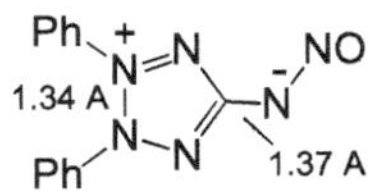

26

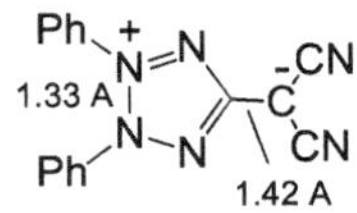

27

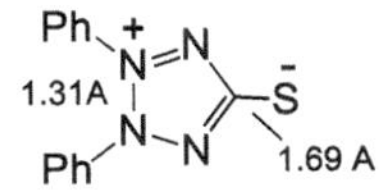

28

Fig. 6 Structures of type B tetrazoles determined by X-ray crystallography.

are shown in Fig. 5. In the 1,2-dithiol-4-one **23** the CO bond is intermediate between a single and double bond and comparable to those in amides.[7,8] This suggests a mixture of structures **19** and **20**, and this is consistent with the measured dipole moment of the 3,5-di-tert-butyl derivative (3.50 D).[8] In an ab initio MP2 study the calculated CO bond length of the 3,5-dimethyl structure (1.253 Å) is shortened to 1.206 Å in the 3,5-diaza analog, suggesting an increase in the contribution of hybrid **19**, as summarized in the form **21**.[9]

The imine **24** is the only reported X-ray structure of a type B pyrazole derivative.[10] It is interesting to compare structure **24** with structurally related tetrazoles (Fig. 6). In the tetrazolium-olate **25**[11] the exocyclic CO bond length is comparable to that in **23** but the N2—N3 bond length (1.33 Å), as in the other tetrazoles **26–28**,[12,13] is shorter than the pyrazole NN length **24** (1.36 Å). This slight shortening may be attributable to a change in electron distribution due to the ring nitrogens (**21** vs **22**) and a decrease in nuclear–nuclear repulsion. The exocyclic bonds of the four tetrazoles **25–28** (Fig. 6) are all intermediate between single and double bond values implying resonance similar to that in amides and related functional groups.

3.3 The aromaticity of type B mesoionic rings

Historically mesoionic rings have been associated with aromaticity mainly because it is convenient to associate both type A and B mesoionic rings with a 6π cationic ring and an exocyclic anion as shown in structure **29**. However, structural studies (Section 3.2) clearly show that the exocyclic bonds have double bond character and resonance structures of the type **30** and **31**, which are associated with seven π electrons in the ring, make a significant contribution to the structure. The index pEDA is the sum of ring p_z atomic orbital populations minus the aromatic sextet value of six. For seven representative type B rings, pEDA values were calculated to be in the range 0.07–0.32, consistent with electron-rich rings.[9]

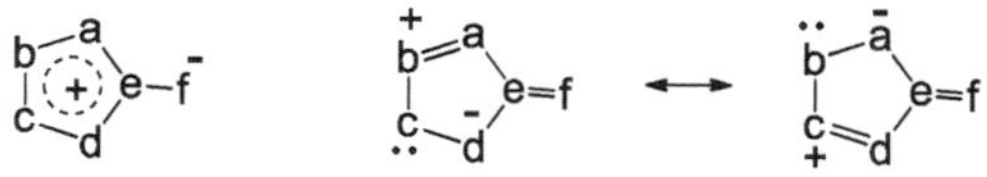

29 **30** **31**

Several indexes of classical aromaticity have been applied to type B rings and the general conclusion is that there is modest stabilization due to cyclic conjugation. Application of the Bird Index (I) of aromaticity to the X-ray structures of nineteen mesoionic rings, including structures **23**, **24**, **25** and **28**, suggested that type B rings have a higher degree of aromaticity than type A rings.[14] In an ab initio MP2 study, the Bird Index was found to decrease with aza substitution of the ring, suggesting lower aromaticity in aza derivatives due to a greater contribution of hybrids **30** and **31**.[9] This conclusion is also supported by aromatic stabilization energies (ASEs) calculated using homodesmotic schemes. The calculated ASEs are lower for aza derivatives but the values for type B rings are greater than those for type A rings, which have values close to zero.[9, 15] Overall, based on their structures and reactions there is little evidence to suggest that type B mesoionic rings are associated with significant classical aromaticity due to cyclic conjugation. They are probably best regarded as heterocycles enjoying some resonance stabilization.

A difference between classical and magnetic aromaticity has been recognized,[16] and for a series of heterocyclic rings different trends are observed.[17] The NICS(1) magnetic aromaticity index for representative type B pyrazole-4-one and tetrazole-5-one derivatives were calculated to be in the range −7.0 to −7.5 (benzene −10.2).[15] These values are consistent with modest magnetic aromaticity.

3.4 General modes of reaction of type B mesoionic rings

Two general modes of reaction of type B mesoionic rings merit discussion. In the first the ring undergoes cycloaddition (Section 3.4.1) leading to a heteropentalene intermediate. In the second mode valence tautomerism to an acyclic form precedes further reaction (Section 3.4.2).

3.4.1 Cycloadditions

This mode of reaction has been observed using electron-deficient alkenes, alkynes and heterocumulenes. Typically, the reagent wx adds to atoms a and f to give a bicyclic intermediate **33** (Scheme 1). Although shown as a concerted mechanism in structure **32**, this may well be a two-step mechanism in which the nucleophilic exocyclic atom f add to the electron-deficient reagent followed by cyclization. The bicyclic products **33** are not stable and ring opening by cleavage of the a—b bond gives the dipolar products **34** which may be stable or lose atom b (e.g., as a nitrene PhN: or S) to give product **35**.

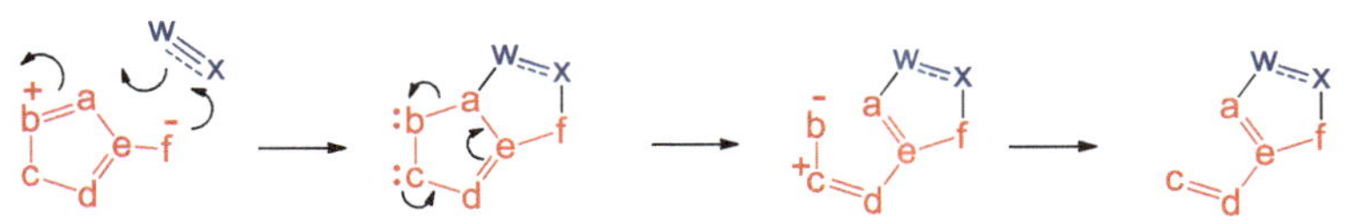

32 **33** **34** **35**

Scheme 1

This type of reaction is illustrated by the work of Boyd, Norris and coworkers who demonstrated, for example, that dehydrodithizone **36** reacts with dimethyl acetylenedicarboxylate (DMAD) to give the thiazole derivative **37** (Scheme 2).[18] In an analogous reaction with benzyne at 150 °C the initial product loses NPh (isolated as azobenzene) giving 2-phenylazobenzothiazole, which corresponds to general structure **35**.[18]

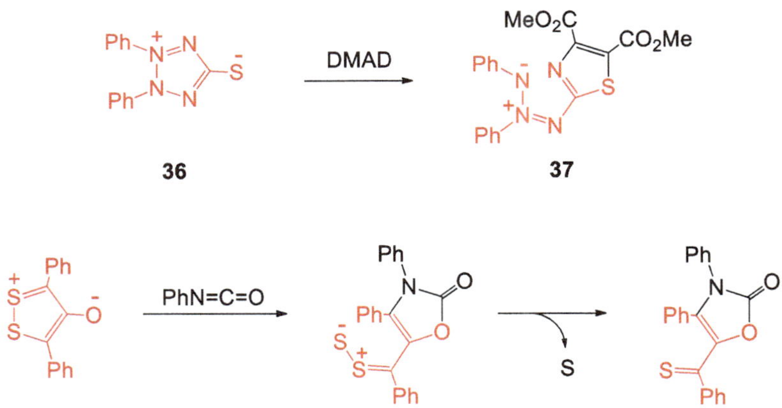

36 **37**

38 **39** **40**

Scheme 2

A similar example is the reaction of the dithiolium-4-olate **38** with phenylisocyanate. In this case the product **39** loses sulfur to give the thioketone **40** (Scheme 2).[19]

3.4.2 Valence tautomerism

In general, type B mesoionic rings **41** may equilibrate with the acyclic valence tautomers **42** (Scheme 3). An MP2 ab initio study suggests that in the gas phase the relative stability of the ring increases with the strength of the single bond b—c (i.e., S—S > RN—NR > O—O). Inclusion of aqueous solvation in the calculations increases the relative stability of the ring, presumably due to the ring polarity.[9] In practice, dioxole derivatives **41** (b–c = O–O) are unstable and exist exclusively as the dicarbonyl valence tautomers **42** (a=b and d=c are C=O).[7] At the other end of the stability range, known dithiole derivatives **41** (b–c = S–S) (Section 4.5) are stable and studies so far have provided only speculative evidence of valence tautomerism. For rings with an N—N bond (**41** b–c = RN–NR) an equilibrium with the acyclic form can be established and, although there is no evidence of this for diazoles (Section 4.3), the known chemistry of type B tetrazoles (Section 4.8) is rich in transformations occurring via valence tautomerism.

41 **42**

Scheme 3

Although the valence tautomer **44** is not detectable, many reactions of dehydrodithizone **43** proceed via initial ring opening (Scheme 4). Acid catalyzed heating results in recyclization to the product **45**.[20] Trapping with enamines gives [4+2] cycloadducts, e.g., **46**. Similar trapping with ynamines occurs but in this case the initial cycloadduct eliminates sulfur giving pyrazoles, **47** (Scheme 4).[21]

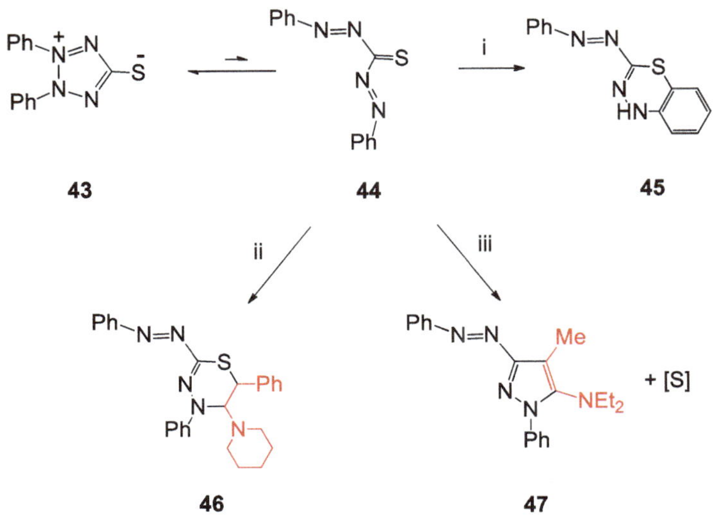

Reagents: i, Δ, AcOH; ii, β-piperidinostyrene; iii, diethylaminopropyne

Scheme 4

An interesting variation occurs when the exocyclic atom carries a nitroso substituent, e.g., **48**. In this case recyclization of the tautomer **49** leads to N-oxide **50** formation (Scheme 5).[22]

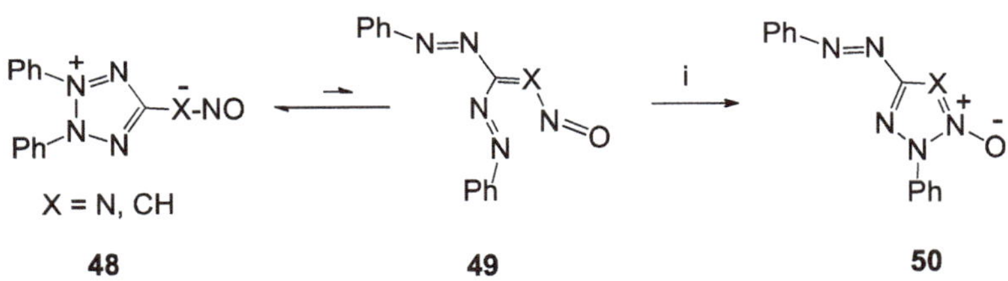

Scheme 5

4. The chemistry of type B mesoionic compounds

4.1 Dioxoles

4.1.1 1,2-Dioxolium-4-olates (51)

51 **52**

The 1,2-dioxolium-4-olates **51** are an example of the acyclic valence tautomers, i.e., **52**, being much more stable than the mesoionic ring (see Section 3.4.2). Ab initio calculations on the derivatives **51** and **52** (R=Me) indicate a free energy difference (ΔG) of 78 kcal mol^{-1} in the gas phase and 75 kcal mol^{-1} in an aqueous environment.[9] A 1978 X-ray crystallography study showed no evidence of a diaryl mesoionic tautomer (**51**; R=4-BrC_6H_4).[7] In a photochemical study in the temperature range +20 to −170°C, irradiation of the diphenyl derivative **52** (R=Ph) in methylcyclohexane solution showed no spectroscopic evidence of formation of the valence tautomer **51** (R=Ph).[23] The chemistry of vicinal polycarbonyl compounds, including the triketones **52**, has been comprehensively reviewed.[24]

4.2 Oxazoles

4.2.1 1,2-Oxazolium-4-olates (53)

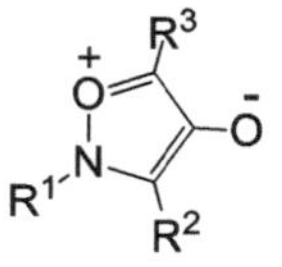

53

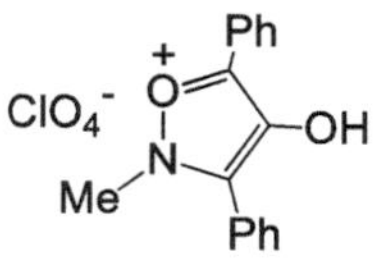

54

In an attempt to prepare the derivative **53** (R^1 =Me, R^2 = R^3 =Ph) Boyd and Norris reported that treatment of the perchlorate **54** with aqueous NaOH resulted in irreversible decomposition with only a transient yellow colouration suggesting initial formation of the desired product.[25] Later using methanolic alkali Friedrichsen and Schmidt isolated the yellow crystalline derivative **53** (R^1 =Me, R^2 = R^3 =Ph) (84%). This study demonstrates that the 1,2-NO bond provides stability relative to the acyclic valence tautomer, although it should be noted that the crystalline product **53** (R^1 =Me, R^2 = R^3 =Ph) was described as unstable.[26] No further studies on this ring system have been reported.

4.2.2 1,2-Oxazolium-4-aminides (55)

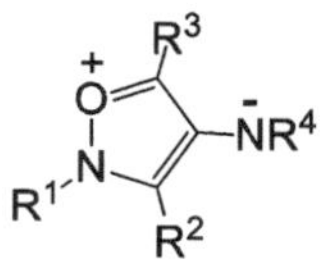

55

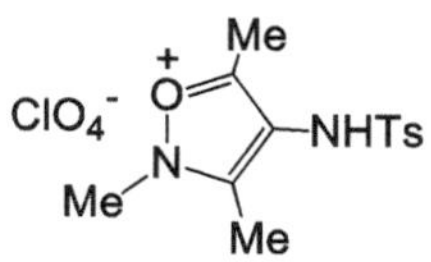

56

Mesoionic 1,2-oxazolium-4-aminides **55** have not been isolated. Treatment of the perchlorate salt **56** with aqueous KOH in acetonitrile resulted in a UV spectrum (λ_{max} 250 nm) that was attributed to the derivative **55** ($R^1 = R^2 = R^3 =$ Me, $R^4 =$ Ts). Attempts to isolate the product resulted in the formation of tars.[25]

4.3 Diazoles

4.3.1 1,2-Diazolium-4-olates (57)

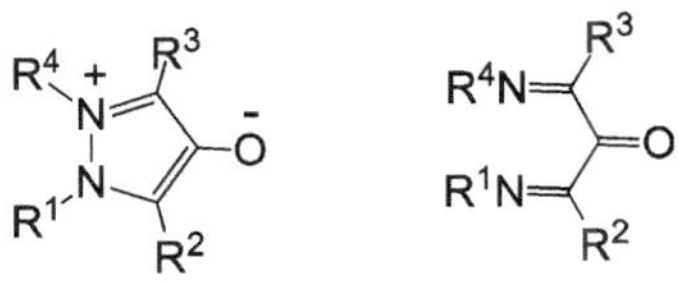

57 **58**

Ab initio calculations on the derivatives **57** and **58** ($R^1 = R^4 =$ Me, Ph, $R^2 = R^3 =$ H, Me) indicate that: (i) in the gas phase the valence tautomers **58** are more stable by ca. 12 kcal mol^{-1}; (ii) aqueous solvation stabilizes the mesoionic tautomers by 12±3 kcal mol^{-1}; (iii) inclusion of a hydrogen bonded water molecule with both tautomers increases the relative stabilization of the mesoionic ring by ~5 kcal mol^{-1}.[9] Synthetic studies have shown that the diazoles **57** do have greater stability than the dioxoles **51** and the oxadiazoles **53** and **55**. It is also notable that most derivatives have been fully characterized as monohydrates.

A ^{13}C and ^{14}N NMR study of the derivatives **57** ($R^1 = R^4 =$ Me, $R^2 = R^3 =$ Ph) and **57** ($R^1 =$ Me, $R^2 = R^3 = R^4 =$ Ph) demonstrated that these derivatives are present as the cyclic tautomers in $CDCl_3$ and D_2O. Acidification of the solvents indicated that the exocyclic oxygen atom is the preferential site of protonation.[27]

The preparation of derivatives **57** either from 4-hydroxypyrazoles or from dimethylhydrazine has been reviewed.[1, 2] A major study involving the preparation of eight fully characterized derivatives has been reported by Fagan and coworkers.[28]

Attempts to achieve cycloaddition reactions of the diazolium-4-olates **57** have been disappointing with no significant results using alkynes or alkenes reported. However, the derivative **59** undergoes an interesting reaction with singlet oxygen leading to 1,2-dibenzoyl-1,2-dimethylhydrazine **62** (89%). A mechanism involving the cycloadduct **60** and the rearranged carbene **61** has been proposed (Scheme 6).[29]

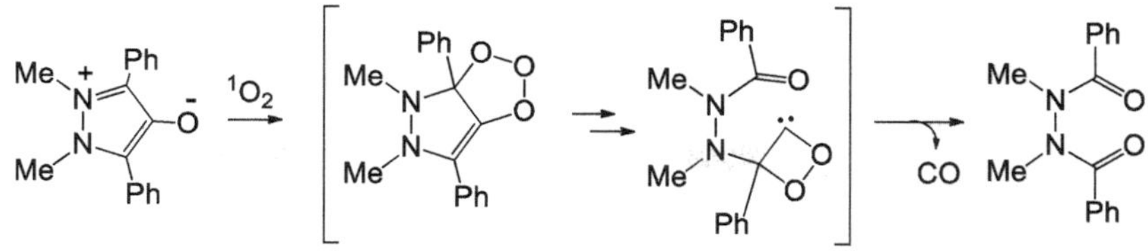

59 60 61 62

Scheme 6

Using homodesmotic reaction schemes the aromatic stabilization energy (ASE) of the dimethyl derivative **63** has been calculated to be 15.8 kcal mol^{-1}. This is significantly greater than that of the type A isomer **64** (ASE 0.5 kcal mol^{-1}), but may reflect symmetrical resonance stabilization rather than true classical aromatic cyclic conjugation. The calculated NICS(1) value of −7.12 for ring **63** indicates moderate magnetic aromaticity. The calculated exocyclic C—O bond lengths (**63** 1.246 Å and **64** 1.231 Å) indicate greater polarity of the CO bond in the type B isomer **63**.[15]

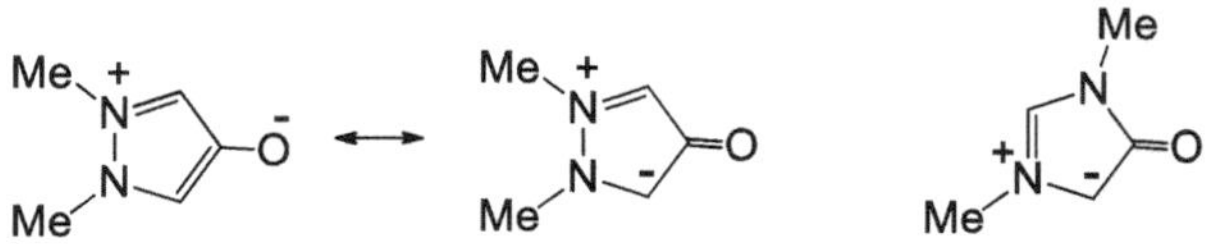

63 64

4.3.2 1,2-Diazolium-4-aminides (65)

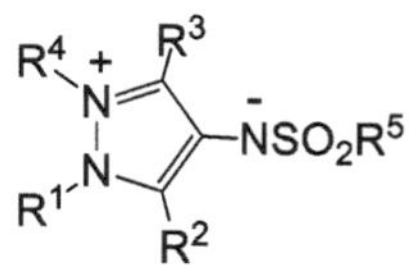

65

Representative examples of the diazolium aminides **65** have been prepared by a sequence of N-mesylation/tosylation, N-methylation and deprotonation of 4-amino-1,3,5-trimethylpyrazole.[25] The structure of derivative **65** ($R^1 = R^2 = R^3 = R^4 =$ Me, $R^5 =$ 4-$MeC_6H_4SO_2$) has been confirmed by X-ray crystallography.[10] No new studies have been reported since the previous reviews.[1, 2]

4.4 Thiazoles

4.4.1 1,2-Thiazolium-4-olates (66)

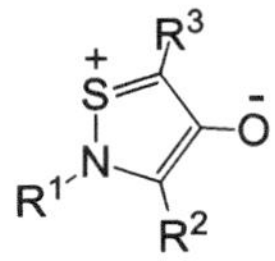

66

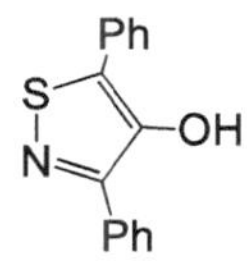

67

Alkylation of the isothiazole **67** followed by base treatment gives the red 1,2-thiazolium-4-olates **66** (R^1 = Me, Et, R^2 = R^3 = Ph) and their UV and ^{13}C NMR spectra have been reported. Their measured dipole moments are 6.23 and 5.26 D, respectively.[8] These studies and related reactions have been reviewed,[2] and no new work has been reported.

4.4.2 1,2-Thiazolium-4-aminides (68)

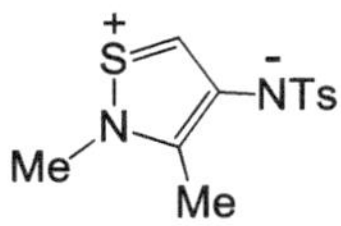

68

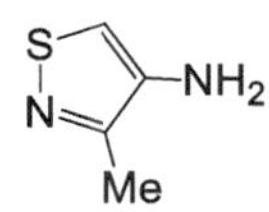

69

Work on this ring system is limited to a single report on the preparation of the derivative **68**, prepared by sequential N-tosylation, N-methylation and deprotonation of 4-amino-3-methylisothiazole **69**. This pale yellow crystalline product decomposes at its melting point (175–180 °C).[25] No further studies have been reported.

4.5 Dithioles

4.5.1 1,2-Dithiolium-4-olates (70)

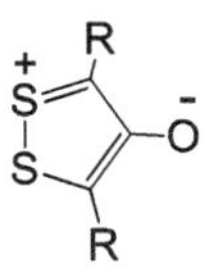

70

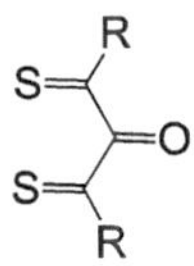

71

The dithiol-4-olates **70** have greater stability than the analogous dioxoles **51** and the diazoles **57** and this is largely attributable to the greater strength of the S—S bond. Ab initio calculations on the valence tautomers **70** (R = Me) and **71** (R = Me) indicate a free energy difference (ΔG) favoring the cyclic form of $-3.9\,\text{kcal mol}^{-1}$ in the gas phase and $-8.3\,\text{kcal mol}^{-1}$ in an aqueous environment.[9] Calculations on the unsubstituted derivatives **70** (R = H) and **71** (R = H) reported similar conclusions.[30] These calculations suggest that for simple derivatives, i.e., R = alkyl, aryl, the valence tautomers **71** may be accessible as reaction intermediates but in solution and the solid state the cyclic tautomers **70** are more stable. However, judicious choice of the substituents R may favor the acyclic form **71** (see below). The cyclic structure of the 3,5-diphenyl derivative **70** (R = Ph) has been confirmed by X-ray crystallography.[7, 8] Methods of preparing mesoionic dithiolones **70** from (i) 1,1,3,3-tetrabromo-1,3-diarylacetones ($ArCBr_2COCBr_2Ar$), (ii) 1,3-diarylpropanetriones ArCOCOCOAr, and (iii) propane-1,3-dione esters [(RCO_2)CHOAc] have been summarized in previous reviews.[1, 2] Treatment of 1,3-dichloroacetone with sulfur followed by methylation (MeI) gives the 3,5-dimethylthio derivative **70** (80%).[31] Treatment of the derivative **72** with either *N,N*-dimethylaniline or morpholine gives the derivatives **73–76** (Scheme 7). This study provides spectroscopic evidence to support the view that the morpholino derivatives **75** and **76** exist as the acyclic valence tautomers, e.g., **71** (R = morpholino).[32]

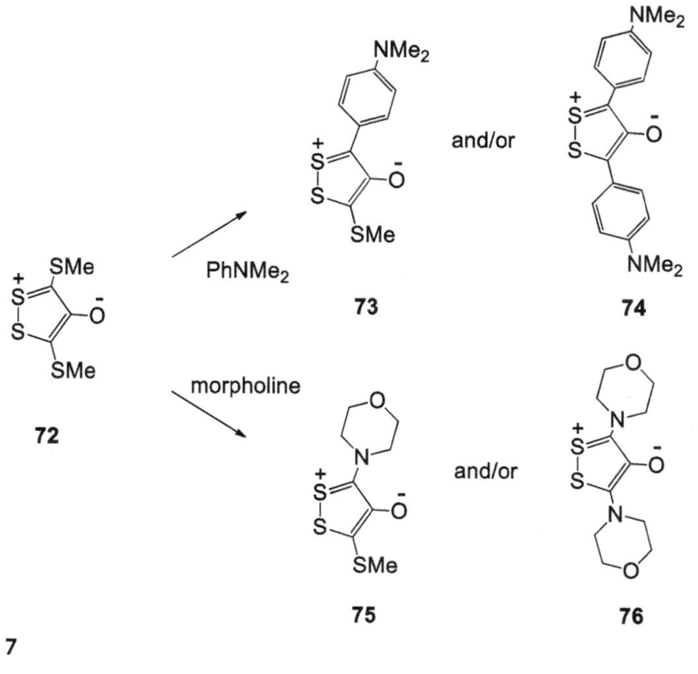

Scheme 7

Reaction of symmetrical derivatives **77** (R^1 = Ph, ^{t}Bu, PhNHCO) (eight examples) with dienes **78** gives the [4 + 3] cycloadducts **79** with yields in the range 34–81% (Scheme 8). Unexpectedly the dimethylthio derivative **72** with 2,3-dimethylbuta-1,3-diene **80** gave the thione 2:1 adduct **81** (35%). The authors suggest that this product may be formed via the valence tautomer **71** (R = SMe).[33]

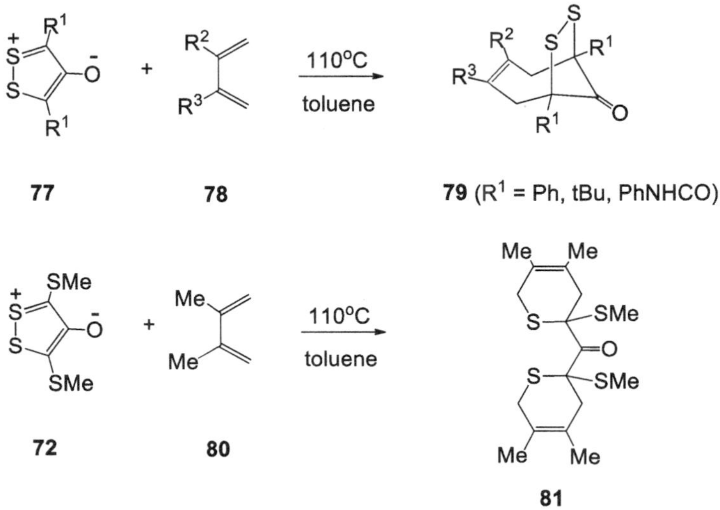

Scheme 8

The 3,5-diphenyl derivative **82** reacts with singlet oxygen in a manner analogous to 1,2-dimethyl-3,5-diphenyl-1,2-diazol-4-one **59** (Section 4.3.1). [3 +2] Cycloaddition of oxygen across the 3-position and the exocyclic oxygen gives an initial molozonide-type adduct that rearranges and loses carbon monoxide leading to 1,2-dibenzoyl disulfide **83** (72%) (Scheme 9).[29]

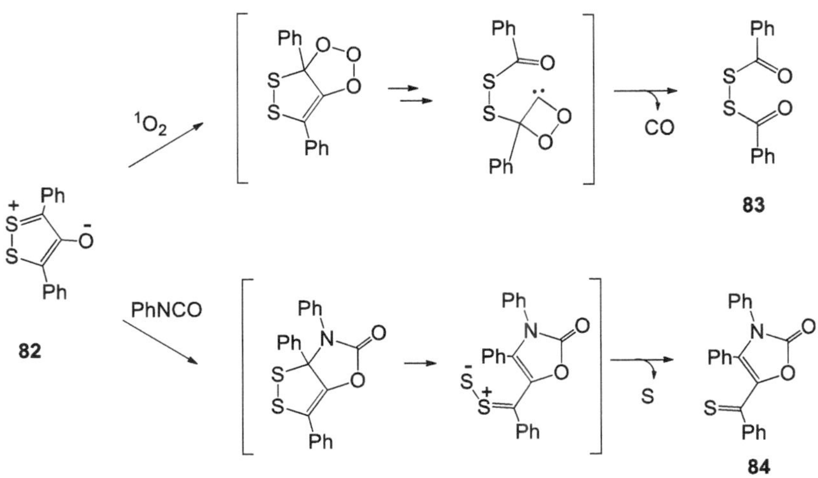

Scheme 9

A similar initial [3 + 2] cycloaddition of phenylisocyanate across the 3-position and the exocyclic oxygen of the diphenyl derivative **82** has been observed (Scheme 9). The structure of the thioketone product **84**, formed by ring opening of the cycloadduct and loss of sulfur, has been determined by X-ray crystallography and the incorrect structural assignment in an earlier publication corrected.[19]

Full details of two independent studies of the reaction of the diphenyl derivative **82** with aniline have been published.[34, 35] Both studies confirm by X-ray crystallography that the initial red product has the 1-phenylimino-2-phenylamino-3-phenylindene structure **85**. McKinnon and coworkers demonstrated that use of redistilled aniline gave only the red product **85**.[34] Use of impure aniline or aniline with H_2S gave an additional blue product (15–24%) to which they assigned the structure **88** on the basis of the formation of an identical product by treatment of 3-phenylindane-1,2-dione **87** with 2-aminobenzenethiol.[34] Ollis and coworkers also established the role of H_2S in the formation of the blue product.[35] They demonstrated reaction conditions that favor either the red product or the blue product depending upon reaction time and H_2S/S_8 content. However, the Ollis group favors the isomeric structure **86** for the blue product on the basis that Raney-nickel desulfurization gave the indole derivative **89**, identical with an authentic sample.[35] The mechanisms of formation of the initial red product **85** and the blue product (**86** or **88**) are still open to speculation.

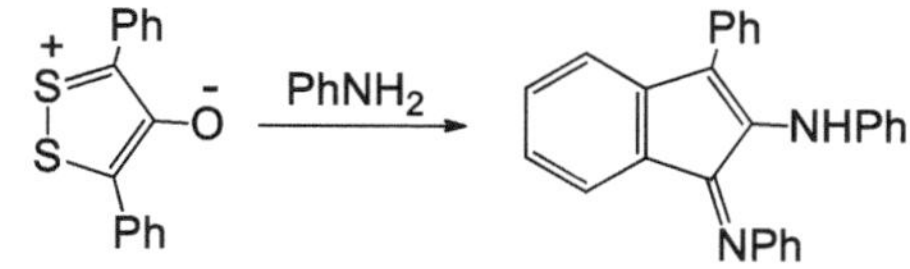

82 **85**

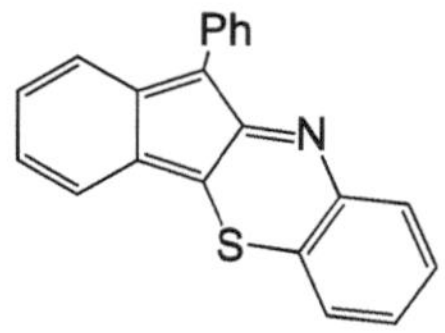

86

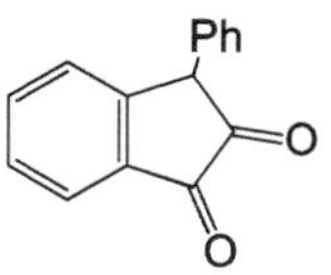

87

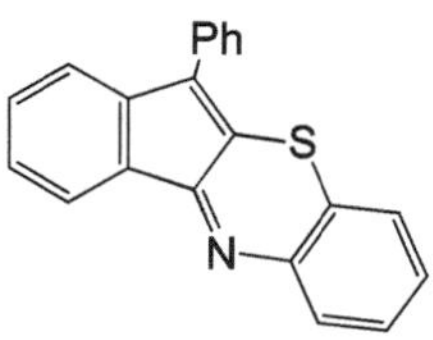

88

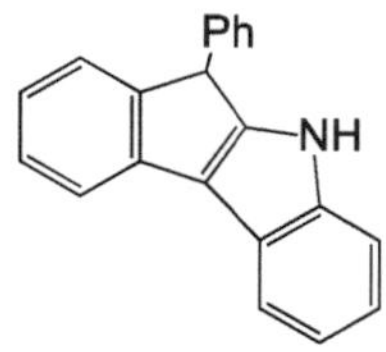

89

4.5.2 1,2-Dithiolium-4-thiolates (*90*)

90 **91**

A study of the structure of the parent molecule **90** (R = H) using ab initio methods calculated the valence tautomers **90** and **91** to be nearly equal in energy in the gas phase. However, the strong polarity of the mesoionic ring suggests that this form may be preferentially stabilized in solution.[30]

Compound **92** (deep violet crystals, mp 157–158 °C) is the only known example of this ring system and was prepared by the sequence shown in Scheme 10. Oxidation by hydrogen peroxide gave the disulfide **93** (72%).[36]

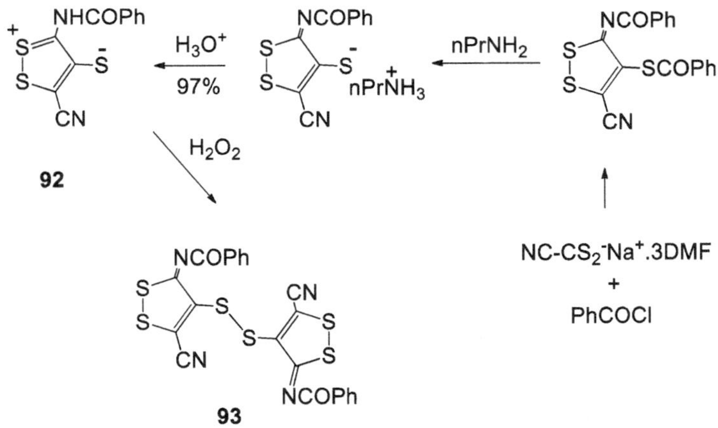

Scheme 10

4.5.3 1,2-Dithiolium-4-methylides (*94*)

As far as we are aware, examples of this ring system are unknown. A study of the structure of the parent molecule **94** ($R^1 = R^2 = H$) using ab initio methods calculated the valence tautomers **95** and **96** to be more than 30 kcal mol^{-1} more stable than the ring. Structure **96** was calculated to be more stable than **95** by 4 kcal mol^{-1}.[30]

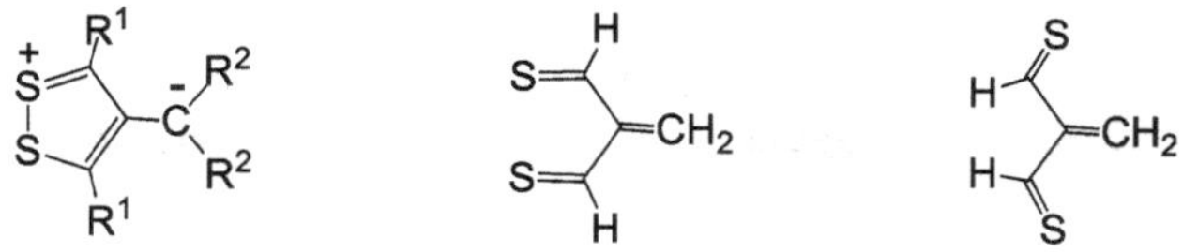

94 95 96

4.6 Thiadiazoles

4.6.1 1,2,5-Thiadiazolium-3-olates (97)

A full account of the preparation of 4-aryl-1,2,5-thiadiazolium-3-olates (**97**; R^2 = Ar) has been published.[37, 38] Reaction of arylacetamide derivatives **99** (R^2 = aryl) with sulfur monochloride in DMF gives the chlorides **98** in good yield. Treatment of these salts with base ($NaHCO_3$ or Et_3N) gives the mesoionic products **97** (R^2 = aryl) as yellow crystals (six examples) in 60–77% yield (Scheme 11). An aryl substituent at position 4 appears to be necessary for stability. The tricyclic derivative **103** was obtained in 65% yield but other precursors lacking a 4-aryl substituent, and also diaryl precursors, did not give the desired products. Attempts to prepare 5-benzyl derivatives resulted in debenzylation and formation of 3-hydroxy-1,2,5-thiadiazoles **104**.

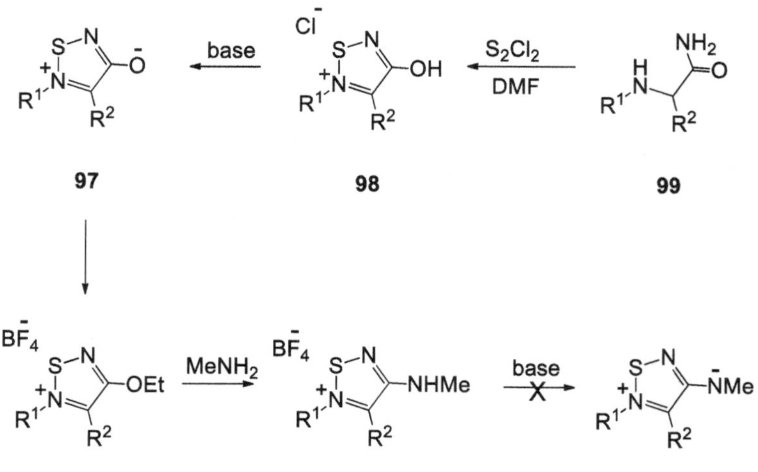

97 98 99

100 101 102

Scheme 11

The derivative **97** (R^1 = Me, R^2 = Ph) with dilute NaOH gave the α-iminoacetamide Ph(C=NMe)$CONH_2$ (38%) whereas reduction with sodium borohydride gave the amide **99** (R^1 = Me, R^2 = Ph) (64.5%). Treatment with triethyloxonium tetrafluoroborate gave the ethoxy salt **100** (R^1 = Me, R^2 = Ph) (58%) which with methylamine formed the 3-methylamine salt **101** (R^1 = Me, R^2 = Ph) (31%). Attempts to form an exo-thio product by treatment with sodium hydrosulfide were unsuccessful as were attempts to achieve reactions with dipolarophiles, such dimethyl acetylenedicarboxylate.[38]

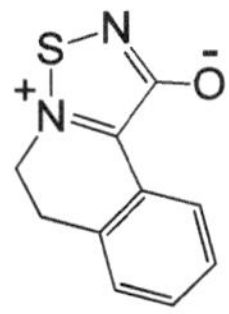

103

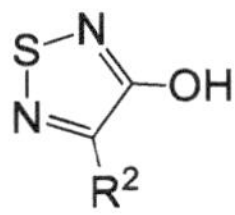

104

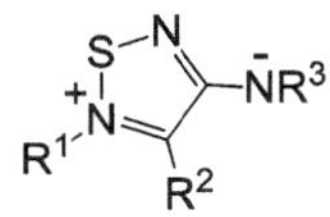

105

A ^{1}H, ^{13}C, ^{14}N and ^{15}N NMR study of the derivative **97** (R^1 = Me, R^2 = Ph) was consistent with the assigned cyclic structure with no evidence of an acyclic valence tautomer. The density of positive charge on N5 was found to be relatively high and protonation was shown to occur on the exo-cyclic oxygen.[39]

4.6.2 1,2,5-Thiadiazolium-3-aminides (105)

Attempts to prepare the derivative **105** (R^1 = R^3 = Me, R^2 = Ph) by reaction of the tetrafluoroborate salt **101** (R^1 = Me, R^2 = Ph) with base (Scheme 11) were unsuccessful.[38] Examples of this ring system remain unknown.

4.7 Dithiadiazoles

4.7.1 1,2,3,5-Dithiadiazolium-4-thiolates (106)

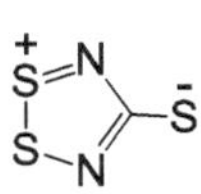

106

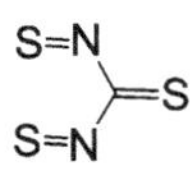

107

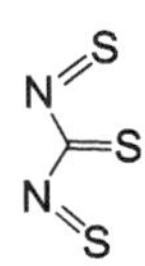

108

An ab initio study calculated the dithiodiazolium-4-thiolate **106** to be 45 kcal mol^{-1} more stable than the valence tautomer **107**. Both ab initio and semi-empirical AM1 studies calculated the fully *cis* valence tautomer **108** to be more stable than structure **107**.[9, 40]

In spite of the calculated stability of structure **106** relative to its valence tautomers, attempts to form this derivative have been unsuccessful. Treatment of the *S*-benzylisothiouronium salts **109** with sulfur monochloride (S_2Cl_2) gives good yields (55–73%) of the deep red crystalline *S*-benzyl-1,2,3,5-dithiadiazolium chlorides **110** (Scheme 12). The structure of the derivative **110** (Ar = 4-ClC_6H_4) was confirmed by X-ray crystallography. Reaction of the 4-chloro derivative with a variety of hard and soft nucleophiles resulted in immediate loss of the red color but no evidence of formation of either the mesoionic product **106** or its decomposition products was obtained. Treatment of the salt **110** (Ar = 4-ClC_6H_4) with thiourea in THF gave the *S*-benzylisothiouronium chloride **109** (Ar = 4-ClC_6H_4) and sulfur suggesting that the expected S_N2 reaction had occurred but the dithiadiazole ring had clearly decomposed.[40]

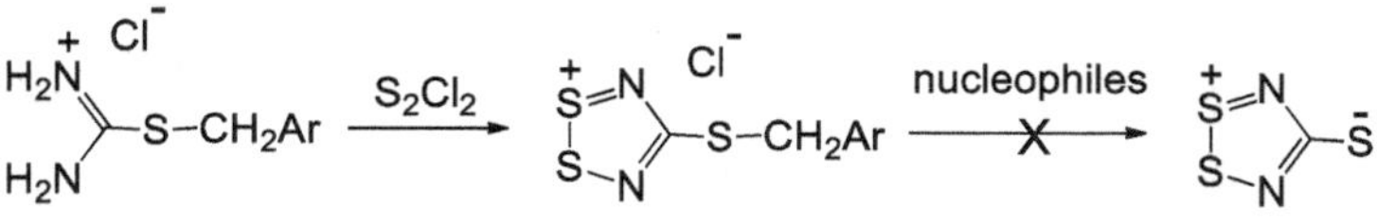

109 **110** **106**

Scheme 12

4.8 Tetrazoles

A review of the chemistry of type B tetrazoles covering the period 1980–2016 has been published.[41] A 2010 review of formazan chemistry contains sections on type B mesoionic tetrazoles.[42]

4.8.1 1,2,3,4-Tetrazolium-5-olates (107)

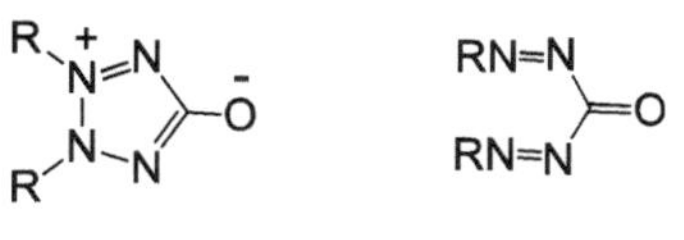

107 **108**

Compared to the corresponding 5-thiolates (Section 4.8.3), the known chemistry of the tetrazolium-5-olates (**107**) is very limited. The calculated energy differences between the valence tautomers **107** and **108** (R = Me, Ph), using the ab initio MP2 method, were found to be in the order of 17 kcal mol^{-1} (gas phase) and 30 kcal mol^{-1} (solvated), which are similar to values calculated for the 5-thiolates.[9] The calculated activation energies for valence tautomerism [ca. 30 kcal mol^{-1} (gas phase) and 40 kcal mol^{-1} (solvated)] for both the 2,3-diphenyl 5-olate and the 5-thiolate are also similar in magnitude.[22] However, in contrast to the 5-thiolates, there is no evidence of the 5-olates **107** reacting via their valence tautomers **108**. Based on these calculations, it is concluded that the tetrazolium-5-olates **107** may undergo thermal ring opening to give small equilibrium concentrations of the valence tautomers **108** but, in contrast to the 5-thiolates, the lack of alternative modes of reaction leads to recyclization to the mesoionic ring. The alternative mode of recyclization observed for the diphenyl 5-thiolate has an activation energy calculated to be much higher for the 5-olate.[22]

In general the known chemistry of the mesoionic 1,2,3,4-tetrazolium-5-olates **107** is centered on the formation of salts and complexes by the diphenyl derivative. Reaction with triethyloxonium tetrafluoroborate gives the 5-ethoxy derivative **109** (R = Et, X = BF_4) but further reactions using C and N nucleophiles were unsuccessful.[43] However, reaction with triflic anhydride gave an intermediate, presumed to be the salt **109** (R = SO_2CF_3, X = CF_3SO_3), which with malononitrile and triethylamine gave a 23% yield of the 5-methylide **110** (Section 4.8.4).[44] Reaction with cobalt halides [$Co(II)X_2 \cdot 6H_2O$] gives the blue complexes **111**.[45] Oxidation using nitrosonium tetrafluoroborate ($NOBF_4$) has been reported to give the peroxide salt **112**.[46, 47]

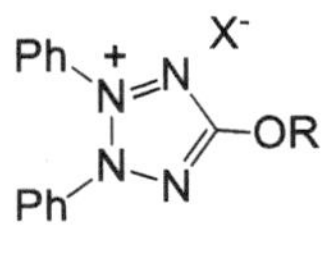

109

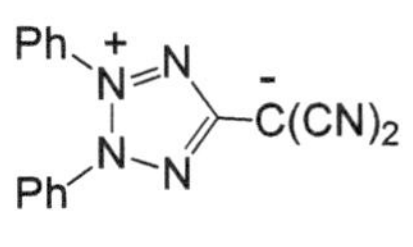

110

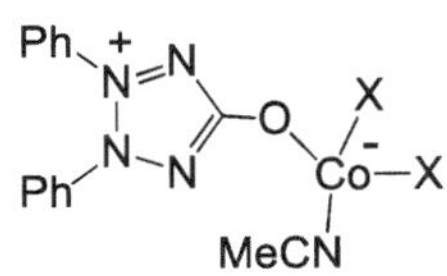

111 (X = Cl, Br)

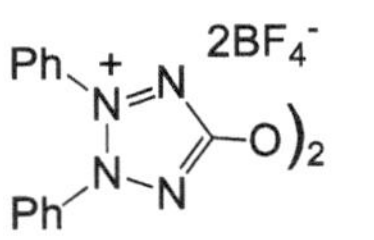

112

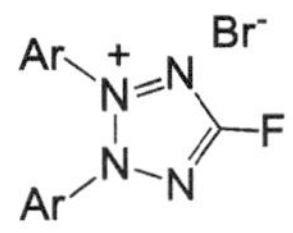

113

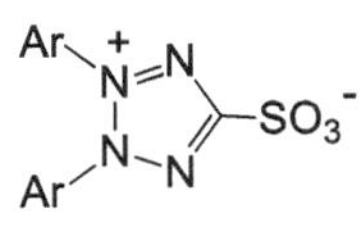

114

The main preparative methods have been covered in previous reviews.[1, 2] The *p*-tolyl derivative has been obtained by hydrolysis of the 5-fluoro salt **113** (Ar = *p*-Tolyl). Treatment of the betaine **114**, obtained by oxidation of the 5-thiolate, with KOH gave **107** (R = Ph) (8.5%), among other products.[47]

^{13}C and ^{15}N NMR studies of the diphenyl derivative support the mesoionic structure and demonstrate that protonation occurs on the exocyclic oxygen atom.[48, 49] The calculated [B3LYP/6-311++G(d,p)] NICS(1) value of −7.5 for the 2,3-dimethyl derivative **107** (R=Me) is consistent moderate magnetic aromaticity.[15]

4.8.2 1,2,3,4-Tetrazolium-5-aminides (115)

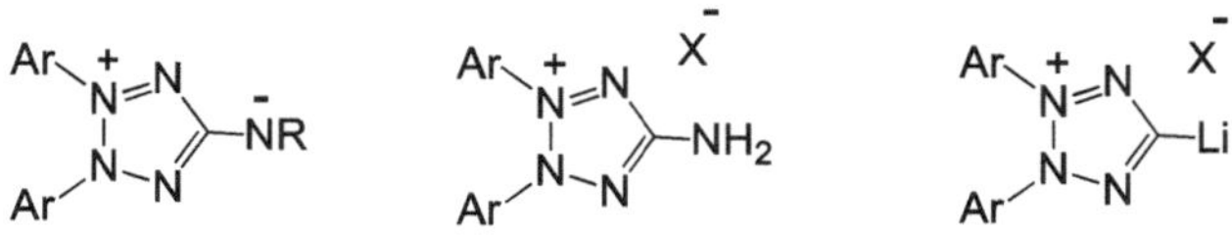

115 **116** **117**

Known stable examples of 1,2,3,4-tetrazolium-5-aminides are limited to 2,3-diaryl rings with the exocyclic nitrogen carrying an electron-stabilizing group **115** (R=NO, Ts). Preparative routes usually start with the 5-amino or 5-lithio salts **116** and **117**. The derivative **115** (Ar=Ph, R=NO) was prepared by Bamberger and coworkers by nitrosation of the chloride **116** (Ar=Ph, X=Cl) in 1926. A reinterpretation of these studies and the revision of the structural assignments was summarized in an earlier review.[1] The crystal structures of the mesoionic ring **115** (Ar=Ph, R=NO) and the precursor salt **116** (Ar=Ph, X=Cl) have subsequently been published.[12,50] ^{13}C and ^{15}N NMR data on the derivatives **115** (Ar=Ph, R=NO, Ts) has been reported.[48,49]

Treatment of the 5-lithio tetrafluoroborate **117** (X=BF_4) with tosyl azide at −90 °C is reported to give a small amount of the derivative **115** (Ar= *p*-anisyl, R=N=NTs) (detected by mass spectrometry). The main product was the 5-azido-tetrazolium tetrafluoroborate (43%).[51]

Interesting mesoionic tetrazole transformations have been reported by Araki and coworkers.[52] Reaction of the salt **118** with primary amines, e.g., benzylamine, using an inorganic base ($NaHCO_3$ or Na_2CO_3) gives the tetrazole products **119** in moderate yield. However, use of triethylamine as base results in formation of the 1,2,4-triazole products **121**. After investigation of possible mechanisms, the authors concluded that in the presence of Et_3N the most plausible mechanism is deprotonation of the initial product **119** leading to transient formation of a tetrazolium-5-aminides **120**. Ring opening to the valence tautomer **123** and a second deprotonation gives the anion **122** which upon recyclization and aerial oxidation gives the observed products **121** (23–50% yield) (Scheme 13).[52] If this mechanism

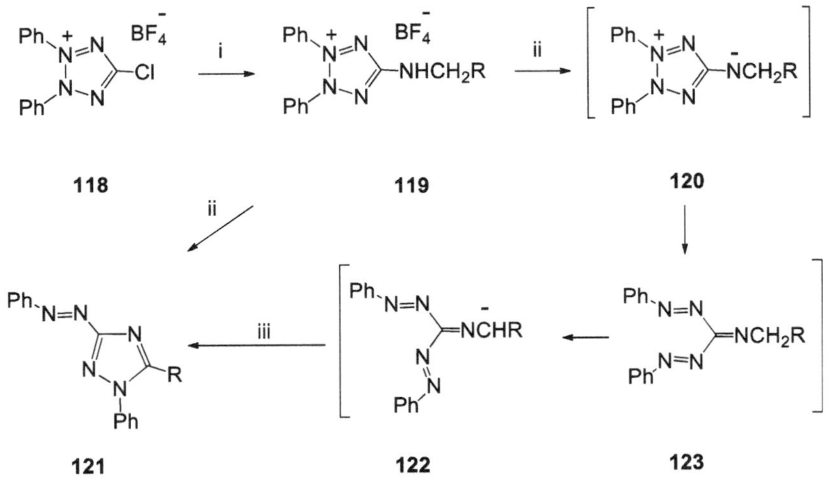

Reagents: i, RCH_2NH_2 + base; ii, Et_3N; iii, cyclisation + O_2.

Scheme 13

is correct, these reactions are a further example of the valence tautomerism of type B mesoionic rings (see Section 3.4.2). The thermal rearrangement of the derivative **115** (Ar = Ph, R = NO) also appears to occur via a similar valence tautomerism.[1,22]

DFT calculations show that the 2-methyl-3*H*-tetrazolium-5-aminides (*E* and *Z*) are more than 100 kJ mol^{-1} higher in energy than the 3-methyl-1*H* isomers.[53]

4.8.3 1,2,3,4-Tetrazolium-5-thiolates (124)

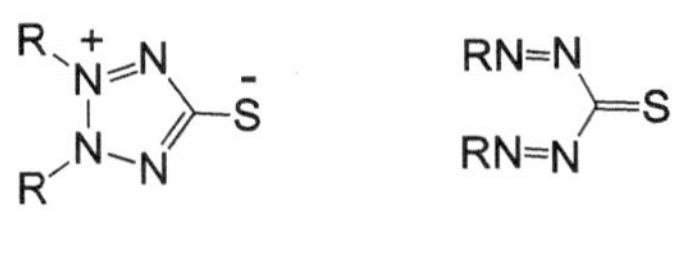

124 **125**

The chemistry of 1,2,3,4-tetrazolium-5-thiolates is dominated by dehydrodithizone **124** (R = Ph), which undergoes addition reactions with a wide variety of unsaturated reagents. In general, electron-rich reagents give products derived from the addition to the valence tautomer **125** (R = Ph) whereas electron-deficient reagents give cycloadducts derived from the mesoionic ring **124** (see Section 3.4). Previous reviews have described many of these reactions.[1,2]

In a more recent study, Potts and coworkers have shown that addition of benzyne to dehydrodithizone **124** (R = Ph) at 70 °C gives the benzothiazole derivative **126** (39%); the structure has been confirmed by X-ray analysis.[54,55] At 150 °C the 2-phenylazo derivative **127** is formed by loss of phenylnitrene.[18] In similar cycloaddition reactions dibenzoylacetylene gives the adduct **128** (77%) and *N*-ethylmaleimide gives the adduct **129** (13%) after loss of nitrene.[54] The product **126–129** are all formed by initial 1,3-addition across the exocyclic sulfur atom and the 1-position of dehydrodithizone, as discussed in Section 3.4.

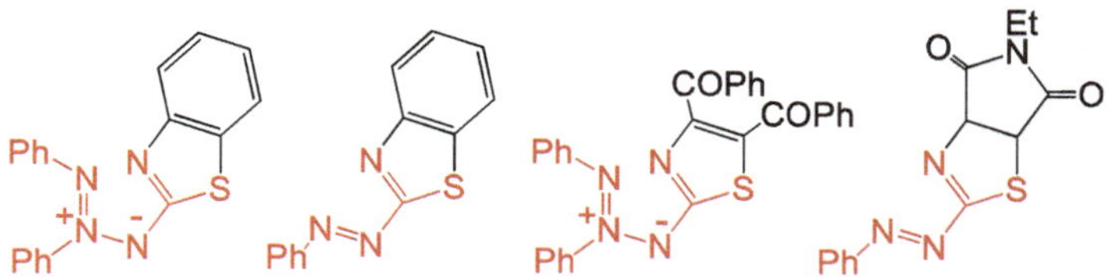

126 **127** **128** **129**

Reaction with diphenylcyclopropenethione is reported to give the 1,3,4-thiadiazepine derivative **130** (21%), presumably by a mechanism involving addition to the valence tautomer **125** (R = Ph) and subsequent rearrangement (Scheme 14).[54] An ab initio study calculated that in the solvated phase the valence tautomer is 32 kcal mol^{-1} higher in energy and with an activation energy of formation of 41 kcal mol^{-1}.[9, 22]

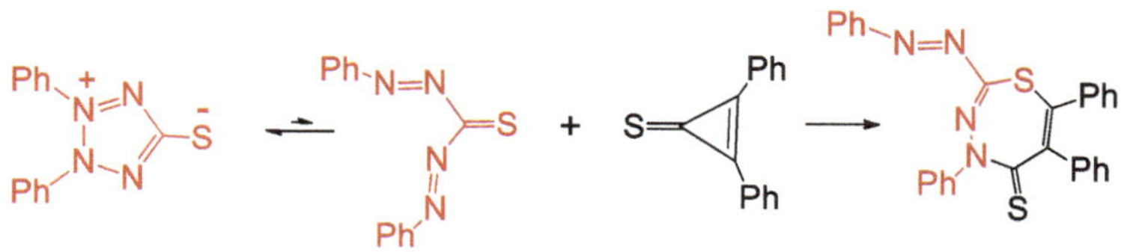

130

Scheme 14

Oxidation of dehydrodithizone gives a variety of structures with yields and product composition depending on reagents and conditions. These products include the salts **131** and **132**, the zwitterion **133** and 1-cyano-2,3-diphenylazimine **134**.[47]

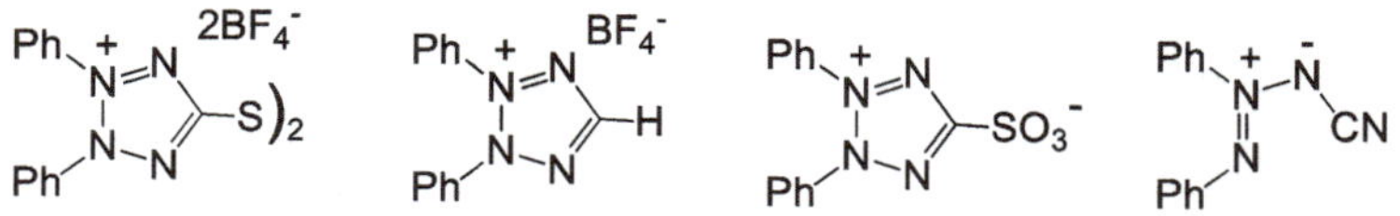

131 **132** **133** **134**

Reaction with one equivalent of bromine gives a yellow crystalline but unstable product formulated as the dibromide **135**. Reaction with the corresponding carbon anions gives the betaines (thiocarbonyl ylides) **136** and **137**.[56]

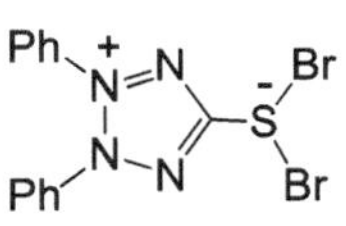

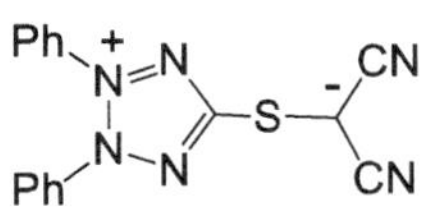

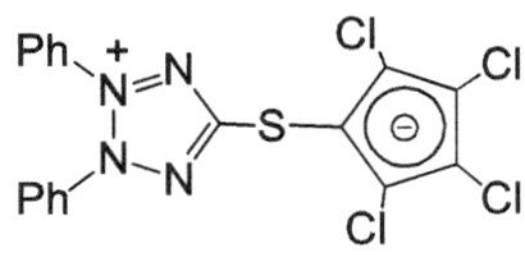

135 **136** **137**

The tetrazolium-5-thiolates react as sulfur bases, nucleophiles and ligands. They are Hammett bases with protonation occurring on the exocyclic sulfur. pK_{BH+} values have been measured for five 3,4-diaryl derivatives with $-pK_{BH+}$ values found to be in the range 1.7–2.1 at 25 °C.[57] The formation of S-CD_3 iodide salts by reaction of diaryl derivatives with trideuteriomethyl iodide has been reported.[58] Dehydrodithizone **124** (R = Ph) reacts as a sulfur ligand and the formation of metal complexes with ruthenium (II),[59] cobalt (II),[45] copper (II),[60] manganese (II),[61] chromium(I),[62] and rhenium (I)[63] has been reported.

Dithizone, the acyclic precursor to dehydrodithizone, is a well-known trace-metal analytical reagent. This has led to the exploration of polymer bound reagents. Reaction of dehydrodithizone **124** (R = Ph) with chloromethylated polystyrene gives the tetrazole derivative **138** (P-TD) which upon reduction gives the immobilized dithizone equivalent **139** (P-D) (Scheme 15). This resin shows selectivity, particularly for gold, platinum and palladium, and has been used for a number of applications.[64–69]

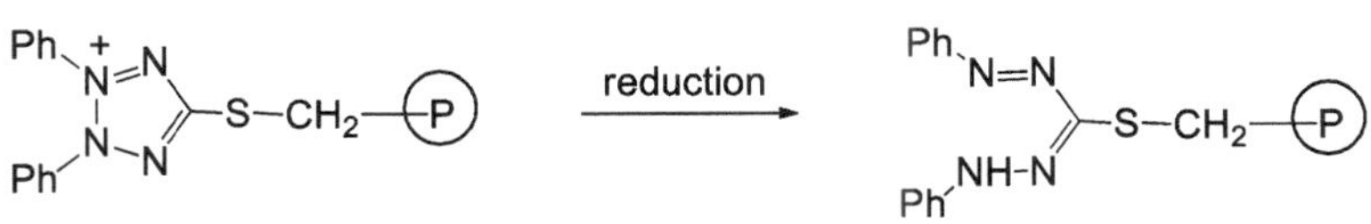

138 (P-TD) **139** (P-D)

Scheme 15

An improved procedure for oxidation of dithizones to dehydrodithizones employs either hydrogen peroxide or air as oxidant. The procedure was used to prepare the derivatives **124** (R = Ph, 4-MeC_6H_4, 2-MeC_6H_4, 4-ClC_6H_4) with yields in the range 95–98%.[70]

Three new X-ray single crystal structure determinations of dehydrodithizone **124** (R = Ph) have been reported.[61,70–72] In addition, X-ray structures of the derivatives **124** (R = 4-MeC_6H_4, 2-$MeOC_6H_4$, 3-FC_6H_4) have been published.[70,72–74] NMR studies with comparisons of ^{13}C and ^{15}N data to that of other mesoionic tetrazoles have been published.[48,49,75] The effect of solvent on the electronic spectra of seven diaryl derivatives has been investigated using thirteen pure solvents and eight water-organic solvent systems.[76]

DFT computational studies of the mesoionic tetrazolium-5-thiolate ring have been reported.[70,77,78] Ab initio MP2 studies of the energy profile of ring opening of dehydrodithizone **124** (R = Ph) to its valence tautomer and an alternative mode of recyclization have been reported. The energy profile of dehydrothizone is compared with those of analogs with different exocyclic groups (O, N—NO and CH—NO).[9,22]

4.8.4 *1,2,3,4-Tetrazolium-5-methylides* (140)

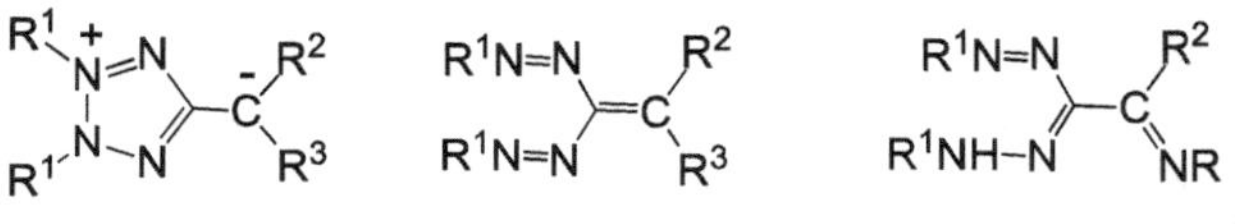

140 **141** **142**

The first examples of mesoionic tetrazolium-5-methylides **140** were reported in 1980. Mild oxidation of the appropriate formazan derivative **142** gives the derivatives **140** (R^1 = Ph, R^2 = R^3 = N=NPh) and **140** (R^1 = Ph, R^2 = H, R^3 = N=O).[79] Although the latter was originally formulated as the valence tautomer **141** (R^1 = Ph, R^2 = H, R^3 = N=O), it is now assumed that it has the mesoionic structure. In subsequent studies, Araki and coworkers prepared the dicyano derivative **140** (R^1 = Ph, R^2 = R^3 = CN) in 23% yield by reaction of the tetrazolium-4-olate **107** (R = Ph) with triflic anhydride followed by malononitrile and triethylamine. NMR studies showed no evidence of equilibrium with the valence tautomer **141** (R^1 = Ph, R^2 = R^3 = CN).[44] A later X-ray crystallography study confirmed that the derivatives **140** (R^1 = Ph, R^2 = CN, R^3 = CN, CO_2Et) have the mesoionic structure.[12] Multiple NMR studies are available.[49,75,80,81]

The formation of the C-nitroso derivative **143** by oxidation of the formazan **140** (R^1 = Ph, R^2 = H R^3 = N=O), as described above, occurs in very low yield (4%) and is accompanied by a greater yield (43%) of the triazole **146**. It is possible that both products are formed by initial oxidative generation of the valence tautomer **144** which can undergo alternative modes of cyclization to either the mesoionic product **143** or the N-oxide

145. Loss of oxygen then gives the observed product **146** (Scheme 16). Analogous formation of a tetrazole N-oxide is observed upon heating the N-nitroso tetrazolium-5-aminides **115** (Ar = Ph, R = NO).[1] The energy profiles of these alternative modes of cyclization have been studied using ab initio MP2 methods.[22]

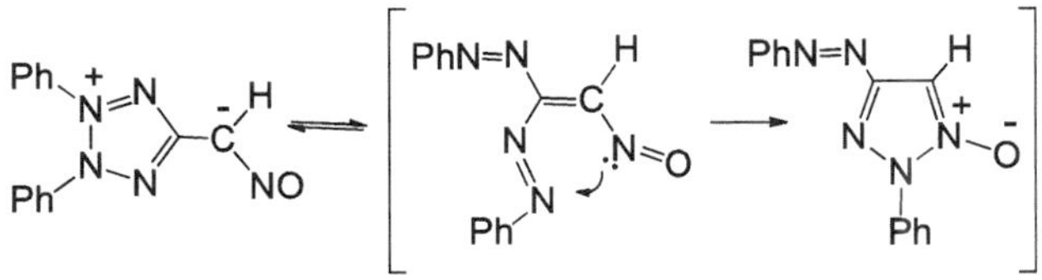

143 **144** **145** **146**

Scheme 16

The methylide **147a** (purple crystals), better represented as the polarized phenolate resonance form **147c**, has been prepared by deprotonation of the corresponding tetrazolium bromide. The *ortho* and *meta* isomers and sterically hindered derivatives were prepared in a similar way. In an anthraquinone derivative the less polar quinone methide structure **147b** is regarded as a better representation.[82,83]

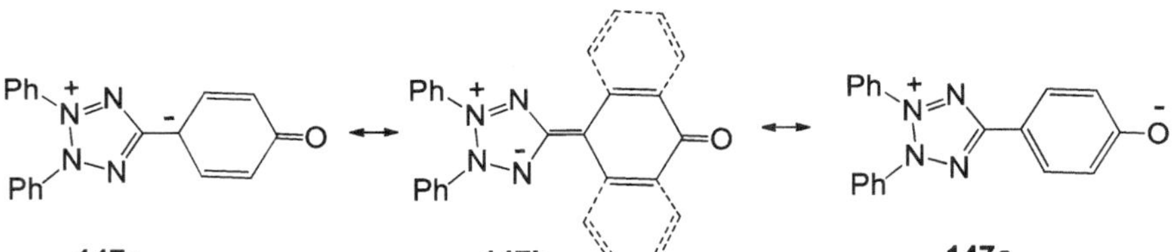

147a **147b** **147c**

5. Conclusions

In the period 1980–2020 type B mesoionic chemistry has not enjoyed the same attention as type A mesoionic chemistry. This is almost certainly because type B derivatives have not, so far, provided useful synthetic applications similar to the 1,3-dipolar cycloadditions of the type A family. In fact since 1980 no new type B ring systems have been revealed, although interesting new reactions of known systems have been reported. Stable type B mesoionic derivatives potentially have useful biological and physical properties and these possibilities are underexploited. They are a significant component of the pool of heterocyclic rings available for investigation as, for example, advanced materials or bioactive products. They also show interesting chemical transformations including cycloadditions and tautomerism which may have synthetic value and merit further study. This review highlights the know chemistry of type B mesoionic rings but also identifies area that have not been explored. We hope that this will encourage and facilitate new research in the next four decades.

References

1. Ollis, W. D.; Ramsden, C. A. *Adv. Heterocycl. Chem.* **1976**, *19*, 1.
2. Newton, C. G.; Ramsden, C. A. *Tetrahedron* **1982**, *38*, 2965.
3. Ramsden, C. A. *Prog. Heterocycl. Chem.* **2016**, *28*, 1.
4. Ramsden, C. A. *Adv. Heterocycl. Chem.* **2022**, 137 (Chapter 1).
5. Dewar, M. J. S.; Dougherty, R. C. *The PMO Theory of Organic Chemistry*; Plenum Press: New York, 1975.
6. Ollis, W. D.; Stanforth, S. P.; Ramsden, C. A. *Tetrahedron* **1985**, *41*, 2239.
7. Cannon, J. R.; Potts, K. T.; Raston, C. L.; Sierakowski, A. F.; White, A. H. *Aust. J. Chem.* **1978**, *31*, 297.
8. Barillier, D. *Phosphorus Sulfur Relat. Elem.* **1980**, *8*, 79.
9. Ramsden, C. A.; Oziminski, W. P. *Tetrahedron Lett.* **2019**, *60*, 150876.
10. Lindley, P. F.; Mahmoud, M. M.; Dodd, C.; Smith, C. H.; Boyd, G. V.; Norris, T. *Acta Crystallogr.* **1977**, *B33*, 2160.
11. King, T. J.; Preston, P. N.; Suffolk, J. S.; Turnbull, K. *J. Chem. Soc., Perkin Trans. 2* **1979**, 1751.
12. Luboradzki, R.; Koźmiński, W.; Stefaniak, L. *J. Crystallogr. Spectrosc. Res.* **1993**, *23*, 133.
13. Kushi, Y.; Fernando, Q. *J. Am. Chem. Soc.* **1965**, *1970*, 92.
14. Bird, C. W. *Tetrahedron* **1985**, *41*, 1409.
15. Oziminski, W. P.; Ramsden, C. A. *Tetrahedron* **2015**, *71*, 7191.
16. Cyranski, M. K.; Krygowski, T. M.; Katritzky, A. R.; Schleyer, P.v. R. *J. Org. Chem.* **2002**, *67*, 1333.
17. Ramsden, C. A. *Tetrahedron* **2010**, *66*, 2695.
18. Boyd, G. V.; Norris, T.; Lindley, P. F.; Mahmoud, M. M. *J. Chem. Soc., Perkin Trans. 1* **1977**, 1612.
19. Probert, M. R.; Ramsden, C. A.; Stanforth, S. P. *Tetrahedron Lett.* **2017**, *58*, 1425.
20. McDonald, W. S.; Irving, H. M. N. H.; Raper, G.; Rupainwar, D. C. *Chem. Commun.* **1969**, 392.
21. Boyd, G. V.; Norris, T.; Lindley, P. F. *J. Chem. Soc., Perkin Trans. 1* **1976**, 1673.
22. Oziminski, W. P.; Ramsden, C. A. *Struct. Chem.* **2021**,. https://doi.org/10.1007/s11224-021-01841-8.
23. Rubin, M. B.; Heller, M.; Monisov, R.; Gleiter, R.; Doerner, T. *J. Photochem. Photobiol. A* **1995**, *87*, 7.
24. Rubin, M. B.; Gleiter, R. *Chem. Rev.* **2000**, *100*, 1121.
25. Boyd, G. V.; Norris, T. *J. Chem. Soc., Perkin Trans. 1* **1974**, 1028.
26. Friedrichsen, W.; Schmidt, R. *Justus Liebigs Ann. Chem.* **1978**, 1540.
27. Kozminski, W.; Jazwinski, J.; Stefaniak, L.; Webb, G. A. *J. Mol. Struct.* **1991**, *243*, 369.
28. Fagan, P. J.; Neidert, E. E.; Nye, M. J.; O'Hare, M. J.; Tang, W.-P. *Can. J. Chem.* **1979**, *57*, 904.
29. Gotthardt, H.; Schenk, K. H. *Chem. Ber.* **1986**, *119*, 762.
30. Mann, M.; Fabian, J. *J. Phys. Org. Chem.* **1995**, *8*, 536.
31. Viola, H.; Jakisch, L.; Mayer, R. *Z. Chem.* **1982**, *22*, 405.
32. Mayer, R.; Fabian, J.; Viola, H.; Jakisch, L. *Phosphorus Sulfur Relat. Elem.* **1987**, *31*, 109.
33. Hantke, K.; Gotthardt, H. *J. Chem. Soc., Chem. Commun.* **1984**, 1682.

34. McKinnon, D. M.; Clark, P. D.; Martin, R. O.; Delbaere, L. T. J.; Wilson, Q. J. *Can. J. Chem.* **1987**, *65*, 2830.
35. Newton, C. G.; Ollis, W. D.; Rowson, G. P.; Hamor, M. J.; Hamor, T. A. *Tetrahedron* **1992**, *48*, 8127.
36. Kibbel, H. U.; Teller, J.; Hansen, P.; Reiter, A. *J. Prakt. Chem.* **1980**, *322*, 769.
37. Masuda, K.; Adachi, J.; Nomura, K. *J. Chem. Soc., Chem. Commun.* **1979**, *331*.
38. Masuda, K.; Adachi, J.; Nomura, K. *J. Chem. Soc., Perkin Trans. 1* **1981**, 1033.
39. Koźmiński, W.; Jaźwińska, J.; Stefaniak, L.; Webb, G. A. *J. Mol. Struct.* **1994**, *323*, 177.
40. Gelbrich, T.; Humphries, M.; Hursthouse, M. B.; Ramsden, C. A. *Arkivoc* **2002**, *vi*, 224.
41. Moderhack, D. *Heterocycles* **2016**, *92*, 185.
42. Buzykin, B. I. *Chem. Heterocycl. Compd.* **2010**, *46*, 379.
43. Hanley, R. N.; Ollis, W. D.; Ramsden, C. A.; Smith, I. S. *J. Chem. Soc., Perkin Trans. 1* **1979**, 744.
44. Araki, S.; Mizuya, J.; Butsugan, Y. *J. Chem. Soc., Perkin Trans. 1* **1985**, 2439.
45. Vaidya, S.; Upadhyay, A.; Singh, S. K.; Gupta, T.; Tewary, S.; Langley, S. K.; Walsh, J. P. S.; Murray, K. S.; Rajaraman, G.; Shanmugam, M. *Chem. Commun.* **2015**, *51*, 3739.
46. Lowack, R. H. *Derivatisierung und Ringöffnung reduzierbarer Tetrazolium-Systeme*; Dissertation, University of Erlangen-Nürnberg: Germany, 1990.
47. Lowack, R. H.; Weiss, R. *J. Am. Chem. Soc.* **1990**, *112*, 333.
48. Koźmiński, W.; Jaźwiński, J.; Stefaniak, L. *Magn. Reson. Chem.* **1993**, *31*, 200.
49. Jaźwiński, J.; Stefaniak, L. *Chem. Heterocycl. Compd.* **1995**, *31*, 1027.
50. Luboradzki, R.; Lipkowski, J.; Koźmiński, W.; Stefaniak, L. *J. Chem. Crystallogr.* **1995**, *25*, 29.
51. Weiss, R.; Lowack, R. H. *Angew. Chem., Int. Ed.* **1991**, *30*, 1162.
52. Araki, S.; Hirose, S.; Konishi, Y.; Nogara, M.; Hirashita, T. *Beilstein J. Org. Chem.* **2009**,. https://doi.org/10.3762/bjoc.5.8.
53. Gómez-Zavaglia, A.; Reva, I. D.; Frija, L.; Cristiano, M. L.; Fausto, R. *J. Phys. Chem. A* **2005**, *109*, 7967.
54. Potts, K. T.; Elliott, A. J.; Titus, G. R.; Al-Hilal, D.; Lindley, P. F.; Boyd, G. V.; Norris, T. *J. Chem. Soc., Perkin Trans. 1* **1981**, 2692.
55. Potts, K. T.; Elliott, A. J.; Titus, G. R.; Al-Hilal, D.; Lindley, P. F.; Boyd, G. V.; Norris, T. *Acta Crystallogr., Sect. B: Struct. Sci., Cryst. Eng. Mater.* **1982**, *38*, 682.
56. Araki, S.; Butsugan, Y. *Chem. Lett.* **1985**, *14*, 1639.
57. Kiwan, A. M.; Wanas, G. A. *J. Chem. Soc., Perkin Trans. 2* **1981**, 1534.
58. Hutton, A. T.; Irving, H. M. N. H. *J. Chem. Soc., Perkin Trans.* **1982**, *2*, 1117.
59. Walsh, J. L.; McCracken, R.; McPhail, A. T. *Polyhedron* **1998**, *17*, 3221.
60. Zhao, Y.; Li, L.; Yan, R.; Wu, F.; Zhang, J.; Guo, Z.; Wu, Y. *J. Mol. Struct.* **2018**, *1156*, 30.
61. Koksharova, T. V. *J. Struct. Chem.* **2004**, *45*, 344.
62. Pasynskii, A. A.; Il'in, A. N.; Shapovalov, S. S.; Torubaev, Y. *Russ. J. Inorg. Chem.* **2007**, *52*, 875.
63. Mukiza, J.; Habarurema, G.; Gerber, T. I.; Hosten, E.; Nkuranga, J. B.; Habinshuti, J. B.; Betz, R. *Inorg. Chem. Commun.* **2020**, *113*, 107804.
64. Grote, M.; Kettrup, A. *Anal. Chim. Acta* **1985**, *172*, 223.
65. Grote, M.; Kettrup, A. *Anal. Chim. Acta* **1985**, *172*, 255.
66. Grote, M.; Kettrup, A. *Anal. Chim. Acta* **1987**, *201*, 95.
67. Grote, M.; Sandrock, M.; Kettrup, A. *React. Polym.* **1990**, *13*, 267.
68. Shah, R.; Devi, S. *React. Funct. Polym.* **1996**, *31*, 1.
69. Shah, R.; Devi, S. *Anal. Chim. Acta* **1997**, *341*, 217.
70. Jian, F.; Zhao, P.; Zhang, L.; Hou, Y. *J. Org. Chem.* **2005**, *70*, 8322.
71. Liu, S.; Zubieta, J. *Polyhedron* **1989**, *8*, 677.

72. Jian, F.; Zhao, P.; Hou, Y.; Bei, F. *Chin. J. Chem.* **2005**, *23*, 548.
73. von Eschwege, K. G.; Muller, A. *Acta Crystallogr.* **2009**, *E65*, o2.
74. von Eschwege, K. G.; Muller, A. *Acta Crystallogr.* **2009**, *E65*, o1864.
75. Koźmiński, W.; Jaźwiński, J.; Stefaniak, L. *J. Mol. Struct.* **1993**, *295*, 15.
76. Taha, A.; Kiwan, A. M. *New J. Chem.* **2001**, *25*, 502.
77. Irshaidat, T. *J. Phys. Org. Chem.* **2010**, *23*, 67.
78. von Eschwege, K. G.; Conradie, J.; Kuhn, A. *J. Phys. Chem. A* **2011**, *115*, 14637.
79. Neugebauer, F. A.; Fischer, H. *Chem. Ber.* **1980**, *113*, 1226.
80. Koźmiński, W.; Jaźwiński, J.; Stefaniak, L. *Magn. Reson. Chem.* **1994**, *32*, 746.
81. Koźmiński, W.; Jaźwiński, J.; Stefaniak, L.; Webb, G. A. *J. Spectrosc.* **1994**, *12*, 21.
82. Araki, S.; Aoyama, N.; Butsugan, Y. *Tetrahedron Lett.* **1987**, *28*, 4289.
83. Araki, S.; Aoyama, N.; Butsugan, Y. *Bull. Chem. Soc. Jpn.* **1989**, *62*, 1612.

APPENDIX 1 (REPRINTED FROM OLLIS, W.D.; RAMSDEN, C.A. ADV. HETEROCYCL. CHEM. 1976, *19*, 1.)

CHAPTER FIVE

Meso-ionic compounds☆

W. David Ollis[a] and Christopher A. Ramsden[b,*]
[a]Department of Chemistry, The University of Sheffield, Sheffield, England
[b]Lennard-Jones Laboratories, School of Physical and Geographical Sciences, Keele University, Staffordshire, United Kingdom
*Corresponding author: e-mail address: c.a.ramsden@keele.ac.uk

Contents

☆The review is based on two lectures given by one of us (W.D.O.): a Plenary Lecture entitled "Meso-ionic Heterocycles—Fact and Fiction" given at the Chemical Society Autumn Meeting (September 26, 1972) and the Second Ronald Slack Memorial Lecture, entitled "Meso-ionic Heterocycles—Serendipity and Systematic Investigation" (February 8, 1974).

Advances in Heterocyclic Chemistry, Volume 137
ISSN 0065-2725
https://doi.org/10.1016/bs.aihch.2021.11.001

Abstract

This review provides a general survey of the chemistry of known type A and type B mesoionic heterocycles up to September 1974 with an appendix extending the literature coverage to the end of 1975. The extensive chemistry of sydnones and sydnone imines is not included as this is reviewed elsewhere. The difference between type A and type B mesoionic rings is demonstrated and the number of possible mesoionic rings of each type is discussed. The preparation and reactions of known systems are then systematically reviewed. Additional sections cover general properties of mesoionic rings including valence tautomerism, photochemistry, X-ray crystallography, molecular orbital calculations, pharmacological activity and the interconversion of mesoionic isomers. In a final section, the case for restricting the term mesoionic to five-membered heterocycles of type A and type B is emphasised. Examples of other dipolar heterocycles that have previously been described as mesoionic are discussed and it is recommended that these be described as mesomeric betaines.

Keywords: Mesoionic, Heterocycle, Mesomeric betaine, Cycloaddition, Valence tautomerism, Five-membered rings, Type A, Type B

1. Introduction

The introduction of the concept of meso-ionic molecules by Baker and Ollis[1,2] in 1949 has proved to be a fruitful development in heterocyclic chemistry, and many new meso-ionic systems have been reported during the past two decades. These studies have been concerned with the synthesis and general chemistry of meso-ionic heterocycles, but an additional incentive has been provided by the discovery that some meso-ionic compounds participate in 1,3-dipolar cycloaddition reactions.[3] These cycloaddition reactions of meso-ionic compounds are of considerable general interest in themselves, and they also provide novel synthetic routes to a large variety of heterocycles. Interest in meso-ionic compounds has also been encouraged by the discovery that some types show various pharmacological activities.[4]

Consideration of the structure of the sydnones (**1**) led to our proposal of the term meso-ionic.[1] When this term was proposed, it included a number of known heterocycles; the prediction was also made that new representatives of this general class might be discovered. Subsequently, the sydnone imines (**2**) were recognized either as their cations (**3**) or as their *N*-acyl derivatives (**4**). The extensive chemistry of the sydnones (**1**) and the sydnone imines (**2**) has been well reviewed elsewhere,[2–9] and these topics are not given detailed consideration in this review. However, recent work on the sydnones and sydnone imines is included in Sections 9, 11, and 12, which deal with photochemistry, theoretical studies, and pharmacological activity.

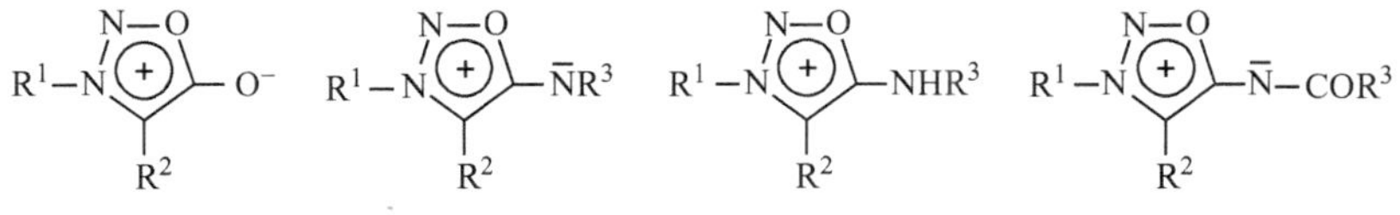

(1) **(2)** **(3)** **(4)**

The purpose of this chapter is to provide a general survey of meso-ionic heterocycles described up to September 1974. Some more recent references are also included, but in view of the extensive current activity in this area of heterocyclic chemistry, it is necessary to be selective rather than comprehensive. No attempt has been made to discuss the chemistry of meso-ionic heterocycles in detail. Our main objective is to present a summary of the presently known types of meso-ionic heterocycles, which can then be considered in relation to the total number of possibilities associated with certain structural conditions.

It is now evident that it is desirable to divide five-membered meso-ionic heterocycles into two general classes: type A and type B. This review is mainly concerned with the presently known 44 members of type A (Table 1) and eight members of type B (Table 2). Brief reference is made at the end of the review to various six-membered heterocycles and polycyclic systems that have been described as meso-ionic, but it is now firmly proposed that this practice should be discontinued.

Table 1 Known five-membered meso-ionic heterocycles of Type A (**19**).

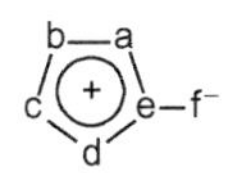

		Atom or group[a]					
Parent system	**Hetero-cycle**	**a**	**b**	**c**	**d**	**e**	**f**
Oxazoles							
1,3-Oxazol-5-ones	**66**	O	CR	NR	CR	C	O
1,3-Oxazol-5-imines	**80**	O	CR	NR	CR	C	NR
1,3-Oxazol-4-imines	**87**	NR	CR	O	CR	C	NR
Diazoles							
1,3-Diazol-4-ones	**91**	NR	CR	NR	CR	C	O
1,3-Diazol-4-imines	**98**	NR	CR	NR	CR	C	NR
1,3-Diazole-4-thiones	**103**	NR	CR	NR	CR	C	S

Table 1 Known five-membered meso-ionic heterocycles of Type A (**19**)—cont'd

Parent system	Hetero-cycle	Atom or group[a]					
		a	b	c	d	e	f
Thiazoles							
1,3-Thiazol-5-ones	**105**	S	CR	NR	CR	C	O
1,3-Thiazol-5-imines	**108**	S	CR	NR	CR	C	NR
1,3-Thiazole-5-thiones	**109**	S	CR	NR	CR	C	S
1,3-Thiazol-4-ones	**114**	NR	CR	S	CR	C	O
1,3-Thiazol-4-imines	**124**	NR	CR	S	CR	C	NR
Dithioles							
1,3-Dithiol-4-ones	**134**	S	CR	S	CR	C	O
1,3-Dithiol-4-imines	**141**	S	CR	S	CR	C	NR
Oxadiazoles							
1,3,4-Oxadiazol-2-ones	**146**	O	CR	NR	N	C	O
1,3,4-Oxadiazol-2-imines	**153**	O	CR	NR	N	C	NR
1,3,4-Oxadiazole-2-thiones	**156**	O	CR	NR	N	C	S
1,3,4-Oxadiazol-2-enes	**164**	O	CR	NR	N	C	CXY[b]
1,2,3-Oxadiazol-5-ones	**1**	O	N	NR	CR	C	O
1,2,3-Oxadiazol-S-imines	**2**	O	N	NR	CR	C	NR
Oxathiazoles							
1,3,2-Oxathiazol-5-ones	**169**	O	N	S	CR	C	O
Triazoles							
1,2,3-Triazol-4-ones	**176**	NR	N	NR	CR	C	O
1,2,3 -T riazol-4-imines	**193**	NR	N	NR	CR	C	NR
1,2,3-Triazole-4-thiones	**196**	NR	N	NR	CR	C	S
1,2,4-Triazol-3-ones	**200**	NR	CR	NR	N	C	O
1,2,4-T riazol-3-imines	**216**	NR	CR	NR	N	C	NR[c]
1,2,4-Triazole-3-thiones	**227**	NR	CR	NR	N	C	S
1,2,4-T riazol-3 -enes	**240**	NR	CR	NR	N	C	CXY[b]

Continued

Table 1 Known five-membered meso-ionic heterocycles of Type A (**19**)—cont'd

Parent system	Hetero-cycle	Atom or group[a]					
		a	b	c	d	e	f
Thiadiazoles							
1,3,4-Thiadiazol-2-ones	**243**	S	CR	NR	N	C	O
1,3,4-Thiadiazol-2-imines	**247**	S	CR	NR	N	C	NR[c]
1,3,4-Thiadiazole-2-thiones	**251**	S	CR	NR	N	C	S
1,3,4-Thiadiazol-2-enes	**261**	S	CR	NR	N	C	CXY[b]
1,2,3-Thiadiazol-4-ones	**265**	NR	N	S	CR	C	O
1,2,4-Thiadiazol-3-imines	**268**	NR	CR	S	N	C	NR
Oxatriazoles							
1,2,3,4-Oxatriazol-5-ones	**271**	O	N	NR	N	C	O
1,2,3,4-Oxatriazol-5-imines	**277**	O	N	NR	N	C	NR
1,2,3,4-Oxatriazole-5-thiones	**286**	O	N	NR	N	C	S
Tetrazoles							
1,2,3,4-Tetrazol-5-ones	**289**	NR	N	NR	N	C	O
1,2,3,4-Tetrazol-5-imines	**293**	NR	N	NR	N	C	NR
1,2,3,4-Tetrazole-5-thiones	**295**	NR	N	NR	N	C	S
1,2,3,4-Tetrazol-5-enes	**296**	NR	N	NR	N	C	CXY[b]
Thiatriazoles							
1,2,3,4-Thiatriazol-5-ones	**297**	S	N	NR	N	c	O
1,2,3,4-Thiatriazol-5-imines	**299**	S	N	NR	N	c	NR
1,2,3,4-Thiatriazole-5-thiones	**300**	S	N	NR	N	c	S
1,2,3,4-Thiatriazol-5-enes	**301**	S	N	NR	N	c	CXY[b]

[a]The groupings a and c each contribute 2 electrons to the π-electron system of the heterocycle; b, d, e, and f each contribute 1 electron.
[b]In these compounds the group f is a stabilized carbanionoid residue. The substituents X and Y include CO_2Et, CN, or a fluorenylidene residue.
[c]Recently, these two classes have been extended to include cases where R is N=CHAr. Thus, the stabilized carbanionoid residue f is associated with the grouping $\overbrace{N-N-CHAr}$.

Table 2 Known five-membered meso-ionic heterocycles of type B (**20**).

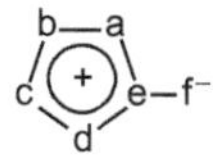

		Atom or group[a]					
Parent system	**Hetero-cycle**	**a**	**b**	**c**	**d**	**e**	**f**
Oxazoles							
1,2-Oxazol-4-imines	**370**	CR	O	NR	CR	C	NR
Diazoles							
1,2-Diazol-4-ones	**373**	CR	NR	NR	CR	C	O
1,2-Diazol-4-imines	**382**	CR	NR	NR	CR	C	NR
Thiazoles							
1,2-Thiazol-4-imines	**386**	CR	S	NR	CR	C	NR
Dithioles							
1,2-Dithiol-4-ones	**388**	CR	S	S	CR	C	O
Tetrazoles							
1,2,3,4-Tetrazol-5-ones	**392**	N	NR	NR	N	C	O
1,2,3,4-T etrazol-5-imines	**399**	N	NR	NR	N	C	NR
1,2,3,4-Tetrazole-5-thiones	**413**	N	NR	NR	N	C	S

[a]The groupings b and c each contribute 2 electrons to the π-electron system of the heterocycle; a, d, e, and f each contribute 1 electron.

2. The representation of meso-ionic heterocycles

Since its introduction in 1949,[1] the term meso-ionic has found almost universal recognition. Its application has depended upon the definition[2] that "a compound may be appropriately called meso-ionic if it is a five- or possibly a six-membered heterocyclic compound which cannot be represented *satisfactorily* by any one covalent or polar structure and possesses a sextet of electrons in association with the atoms comprising the ring. The ring bears a fractional positive charge balanced by a corresponding negative charge located on a covalently attached atom or group of atoms." When this definition was proposed,[2] it was recognized that an inevitable ambiguity was associated with the word "satisfactorily." The adoption of the term meso-ionic and its frequent use in accord with the above definition has demonstrated its usefulness.

However, several symbols involving positive and negative signs continue to be used. These are shown in the general type formulas (**5–8**), in which a, b, c, d, e, and f refer to atoms or groups derived from suitably substituted carbon- or heteroatoms. The arguments in favor of the use of the large, full circle[10] as in formula **5** to represent delocalization in association with a partial positive charge have been clearly presented.[10] These arguments were subsequently reinforced by a general policy adopted by the Chemical Society for the structural representation of aromatic compounds.[10] Because meso-ionic heterocycles cannot be represented by a covalent formulation, the use of the symbolism **5** is still advocated; **5** is preferred to **6**, **7**, or **8**. This symbolism (**5**) is intended to convey the general attributes of their structure and electronic makeup so that the five-membered heterocyclic ring is associated with a partial positive charge and the exocyclic substituent bears a corresponding partial negative charge. It is incorrect to conclude that the general formula **5** necessarily implies extensive polarization or the exhibition of properties or stability that can be characteristically associated with classical aromaticity.

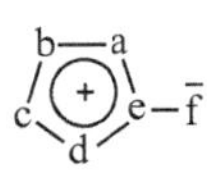

(5)

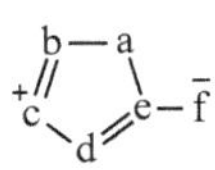

(6)

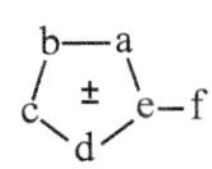

(7)

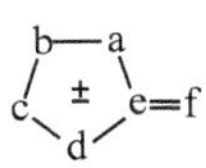

(8)

The betaine-type formula (**6**) has not been found to be acceptable for the general representation of meso-ionic heterocycles, but in view of the recent advocacy[9b] of the ± symbolism depicted in formula **7** some further comment is necessary. Although we originally used this ± symbolism in the representation of the sydnones, we subsequently recommended[2] its replacement by **5** because this is in accord with current practice.[10] Our present view is that the use of a special ± symbol (**7**) is not justified: it is too vague to be useful, and in any case it can cause misunderstanding if formulas of type **8** with e=f are used as well as those of type **7** with e—f. The general formula **5** has much to commend it for discussion of type A and type B five-membered meso-ionic heterocycles and for the correlation of corresponding structural features of these two types (Tables 1 and 2).

3. The two types of five-membered meso-ionic heterocycles

Six general types (Fig. 1) of five-membered heterocycles can be depicted (**9–14**) in which a, b, c, d, e, and f represent suitably substituted

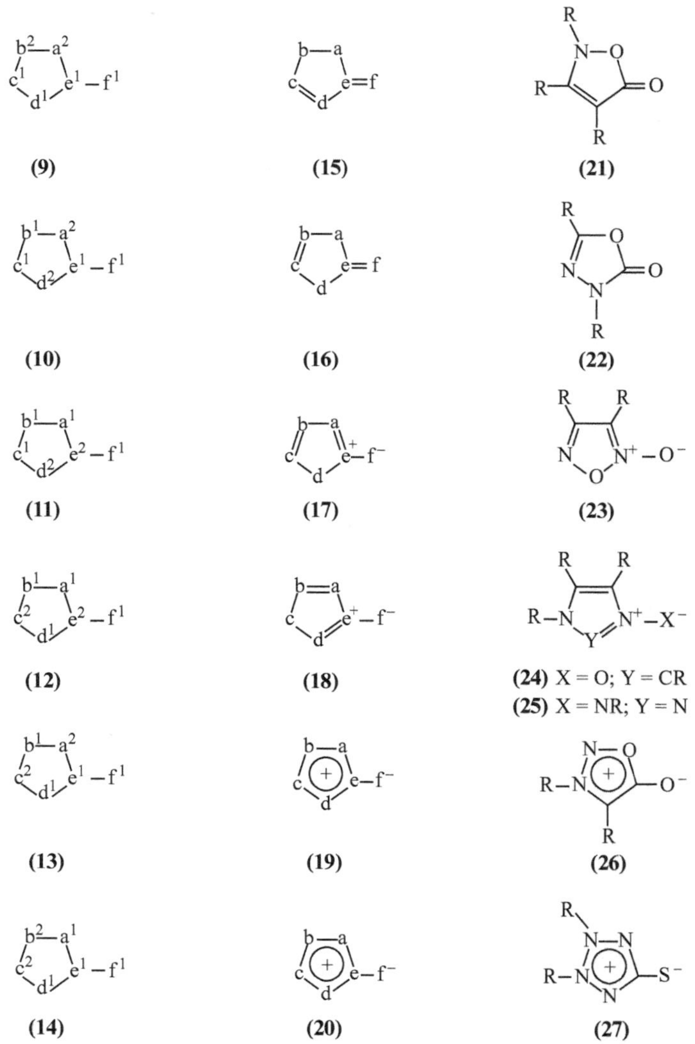

Fig. 1 Six general types of five-membered heterocycles (**9–14**), with examples. Inspection of figure shows that the formulas **19** and **20** as printed are identical. A difference becomes apparent only when they are considered in relation to the corresponding formulas (**13** and **14**), which indicate the origin of the electrons associated with their π-systems.

carbon or heteroatoms. These are chosen to permit a mutually conjugated system associated with all the component atoms of the five-membered ring as well as the exocyclic substituent. The conjugated systems are associated with eight π-electrons whose origin is indicated by superscripts (see **9–14**). Each of these representations (**9–14**) is then developed, giving the corresponding formulas (**15–20**).

The six general heterocyclic types (**15–20**) in association with the examples (**21–27**) illustrate particularly well the need for the term meso-ionic and the requirements that should be observed in meeting the condition of *satisfactory representation*. The compounds **21**[11] and **22**[12] are certainly acceptably represented by their covalent formulations. The types **17** and **18** correspond with the *N*-oxides **23**[13] and **24**,[14] and these are also satisfactorily represented by the single dipolar structures **23** and **24**. Similarly, the compounds **25** are not meso-ionic; they are acceptably described by dipolar structures of the general type **18**.[3a] Compounds belonging to the general types **19** and **20** cannot be satisfactorily represented by covalent or single dipolar structures, and they are best regarded as five-membered meso-ionic heterocycles. It is convenient to associate the two general formulas **19** and **20** with type A and type B meso-ionic systems. The origin of the π-electron system of type A meso-ionic compounds is given in the cipher **13**, and they are exemplified by the sydnones (**26**). The origin (**14**) of the electrons constituting the π-electron system of type B meso-ionic compounds (**20**) is different; this class is exemplified by dehydrodithizone (**27**, R = Ph).

For the five-membered heterocycles (Fig. 1), it is important that the term meso-ionic is reserved for members of the general types **19** and **20**. This point needs to be emphasized because recently several structures have been described as meso-ionic although they were in fact ylides **17** and **18**. Petersen and Heitzer[15] have referred to the intermediate isolated by the oxidation of butane-2,3-dione bisbenzoylhydrazone (**28**) as the "meso-ionic" compound **29**, whereas this structure (**29**) belongs to the ylide class (**17**). Further investigation[16] has established that this isolable intermediate is in fact the amidotriazole enol benzoate (**30**). The postulated intermediate in the thermal and photochemical transformation of bisphenylazostilbene (**31**) into 2,4,5-triphenyl-1,2,3-triazole (**32**) has been called "meso-ionic,"[17a] whereas it should have been described as an ylide (**33**) since it belongs to the general class **17**. Another example of incorrect terminology is the description of the adducts of dicyanoketene ethylene acetals with pyridine thioamides and sulfides as "meso-ionic."[17b] This is not correct since these adducts are acceptably described as enol betaines. Some benztriazole *N*-oxides have also been incorrectly described as "meso-ionic"[17c]; this is unnecessary since they clearly belong to the ylide general class (**17**). The use of the term "meso-ionic zwitterion"[17d] cannot be recommended. An amusingly confused situation has been provided by a recent abstract. The term "meso-ionic" does not appear in the report of the crystal structure of two *N*-ammonia amidates[17e] and neither does it appear in the abstract.[17f] However, this abstract is indexed under "meso-ionic ammonium crystal structure"![17g]

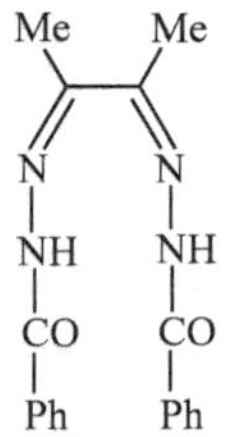
(28)

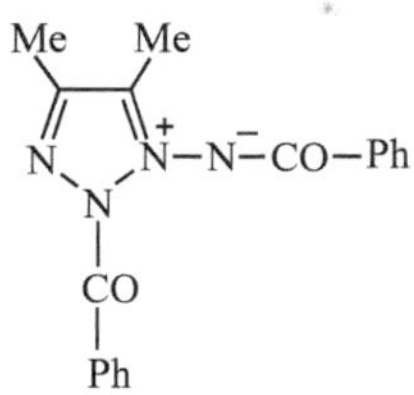
(29)

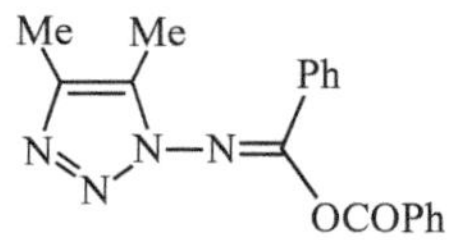
(30)

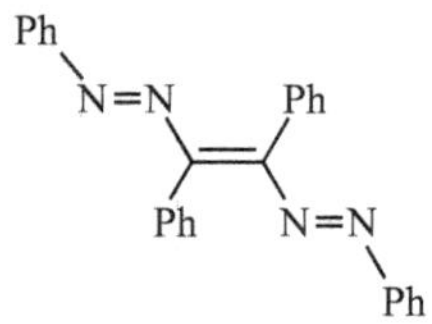
(31)

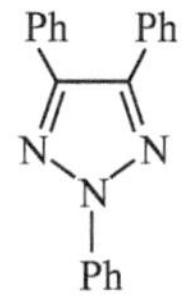
(32)

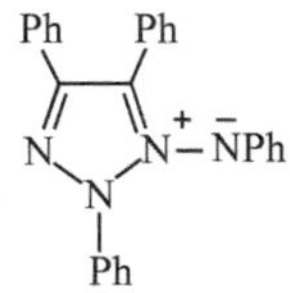
(33)

Having emphasized the need to discriminate between the ylides **17** and **18** and the five-membered meso-ionic heterocycles **19** and **20**, it is necessary to mention that the systematic naming of meso-ionic compounds is best done using the same nomenclature as that used to name betaines. Thus, *NC*-diphenylsydnone (**26**, R = Ph) may be systematically described as anhydro-5-hydroxy-3,4-diphenyl-1-2,3-oxadiazolium hydroxide. However, this nomenclature is somewhat cumbersome, so it is often convenient to refer to the various types of meso-ionic compounds in terms of the parent heterocyclic system and then qualify such names by the adjective meso-ionic. Thus, the sydnones (**26**) are conveniently described as meso-ionic 1,2,3-oxadiazol-5-ones. This terminology is frequently used throughout this review, and the relation between structural formulae and names is given in Tables 1 and 2.

4. Five-membered meso-ionic heterocycles of type A

Forty-four five-membered heterocycles of type A (**13**, **19**) have been described (Table 1). If the atoms or groups a, b, c, d, e, and f are selected from suitably substituted carbon, nitrogen, oxygen, and sulfur atoms, then with these conditions it can be shown that 144 structural possibilities are provided by the general formula **19**. The number of structural possibilities can be deduced in various ways, but a very useful approach is to regard type A meso-ionic molecules (**19**) as being derived by the *union* (←u→) of 1,3-dipoles (**34**) and heterocumulenes (**35**).

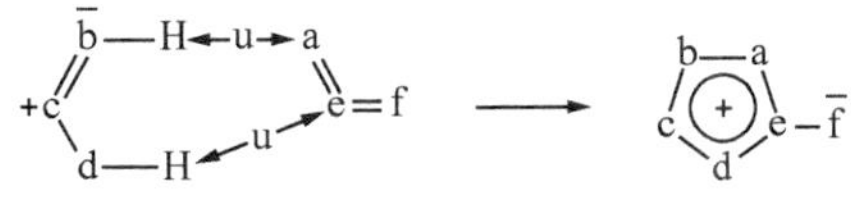

(34) **(35)** **(19)**

In this approach, the term union is used in the sense defined by Dewar[18]: *union* is a process in which two conjugated molecules combine in such a way that their two π-systems unite into one larger one. For example, biphenyl can be formed by the *union* of two molecules of benzene:

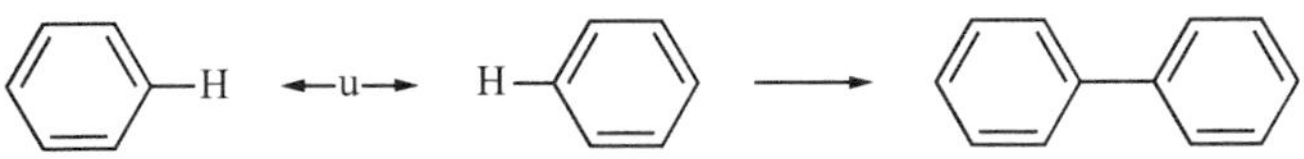

The conditions given above for a, b, c, d, e, and f provide nine 1,3-dipoles (**36–44**), which can be involved in union with 10 heterocumulenes (e = C; a and f variously chosen from >O, >S, >NR, and >CXY). This provides 144 structural possibilities, of which 44 are at present known (Table 1).

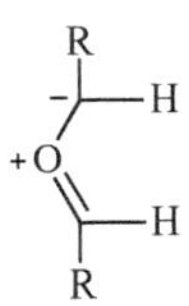

(36) Carbonyl ylides

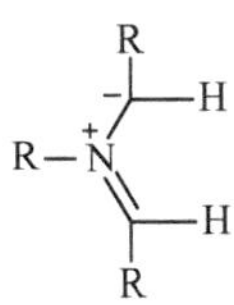

(37) Azomethine ylides

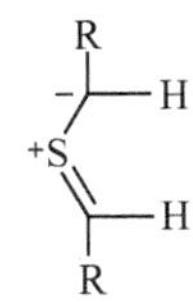

(38) Thiocarbonyl ylides

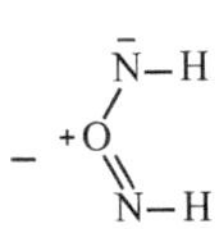

(39) Nitrosoimines

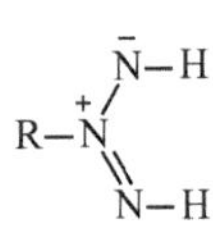

(40) Azonium imines

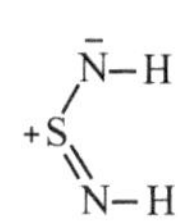

(41) Thionitrosoimines

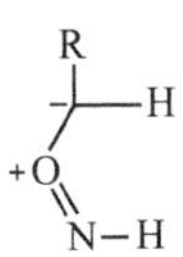

(42) Carbonyl imines

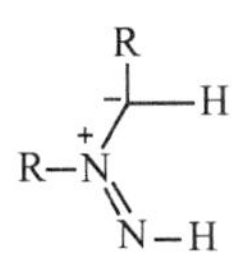

(43) Azomethine imines

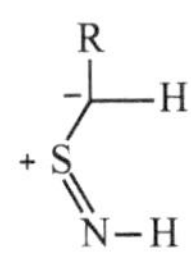

(44) Thiocarbonyl imines

The relation between the structures of the 44 known five-membered meso-ionic heterocycles (Table 1) and the 1,3-dipoles (**36–44**) from which they can be formally derived by *union* is as follows:

(**36**) Carbonyl ylides: **87**

(**37**) Azomethine ylides: **66, 80, 91, 98, 103, 105, 108, 109**

(**38**) Thiocarbonyl ylides: **114, 124, 134, 141**

(**40**) Azonium imines: **271, 277, 286, 289, 292, 294, 296, 297, 299, 300, 301**

(**43**) Azomethine imines: **1, 2, 146, 153, 156, 164, 176, 193, 196, 200, 216, 227, 240, 243, 247, 251, 261**

(**44**) Thiocarbonyl imines: **169, 265, 268**

This list shows that the five-membered meso-ionic heterocycles related to the azomethine imine 1,3-dipole (**43**) comprise the largest group and that meso-ionic compounds related to the nitrosoimines (**39**), thionitrosoimines (**41**), and carbonyl imines (**42**) are not yet known. Clearly synthetic challenges still exist in the field of meso-ionic compounds.

5. Five-membered meso-ionic heterocycles of type B

This second class of five-membered heterocyclic meso-ionic compounds is represented by the type formula **14; 20**. So far, only eight representatives (Table 2) of type B have been described, whereas acceptable combinations of the groupings a, b, c, d, e, and f derived from suitably substituted carbon, nitrogen, oxygen, or sulfur atoms lead, in principle, to 84 possibilities. However, not all these 84 possible structures are necessarily well based because valence tautomerism (see Section 6) might well be associated with thermodynamic preference for the acyclic covalent valence isomer (**46**) rather than the cyclic meso-ionic alternative (**20**).

6. The valence tautomerism of five-membered meso-ionic compounds of type A and type B

This possibility can be depicted generally by the situations **19** ⇆ **45** for type A and **20** ⇆ **46** for type B five-membered meso-ionic heterocycles.

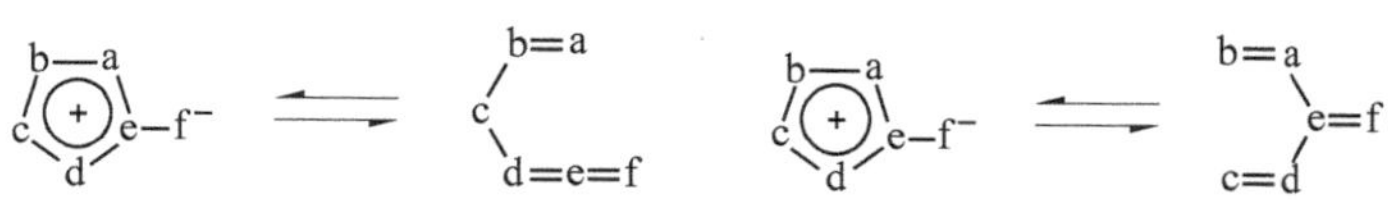

(**19**) Type A (**45**) (**20**) Type B (**46**)

For the type A meso-ionic heterocycles the valence tautomerism **19** ⇆ **45** is now considered. Examples include the possibility that the sydnones (1) could be associated with a valence tautomerism involving the nitrosoketene isomer (**47**).[1,2] Similarly, the relation between the isosydnones (**48**) and their acylaminoisocyanate isomers (**49**) have been noted.[19] Although there is no definitive evidence that the sydnones (**1**) are involved in the valence tautomerism **1** ⇆ **47**, some cycloaddition reaction of isosydnones (**48**) have been considered as involving the acylaminoisocyanate (**49**).[20]

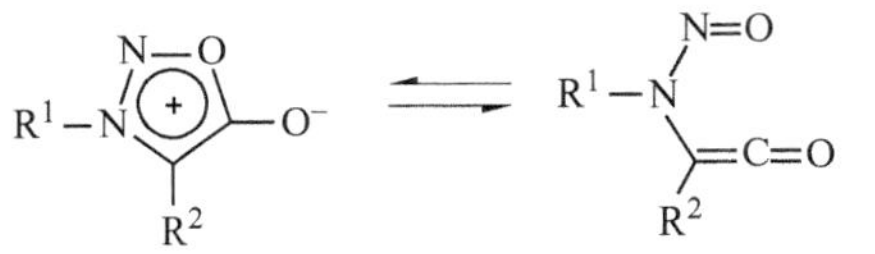

(1) (47)

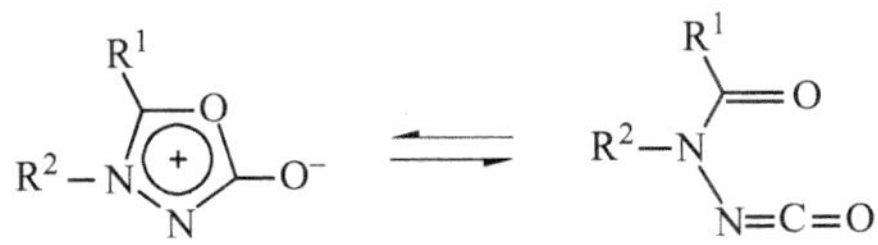

(48) (49)

Some highly relevant results have been obtained with the meso-ionic 1,3-oxazol-5-one (**50**), whose cycloaddition reactions are most directly interpreted[3,21] as involving the valence tautomer **51**. Thus, azomethines gave azetidinones (**52**), carbodiimides gave iminoazetidones (**53**), and enamines [e.g., *N*-(1-cyclopentenyl)morpholine] gave the intermediate cycloadducts (e.g., **54**) which are then transformed to the products (e.g., **55**; this product is incorrectly formulated in a review[9b]). The 1,3-oxazol-5-one (**50**) on heating in xylene gives the allene (**57**), which is probably formed from the dimer (**56**) of the valence tautomer (**51**). The probable involvement of the acylaminoketene (**51**) as a valence tautomer in these thermal reactions is clear, but its spectroscopic detection was not achieved: presumably its equilibrium concentration must be very small. In contrast the valence tautomer (**59**) has been detected spectroscopically (ν_{max}2260 cm^{-1}; N=C=0) during the photolysis of the meso-ionic 1,3,4-thiadiazol-2-one (**58**).[22]

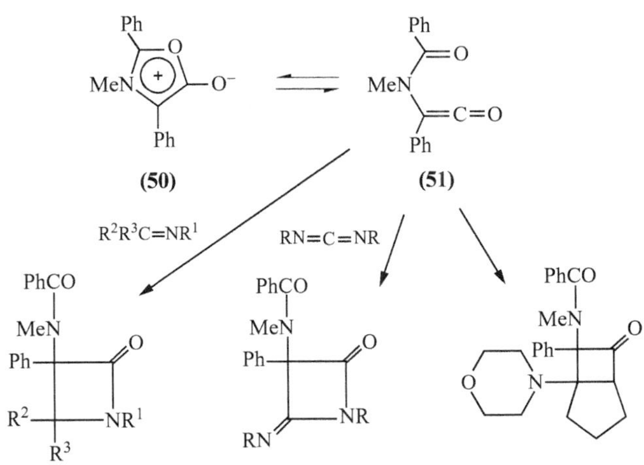

(52) (53) (54)

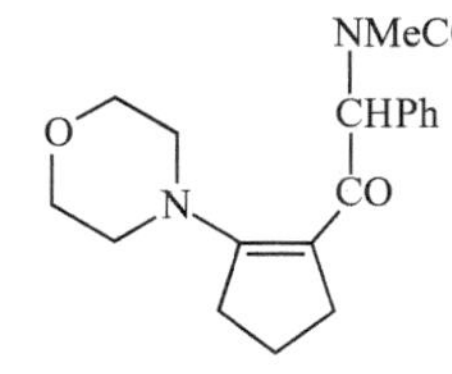

(55)

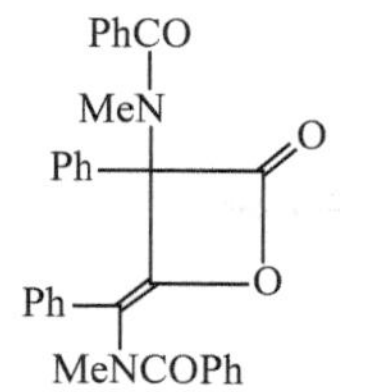

(56)

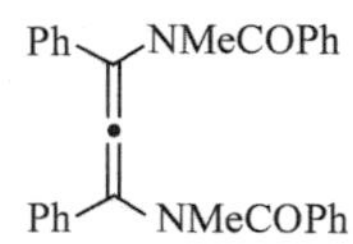

(57)

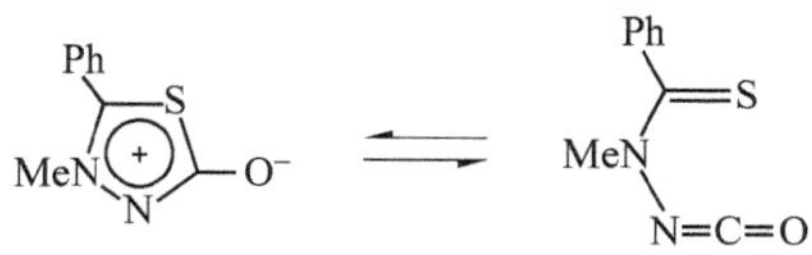

(58) (59)

Regarding the valence tautomerism **20** ⇆ **46**, which is, in principle, possible for five-membered meso-ionic heterocycles of type B, it may be noted that whereas the triketone (**60**) is obviously the more stable partner with respect to its meso-ionic valence tautomer (**61**), the converse situation exists in the case of the bisthiotriketone (**62**). This has the meso-ionic 1,2-dithiol-4-one structure (**63**).[23,24] Similarly, dehydrodithizone has the meso-ionic structure **65**[25,26] rather than the covalent alternative **64**.

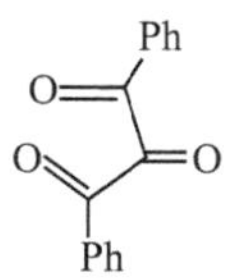

(60)

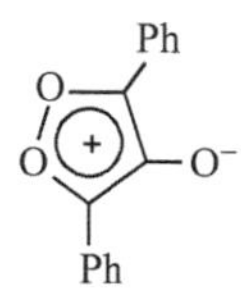

(61)

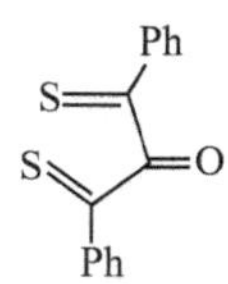

(62)

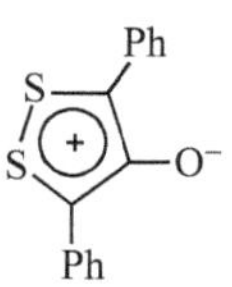

(63)

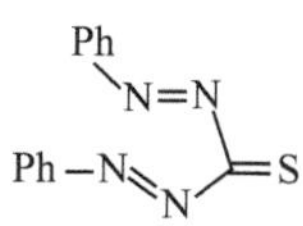

(64)

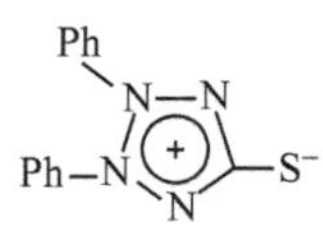

(65)

7. The chemistry of meso-ionic compounds of type A

These are listed under the parent heterocyclic systems in the order given in Table 1 (Section 4).

7.1 Oxazoles

7.1.1 1,3-Oxazol-5-ones (anhydro-5-hydroxy-l,3-oxazolium hydroxides) (66)

Anhydro-acylation of 1,2-dihydro-2-oxo-l-pyridylacetic acid with acid anhydrides yields bicyclic meso-ionic 4-acyl-1,3-oxazol-5-ones (**67**).[27,28]

Huisgen and his coworkers have reported[3,21,29,30] extensive studies of the monocyclic meso-ionic 1,3-oxazol-5-ones (**66**), which they have lightheartedly named miinchnones.[3] Cyclodehydration of *N*-benzoyl-*N*-methyl-*C*-phenylglycine using acetic anhydride at 55° gives the 3-methyl-2,4-diphenyl derivative (**66**, $R^1=R^3=Ph$; $R^2=Me$) as orange-yellow needles, m.p. 151°–152° (decomp.). This compound is very reactive[30]: exposure to moist air gives back the acylamino acid, alcoholysis yields the ester, and aminolysis the amide. The meso-ionic oxazolone (**66**, $R^1=R^3=Ph$; $R^2=Me$) reacts with acetic anhydride at 90° giving the methyl ketone (**68**, $R^1=R^3=Ph$; $R^2=Me$) directly. Based upon this result, a mechanism for the Dakin–West reaction involving a meso-ionic intermediate has been proposed.[31]

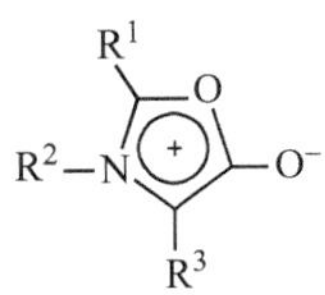

(66)

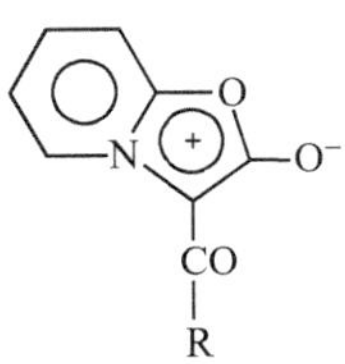

(67)

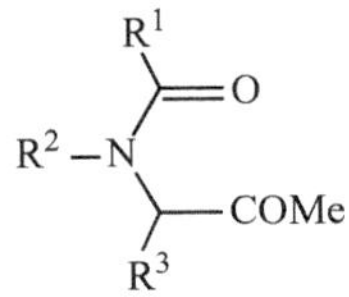

(68)

N-Benzoyl-*C*-phenylglycine with trifluoroacetic anhydride at room temperature yields the bright yellow 4-trifluoroacetyl-1,3-oxazol-5-one (**66**, $R^1=R^2=Ph$; $R^3=COCF_3$), but the corresponding reaction with acetic anhydride gave the Dakin–West reaction product (**68**, $R^1=R^2=Ph$; $R^3=H$).[32] Other cyclodehydrating agents used to prepare meso-ionic 1,3-oxazol-5-ones (**66**) include oxalyl chloride[33,34] or dicyclohexylcarbodiimide.[35]

N-Alkyl- and *N*-aryl-α-acylamino acids with a mixture of acetic anhydride and perchloric acid form 1,3-oxazolonium perchlorates (**69**),[36] which with trifluoroacetic anhydride and base (triethylamine or solid sodium carbonate) yield the known 4-trifluoroacetyl derivative of the meso-ionic 1,3-oxazol-5-one (**66**, $R^3=COCF_3$). The 4-unsubstituted meso-ionic 1,3-oxazol-5-ones (**66**, $R^3=H$) could not be isolated, but treatment of the 1,3-oxazolonium perchlorate (**69**, $R^1=R^2=Ph$) with base and dimethyl acetylenedicarboxylate gave the pyrrole (**70**), presumably by 1,3-cycloaddition to the meso-ionic 1,3-oxazol-5-one (**66**, $R^1=R^2=$ Ph; $R^3=H$).[36]

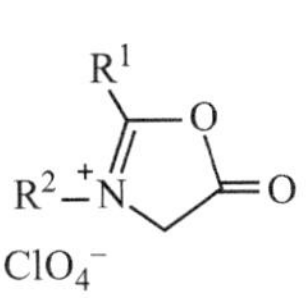

(69)

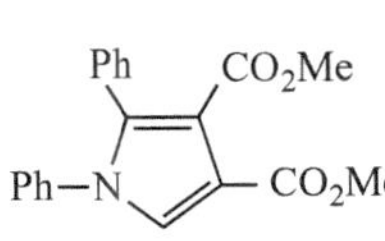

(70)

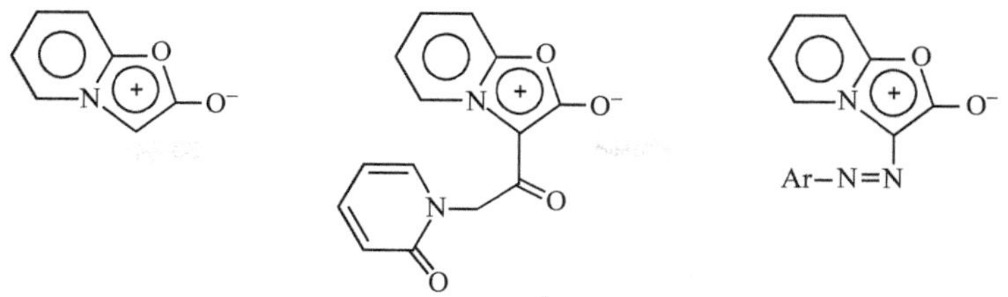

(71) (72) (73)

Cyclodehydration of l-carboxymethyl-2-pyridone with acetic anhydride and perchloric acid gave the corresponding 1,3-oxazolonium perchlorate which with triethylamine gave the dimer (**72**) directly. Similar dimerization of other meso-ionic 1,3-oxazol-5-ones (**66**) have been reported.[36] The monomer (**71**) could not be isolated, but acetyl (**67**, R=Me) and benzoyl (**67**, R=Ph) derivatives were prepared from a solution of the monomer (**71**). Treatment of the monomer solution with aryldiazonium fluoroborates gave the azo derivatives (**73**).[37,38]

The meso-ionic 1,3-oxazol-5-ones show an incredible array of cycloaddition reactions. Reference has already been made to the cycloaddition reactions of the derivative **50**, which are interpreted as involving cycloaddition to the valence tautomer **51**.[21] In addition, an extremely comprehensive study of the 1,3-dipolar cycloaddition reactions of meso-ionic 1,3-oxazol-5-ones (**66**) has been undertaken by Huisgen and his coworkers.[3,39–45] The 1,3-dipolarophiles that have been examined include alkenes, alkynes, aldehydes, *α*-keto esters, *α*-diketones, thiobenzophenone, thiono esters, carbon oxysulfide, carbon disulfide, nitriles, nitro-, nitroso-, and azo-compounds, and cyclopropene and cyclobutene derivatives.[46] In these reactions, the 1,3-oxazol-5-ones (**66**) may be generated in the presence of the 1,3-dipolarophile,[46] and an example[3] of this *in situ* technique is the direct formation of the 2-pyrroline (**75**) in the reaction of *N*-formyl-*N*-phenylglycine (**74**) with acetic anhydride and dimethyl fumarate.

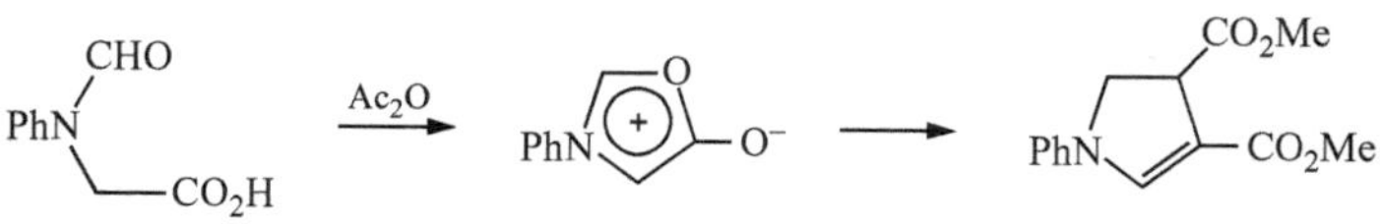

(74) (75)

These results are relevant to the tautomerism which is, in principle, possible between Δ^2-oxazolin-5-ones (azlactones) (**76**) and their meso-ionic isomers (**77**). Kille and Fleury[47] interpreted the spectroscopic properties of the copper-red anhydro derivative (77, R^1=Ph; R^2=p-$O_2N \cdot C_6H_4$) as supporting the meso-ionic structure (**77**). Huisgen and his coworkers[48] have examined the solvent dependence for the equilibrium **76** ⇆ **77**, $R^1=R^2=Ph$, and have given the following estimates of the equilibrium concentration of the meso-ionic partner (77, $R^1=R^2=Ph$): dimethylformamide 49%, dimethyl sulfoxide 32%, acetone 0.26%, chloroform 0.007%. They have also shown that azlactones **76** ⇆ **77** participate in 1,3-dipolar cycloaddition with alkynes[49,50] and alkenes.[50] In view of the extensive earlier studies of azlactones (**76**), it is remarkable that their tautomerism with meso-ionic 1,3-oxazol-5-ones (**77**) should have been recognized so recently. It is clear that this tautomerism (**76** ⇆ **77**) may well be involved in the racemization of optically active amino acids during *N*-acylation.

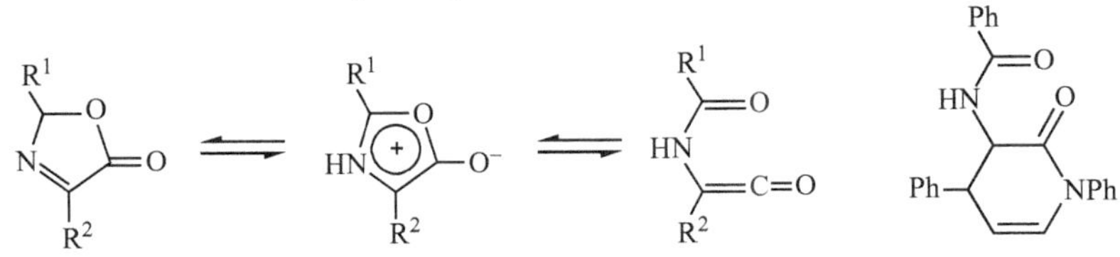

(76) **(77)** **(78)** **(79)**

Recently, a novel reaction between the azlactone **76**, $R^1=Ph$; $R^2=Me$, and cinnamylidene aniline yielding the adduct **79** has been described.[51a] This result has been interpreted as involving a Diels–Alder cycloaddition between cinnamylidene aniline and the valence tautomeric ketene **78**, $R^1=Ph$; $R^2=Me$.

The cycloaddition reactions of 2-phenyloxazol-4(5*H*)-one with acetylenic dipolarophiles has been briefly reported.[51b] The formation of 2-phenylfurans may well involve a tautomerism analogous to that exhibited by azlactones (**76** ⇆ **77**).

7.1.2 1,3-Oxazol-5-imines (anhydro-5-amino-1,3-oxazolium hydroxides) (*80*)

A mechanism has been proposed[52] involving a meso-ionic 1,3-oxazol-5-imine (münchnone imine) intermediate for the acid-catalyzed formation of aldehydes from Reissert compounds. Subsequently[53] this meso-ionic intermediate was trapped using 1,3-dipolarophiles. 2-Benzoyl-1,2-dihydroisoquinaldonitrile (**81**) and tetrafluoroboric acid gave the Reissert salt (**82**), which with dimethyl acetylenedicarboxylate gave the product **83**, $R=R^2=CO_2Me$, and with ethyl phenyl propiolate yielded the primary adduct (**84**) and a secondary product (**83**, $R^1=CO_2Et$; $R^2=Ph$).

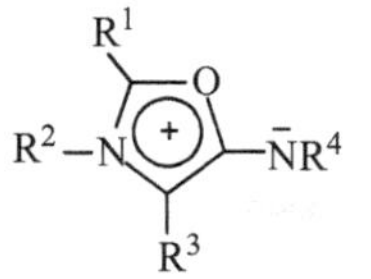

(80)

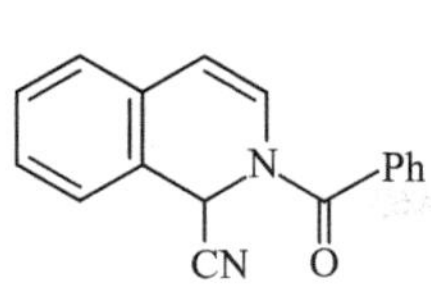

(81)

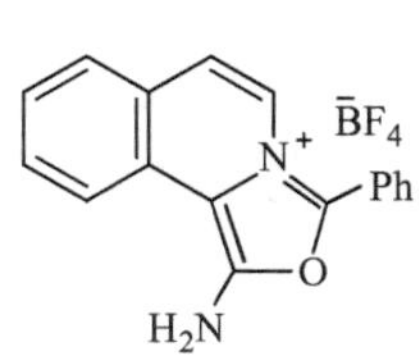

(82)

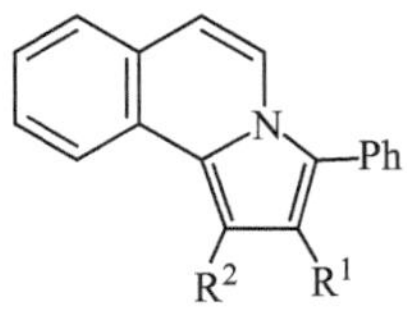

(83)

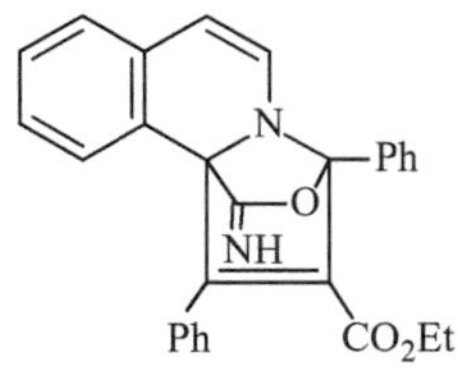

(84)

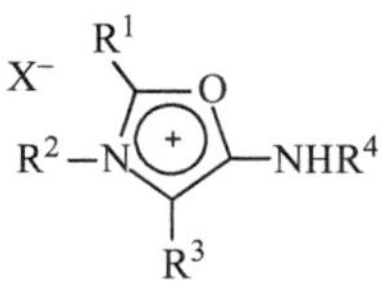

(85)

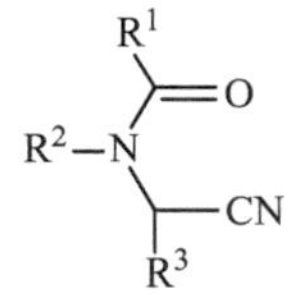

(86)

The 1,3-oxazolium chloride (**85**, $R^1=R^2=Ph$; $R^3=R^4=H$; $X=Cl$) is formed from *N*-benzoylanilinoacetonitrile (**86**, $R^1=R^2=Ph$; $R^3=H$) and hydrogen chloride.[54] Similarly, this nitrile (**86**, $R^1=R^2=Ph$; $R^3=H$) and acetyl or benzoyl perchlorate give the 1,3-oxazolium perchlorates (**85**, $R^1=R^2=Ph$; $R^3=H$; $R^4=MeCO$ or Ph.CO; $X=ClO_4$),[54] but nitriles (**86**) and acid anhydrides give meso-ionic 1,3-oxazol-5-imines (**80**, $R^4=RCO$) directly.[55,56]

7.1.3 1,3-Oxazol-4-imines (anhydro-4-amino-1,3-oxazolium hydroxides) (87)

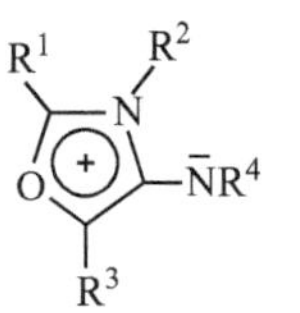

(87)

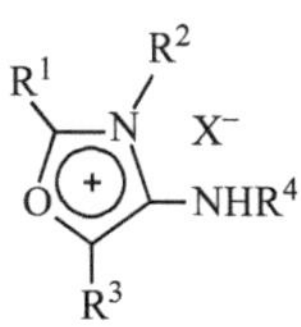

(88)

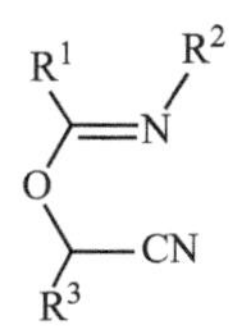

(89)

Meso-ionic 1,3-oxazol-4-imines (**87**) have not been isolated by basic treatment of the corresponding salts (**88**). These salts (**88**) are formed by acid-catalyzed cyclization of the nitriles (**89**).[57]

7.2 Diazoles

7.2.1 1,3-Diazol-4-ones (anhydro-4-hydroxy-1,3-diazolium hydroxides) (91)

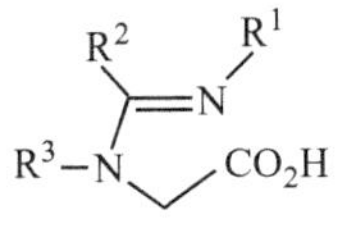

(90)

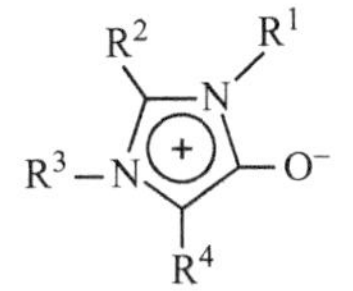

(91)

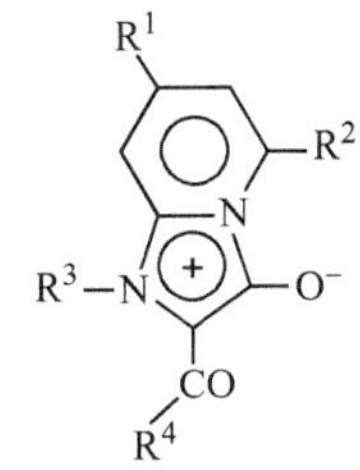

(92)

Anhydro-acylation of *N*-(*N*-phenylbenzimidoyl)amino acids (**90**) and the corresponding nitriles yield the meso-ionic 5-acyl-1,3-diazol-4-ones (**91**, R^4=COR); derivatives (**92**) are analogously formed from *N*-2-pyridylglycines.[28,58a] The meso-ionic compounds **91** and **92** are remarkably stable; they are unreactive toward alkalis, acids, and amines.

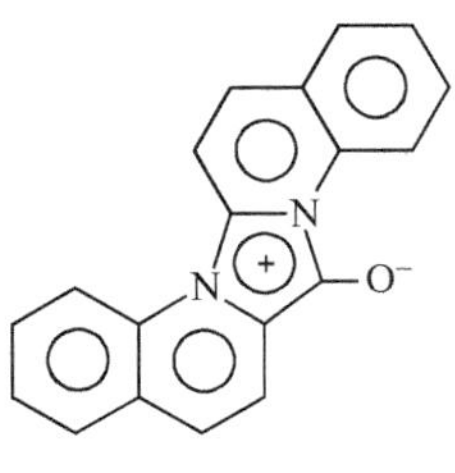

(93a)

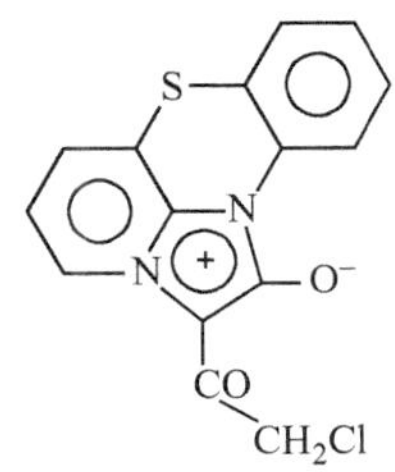

(93b)

Polycyclic meso-ionic 1,3-diazol-4-ones (**93**) are also known. These include Besthorn's Red (**93a**) which was first described in 1904. The discovery of Besthorn's Red and its subsequent formulation (**93a**) forms an interesting part of the history of meso-ionic compounds.[2,9b]

Recently,[58b] an analogous meso-ionic 1,3-diazol-4-one (**93b**) was unexpectedly encountered as the product of the reaction between 1-azaphenothiazine and chloroacetic anhydride.

The triphenyl derivative (91, R^1=R^3=R^4=Ph, R^2=H) is formed[59c] in a mechanistically interesting reaction between benzoyl formic acid anil (Ph–N=C Ph–CO_2H), trifluoroacetic anhydride, and pyridine. Its 1,3-dipolar cycloaddition reactions with alkynes and alkenes have been reported.[59c]

Two synthetic routes[59] to meso-ionic 1,3-diazol-4-ones (**91**) have recently been reported. 1,3-Cycloaddition[59a] of phenyl isocyanate to the meso-ionic 1,3-oxazol-5-one (**94**) yields the intermediate (**95** or **96**) from which carbon dioxide was extruded in boiling xylene, giving the meso-ionic 1,3-diazol-4-one (**97**).

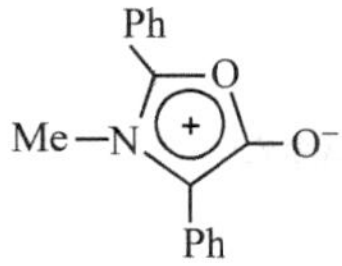

(94)

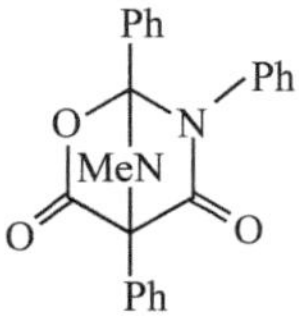

(95)

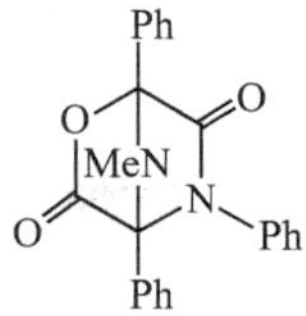

(96)

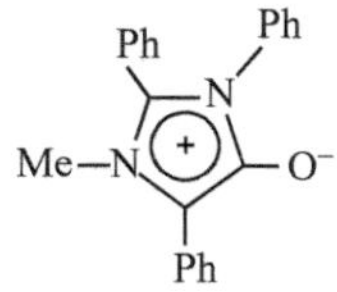

(97)

A general route[59b] to meso-ionic 1,3-diazol-4-ones (**91**) involves the reaction of amino amides ($R^1NHCOCH_2NHR^3$) either with triethyl orthoformate or with *N*-phenylbenzimidoyl chloride.

7.2.2 1,3-Diazol-4-imines (anhydro-4-amino-1,3-diazolium hydroxides) (98)

The meso-ionic 1,3-diazol-4-imine (**98**, $R^1=R^2=Ph$, $R^3=Me$, $R^4=H$, $R^5=COPh$) has been prepared[60] from the nitrile (**99**, $R^3=Me$) and benzoyl chloride followed by treatment with sodium hydrogen carbonate. This compound (**98**, $R^1=R^2=Ph$, $R^3=Me$, $R^4=H$, $R^5=COPh$) undergoes a 1,3-dipolar cycloaddition with dimethyl acetylenedicarboxylate in boiling benzene during 10 min, yielding the pyrrole **(100)**.[60] A similar 1,3-dipolar cycloaddition occurs with diethyl azodicarboxylate, and the intermediate 1,3-cycloadduct was isolated.[60]

Amidines **(101)** and halogenoacetonitriles (XCH_2CN) yield 4-amino-1,3-diazolium salts **(102)** directly.[61] These salts **(102)** could not be transformed into the corresponding meso-ionic heterocycles (**98**) with base; they were unchanged by treatment with triethylamine or sodiumhydrogen carbonate. Sodium hydroxide transformed the salt (**102**, $R^1=R^2=R^3=Ph$, $X=Br$) into the amidinonitrile (**99**, $R^3=Ph$). The salts **(102)** and acetic anhydride gave *N*-acetyl derivatives.

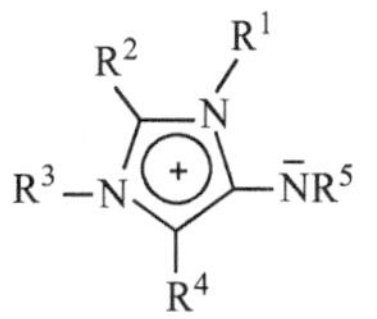

(98)

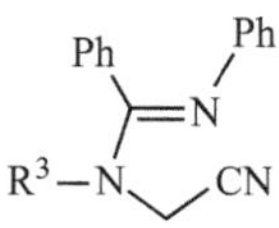

(99)

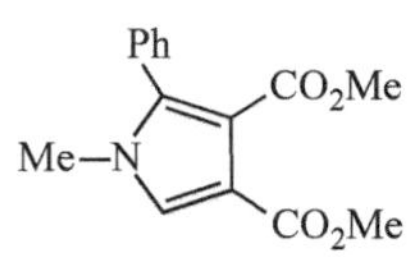

(100)

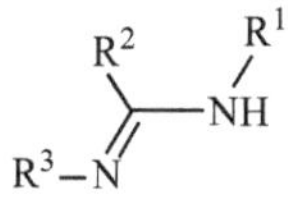

(101)

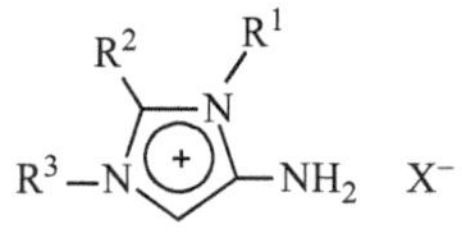

(102)

7.2.3 1,3-Diazole-4-thiones (anhydro-4-mercapto-1,3-diazolium hydroxides) *(103)*

Meso-ionic 1,3-diazole-4-thiones (**103**, R^1=Ph) have been prepared by 1,3-dipolar cycloaddition of phenylisothiocyanate to meso-ionic 1,3-oxazol-5-ones (**66**)[62] and 1,3-thiazol-5-ones **(105)**.[63] Another route to diazole-4-thiones is exemplified by the formation of the derivative (**103**, R^1=R^3=Me, R^2=R^4=Ph) from the methiodide **(104)** and methylamine.[59]

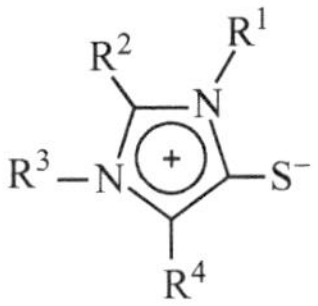

(103)

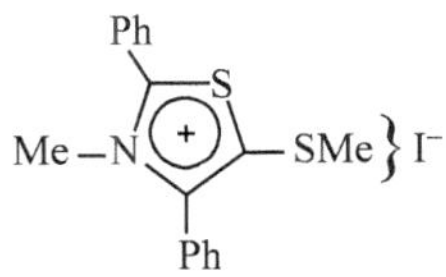

(104)

7.3 Thiazoles

7.3.1 1,3-Thiazol-5-ones (anhydro-5-hydroxy-1,3-thiazolium hydroxides) *(105)*

N-Alkyl- and *N*-aryl-*N*-thiobenzoylglycines **(106)** and hot acid anhydrides or acid chlorides in pyridine yield[28,64] meso-ionic 4-acyl-1,3-thiazol-5-ones (**105**, R^3=R–CO), but at room temperature 4-acylation is avoided.[65]

A second route to meso-ionic 1,3-thiazol-5-ones **(105)** is provided by 1,3-dipolar cycloaddition. Thus, the meso-ionic 1,3-oxazol-5-one (**66**, R^1=R^3=Ph, R^2=Me), and carbon oxysulfide yields the corresponding 1,3-thiazol-5-one (**105**, R^1=R^3=Ph, R^2=Me).[44,62]

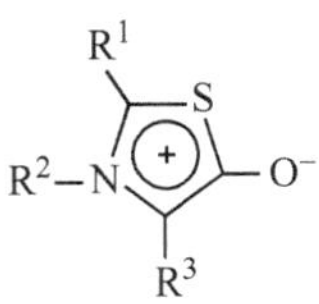

(105)

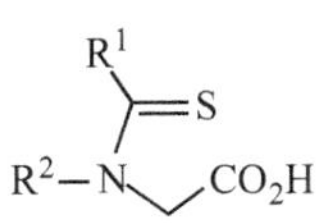

(106)

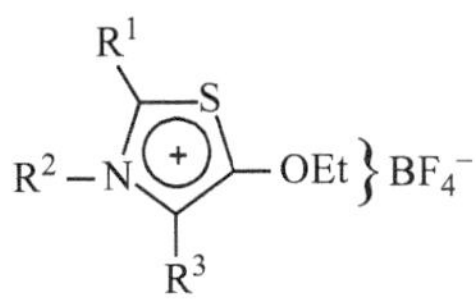

(107)

The meso-ionic 1,3-thiazol-5-ones **(105)** show an interesting range of reactions. Electrophilic substitution reactions[65] have yielded the 4-substituted derivatives (**105**, R^3=Br, I, HgCl, HgOAc), and triethyloxonium tetrafluoroborate gives the corresponding thiazolium salts **(107)**.[66] Meso-ionic 1,3-thiazol-5-ones (**105**, R^3=H or Ph) undergo 1,3-dipolar cycloaddition with cyclopropene derivatives,[46b,46c] diethyl azodicarboxylate,[62] or dimethyl acetylenedicarboxylate,[67a] but the presence of a 4-acyl substituent (**105**, R^3=COR) prevents the cycloaddition reaction.[67a] Good yields of adducts are also obtained if the meso-ionic 1,3-thiazol-5-ones are generated in the presence of the 1,3-dipolarophile.[46]

7.3.2 1,3-Thiazol-5-imines (anhydro-5-amino-1,3-thiazolium hydroxides) (108)

The claim[62] that the compound **103**, R^1=R^2=R^4=Ph, R^3=Me, is rearranged by hydrogen chloride in chloroform, yielding the isomer **108**, R^1=R^3=R^4=Ph, R^2=Me, has not been confirmed.[59,67b]

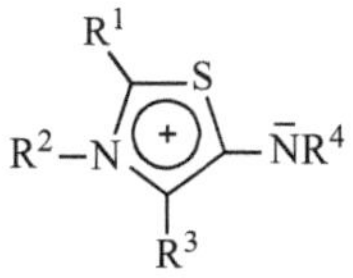

(108)

A general synthetic route to meso-ionic *N*-acyl-1,3-thiazol-5-imines (**108**, R^1=Ph, R^2=Me, R^3=Me or Ph, R^4=COPh) is provided by treatment of thiobenzamidoaminoacetonitriles, Ph–CS–NMe–CHR–CN, first with benzoyl chloride and then with aqueous alkali. Alternatively, the thiobenzamidoaminoacetonitriles and hydrogen chloride give the corresponding salts, which with benzoyl chloride and aqueous alkali give *N*-acyl-1,3-thiazol-5-imines **(108)**.[59b,67b]

Derivatives (**108**, R^1=SMe) have been obtained by treating 5-acylamino-4-thiazoline-2-thiones with methyl iodide and aqueous alkali.[67c]

7.3.3 1,3-Thiazole-5-thiones (anhydro-5-mercapto-1,3-thiazolium hydroxides) (109)

Monocyclic 1,3-thiazole-5-thiones **(109)** are easily prepared by 1,3-dipolar cycloaddition of carbon disulfide to meso-ionic 1,3-oxazol-5-ones (**66**)[44,62] and 1,3-thiazol-5-ones **(105)**.[68] Polycyclic meso-ionic 1,3-thiazole-5-thiones **(110)**[69–71] are formed from carbon disulfide and azomethine ylides **(111)** derived from isoquinolinium iodides and base.

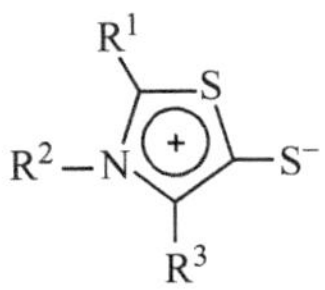

(109)

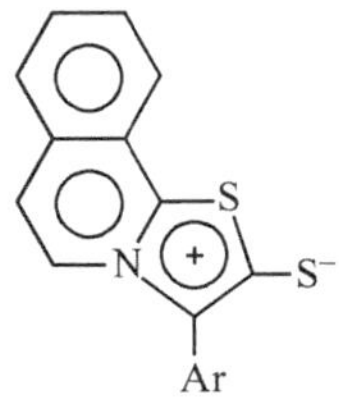

(110)

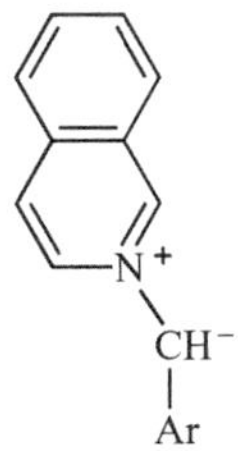

(111)

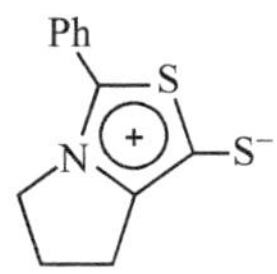

(112)

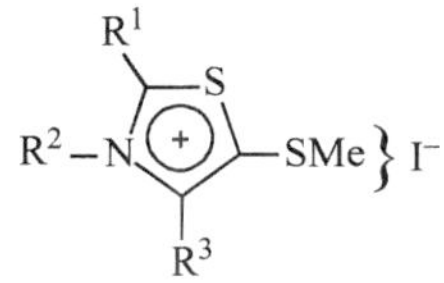

(113)

The meso-ionic 1,3-thiazole-5-thione **(112)**[68] is formed from *N*-thiobenzoylproline and warm thiolactic acid.

The constitution (**110**, Ar=*p*-Br.C_6H_4) was firmly established by X-ray crystallography.[72a] The corresponding derivative (**110**, Ar=*p*-$O_2N.C_6H_4$) yields a methiodide which gives meso-ionic 1,3-diazole-4-thiones **(103)** with primary aliphatic amines.[72b] The meso-ionic 1,3-thiazole-5-thiones **(109)** give the salts **(113)** with methyl iodide.[59]

7.3.4 1,3-Thiazol-4-ones (anhydro-4-hydroxy-1,3-thiazolium hydroxides) (114)

The product obtained by reaction between the acid (**115**, $R^1=R^2=$Ph, $R^3=$H) and acetic anhydride-triethylamine was initially thought[73] to be anhydro-2,3-diphenyl-4-hydroxy-1,3-thiazolium hydroxide, $C_{15}H_{11}NOS$ (**114**, $R^1=R^2=$Ph, $R^3=$H), but later studies[74] established that the product had the molecular formula $C_{30}H_{24}N_2O_2$ and the constitution **116**. The synthesis of the meso-ionic 1,3-thiazol-4-one (**114**, $R^1=R^2=$Ph, $R^3=$H), orange-yellow needles, m.p. 113°–115°, was successfully achieved[74] by dehydration of the acid (**115**, $R^1=R^2=$Ph, $R^3=$H) using acetic anhydride-triethylamine for a few minutes at room temperature. The acids (**115**, $R^1=NR_2$, $R^2=$Ph, $R^3=$H) and acid anhydrides similarly yield the corresponding meso-ionic 1,3-thiazol-4-ones (**114**, $R^1=NR_2$, $R^2=$Ph, $R^3=$H).[75a] Analogous polycylic meso-ionic 1,3-thiazol-4-ones have also been prepared.[75]

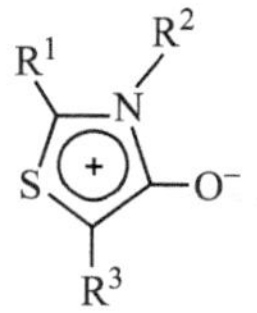

(114)

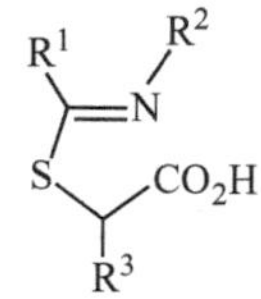

(115)

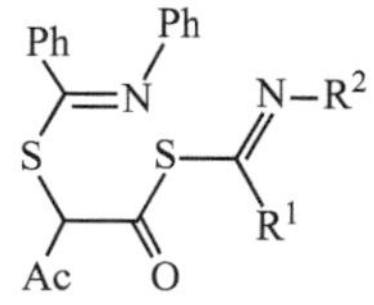

(116)

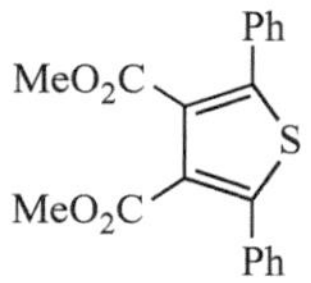

(117)

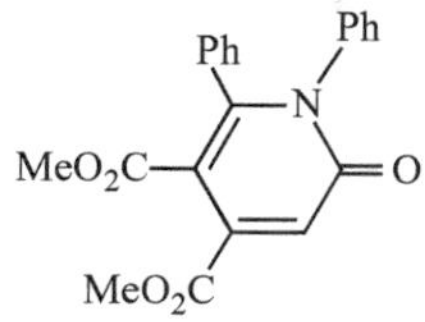

(118)

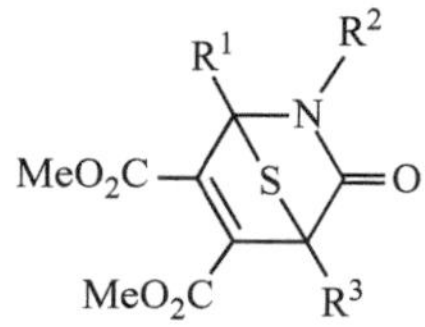

(119)

The meso-ionic 1,3-thiazol-4-ones **(114)** undergo a so-called 1,4-cycloaddition reaction[76] which contrasts with the 1,3-cycloadditions most frequently exhibited by meso-ionic heterocycles.[3] The course of the reaction is also determined by the nature of the 5-substituent (**114**, R^3 = Ph or H). Thus, dimethyl acetylenedicarboxylate and the triphenyl derivative (**114**, R^1 = R^2 = R^3 = Ph) yields the thiophene **(117)**, whereas the diphenyl derivative (**114**, R^1 = R^2 = Ph, R^3 = H) gives the pyridone **(118)**. The triphenyl derivative (**114**, R^1 = R^2 R^3 = Ph) and diben-zoylacetylene in refluxing benzene by a 1,3-cycloaddition yields 3,4-dibenzoyl-2,5-diphenylthiophene, which is a useful intermediate for the synthesis of 1,3,4,6-tetraphenylthieno [3,4-*c*]thiophene.[76] These reactions have been discussed in terms of an intermediate **(119)** formed by a 1,3-cycloaddition which yields the thiophene **(117)** by elimination of phenylisocyanate, whereas the pyridone **(118)** is formed by sulfur extrusion. The triphenyl derivative (**114**, R^1 = R^2 = R^3 = Ph) undergoes a similar cycloaddition reaction with benzyne, yielding the intermediate adduct **120**.[77a] Thermal transformation of the adduct **(120)** occurred in boiling xylene giving diphenylbenzo[*c*]thiophene by elimination of phenylisocyanate. In striking contrast, photochemical irradiation of the adduct gave 1,2,4-triphenyl-3-isoquinolone by sulfur extrusion. The final sentence of this communication[77a] is worthy of quotation: "Such complete selectivity in thermal and photochemical fragmentation of a compound bearing two potentially extrudable groups, *though not without precedent, is unique.*"

The diphenyl derivative (**114**, R^1 = R^2 = Ph, R^3 = H)[74] is extremely sensitive to moisture and is decomposed by atmospheric exposure. With hot acetic anhydride this compound (**114**, R^1 = R^2 = Ph) undergoes 5-acylation (**114**, R^3 = H → R^3 = Ac). Reduction of the triphenyl derivative

(**114**, $R^1=R^2=R^3=Ph$) with sodium borohydride yields 2,3,5-triphenylthiazolidin-4-one **(121)**.[77b] This compound (**114**, $R^1=R^2=R^3=Ph$) forms the corresponding thiazolium salts by treatment with acetyl or benzoyl perchlorate.[77b]

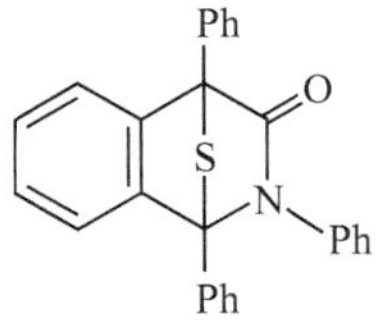

(120)

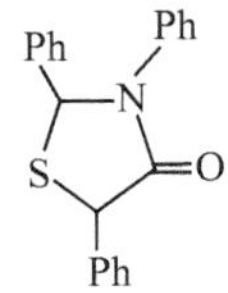

(121)

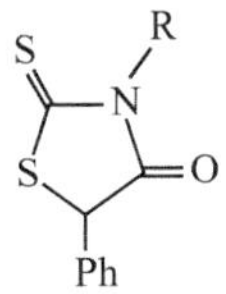

(122)

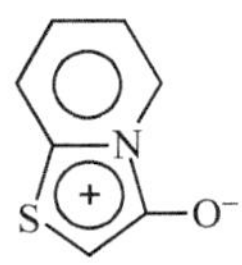

(123)

Alkylation of 5-phenylrhodanines **(122)** with alkyl iodides and sodium ethoxide yields the meso-ionic 2-alkylthio-1,3-thiazol-4-ones (**114**, $R^1=SR$).[78a] The structure of the compound **114**, $R^1=SMe$, $R^2=NH_2$, $R^3=Ph$, was established by X-ray analysis.[78a] The dipole moment of the compound **114**, $R^1=R^2=Me$, $R^3=Ph$, in benzene solution is 5.21 D.[78a] Irradiation of the meso-ionic 1,3-thiazole-4-thione (**114**, $R^1=SMe$, $R^2=Me$, $R^3=Ph$) in ethanol yields the isomeric 3-methyl-4-methylthio-5-phenyl-1,3-thiazol-2-one. A novel mechanism for this photochemical rearrangement has been proposed.[78b]

The bicyclic compound **123** is well known.[75b,79–81] It was first prepared by cyclodehydration of (2-pyridylthio)acetic acid. It is noteworthy that **123** is formed directly from *S*-acetyl-2-mercaptopyridine and chloroacetic acid.[82,83]

7.3.5 1,3-Thiazol-4-imines (anhydro-4-amino-1,3-thiazolium hydroxides) (124)

Acid-catalyzed cyclization of the nitriles **125** yields the salts **126**, $R^3=H$, but treatment of the salts **126**, $R^3=H$, with base results in decomposition and does not lead to the meso-ionic 1,3-thiazol-4-imines (**124**, $R^3=H$).[84–86] However, *N*-acetyl derivatives (**126**, $R^3=Ac$) are formed by acetylation of the salts **126**, $R^3=H$, and then treatment with base does yield *N*-acetyl derivatives of the meso-ionic 1,3-thiazol-4-imines (**124**, $R^3=Ac$).[86] The salts **126**, $R^3=H$ and potassium cyanate yield the *N*-carbamoyl derivatives **124**, $R^3=CONH_2$.[86] Acid hydrolysis of the salts **126**, $R^3=H$ or Ac, X=Cl, gives hydrochlorides of the amides **127**.[86]

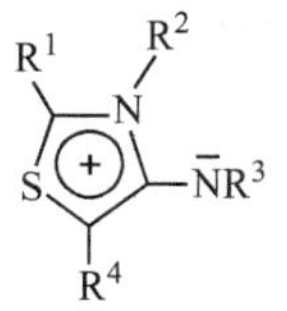

(124)

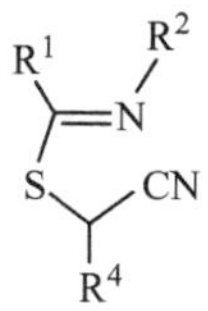

(125)

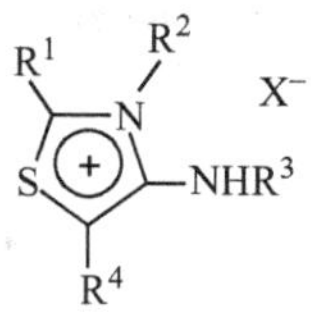

(126)

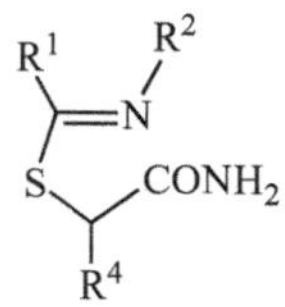

(127)

The *N*-acetyl derivatives of the 2-alkylthio-1,3-thiadiazol-4-imines (**124**, $R^1=SR$, $R^3=Ac$) undergo nucleophilic displacement reaction with amines (benzylamine, cyclohexylamine, morpholine, or aniline) giving the 2-amino derivatives (**124**, $R^1=NR_2$, $R^3=Ac$).[86] The salt (**126**, $R^1=R^2=Ph$, $R^3=R^4=H$, X=Cl) reacts with aniline at room temperature giving 4-anilino-2-phenyl-1,3-thiazole **(128)**, presumably by a mechanism involving cleavage of the heterocyclic ring.[87]

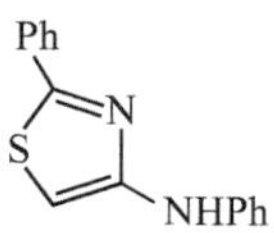

(128)

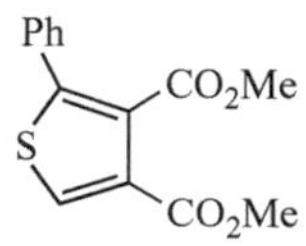

(129)

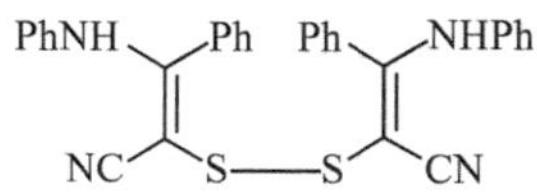

(130)

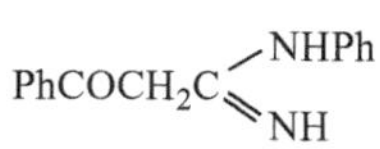

(131)

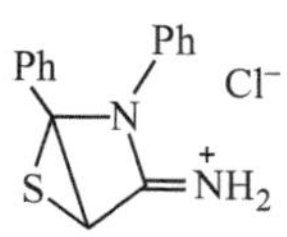

(132)

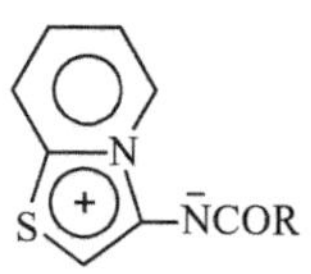

(133)

Most of the known 1,3-dipolar cycloaddition reactions of meso-ionic heterocycles[3] have been concerned with those types in which the exocyclic substituent is either an oxygen or a sulfur atom (**19**, f=O or S). In these cases, the formation of the heterocyclic product involves extrusion of carbon dioxide or carbonyl sulfide from the intermediate cycloadduct. An interesting extension of this reaction type is provided[60] by the formation of the thiophene **129** from anhydro-4-*N*-benzoylamino-2,3-diphenyl-1,3-thiazolium hydroxide (**124**, $R^1=R^2=Ph$, $R^3=COPh$, $R^4=H$) and dimethyl acetylenedicarboxylate. In this case the exocyclic substituent (**19**, f) is NCOPh and the transformation involves the elimination of *N*-benzoyl-*N*′-phenylcarbodiimide, Ph–CON=C=NPh, from the intermediate cycloadduct.

Interest in the photochemistry of meso-ionic compounds is now developing, and an interesting result[88] has been obtained by the photolysis of 5-amino-1,3-thiazolium salts (**126**, R^3=H). For example, irradiation of the salt **126**, R^1=R^2=Ph, R^3=H, X=Cl, in aqueous solution yields the disulfide **130** (23%) and the keto amidine **131** (70%). It is proposed that this reaction involves a bicyclic intermediate **(132)**.[88]

Bicyclic derivatives **(133)** of 1,3-thiazol-4-imines have been prepared[82,89] by treatment of (2-pyridylthio)acetonitrile with acid chlorides. The same compounds **(133)** are also formed from *S*-acyl-2-mercaptopyridine, chloroacetonitrile, and acid chlorides.

7.4 Dithioles

7.4.1 1,3-Dithiol-4-ones (anhydro-4-hydroxy-1,3-dithiolium hydroxides) (134)

The dehydration of thiobenzoylthioglycolic acid (**135**, R^1=Ph, R^2=H) with acetic anhydride–boron trifluoride was initially described[90] as yielding the meso-ionic 1,3-dithiol-4-one (**134**, R^1=Ph, R^2=H), but subsequent studies[91] showed that the product was in fact the 5-substituted derivative **134**, R^1=Ph, R^2=$COCH_2SCSPh$. Authentic meso-ionic 1,3-dithiol-4-ones **(134)** have recently been prepared[91,92] (85–90% yield) by the cyclodehydration of the acids **(135)** with acetic anhydride–triethylamine at 0°–10°. Examples include anhydro-4-hydroxy-2-phenyl-1,3-dithiolium hydroxide (**134**, R^1=Ph, R^2=H)[91] described as scarlet needles, m.p. 113°–115°; this compound is sensitive to moisture. Anhydro-4-hydroxy-2,5-diphenyl-1,3-dithiolium hydroxide (**134**, R^1=R^2=Ph)[92] was obtained as gold lustered, deep violet needles, m.p. 150°–151°; its dipole moment (4.96 D, benzene) is in accord with its meso-ionic formulation.

The meso-ionic 1,3-dithiol-4-ones **(134)** participate[46a,77a,91–94] in 1,3-dipolar cycloaddition reactions giving adducts of the general type **136**. They show a remarkable degree of reactivity toward simple alkenes[94] including tetramethylethylene, cyclopentene, norbornene, and norbornadiene as well as toward the more reactive 1,3-dipolarophilic olefins: dimethyl maleate, dimethyl fumarate, methyl cinnamate, diben-zoylethylene, *N*-phenylmaleimide, and acenaphthylene. Alkynes[91,93] such as dimethyl acetylenedicarboxylate also add to meso-ionic 1,3-dithiol-4-ones **(134)**, but the intermediate cycloadducts are not isolable: they eliminate carbonyl sulfide and yield thiophenes **(137)** directly.[91,93]

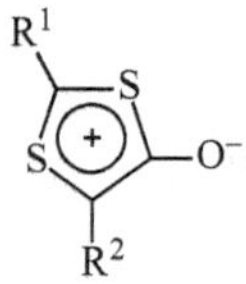

(134)

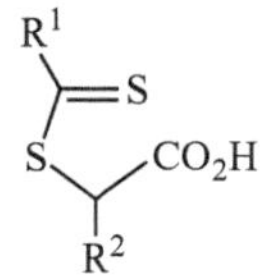

(135)

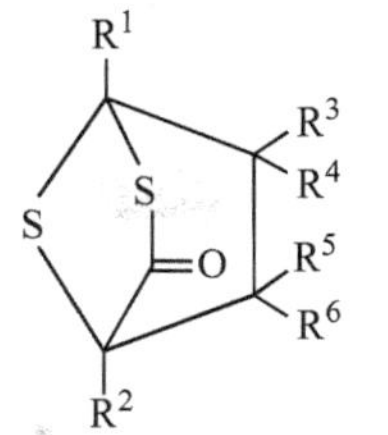

(136)

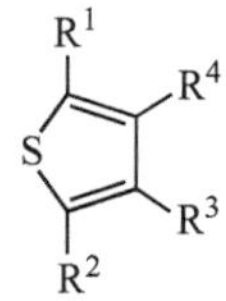

(137)

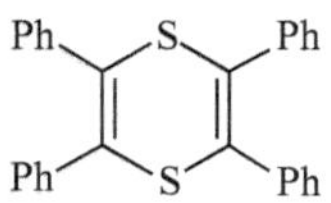

(138)

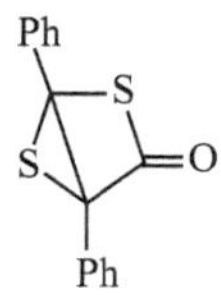

(139)

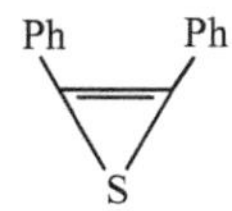

(140)

Photolysis of anhydro-4-hydroxy-2,5-diphenyl-1,3-dithiolium hydroxide (**134**, $R^1 = R^2 = Ph$) in benzene solution yields tetraphenyl-1,4-dithiin (**138**, 19% yield), diphenylacetylene (16% yield), and sulfur. This result has been interpreted[95] as involving the bicyclic photointermediate **139**, which loses carbonyl sulfide to give the 4π-antiaromatic diphenyl thiirene **(140)**, which is the precursor of diphenylacetylene and tetraphenyl-1,4-dithiin **(138)**.

7.4.2 1,3-Dithiol-4-imines (anhydro-4-amino-1,3-dithiolium hydroxides) (141)

The cyanomethyl dithio ester (**142**, $R^1 = Ph$, $R^2 = H$) and an ethereal solution of hydrogen chloride yields the salt **143**, $R^1 = Ph$, $R^2 = R^3 = H$, $X = Cl$.[96] This hydrochloride (**143**, $R^1 = Ph$, $R^2 = H$) could not be acylated directly, but the cyanomethyl dithio ester (**142**, $R^1 = Ph$, $R^2 = H$) with benzoyl or acetyl chloride gave the salts **143**, $R^1 = Ph$, $R^2 = H$, $R^3 = COPh$ or COMe, which with aqueous sodium hydrogen carbonate gave the red meso-ionic derivatives **141**, $R^1 = Ph$, $R^2 = H$, $R^3 = COPh$ or COMe.[90,96]

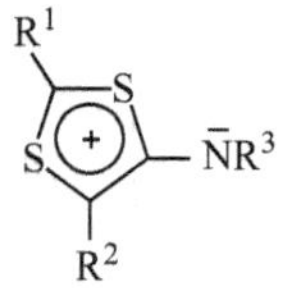

(141)

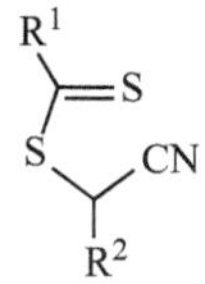

(142)

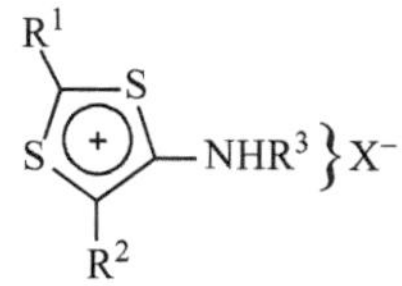

(143)

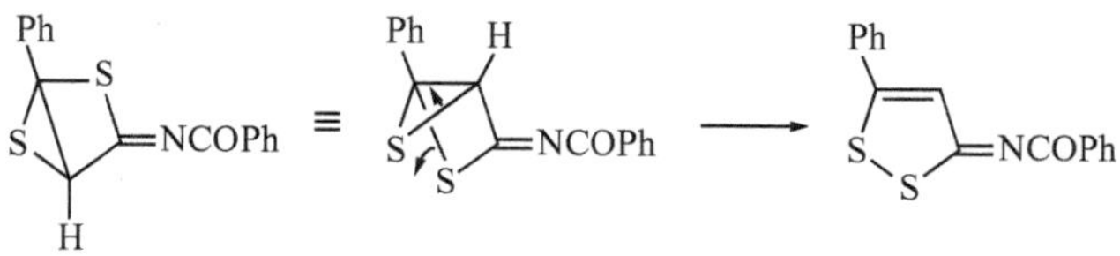

(144) (144) (145)

Irradiation of a benzene solution of anhydro-4-benzoylamino-2-phenyl-1,3-dithiolium hydroxide (**141**, R^1=Ph, R^2=H, R^3=COPh) yields the 1,2-dithiole (**145**, 80%). This photoisomerization has been interpreted as involving the bicyclic intermediate **144**.[97]

7.5 Oxadiazoles

7.5.1 1,3,4-Oxadiazol-2-ones (anhydro-2-hydroxy-1,3,4-oxadiazolium hydroxides) (146)

These meso-ionic compounds are often referred to as isosydnones **(146)**. They are prepared from *N*-acyl-*N*-(alkyl or aryl)-hydrazines (**147**, X=O) or their hydrochlorides and carbonyl chloride.[98–101] The isosydnones **(146)** are colorless, reasonably stable compounds, but they are clearly more reactive toward nucleophilic reagents[102] than the sydnones (1). On heating with the indicated reagents, the following products are formed[20,98,103]: water gives *N*-acylhydrazines (**147**, X=O) and the corresponding urea derivative, alcohols give urethans **(148)**, and aniline gives *N*-phenylureas **(149)**. The reaction of isosydnones **(146)** with hydrogen sulfide in the presence of the tertiary bases, pyridine or triethylamine, provides a useful synthetic route of *N*-thioacylhydrazines (**147**, X=S).[103]

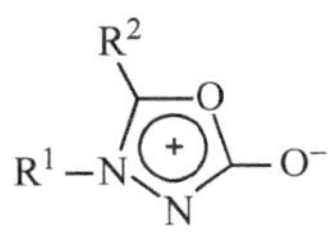

(146)

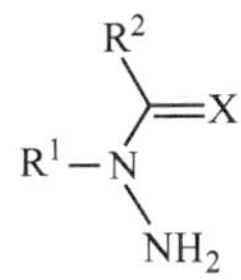

(147)

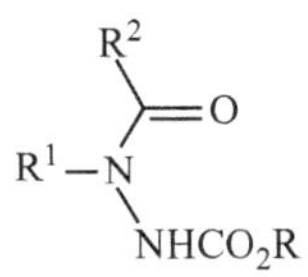

(148)

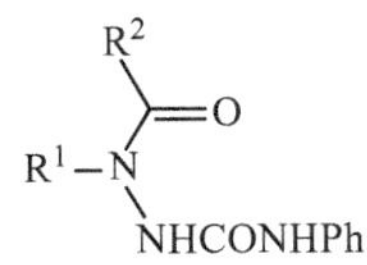

(149)

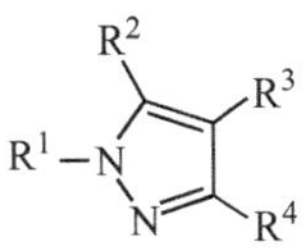

(150)

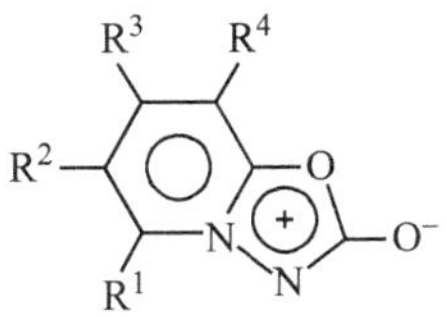

(151)

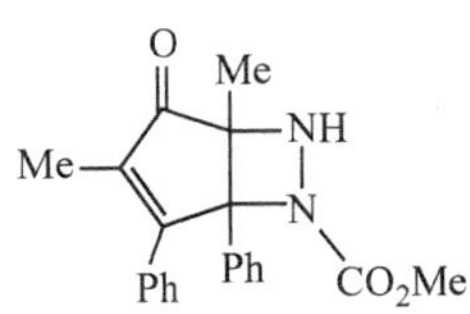

(152)

The dipole moment of 4,5-diphenyl isosydnone (**146**, R^1=R^2=Ph) is 7.82 D (benzene), and comparison with the dipole moments of other diaryl isosydnones has given a value of 7.3 D for the isosydnone group moment.[19] This and the spectroscopic properties of isosydnones are in full accord with their meso-ionic formulation **(146)**.[19]

Isosydnones **(146)** react with alkynes to give pyrazoles **(150)**, For example, 4,5-diphenylisosydnone (**146**, R^1=R^2=Ph) and ethyl phenyl propiolate gives 4-ethoxycarbonyl-1,3,5-triphenylpyrazole (**150**, $R^1=R^2=R^4=Ph$, $R^3=CO_2Et$) identical with the product from 4,5-diphenylsydnone (1, $R^1=R^2=Ph$).[104] The rate of 1,3-cycloaddition for isosydnones **(146)** is relatively slow in comparison with sydnones (**1**).[20,104] A number of other cycloaddition reactions of isosydnones with alkenes, alkynes, and carbonyl compounds have been reported.[20]

The first isosydnone to be prepared[105a] was the bicyclic derivative **151**, $R^1=R^2=R^3=R^4=H$, formed from l-aminopyrid-2-one and carbonyl chloride. This substance (**151**, $R^1=R^2=R^3=R^4=H$) is unchanged by irradiation at 350 and 300 nm.[105b] Recently, the formation of the bicyclic derivative **151**, $R^1=R^4=Me$, $R^2=R^3=Ph$, by an unusual rearrangement has been reported.[106] The bicyclic diazetidine ester **(152)** and trifluoroacetic acid gave an isomeric α-pyridone, which yielded the bicyclic meso-ionic compound **151**, $R^1=R^4=Me$, $R^2=R^3=Ph$, on heating.[106]

7.5.2 1,3,4-Oxadiazol-2-imines (anhydro-2-amino-1,3,4-oxadiazolium hydroxides) (153)

This new class of meso-ionic compounds have been recently synthesized.[107a] *N*-methyl-*N*-benzoylhydrazine (**147**, $R^1=Me$, $R^2=Ph$) and aryl isocyanide dichlorides ($ArN{\equiv}CCl_2$) yield 1,3,4-oxadiazolium chlorides (**154**, $R^1=Me$, $R^2=Ph$, $R^3=Ar$), which with diazomethane give the meso-ionic 1,3,4-oxadiazol-2-imines **(153)** as yellow crystalline compounds. Subsequently, this synthetic method[107] has been extended by the use of acylisocyanide dichlorides. These reagents yield the 1,3,4-oxadiazol-2-imines **153**, $R^3=SO_2R$,[107b] and **153**, $R^3=COR$.[107c]

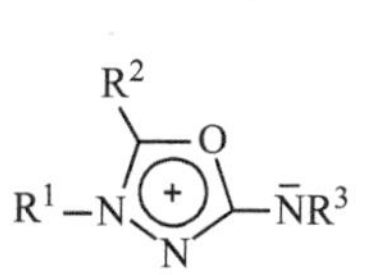

(153)

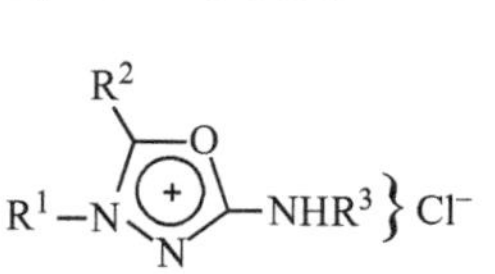

(154)

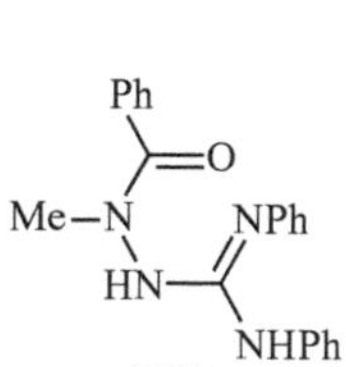

(155)

The derivative **153**, $R^1=Me$, $R^2=R^3=Ph$, has a dipole moment ($\mu=7.56$ D) in accord with its meso-ionic structure.[108] The meso-ionic 1,3,4-oxadiazol-2-imine (**153**, $R^1=Me$, $R^2=R^3=Ph$) reacts with aniline giving the guanidine derivative **155**.[107a]

7.5.3 1,3,4-Oxadiazole-2-thiones (anhydro-2-mereapto-1,3,4-oxadiazolium hydroxides) (156)

A representative of this class (**156**, $R^1=Me$, $R^2=Ph$) was first obtained[109] by heating in pyridine the methiodide (**157**, $R^1=Me$, $R^2=Ph$) derived from 2-phenyl-4-methylthio-1,3,4-oxadiazole **(158)**. Subsequently, three general methods for the synthesis of meso-ionic 1,3,4-oxadiazole-2-thiones **(156)** have been described. These involve (i) the cyclization of the dithiocarbamic acid ammonium salts **(159)** with phosphorus oxychloride-triethylamine,[110,111] (ii) the reaction between *N*-acylhydrazines **(147)** and thiophosgene,[112] and (iii) the condensation of *N*-acylhydrazines **(147)** and carbon disulfide with diethyl carbo-diimide.[112] The reactions of the diphenyl derivative **156**, $R^1 = R^2 = Ph$, with acid, alkali, ethanol, cyclohexylamine, and triethyloxonium tetrafluoroborate are straightforward.[112b]

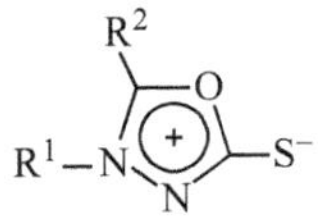
(156)

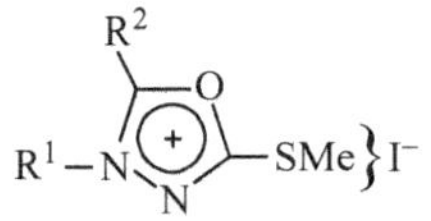
(157)

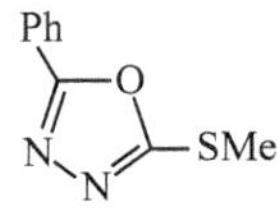
(158)

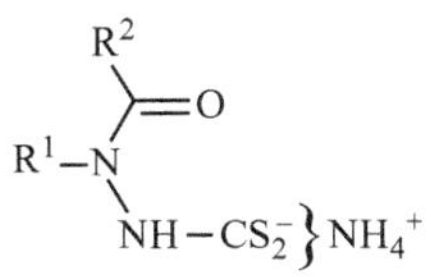
(159)

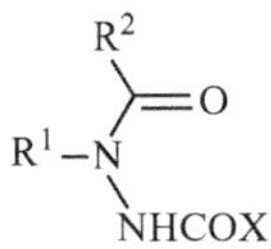
(160)

(a) X = SMe
(b) X = NHPh
(c) X = NEt_2

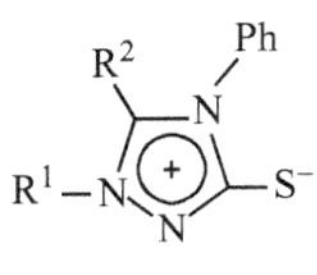
(161)

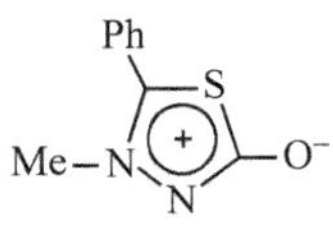
(162)

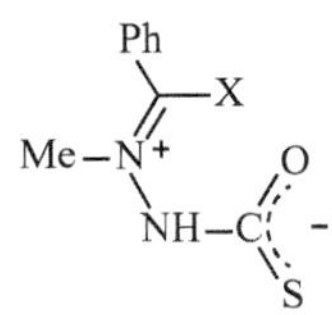
(163) X = OEt or SEt

The 1,3,4-oxadiazole-2-thione (**156**, $R^1 = Me$, $R^2 = Ph$) has a dipole moment of 9.10 D, which obviously supports its meso-ionic formulation.[113] The meso-ionic 1,3,4-oxadiazole-2-thiones **(156)** form methiodides **(157)**[112] which are easily hydrolyzed, yielding the acyclic esters (**160a**). The meso-ionic 1,3,4-oxadiazole-2-thione (**156**, $R^1 = Me$, $R^2 = Ph$) is transformed by reaction with aniline into the 1,2,4-triazole-3-thione (**161**, $R^1 = Me$, $R^2 = Ph$)[103]; the presumption that this transformation involves the thiosemicarbazide **160b** as an intermediate is supported by the isolation of the derivative **160c** by a similar reaction with diethylamine.[103]

The existence of the isomeric meso-ionic compounds **156** and **162** has been recently established.[102,103] The transformation **156** → **162**, $R^1 = Me$, $R^2 = Ph$, is achieved by heating in ethanolic solution or with ethanethiol in benzene; this isomerization could well involve the betaine intermediate **(163)**.[102,103]

7.5.4 1,3,4-Oxadiazol-2-enes (anhydro-2-alkyl-1,3,4-oxadiazolium hydroxides) *(164)*

Representatives of this novel class of meso-ionic compounds in which the exocyclic substituent f (see Table 1) is a stabilized carbanionoid residue [$-\overline{C}(CN)CO_2Me$ or $-\overline{C}(CN)_2$] have been recently synthesized. Base-catalyzed (potassium carbonate or triethylamine) condensation of *N*-acylhydrazines **(147)** with 3,3-dichloroacrylonitriles **(165)** yield the greenish-yellow meso-ionic 1,3,4-oxadiazol-2-enes **(164)** directly.[114]

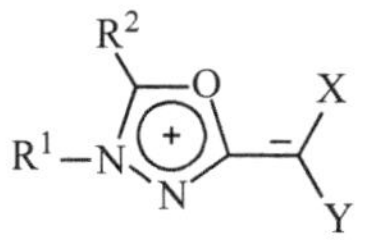

(164)

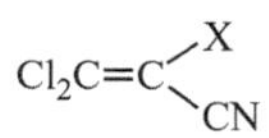

(165)

X and Y = CN or CO_2Me

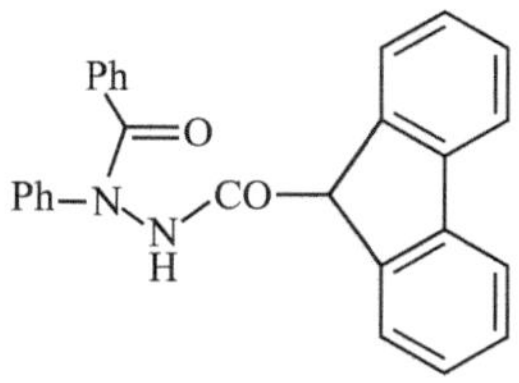

(166)

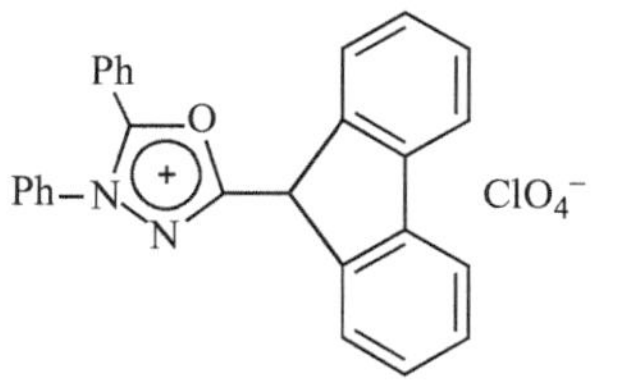

(167)

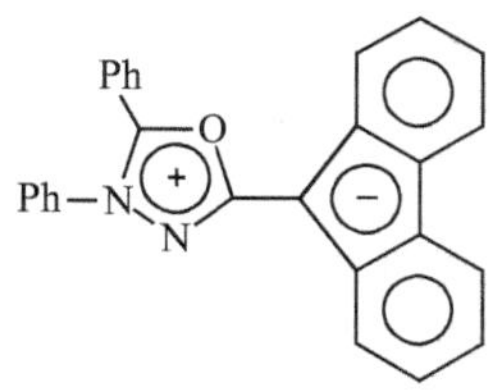

(168)

An analogous meso-ionic 1,3,4-oxadiazol-2-ene **(168)** in which the negative charge is associated with a fluorenyl residue has also been recently described.[115] The *N,N'*-diacylhydrazine **(166)** and acetic anhydride-perchloric acid yield the 1,3,4-oxadiazolium perchlorate **(167)**, which by addition of triethylamine to its solution in acetonitrile yields the purple meso-ionic compound **168**, m.p. 105°–107°.

7.5.5 1,2,3-Oxadiazol-5-ones (sydnones) *(1)*

Comprehensive reviews of the chemistry of sydnones (**1**) have been published elsewhere,[2–9] and a general discussion of this subject is not included here. Recent work on the photochemistry (Section 9), physical and theoretical studies (Section 11), and pharmacological activity (Section 12) of sydnones (**1**) is included in later sections.

7.5.6 1,2,3-Oxadiazol-5-imines (sydnone imines) *(2)*

The chemistry of sydnone imines (**2**) has been reviewed,[2–9] and a discussion of their chemistry is not duplicated here. Recent photochemical and pharmacological studies are included in Sections 9 and 12.

7.6 Oxathiazoles

7.6.1 1,3,2-Oxathiazol-5-ones (anhydro-5-hydroxy-1,3,2-oxathiazolium hydroxides) (169)

1-Mercaptophenylacetic acid, acetic anhydride, and sodium nitrite yielded the first example[116] (**169**, R = Ph) of a meso-ionic 1,3,2-oxathiazol-5-one; this compound has a dipole moment of 4.5 D. An X-ray crystallographic investigation of the 4-phenyl-1,3,2-oxathiazol-5-one (**169**, R = Ph) has been reported.[117] This compound is incorrectly described as a "thiosydnone." The bond distances support the structure with tetracovalent sulfur, and on this evidence alone the compound **169**, R = Ph, could be judged not to be "meso-ionic" because it can, on the basis of the X-ray evidence,[117] be more satisfactorily represented by a covalent formulation.

More extensive studies[118] have shown that α-mercaptocarboxylic acids, ethyl nitrite, and *N,N'*-dicyclohexylcarbodiimide with a catalytic amount of sulfuric acid yield meso-ionic 1,3,2-oxathiazole-5-ones **(169)**, presumably via intermediate *S*-nitroso derivatives **(170)**.

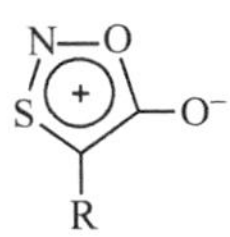

(169)

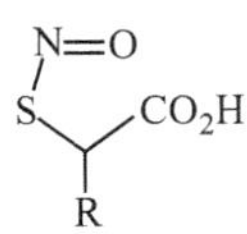

(170)

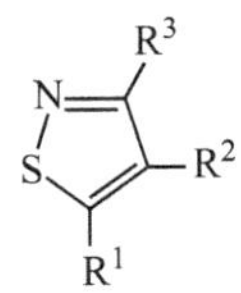

(171)

The meso-ionic 1,3,2-oxathiazol-5-ones **(169)** show an interesting range of reactions with nucleophiles including ammonia, primary amines, and aqueous alkali. They also react with 1,3-dipolarophiles,[118] including dimethyl acetylenedicarboxylate and methyl propiolate, yielding isothiazoles **(171)** and carbon dioxide. 1,3-Dipolar cycloaddition reactions with alkenes such as styrene, dimethyl maleate, and methyl cinnamate also lead to isothiazoles **(171)** directly. Bicyclic intermediates (cf. **136**) were not isolable; these cycloaddition reactions with alkenes giving isothiazoles involve an additional dehydrogenation step.

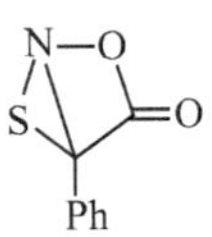

(172)

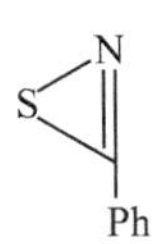

(173)

(174)

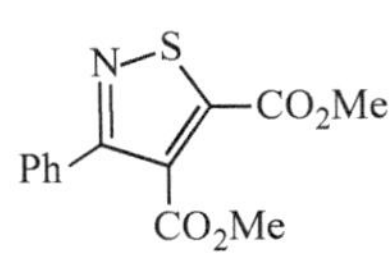

(175)

Photolysis (4045–4078 Å) of anhydro-5-hydroxy-1,3,2-oxathiazolium hydroxide (**169**, R = Ph) gives benzonitrile (77%), sulfur (>90%), and carbon dioxide.[118] This reaction has been interpreted[118] as involving the bicyclic intermediate **172**, which then transforms via the thiazirin **173** into the thiobenzonitrile oxide **174**. The postulated intermediates **172** and **173** are analogous to those proposed (**139** and **140**) for another photoreaction.[95] The possible involvement of the thionitrile oxide **174** is indicated by its trapping with dimethyl acetylenedicarboxylate: this yields 4,5-dimethoxycarbonyl-3-phenylisothiazole **(175)**.[118]

7.7 Triazoles

7.7.1 *1,2,3-Triazol-4-ones (anhydro-4-hydroxy-1,2,3-triazolium hydroxides)* (176)

Four methods have been described for the synthesis of meso-ionic 1,2,3-triazol-4-ones **(176)**. Alkylation of either 1-substituted 4-hydroxy-1,2,3-triazoles **(177)** or 5-hydroxy-1,2,3-triazoles **(178)** gives mixtures of products including meso-ionic 1,2,3-triazol-4-ones **(176)**.[119–124] This *N*-alkylation has been achieved either with diazomethane or by treating the sodium salts with alkyl halides. 1-Substituted 4- or 5-methoxy-1,2,3-triazoles are rearranged to their meso-ionic isomers in the presence of methyl iodide.[121,122] Thus, 1-methyl-5-methoxy-1,2,3-triazole **(179)** and methyl iodide in chloroform give anhydro-4-hydroxy-1,3-dimethyl-1,2,3-triazolium hydroxide **(180)** in high yield; this reaction has been monitored by nuclear magnetic resonance (NMR) spectroscopy and has been shown to involve the cation **181** as an intermediate.[121,122] The formation of the meso-ionic 5-bromo-1,2,3-triazol-4-one **(182)** from the triazolium *p*-toluenesulfonate (**183a**) and aqueous sodium hydroxide is interesting mechanistically and is believed to involve an ylide intermediate.[125,126] Analogously, 1,3-dimethyl-1,2,3-triazolium tosylate (**183b**), *N*-bromoacetamide, and aqueous sodium hydroxide give the meso-ionic 1,2,3-triazole-4-one **(182)** (71% yield).[125]

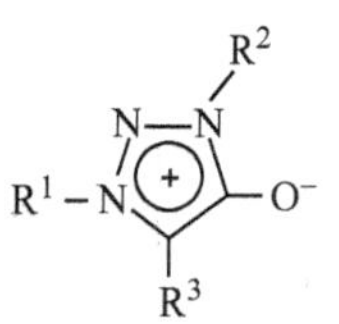

(176)

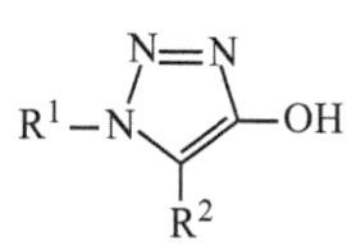

(177)

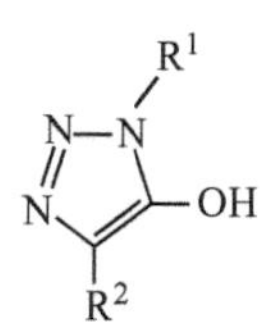

(178)

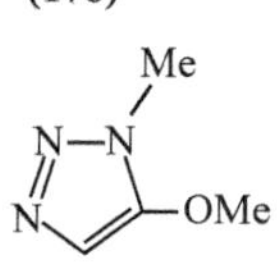

(179)

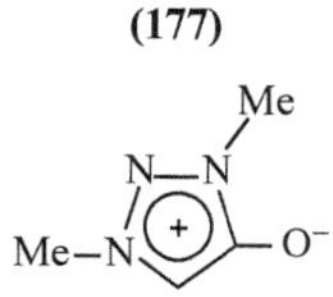

(180)

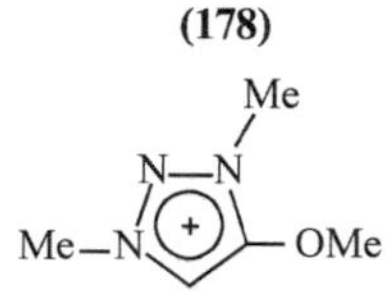

(181)

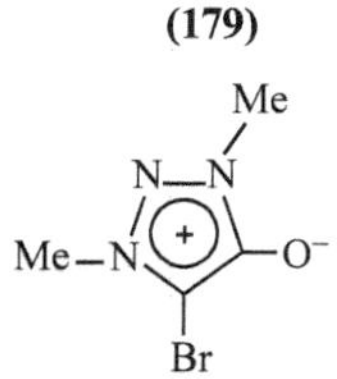

(182)

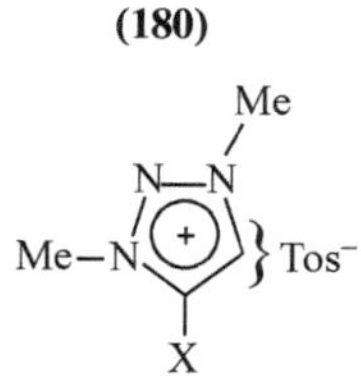

(183a) X = Br
(183b) X = H

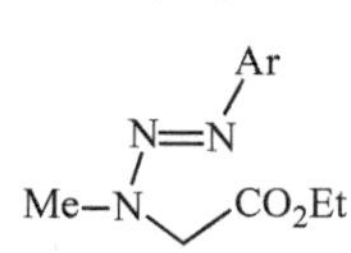

(184)

Cyclization of arylazoaminoacetates **(184)** with thionyl chloride–pyridine gives anhydro-4-hydroxy-3-aryl-1-methyl-1,2,3-triazolium hydroxides (**176**, R^1=Me, R^2=H, R^3=Ar).[127] These compounds are usually basic and yield the corresponding triazolium chlorides (**185**, R=H, X=Cl) with hydrogen chloride in benzene solution; corresponding *O*-ethylation yielding the salts **185**, R=Et, X=BF_4, takes place with triethyloxonium tetrafluoroborate.[127] They undergo 1,3-cycloaddition reactions.[127] The meso-ionic compounds **176**, R^1=Me, R^2=*p*-Me. C_6H_4, R^3=H, and dimethyl acetylenedicarboxylate give the pyrazole **186**, whereas with diethyl azodicarboxylate the 1,3-cycloadduct (**187**, Ar=*p*-Me.C_6H_4) could be isolated.[127] In contrast, tetracyanoethylene yields the product **188**, Ar=*p*-Me.C_6H_4.[127] It is noteworthy that the heterocumulenes phenyl isocyanate and phenyl isothiocyanate did not react with the meso-ionic 1,2,3-triazol-4-one (**176**, R^1=Me, R^2=Ph, R^3=H), but with these two heterocumulenes and the compound **176**, R^1=R^2=Me, R^3=H, the expected 1,3-dipolar cycloaddition was observed.[127]

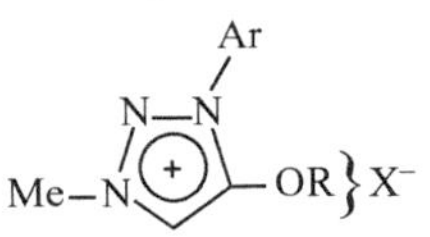

(185)

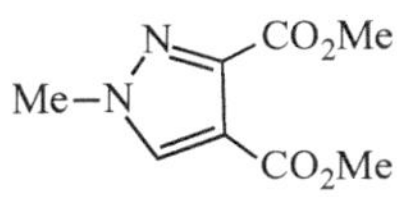

(186)

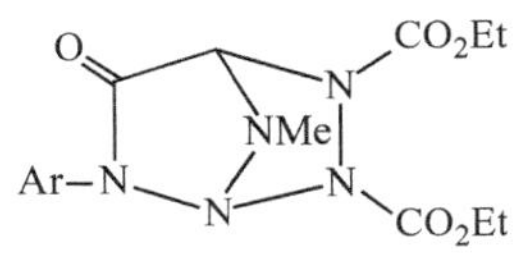

(187)

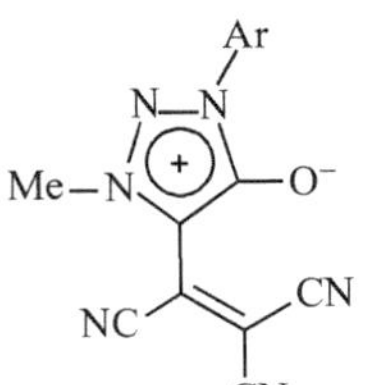

(188)

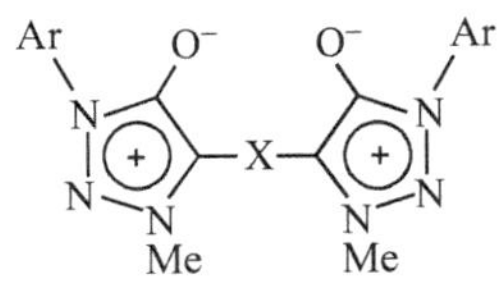

(189)

The 5-unsubstituted-1,2,3-triazol-4-ones (**176**, R^3=H) participate in electrophilic substitution reactions. Bromination in chloroform of anhydro-4-hydroxy-1,3-dimethyl-1,2,3-triazolium hydroxide **(180)** gave its 5-bromo derivative **(182)**.[121] The meso-ionic 3-aryl-1,2,3-triazol-4-ones (**176**, R^1=Me, R^2=Ar, R^3=H) gave 5-bromo derivatives (**176**, R^1=Me, R^2=Ar, R^3=Br) with bromine in acetic acid. Their reaction with sulfur monochloride gave the sulfide (**189**, X=S), and with thionyl chloride they gave the sulfoxide (**189**, X=SO).[127]

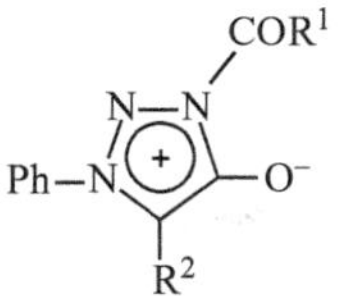

(190)

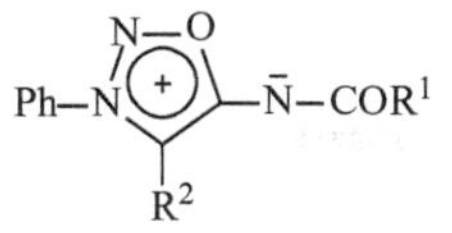

(191)

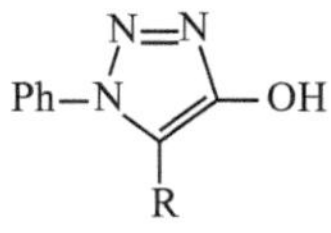

(192)

Meso-ionic 3-acyl-1-phenyl-1,2,3-triazol-4-ones (**190**, R^1=Ph or Me, R^2=H) have been postulated as intermediates in the interesting base-catalyzed transformation of *N*-acylsydnone imines (**191**, R^1=Ph or Me, R^2=H) into 4-hydroxy-1-phenyl-1,2,3-triazole (**192**, R=H).[128a] A similar photochemical transformation (**191**→**190**, R^1=Me or Ph, R^2=Ph) has also been reported.[128b] A different base-catalyzed transformation of sydnone imine derivatives into pyrazolones has been recently reported.[129]

7.7.2 1,2,3-Triazol-4-imines (anhydro-4-amino-1,2,3-triazolium hydroxides) (193)

The *N*-methyl-*N*′-arylazoaminoacetonitriles (**194**) and hydrogen chloride in ether yield the 1,2,3-triazolium chlorides (**195**). These salts (**195**) and base did not give the corresponding meso-ionic compounds (**193**, R^1=Me, R^2=Ar, R^3=R^4=H), but the corresponding *N*-acetyl derivatives (**193**, R^1=Me, R^2=Ar, R^3=H, R^4=Ac) were produced either from the nitriles (**194**) and acetyl chloride or from the 1,2,3-triazolium chlorides (**195**) and acetic anhydride, followed by treatment with ammonium hydroxide.[127]

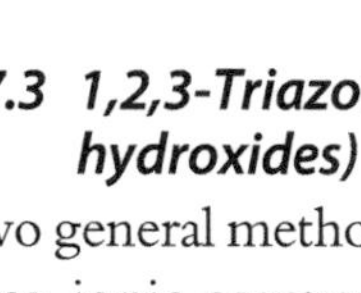

(193)

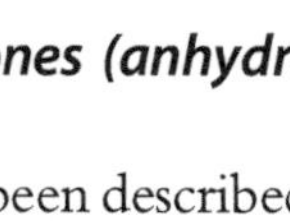

(194)

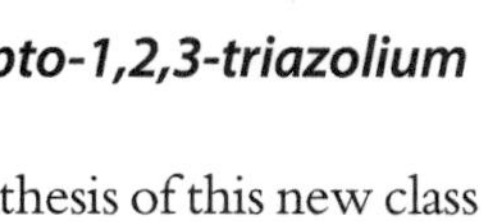

(195)

7.7.3 1,2,3-Triazole-4-thiones (anhydro-4-mercapto-1,2,3-triazolium hydroxides) (196)

Two general methods have been described for the synthesis of this new class of meso-ionic compounds (**196**).[126,130] The most convenient method is by the treatment of 4-bromo-1,2,3-triazolium salts (**197**, X=Br) with sodium sulfide in dimethylformamide. Alternatively, *N*-methylation of the isomeric 4- or 5-alkylmercapto-1,2,3-triazoles **198** or **199** with methyl tosylate gave intermediate triazolium salts (**197**, X=SR, Y=Tos), which yielded meso-ionic 1,2,3-triazole-4-thiones (**196**) by 5-dealkylation by heating with piperidine.

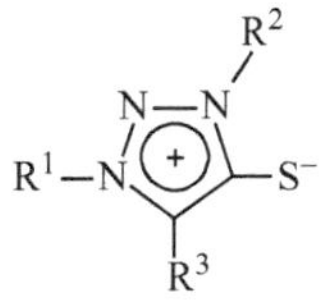

(196)

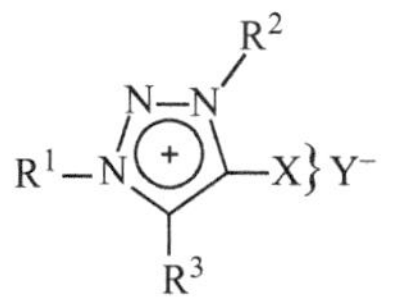

(197)

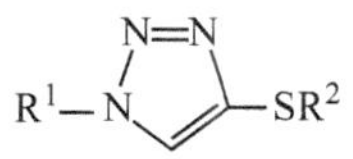

(198)

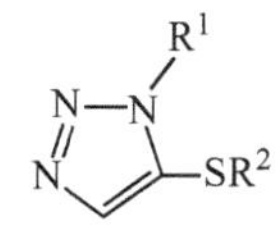

(199)

In contrast with the meso-ionic 1,2,3-triazol-4-ones (**176**) which are relatively stable to heat, the meso-ionic 1,2,3-triazole-4-thiones (**196**) rearrange when heated in benzene solution at 180°. Under these conditions, anhydro-1-benzyl-3-methyl-4-mercapto-1,2,3-triazolium hydroxide (**196**, R^1=$PhCH_2$, R^2=Me, R^3=H) gives a mixture of the 5-alkylmercapto-1,2,3-triazoles (**199**, R^1=Me, R^2=$PhCH_2$, 82% yield; **199**, R^1=R^2=$PhCH_2$, 2% yield; and **199**, R^1=R^2=Me, 13% yield) formed by intermolecular alkyl group transfer reactions.[130] The meso-ionic 1,2,3-triazole-4-thiones (**196**) and methyl iodide give methiodides (**197**, X=SMe, Y=I). Anhydro-1-benzyl-4-mercapto-3-methyl-1,2,3-triazolium hydroxide (**196**, R^1=$PhCH_2$, R^2=Me, R^3=H) on heating with benzoyl chloride–pyridine gives a mixture of 1,2,3-triazoles (**198**, R^1=$PhCH_2$, R^2=PhCO, 32% yield; and **199**, R^1=Me, R^2=PhCO, 49% yield). This reaction presumably involves the *S*-benzoyl-1,2,3-triazolium salt (**197**, R^1=$PhCH_2$, R^2=Me, X=SCOPh, Y=Cl) as an intermediate.[130]

7.7.4 1,2,4-Triazol-3-ones (anhydro-3-hydroxy-1,2,4-triazolium hydroxides) (200)

The "endoxytriazolines" originally described by Busch were later formulated[2b] either as meso-ionic 1,2,4-triazol-3-ones (**200**) or as meso-ionic 1,3,4-oxadiazol-2-imines (**153**). This ambiguity has now been resolved by the specific synthesis[107] of the meso-ionic 1,3,4-oxadiazol-2-imines (**153**), and recent studies[131,132] have firmly established that Busch's "endoxytriazolines" are meso-ionic 1,2,4-triazol-3-ones (**200**).

A number of general methods for the synthesis of meso-ionic 1,2,4-triazol-3-ones are available. Sodium ethoxide-catalyzed cyclization of 1-benzoyl-1,4-diphenylsemicarbazide (**201**, R^1=R^2=R^3=Ph, X=O) yielded anhydro-3-hydroxy-1,4,5-triphenyl-1,2,4-triazolium hydroxide (**200**, R^1=R^2=R^3=Ph).[133] A general route to meso-ionic 1,2,4-triazol-3-ones (**200**) is exemplified by the formation of the 1,4,5-triphenyl derivative (**200**, R^1=R^2=R^3=Ph) from *N*-amino-*N*,*N*′-diphenylbenzamidine (**202**, R^1=R^2=R^3=Ph) and phosgene.[111,131,132,134] In contrast with this ready meso-ionic compound formation, the corresponding reaction of the *N*-methylbenzamidine (**202**, R^1=Me, R^2=R^3=Ph) did not yield the meso-ionic 1,2,4-triazol-3-one (**200**, R^1=Me, R^2=R^3=Ph).[132]

The product was in fact 3,4-diphenyl-2-methyl-1,2,4-triazol-5-onium chloride (**203**), which on heating gave 3,4-diphenyl-1,2,4-triazol-5-one (**204**, R=Ph).[132] The formation of the *N*-methyl derivative (**200**, R^1=Me, R^2=R^3=Ph, yield 79%) by heating the *N*-thiobenzoyl semicarbazide (**201**, R^1=Me, R^2=R^3=Ph, X=S) with potassium carbonate in methyl cyanide has been reported.[135] Another synthesis of *N*-methyl derivatives (**200**, R^1=Me) involves méthylation of 3-methyl-4-phenyl-1,2,4-triazol-5-one[136] (**204**, R=Me). This yields a mixture including the meso-ionic compound **200**, R^1=R^2=Me, R^3=Ph. The 1,3-dipolar cycloaddition of phenyl isocyanate to *N*-phenylsydnone (**205**) yielding the meso-ionic 1,4-diphenyl-1,2,4-triazol-3-one (**200**, R^1=R^3=Ph, R^2=H)[137,138]provides a comparatively rare example of the interconversion of one type of meso-ionic compound into another.

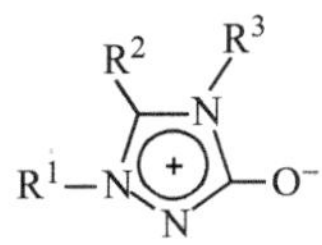
(200)

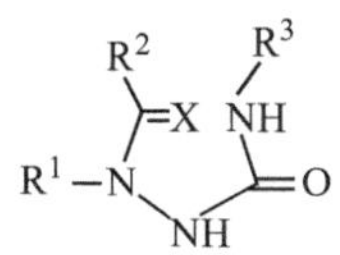
(201)

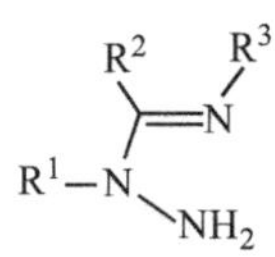
(202)

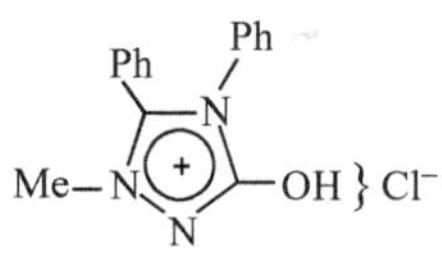
(203)

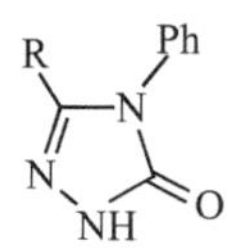
(204)

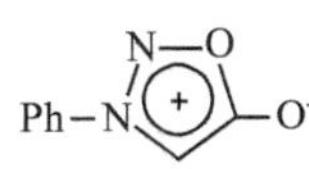
(205)

The bicyclic meso-ionic 3-oxo-1,2,4-triazolo[4,5-*a*]pyridines (**206**) have been prepared by the following methods: (i) the reaction of the hydrazines (**207**, X=H) with phosgene,[136,139] (ii) heating the amide (**207**, X=$CONH_2$) or the carbamate (**207**, X=CO_2Et),[136] and (iii) alkylation or acylation of 3-oxo-1,2,4-triazolo[4,5-*a*]pyridine (**208**).[133] The isomeric meso-ionic 2-oxo-1,3,4-triazolo[4,5-*a*]pyridines (**209**) are formed from the carbamoyl chlorides (**210**) and sodium azide.[140]

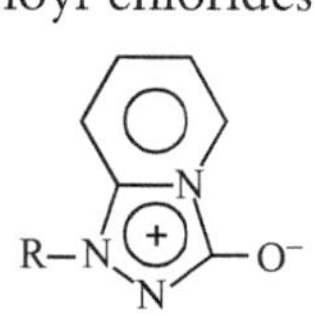
(206)

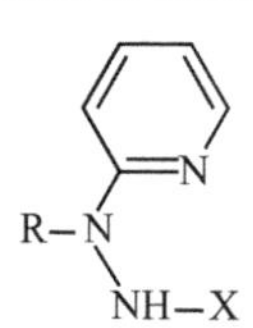
(207)

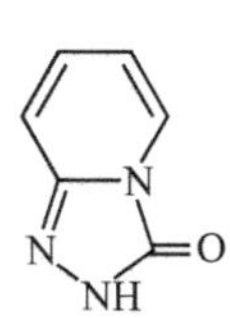
(208)

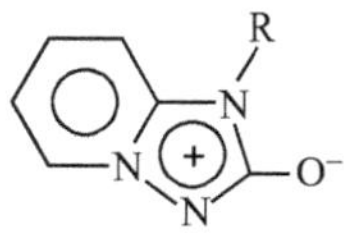
(209)

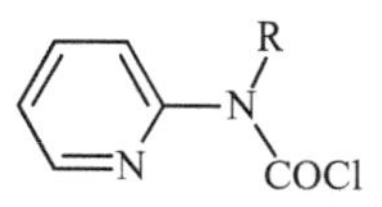
(210)

The meso-ionic 1,2,4-triazol-3-ones (**200**) are stable to acid, but alkaline hydrolysis gives 1,4-disubstituted semicarbazides. They do not normally participate in 1,3-dipolar cycloadditions,[131,132] but the meso-ionic 1,4-diphenyl-1,2,4-triazol-3-one (**200**, $R^1=R^3=Ph$, $R^2=H$) and benzyne yielded 2-phenylindazole.[77a] 1,2,4-Triazolium salts (**211**) are formed with triethyloxonium tetrafluoroborate.[66] Reduction of the meso-ionic compound **200**, $R^1=Me$, $R^2=R^3=Ph$, with lithium aluminum hydride gives the triazolidinone **212**.[107]

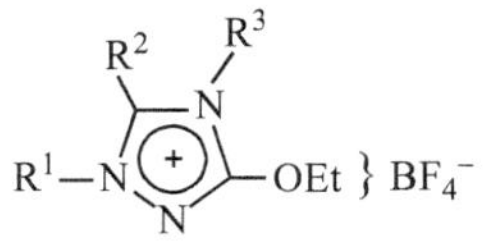

(211)

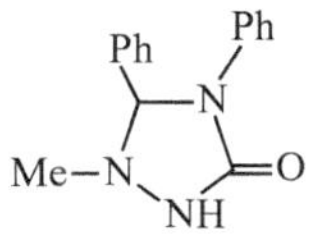

(212)

Nuclear magnetic resonance studies[141] on meso-ionic 1,2,4-triazol-3-ones (**200**) were used to examine their relationship to the alternative 1,3,4-oxadiazol-2-imine structure (**153**). The effect of solvent polarity upon the ultraviolet spectrum of anhydro-3-hydroxy-1,4-diphenyl-1,2,4-triazolium hydroxide (**200**, $R^1=R^3=Ph$, $R^2=H$) has been discussed in terms of its meso-ionic structure.[142]

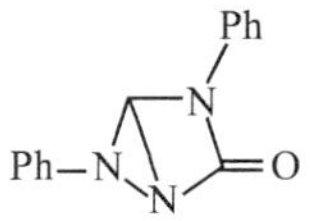

(213)

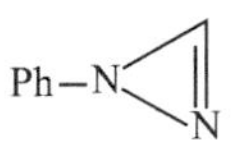

(214)

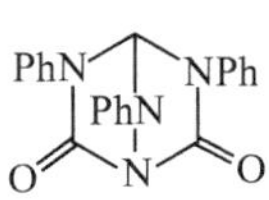

(215)

The photolysis of the meso-ionic 1,4-diphenyl-1,2,4-triazol-3-one (**200**, $R^1=R^3=Ph$, $R^2=H$) was stated[97a] to yield phenyl isocyanate (13%), *N,N'*-diphenylurea (23%), and the bicyclic compound **215** (49%). These results were interpreted in terms of the fragmentation of the photo-intermediate **213** yielding the *N*-phenyldiazirine (**214**).[97] A later publication by the same group[97b] reports different results. Photolysis of meso-ionic 1,4-diphenyl-1,2,4-triazol-3-one (**200**, $R^1=R^{3=}Ph$, $R^2=H$) was stated to yield[97a] phenylisocyanate and the bicyclic compound **215**. Later studies[97b] have shown that the bicyclic compound **215** is not produced by the photolysis of **200**, $R^1=R^3=Ph$, $R^2=H$. Irradiation in methanol-methylene chloride gave methyl phenylcar-bamate (25%), benzimidazole (18%), and azobenzene (7%).

7.7.5 1,2,4-Triazol-3-imines (anhydro-3-amino-1,2,4-triazolium hydroxides) (216)

The meso-ionic 1,2,4-triazol-3-imines can in principle exist as isomers **216** and **217**, but only recently has a specific synthesis of these two isomers been described.[143] Aryl isocyanide dichlorides (**218**)[144] and *N*-aminoamidines (**219**) in boiling toluene give the intermediate 2-arylamino-1,3,4-triazolium chlorides (**220**, R^5=H), which with ammonium hydroxide yield the yellow crystalline meso-ionic products (**216**). This method has been used to synthesize pairs of isomers (**216** and **217**), and their equilibration (**216** ⇆ **217**) in boiling ethanol has been reported.[143] The original syntheses by Busch of the "endoiminotriazolines" later recognized as meso-ionic 1,2,4-triazol-3-imines (**216** and **217**)[2b] involved either the reaction of *N*-aminoguanidines (**221**) with acid chlorides or the oxidation of triazolidines (**222**) obtained from *N*-aminoguanidines (**221**) and aldehydes. These two methods presumably yielded one of two possible products (**216** or **217**) or mixtures (**216** and **217**).

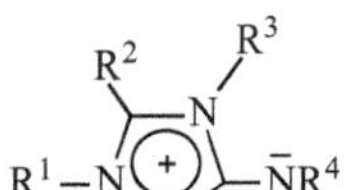

(216)

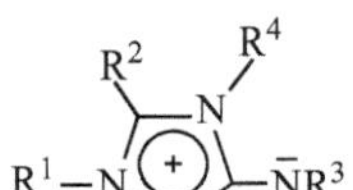

(217)

$Ar-N=CCl_2$

(218)

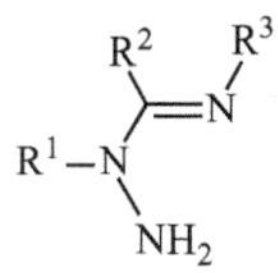

(219)

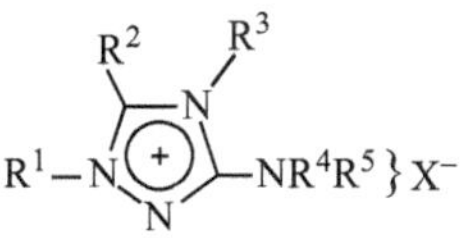

(220)

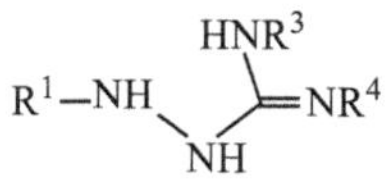

(221)

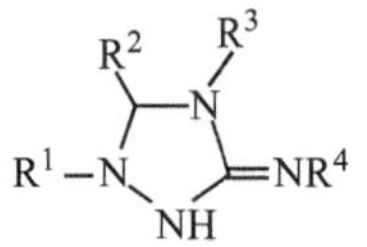

(222)

The dipole moments in benzene solution of the compounds **216**, R^1=Me, R^2=R^3=R^4=Ph (μ=8.2 D); **216**, R^1=Me, R^2=R^3=Ph, R^4=*p*-Cl.C_6H_4 (μ=9.9 D), and its isomer **217**, R^1=Me, R^2=R^3=Ph, $R^{4=}$*p*-Cl.C_6H_4 (μ=8.2 D) clearly support their meso-ionic structure.[108] Like the analytical reagent nitron (**216**, R^1=R^3=R^4=Ph, R^2=H), the meso-ionic 1,2,4-triazol-3-imines (**216**, R^1=Me, R^2=Ph, R^3=R^4=Ar) give precipitates of the nitrates (**220**, R^1=Me, R^2=Ph, R^3=R^4=Ar, R^5=H, X=NO_3) with dilute nitric acid. With methyl iodide they form methiodides (**220**, R^1=Me, R^2=Ph, R^3=R^4=Ar, R^3=Me, X=I), and they are reduced to triazolidines (**222**, R^1=Me, R^2=Ph, R^3=$R^{4=}$Ar) by lithium aluminum hydride.[143]

$BR_2C{=}N{-}N{=}CHAr$

(223)

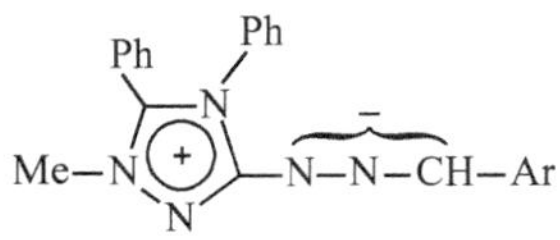

(224)

4-Aryl-1,1-dibromo-2,3-diazabutadienes (**223**)[145] and the *N*-amino-amidine (**219**, $R^1{=}Me$, $R^2{=}R^3{=}Ph$) yield 1,3,4-triazolium bromides (**220**, $R^1{=}Me$, $R^2{=}R^3{=}Ph$, $R^4{=}H$, $R^5{=}N{=}CHAr$), which with ammonia in chloroform solution give novel derivatives (**224**) of meso-ionic 1,2,4-triazol-3-imines (**216**). These compounds are of interest in that they belong to a new type of meso-ionic heterocycle in which the exocyclic substituent f (see Table 1) is a stabilized carbanionoid residue, $\widehat{N - N - C}HAr$.[146]

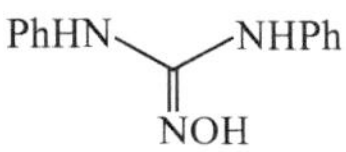

(225)

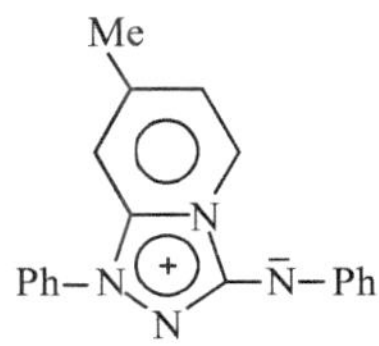

(226)

An interesting rearrangement is involved in the reaction between *N*-hydroxy-*N'*,*N*" diphenylguanidine (**225**), toluenesulfonyl chloride, and *γ*-picoline in benzene at 0°. This yields a colorless toluenesulfonate which with alkali gives red anhydro-*s*-triazolo[4,3-*a*]pyridinium hydroxide (**226**), which is a derivative of meso-ionic 1,2,4-triazol-3-imine (**216**).[147]

7.7.6 1,2,4-Triazole-3-thiones (anhydro-3-mercapto-1,2,4-triazolium hydroxides) (227)

When Busch's "endo-thiotriazolines" were formulated as meso-ionic compounds,[2b] it was recognized that two constitutional possibilities (**227** or **228**) required consideration. Later studies[141,148] established that the "endo-thiotriazolines" were in fact the meso-ionic 1,2,4-triazole-3-thiones (**227**). Recently, a specific synthesis of their meso-ionic isomers (**228**) has been reported,[149] and under equilibration conditions in hot ethanol the rearrangement **228** → **227** occurs.[149a] This explains why the meso-ionic 1,2,4-triazole-3-thiones (**227**) are the products of synthetic routes that could, in principle, yield either the meso-ionic 1,2,4-triazole-3-thiones (**227**) or their isomers (**228**).

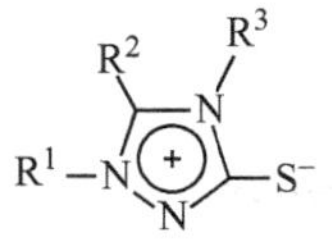

(227)

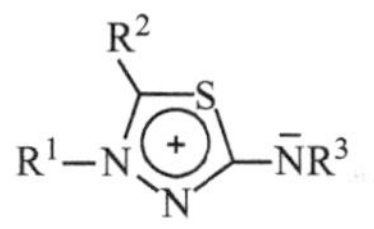

(228)

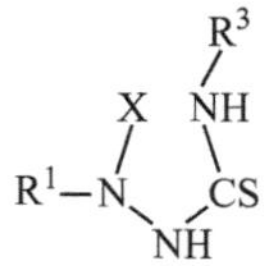

(229) X = H
(229) X = COR^2

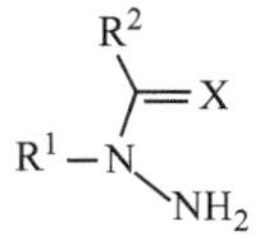

(231) X = NR^3
(232) X = S

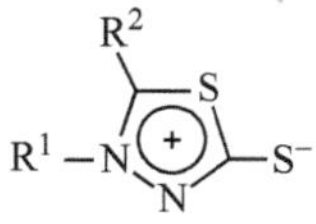

(233)

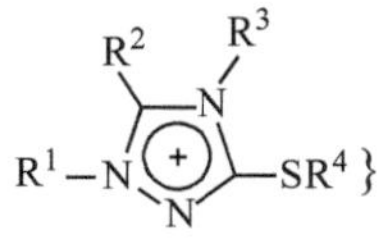

(234)

A number of methods are available for the synthesis of meso-ionic 1,2,4-triazole-3-thiones (**227**). These include (i) anhydro-acylation of 1,4-disubstituted thiosemicarbazides (**229**),[2b,9b,131,132,150] (ii) heating of 4-acyl-1, 4-disubstituted thiosemicarbazides (**230**),[132,151] (iii) reaction of *N*-aminoamidines (**231**) with thiophosgene,[9b,131,132] (iv) reaction of *N*-aminoamidines or *N*-thioacylhydrazines (**232**) with isothiocyanates,[149,152] or carbon disulfide–dicyclohexylcarbodiimide,[152b] and (v) reaction of meso-ionic 1,3,4-thiadiazoles (**233**), or their corresponding methiodides, with primary amines.[2b,9b]

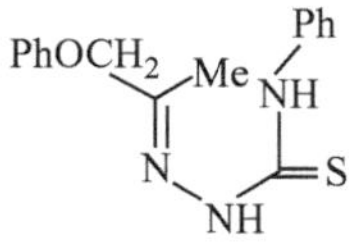

(235)

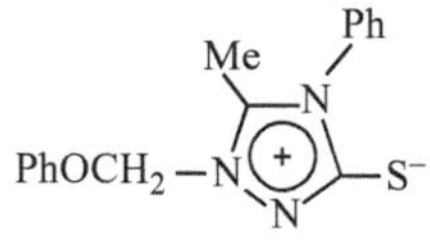

(236)

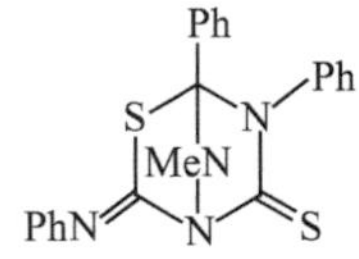

(237)

The oxidative cyclization of the hydrazine **235** to the meso-ionic 1,2,4-triazole-3-thione (**236**) is interpreted as involving a 1,2-shift of the phenoxymethyl group.[153] The isomerization **228** → **227** in boiling ethanol has been mentioned above.[149a] This transformation (**228** → **227**, R^1 = Me, R^2 = R^3 = Ph) may also be achieved with phenyl-isothiocyanate in benzene solution at room temperature and presumably involves the 1,3-dipolar cycloadduct (**237**) as an intermediate.[149a] This 1,3-cycloaddition reaction of meso-ionic 1,3,4-thiadiazol-2-imines (**228**) contrasts with the behavior of their isomers (**227**); which do not apparently react with 1,3-dipolarophiles.[132]

An interesting proposal has been put forward[132] to account for an unexpected difference in the acylation reactions of 1,4-disubstituted thiosemicarbazides (**229**). The *N*-methyl derivative (**229**, R^1 = Ph, R^3 = Me) and acetic anhydride yields the expected meso-ionic product (**227**, $R^1=Ph$, $R^2=R^3=Me$), whereas the corresponding reaction of the *N*-phenyl derivative (**229**, $R^1=R^3=Ph$) gives 1,4-diphenyl-3-methyl-1,2,4-triazoline-5-thione (**238**) rather than the expected meso-ionic isomer (**227**, $R^1=R^3=Ph$, $R^2=Me$). This reaction does not involve a phenyl migration![132] This exclamation mark seemed to be appropriate at the time when this sentence was first written, but subsequently the proposal put forward by Potts et al.[132a] for the transformation (**229**, $R^1=R^3=Ph \rightarrow$ **238**) was shown to be incorrect. The alternative that the transformation is, in fact **229** → **227**, $R^1=R^3=Ph$, $R^2=Me$, has been clearly demonstrated by Grashey and Baumann.[132b] Presumably, the mystery[132a] is now settled.[132b] The latest interpretation[132b] is in accord with the related suggestions made by Evans and Milligan[141b] regarding the thermal dehydration product[2b] of 1-formyl-1,4-diphenylthiosemicarbazide (**230**, $R^1=R^3=Ph$, $R^2=H$).

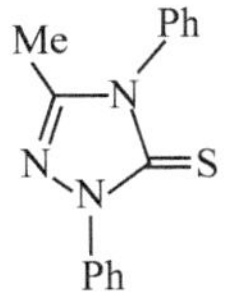

(238)

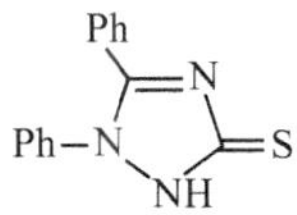

(239)

The NMR spectra[141] and dipole moments[108] of 1,2,4-triazole-3-thiones support their meso-ionic formulation (**227**). For example, the compound (**227**, $R^1=Me$, $R^2=R^3=Ph$) has a dipole moment of 9.1 D in dioxan solution.[108] The effect of solvent polarity upon the ultraviolet and visible spectra of the triphenyl derivative (**227**, $R^1=R^2=R^3=Ph$) has been reported but no direct interpretation was made.[142] The meso-ionic 1,2,4-triazole-3-thiones (**227**) form hydrochlorides[131] (**234**, $R^4=H$, $X=Cl$) and methiodides[132] (**234**, $R^4=Me$, $X=I$); they yield 1,2,4-triazolidine-3-thiones by lithium aluminum hydride reduction.[149a]

Recently, a novel member of the meso-ionic class of 1,2,4-triazole-3-thiones (**227**) has been described.[154] Heating the sodium salt, Ph–CO–NPh–NH–CS–S–CH_2–CO_2Na, with hydrazine gave the compound **227**, $R^1=R^2=Ph$, $R^3=NH_2$, which was characterized as a monobenzylidene derivative. The meso-ionic hydrazine derivative **227**, $R^1=R^2=Ph$, $R^3=NH_2$, undergoes an interesting reaction with nitrous acid, yielding the 1,2,4-triazoline-3-thione (**239**).[154]

7.7.7 1,2,4-Triazol-3-enes (anhydro-3-alkyl-1,2,4-triazolium hydroxides) (*240*)

These heterocycles (**240**) are the first representatives of meso-ionic compounds to be synthesized in which the exocyclic substituent (f, Table 1) is a stabilized carbanionoid group $[-\overline{C}(CN)CO_2Me$ or $-\overline{C}(CN)_2]$. Their synthesis[132b,155] involves the reaction between (i) *N*-aminoamidines (**231**) and bis(methylthio)acrylonitriles (**241**), (ii) *N*-thioacylhydrazines (**232**), and 3-alkylamino-3-methylthioacrylonitriles (**242**), and (iii) 1,2,4-triazolium iodides (**234**, $R^4=Me$, $X=I$) and malononitrile.

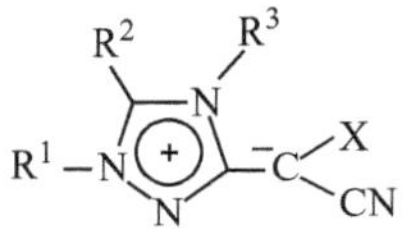

(240)

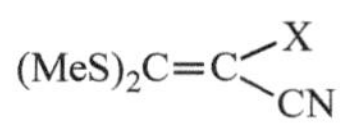

(241)

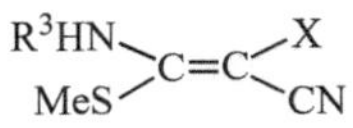

(242)

In **(240)-(242)**, X = CN or CO_2Me

The meso-ionic 1,2,4-triazol-3-enes (**240**) are highly stable, colorless to greenish-yellow compounds.[155]

7.8 Thiadiazoles

7.8.1 1,3,4-Thiadiazol-2-ones (anhydro-2-hydroxy-1,3,4-thiadiazolium hydroxides) (*243*)

The meso-ionic 1,3,4-thiadiazol-2-ones (**243**) have been prepared from *N*-thioacylhydrazines (**232**) or their hydrochlorides[102,103] by reaction with carbonyl chloride[102,103,110,111,135,156] or methyl chlorocarbonate.[135] They are also formed by the isomerization of meso-ionic 1,3,4-oxadiazole-2-thiones (**156**).[102,103]

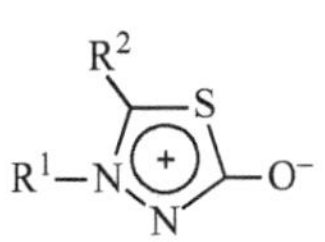

(243)

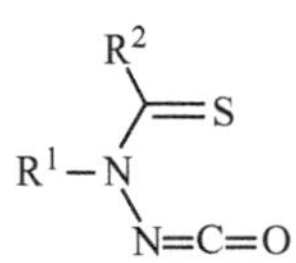

(244)

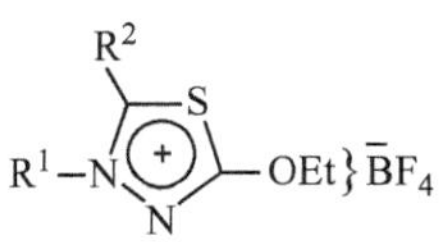

(245)

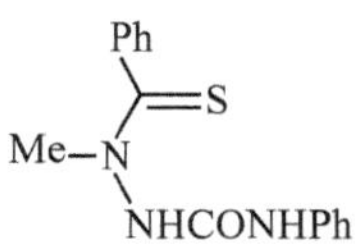

(246)

Photolysis (2537 Å) in methyl cyanide solution of the compound **243**, R^1=Me, R^2=Ph, gives *N*-methylthiobenzamide.[157] This reaction has been interpreted in terms of the homolysis of the valence tautomer **244**, which has been detected spectroscopically (ν_{max}2260 cm^{-1}; N=C=O).[22] The 1,3,4-thiadiazol-2-ones have dipole moments (**243**, R^1=R^2=Ph, μ=7.75 D; **243**, R^1=Me, R^2=Ph, μ=7.7 D) consistent with their meso-ionic formulation.[113] They react with triethyloxonium tetrafluoroborate, yielding the salts **245**.[66] The compound **243**, R^1=Me, R^2=Ph, and aniline gives the semicarbazide **246**.[135]

7.8.2 1,3,4-Thiadiazol-2-imines (Anhydro-2-amino-1,3,4-thiadiazolium Hydroxides) (247)

The novel 1,3,4-thiadiazol-2-imines (**247**) have been prepared from *N*-thioacylhydrazines (**232**) and aryl isocyanide dichlorides (ArN=CCl_2).[149a] This yields the intermediate salts (**248**) from which the meso-ionic heterocycles (**247**) are obtained as deep red oils by treatment of their chloroform solutions with anhydrous ammonia.[149a] Analogous transformations have subsequently been achieved using similar reactions with benzoyl isocyanide dichloride (PhCON=CCl_2), which yield the corresponding meso-ionic 1,3,4-thiadiazol-2-imines (**247**, R^3=Ph.CO)[107b] and aryl sulfonyl isocyanide dichlorides ($ArSO_2N$=CCl_2), which similarly yield the compounds **247**, R^3=$ArSO_2$.[107b] The corresponding reaction of *N*-thioacylhydrazines (**232**) with dibromodiazabutadienes (**223**)[145] has provided a new type of meso-ionic heterocycle (**249**).[146] Thus, within the 1,3,4-thiadiazol-2-imine class of meso-ionic heterocycle (**247**) there are four subclasses in which the exocyclic substituent (f, see Table 1) may be associated with $\overline{N}-Ar$,[149a] $\overline{N}-CO-Ar$,[107b] $\overline{N}-SO_2-Ar$,[107b] or $\overbrace{N-N-C}HAr$.[146] The compound **247**, R^1=Me, R^2=R^3=Ph, has a dipole moment μ=6.7 D.[108]

A second type of synthetic route to meso-ionic 1,3,4-thiadiazol-2-imines (**247**) is based on the acid-catalyzed reaction of *N*-thioacylhydrazines (**232**) with aryl isothiocyanates (Ar–NCS).[152b] This reaction yields the salts (**248**) as precursors of the meso-ionic heterocycles (**247**). An interesting variant upon this route involves the reaction between *N*-thioacylhydrazines (**232**) and acyl isothiocyanates (RCO–NCS).[107c,149b] This leads to the meso-ionic heterocycles **247**, R^3=CO_2Et, $CONMe_2$, COMe, $COCMe_3$, COAr, and SO_2Ph. The investigation of these compounds by X-ray photoelectron spectroscopy is a good example of the application of this physical method for the examination of meso-ionic compounds.[149b]

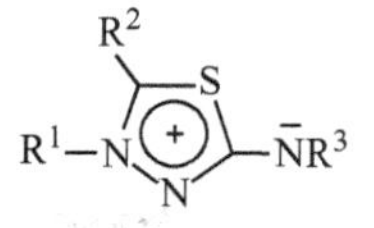
(247)

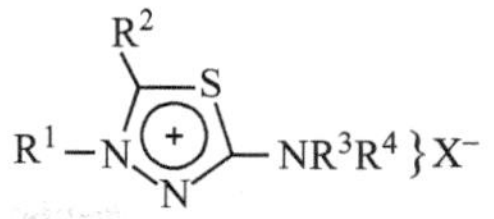
(248)

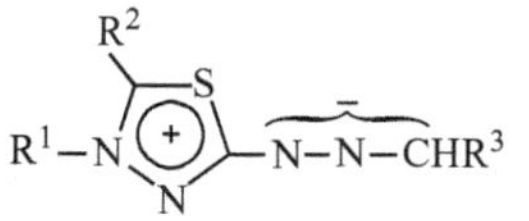
(249)

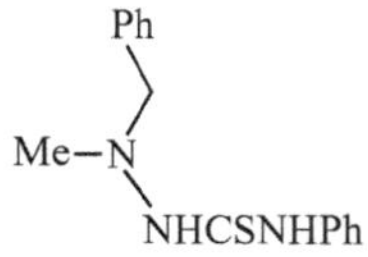
(250)

The isomerization **247** → **227** has been discussed (see Section 7.7.6); this type of reaction is not observed when the compound **249**, R^1=Me, R^2=R^3=Ph, is heated in ethanol.[146] The meso-ionic 1,3,4-thiadiazol-2-imine (**247**, R^1=Me, R^2=R^3=Ph) gives the methiodide (**248**, R^1=R^4=Me, R^2=R^3=Ph, X=I) with methyl iodide, the nitrate (**248**, R^1=Me, R^2=R^3=Ph, R^4=H, X=NO_3) with dilute nitric acid; its reduction with lithium aluminum hydride gives the thiosemicarbazide (**250**).[149]

7.8.3 1,3,4-Thiadiazole-2-thiones (anhydro-2-mercapto-1,3,4-thiadiazolium hydroxides) (251)

Although these substances have been known since the early investigations by Busch,[2b,9b] recent interest has been stimulated by their reported activity as antimicrobial agents.[4,158–160] Methods for the synthesis of meso-ionic 1,3,4-thiadiazole-2-thiones (**251**) include reactions between (i) the salts (**252**) and acid chlorides,[2b,9b,148,161,162] (ii) the salts (**252**) and sodium dithioformate,[162] (iii) *N*-acylhydrazines or *N*-thioacylhydrazines (**232**) and carbon disulfide,[152,163,164] (iv) *N*-thioacylhydrazines (**232**) and thiocarbonyl chloride, (v) *N*-aminoamidines and carbon disulfide.[152] Another route[148] to meso-ionic 1,3,4-thiadiazole-2-thiones (**251**) involves the condensation of the salts (**252**) with aldehydes giving thiadiazolidine thiones (**253**), which give disulfides (**254**) by oxidation (I_2 or $FeCl_3$). The disulfides (**254**) when heated in high-boiling solvents give the meso-ionic 1,3,4-thiadiazole-2-thiones (**251**) and thiadiazolidine thiones (**253**). The polycyclic 1,3,4-thiadiazole-2-thiones (**255**) are formed from quinoline-*N*-imides and carbon disulfide.[165]

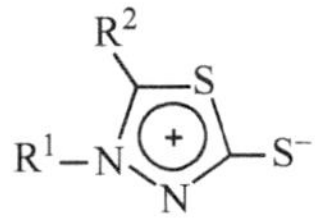

(251)

$R^1-NHNHCS_2^-\} K^+$

(252)

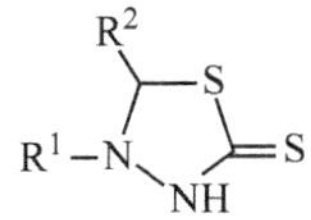

(253)

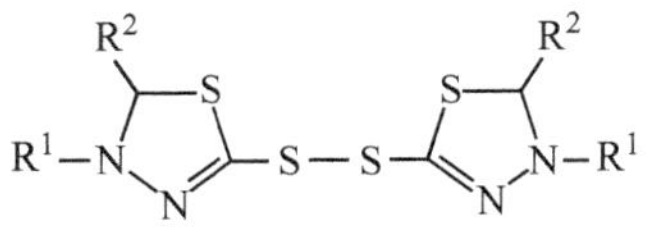

(254)

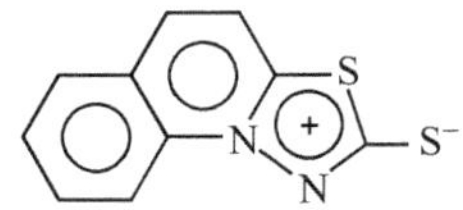

(255)

It has been shown that the transformation **254** → **251** + **253** proceeds with first-order kinetics and that increase in solvent polarity is associated with an increase in the rate. It is proposed that the reaction involves a novel unimolecular heterolytic scission of the sulfur–sulfur bond in the symmetrical disulfide (**254**).[166] The kinetics of the alkaline hydrolysis of anhydro-2-mercapto-4,5-diphenyl-1,3,4-thiadiazolium hydroxide (**251**, $R^1 = R^2 = Ph$) yielding *N*-benzoyl-*N*-phenylhydrazine has been examined: minor variants upon the well-known[2b,9b] chemistry of meso-ionic 1,3,4-thiadiazole-2-thiones have been reported.[167]

The dipole moments of a series of meso-ionic 1,3,4-thiadiazole-2-thiones (**251**) have been analyzed, and it has been shown that their large dipole moments are associated with a heterocyclic ring moment of 7.7 D acting at an angle of 65° to the Ph—N bond direction.[113] The influence of solvent polarity upon their ultraviolet spectra[168,169,170] has been interpreted as involving an $n \rightarrow n^*$ transition associated with the meso-ionic structure **251**. Comparison of the ultraviolet spectra of 5-aryl derivatives (**251**, $R^2 = Ar$) indicates appreciable internuclear conjugation of the biphenyl type.[169] Recently, the X-ray photoelectron (ESCA) spectra[171] of two meso-ionic 1,3,4-thiadiazole-2-thiones (**251**, $R^1 = Me$, $R^2 = Ph$ and *p*-Cl.C_6H_4) have been determined.[172] This important study has shown that the charge densities on the nitrogen and sulfur atoms are in satisfying accord with the charge distributions expected for the meso-ionic structure **251**.

The thermal isomerization **251** → **256** has been shown to occur in cases where $R^1 = Me$.[164]

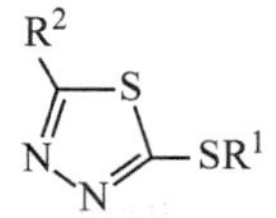

(256)

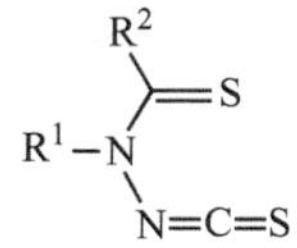

(257)

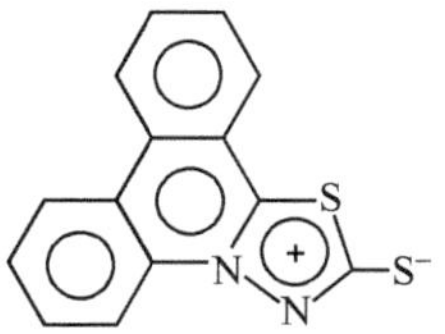

(258)

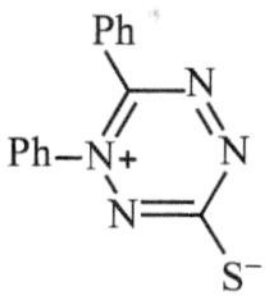

(259)

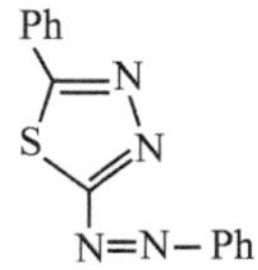

(260)

The valence tautomerism **251** ⇆ **257** has been proposed[22,157] to account for the formation of *N*-methylthiobenzamide by irradiation (2537 Å) in methyl cyanide solution. The valence tautomer **257**, R^1=Me, R^2=Ph, has been detected spectroscopically (ν_{max}2060 cm^{-1}; —N=C=S) either during photolysis at 25° or by heating at 160°.[22] Photochemical oxidative cyclization of 4,5-diaryl derivatives (**251**, R^1=R^2=Ar) analogous to the formation of phenanthrene from stilbene has been reported. Thus, irradiation of the 4,5-diphenyl derivative **251**, R^1=R^2=Ph, yields the tetracyclic meso-ionic compound (**258**).[168]

The product $C_{14}H_{10}N_4S$, obtained from the meso-ionic 1,3,4-thiadiazole-2-thione (**251**, R^1=R^2=Ph) and dimethylazodicarboxylate was initially regarded as the six-membered "meso-ionic" heterocycle **259**.[173] However, subsequent studies have established that the product is, in fact, 2-phenyl-5-phenylazo-1,3,4-thiadiazole (**260**).[174]

7.8.4 1,3,4-Thiadiazol-2-enes (anhydro-2-alkyl-1,3,4-thiadiazolium hydroxides) (261)

N-Thioacylhydrazines (**232**) and either 3,3-dichloroacrylonitriles (**262**, Y=Cl), 3,3-bis(methylthio)acrylonitriles (**262**, Y=SMe), or cyanomethoxycarbonylthioketene prepared *in situ* from the dithietanone form meso-ionic 1,3,4-thiadiazol-2-enes (**261**).[114]

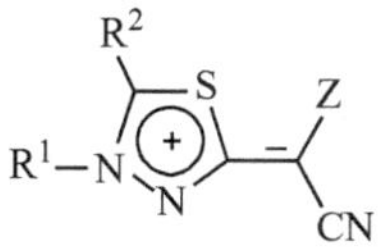

(261)

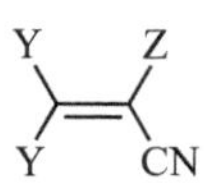

(262)

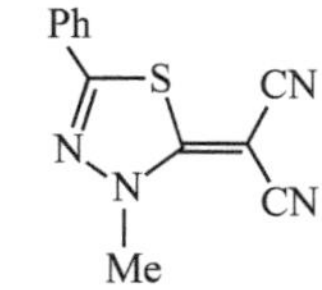

(263)

In **(261)** and **(262)** Z = CN or CO_2Me

The orange-yellow meso-ionic thiadiazol-2-enes (**261**) are reasonably stable, but the isomerization (**261**, R^1 = Me, R^2 = Ph, Z = CN) → **263** occurs at 250°.[114]

The reaction between *N*-thiobenzoyl-*N*-methylhydrazine (**232**, R^1 = Me, R^2 = Ph), and phenylthioacetylthioglycolic acid, Ph–CH_2–CS–S–CH_2–CO_2H, did not yield the expected product. A deep red substance, m.p. 238°–239°, has been isolated, and its X-ray crystallographic examination has shown that it has the meso-ionic structure **264**.[175] This is the first example of the exocyclic group (f; Table 1) being associated with the carbanionoid group, $-\overline{C}$Ph. CS.CH_2Ph.

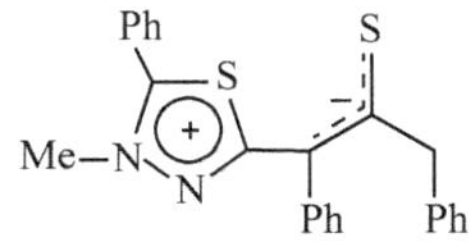

(264)

7.8.5 1,2,3-Thiadiazol-4-ones (anhydro-4-hydroxy-1,2,3-thiadiazolium hydroxides) (265)

Cyclodehydration of arylazomercaptoacetic acids (**266**) by acetic anhydride–pyridine yields the meso-ionic 1,2,3-thiadiazol-4-ones (**265**, R^1 = Ar).[176–178]

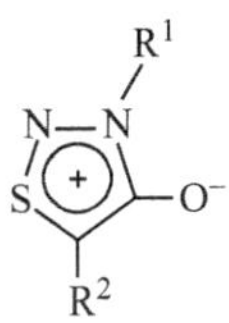

(265)

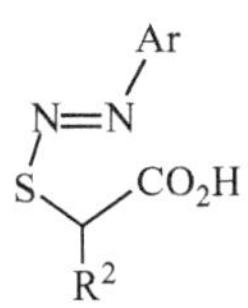

(266)

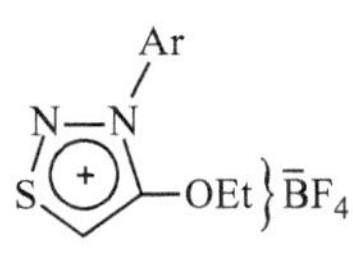

(267)

The electrophilic substitution of the 3-aryl compounds (**265**, R^1 = Ar, R^2 = H) exemplified by the formation of 5-bromo- (**265**, R^1 = Ar, R^2 = Br) and 5-nitro derivatives (**265**, R^1 = Ar, R^2 = $N0_2$) has been put forward as evidence against the meso-ionic formulation **265**.[176] This approach is unacceptable[2b] since ground state charge distribution cannot be deduced from reaction products. The aluminum–amalgam reduction[179] of meso-ionic 1,2,3-thiadiazol-4-ones (**265**) yields either *N*-mercaptoacetyl-*N*-arylhydrazines or *N*-acyl-*N*-arylhydrazines. Triethyloxonium tetrafluoroborate and meso-ionic 1,2,3-thiadiazol-4-ones (**265**) yield 1,2,3-thiadiazolium tetrafluoroborates (**267**).[66] The effect of solvent on the ultraviolet spectra of meso-ionic 1,2,3-thiadiazol-4-ones (**265**) has been reported.[180]

7.8.6 1,2,4-Thiadiazol-3-imines (anhydro-3-amino-1,2,4-thiadiazolium hydroxides) (268)

Only one representative of this class has been described.[181] 3-Amino-5-phenyl-1,2,4-thiadiazole (**269**) and methyl iodide gave the iodide (**270**) which with silver oxide gave the meso-ionic 1,2,4-thiadiazol-3-imine (**268**, R^1 = Ph, R^2 = Me, R^3 = H).[181]

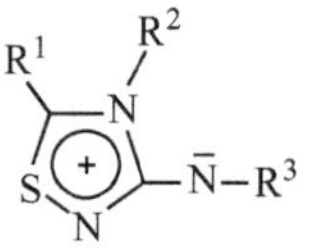

(268)

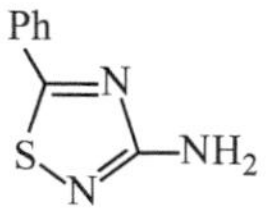

(269)

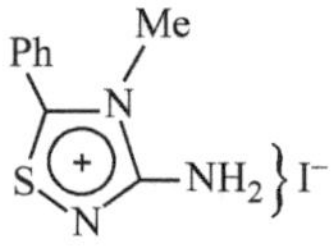

(270)

7.9 Oxatriazoles

7.9.1 1,2,3,4-Oxatriazol-5-ones (anhydro-5-hydroxy-1,2,3,4-oxatriazolium hydroxides) (271)

Meso-ionic 3-alkyl-1,2,3,4-oxatriazol-5-ones (**271**) are obtained[101,182,183] by the nitrosation of 1-alkyl semicarbazides, $RNHNHCONH_2$. At low temperatures, the intermediate *N*-nitroso derivatives (**272**) can be isolated, which cyclize on heating.[184] An alternative synthetic route is illustrated by the formation of meso-ionic 3-methyl-1,2,3,4-oxatriazol-5-one (**271**, R = Me) from *N*-nitroso-*N*-methylhydrazine (**273**) and phosgene.[185]

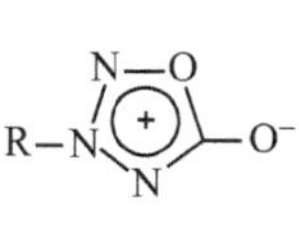

(271)

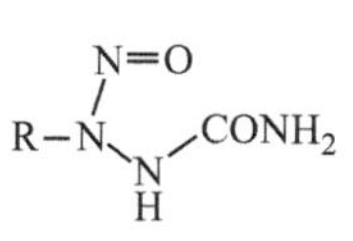

(272)

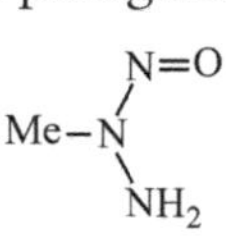

(273)

Analogous routes cannot be successfully employed for the synthesis of meso-ionic 3-aryl-1,2,3,4-oxatriazol-5-ones (**271**). Thus, nitrosation of 1-arylsemicarbazides yields arylazocarbonamides with loss of nitroxyl.[186] Meso-ionic 3-aryl-1,2,3,4-oxatriazol-5-ones are formed by (i) nitrosation of potassium arylhydrazonomethane disulfonates (**274**),[187,188] (ii) coupling of diazonium salts with nitroform,[189,190] and (iii) coupling of diazonium salts with the potassium salt of dinitromethane.[191] Reactions (ii) and (iii) presumably involve intermediates of the type **275**. The addition of nitrous acid is not required because intermediate *N*-nitroso derivatives are presumably formed *in situ* by hydrolysis of the compounds **275**. This view is supported by the isolation of the compound **275**, Ar = *p*-O_2N . C_6H_4, and its thermal transformation into the meso-ionic derivative (**271**, Ar = *p*-O_2N . C_6H_4).[191]

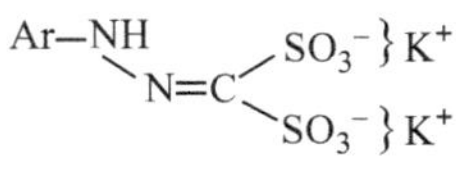

(274)

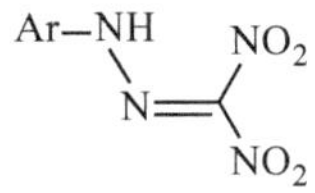

(275)

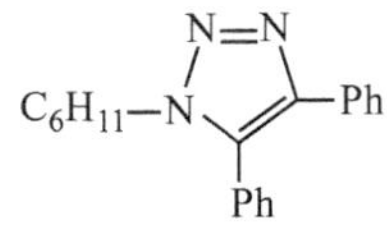

(276)

Information concerning the chemistry of meso-ionic 3-alkyl- and 3-aryl-1,2,3,4-oxatriazol-5-ones (**271**) is limited, but further investigation may well be encouraged by reports of pronounced hypotensive activity.[192] 3-Cyclohexyl-1,2,3,4-oxatriazol-5-one (**271**, R = cyclohexyl) is resistant to attack by dilute mineral acid, but warm concentrated sulfuric acid gives cyclohexanol and carbon dioxide.[184] In contrast, acid hydrolysis of 3-phenyl-1,2,3,4-oxatriazol-5-one (**271**, R = Ph) yields phenyl azide.[189] Meso-ionic 3-cyclohexyl-1,2,3,4-oxatriazol-5-one shows two unusual reactions: its photoirradiation in benzene gives cyclohexanone[97a] and heating with diphenylacetylene yields 1-cyclohexyl-4,5-diphenyl-1,2,3-triazole (**276**)[104] rather than the expected 2-cyclohexyl-4,5-diphenyl-1,2,3-triazole.

The dipole moment of the *N*-phenyl derivative (**271**, R = Ph) in benzene solution is 6.14 D,[108] which may be compared with the molecular orbital calculations by Sundaram and Purcell[193] which give a value of 5.3 D for the meso-ionic 1,2,3,4-oxatriazol-5-one heterocyclic ring moment.

7.9.2 1,2,3,4-Oxatriazol-5-imines (anhydro-5-amino-1,2,3,4-oxatriazolium hydroxides) (277)

Although this was not appreciated at the time, these compounds were first handled by Busch and Becker in 1896.[194] Their study continued until a more extensive report by Busch and Schmidt was published in 1929.[195] However, Busch and his coworkers were apparently unaware of some closely related work reported by von Pechmann in 1896.[187] These earlier studies have been interpreted by Farrar,[188] extended by Christopherson and Treppendahl,[196] and brought to a definitive conclusion by the Sheffield group.[197] This history is an interesting facet of the development of the chemistry of meso-ionic heterocycles.

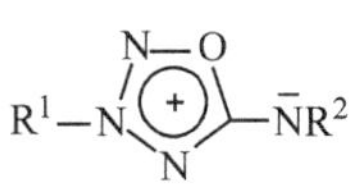

(277)

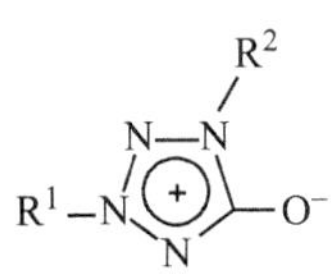

(278)

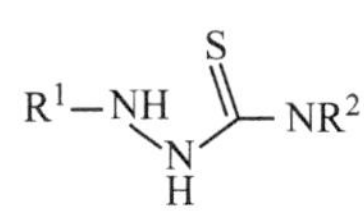

(279)

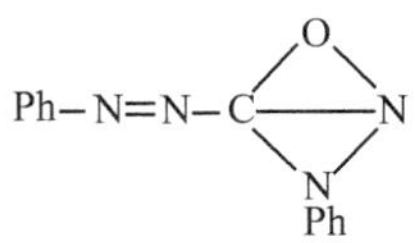

(280)

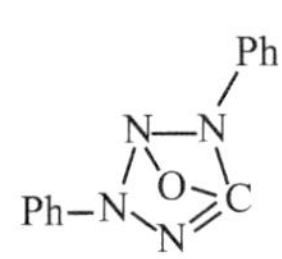

(281)

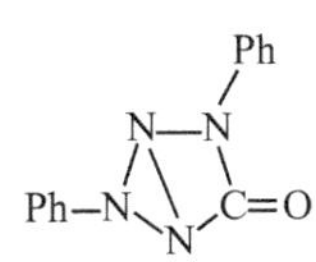

(282)

In the early investigation[194,195] of the nitrosation of 1,4-diphenylthiosemicarbazide (**279**, $R^1=R^2=Ph$), the product was obtained as deep red needles, m.p. 110°, and was allocated the remarkable "1,3-endoxyhydrazomethylene" structure (**280**). Treatment of this deep red compound with warm aqueous alkali gave a colorless isomer, m.p. 157°, which was formulated as having one of the equally unacceptable structures **281** or **282**.[195] Recent investigations[196,197] have established that nitrosation of 1,4-diarylthiosemicarbazides (**279**) yield the meso-ionic 1,2,3,4-oxatriazol-5-imines (**277**), which are transformed by base into the meso-ionic isomers (**278**). These alkaline transformation products formulated as meso-ionic 1,2,3,4-tetrazol-5-ones (**278**) are in fact identical (see Section 7.10.1) with substances prepared, but not formulated, by von Pechmann many years ago.[187]

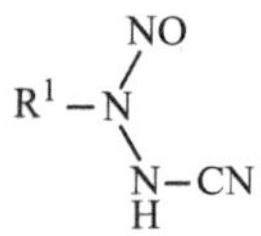

(283)

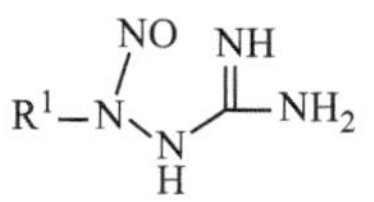

(284)

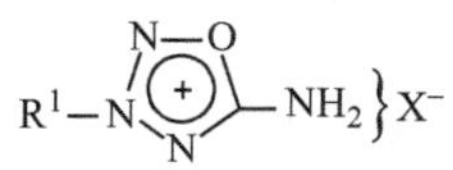

(285)

Treatment of the *N*-nitrosocyanides (**283**)[198] or the *N*-nitrosoguanylhydrazines (**284**)[199] with hydrogen chloride yielded the 1,2,3,4-oxatriazolium chlorides (**285**, X=Cl), which with aqueous sodium bicarbonate gave the meso-ionic 1,2,3,4-oxatriazol-5-imines (**277**, $R^2=H$).[198] Nitrosation of these compounds (**277**, $R^2=H$) gave the interesting *N*-nitroso compounds (**277**, $R^2=NO$), and acylation gave the *N*-acyl derivatives (**277**, $R^2=RCO$).[198]

Their formulation (**277**) as meso-ionic heterocycles is supported by their dipole moments in benzene[108]: **277**, $R^1=R^2=Ph$, $\mu=5.42$ D; **277**, $R^1=p\text{-Cl} . C_6H_4$, $R^2=Ph$, $\mu=4.17$ D; **277**, $R^1=Ph$, $R^2=p\text{-C1} . C_6H_4$, $\mu=6.29$ D. Acid hydrolysis of the diphenyl derivative (**277**, $R^1=R^2=Ph$) gives the meso-ionic 1,2,3,4-oxatriazol-5-one (**271**, R=Ph).[196] An interesting aryl group exchange is achieved by a 1,3-dipolar cycloaddition reaction between meso-ionic 1,2,3,4-oxatriazol-5-imines (**277**) and aryl isocyanates. Thus, the diphenyl derivative (**277**, $R^1=R^2=Ph$) and *p*-chlorophenyl isocyanate yields the product **277**, $R^1=Ph$, $R^2=p\text{-Cl}.C_6H_4$.[197]

7.9.3 1,2,3,4-Oxatriazole-5-thiones (anhydro-5-mercapto-1,2,3,4-oxatriazolium hydroxides) (286)

Meso-ionic 3-aryl-1,2,3,4-oxatriazole-5-thiones (**286**) are obtained by nitrosation or aryl dithiocarbazinic acid salts (**287**) formed from arylhydrazines and carbon disulfide.[200] Their meso-ionic formulation (**286**) is supported by their mass spectra[201] and their dipole moments: **286**, R = Ph, μ = 6.83 D; **286**, R = *p*-Cl.C_6H_4, μ = 5.08 D; **286**, R = *p*-Me.C_6H_4, μ = 7.08 D.[108]

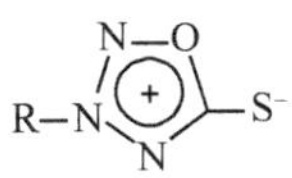

(286)

$R{-}NH{-}NH{-}CS_2^- \} H_3\overset{+}{N}NHR$

(287)

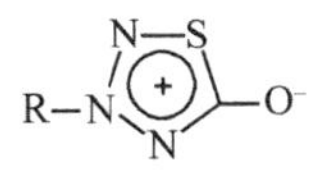

(288)

The meso-ionic 1,2,3,4-oxatriazoles (**286**)[200] yield phenol by acidic hydrolysis, phenyl azide by alkaline hydrolysis, and the *S*-ethyl 1,2,3,4-oxatriazolium cation with triethyloxonium tetrafluoroborate. The rearrangement **286** → **288** is achieved with boiling ethanolic ammonia.

7.10 Tetrazoles

7.10.1 1,2,3,4-Tetrazol-5-ones (anhydro-5-hydroxy-1,2,3,4-tetrazolium hydroxides) (289)

Although representatives of this class of meso-ionic heterocycles were first prepared in 1896[187] and were further studied in 1929,[195] their formulation as meso-ionic compounds was not made until 1966,[188] and this was confirmed only recently by detailed study.[197]

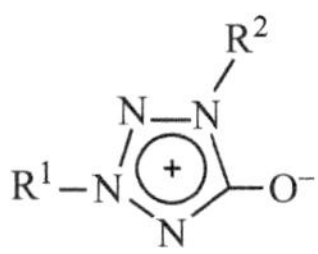

(289)

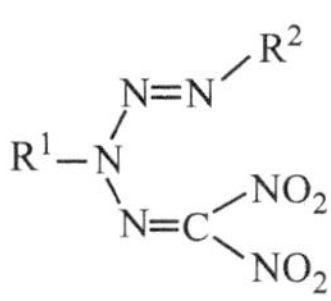

(290)

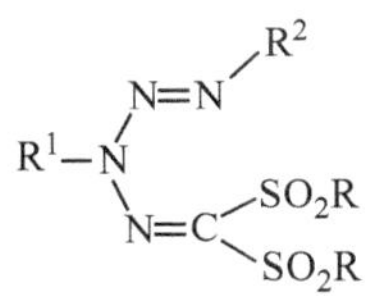

(291)

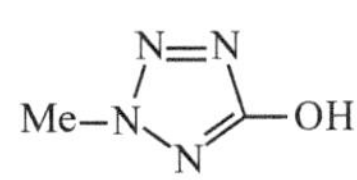

(292)

Mixtures of meso-ionic 1,3-diaryl-1,2,3,4-tetrazol-5-ones (**289**) and 1,2,3,4-oxatriazol-5-ones (**271**) are produced by the base-catalyzed coupling of aryl diazonium salts with dinitromethane[191] or bisalkylsulfonylmethanes.[191,202,203] Presumably these synthetic routes involve a Japp–Klingemann reaction[204] leading eventually to the tetra-aza intermediates **290** and **291**. The second type of synthesis[197] of meso-ionic 1,2,3,4-tetrazol-5-ones (**289**) by the rearrangement **277** → **289** with aqueous ethanolic sodium hydroxide has been discussed (Section 7.9.2).

2-Methyl-5-hydroxytetrazole (**292**) and diazomethane yield 2-methyl-5-methoxytetrazole and an isomer which could be meso-ionic 1,3-dimethyl-1,2,3,4-tetrazol-5-one (**289**, $R^1 = R^2 = Me$).[205] Treatment of the sweetening agent dulcin (*p*-ethoxyphenylurea) with nitrous acid gives a fluorescent product identified as the meso-ionic 1,2,3,4-tetrazol-5-one (**289**, $R^1 = R^2 = p$-EtO.C_6H_4).[206]

Meso-ionic 1,2,3,4-tetrazol-5-ones (**289**) react with triethyloxonium tetrafluoroborate, yielding salts (**294**)[197] that react with sodium sulfide in dimethylformamide, giving meso-ionic 1,2,3,4-tetrazole-5-thiones (**295**).[207]

7.10.2 1,2,3,4-Tetrazol-5-imines (anhydro-5-amino-1,2,3,4-tetrazolium hydroxides) (293)

The first representative of this class of meso-ionic heterocycles was the 1,3-dimethyl derivative (**293**, $R^1=R^2=Me$, $R^3=H$) obtained by methylation of 5-amino-2-methyltetrazole with methyl benzene-sulfonate.[208] Meso-ionic 1,2,3,4-tetrazol-5-acylimides (**293**, $R^1=R^2=Me$, $R^3=Me.CO$, p-$Me.C_6H_4.SO_2$, p-$Cl.C_6H_4.SO_2$) have been prepared by methylation of the corresponding 1-methyltetrazole using dimethyl sulfate.[209,210] The sequence **293**, $R^1=R^2=Me$, $R^3=Me.CO \rightarrow R^3=H \rightarrow R^3=p\text{-}Me.C_6H_4SO_2$, was achieved by acid hydrolysis followed by tosylation. The substances (**293**, $R^1=R^2=Me$, $R^3=H$) prepared by the two routes are identical.[208,210]

Comparison has been made[211] of the relative rates of hydrolysis of the meso-ionic compounds **293**, $R^1=R^2=Me$, $R^3=p\text{-}Me.C_6H_4.SO_2$, and p-$Cl.C_6H_4.SO_2$, with their isomers, which can be represented by a conventional covalent structure. The meso-ionic compounds hydrolyze relatively more slowly, and this has been related to π-electron distribution.

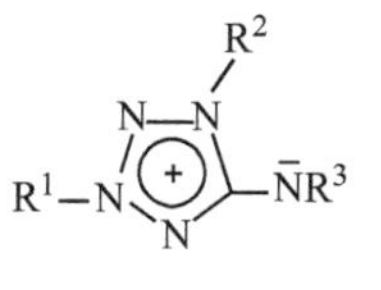

(293)

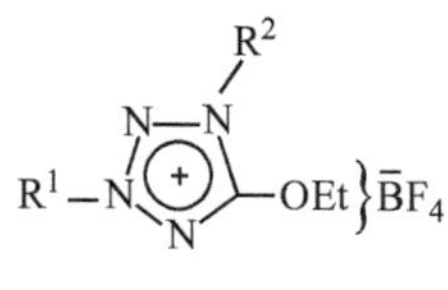

(294)

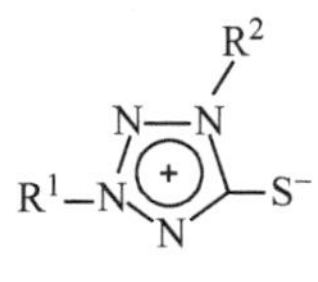

(295)

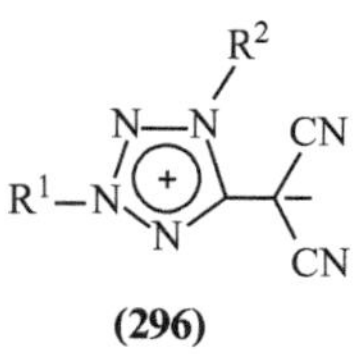

(296)

7.10.3 1,2,3,4-Tetrazole-5-thiones (anhydro-5-mercapto-1,2,3,4-tetrazolium hydroxides) (295)

The salts (**294**) (Section 7.10.1) are useful intermediates[197] for the preparation of other meso-ionic systems by reaction with appropriate nucleophiles. Thus, reaction with sodium sulfide in dimethyl formamide yields the 1,2,3,4-tetrazole-5-thiones (**295**), a new class of meso-ionic heterocycle.[207]

The 1,3-diphenyl derivative (**295**, $R^1 = R^2 = Ph$) has a dipole moment of 6.5 D in benzene solution in accord with its meso-ionic formulation.[108]

7.10.4 1,2,3,4-Tetrazol-5-enes (anhydro-5-alkyl-1,2,3,4-tetrazolium hydroxides) (296)

The salts (**294**)[197] with malononitrile and triethylamine in boiling acetonitrile yield the new class of meso-ionic heterocycles[212] in which the exocyclic carbanionoid residue is a biscyanomethylene group. The diphenyl derivative (**296**, $R^1 = R^2 = Ph$) has a dipole moment of 9.54 D in benzene solution.[108]

7.11 Thiatriazoles

7.11.1 1,2,3,4-Thiatriazol-5-ones (anhydro-5-hydroxy-l2,3,4-thiatriazolium hydroxides) (297)

Treatment of meso-ionic 1,2,3,4-oxatriazole-5-thiones (**286**) (Section 7.9.3) with boiling ethanolic ammonia yields the isomers **297**.[200] These belong to a new class of meso-ionic heterocycle, which by *O*-alkylation with triethyloxonium tetrafluoroborate yield the salts **298**.[200] These are useful intermediates for the synthesis of a number of novel types of meso-ionic 1,2,3,4-thiatriazoles (**299**,[207] **300**,[200] and **301**).[212]

The 1,2,3,4-thiatriazol-5-ones (**297**) have dipole moments in accord with their meso-ionic formulation.[108] They are remarkably stable to acidic hydrolysis, and 1,3-dipolar cycloaddition reactions have not been observed: alkaline hydrolysis yields aryl azides.

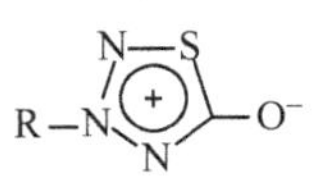

(297)

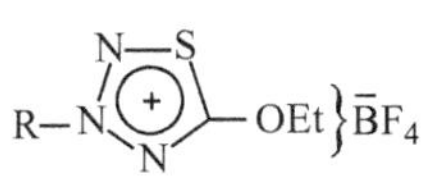

(298)

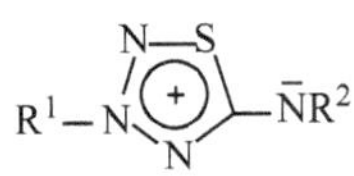

(299)

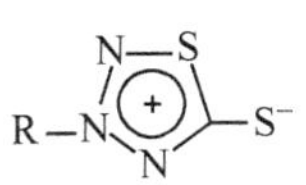

(300)

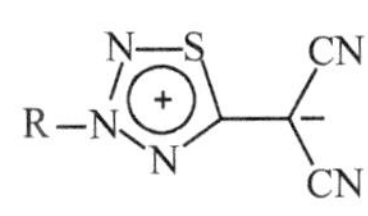

(301)

7.11.2 1,2,3,4-Thiatriazol-5-imines (anhydro-5-amino-1,2,3,4-thiatriazolium hydroxides) (299)

The salts (**298**)[200] undergo a nucleophilic displacement with arylamines, yielding intermediate tetrafluoroborates, which with ethanolic sodium hydroxide yield red meso-ionic 1,2,3,4-thiatriazol-5-imines (**299**).[207] The dipole moment of the diphenyl derivative in benzene is 3.7 D.[108]

7.11.3 1,2,3,4-Thiatriazole-5-thiones (anhydro-5-mercapto-1,2,3,4-thiatriazolium hydroxides) (300)

The salts (**298**) and sodium sulfide in dimethyl formamide yield the orange 1,2,3,4-thiatriazole-5-thiones (**300**).[200]

7.11.4 1,2,3,4-Thiatriazol-5-enes (anhydro-5-alkyl-1,2,3,4-thiatriazolium hydroxides) (301)

The salts (**298**) and malononitrile yield the orange biscyanomethylene derivatives (**301**). The compound **301**, R=Ph, has a dipole moment of 8.84 D in benzene.[212]

8. The existence and interconversion of type A meso-ionic isomers

When the term "meso-ionic" was put forward by Baker, Ollis, and Poole in 1949,[1] the possible existence of meso-ionic isomers **302** and **303** was recognized in principle: the symbols a, b, c, d, e, and f have the same meaning as that given earlier (Section 2). A typical pair of meso-ionic isomers can be more specifically represented by **304** and **305**. Single compounds having the type structures **304** or **305** have been known for a number of years, but it is only recently that (i) of the existence of meso-ionic isomers **304** and **305** has been demonstrated,[103,107,143,149,197,200,207] (ii) methods have been developed for the specific synthesis of both isomers,[103,107,143,149,197,200,207] (iii) physical[108,201] and chemical methods for the characterization of each isomer have been developed, and (iv) the interconversion between meso-ionic heterocyclic isomers has been studied. This has led to assessments of the relative thermodynamic stability of pairs of isomers.

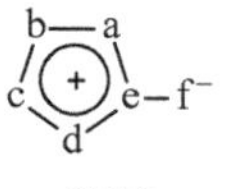

(302)

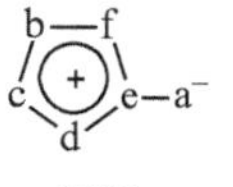

(303)

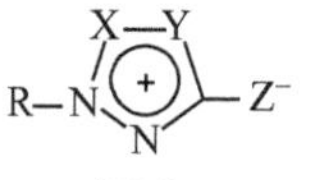

(304)

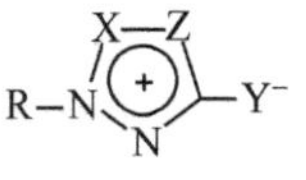

(305)

Methods for the synthesis of meso-ionic isomers **304** and **305**, their characterization and interconversion have been discussed in appropriate sections of the review. These results are summarized in Fig. 2, and in order to assist in the location of compounds they are associated with the formula numbers used in the text.

In Fig. 2, the results of equilibration studies are indicated. In only one case (**216** $\leftrightarrows$ **217**), as might be expected, is an equilibrium established,

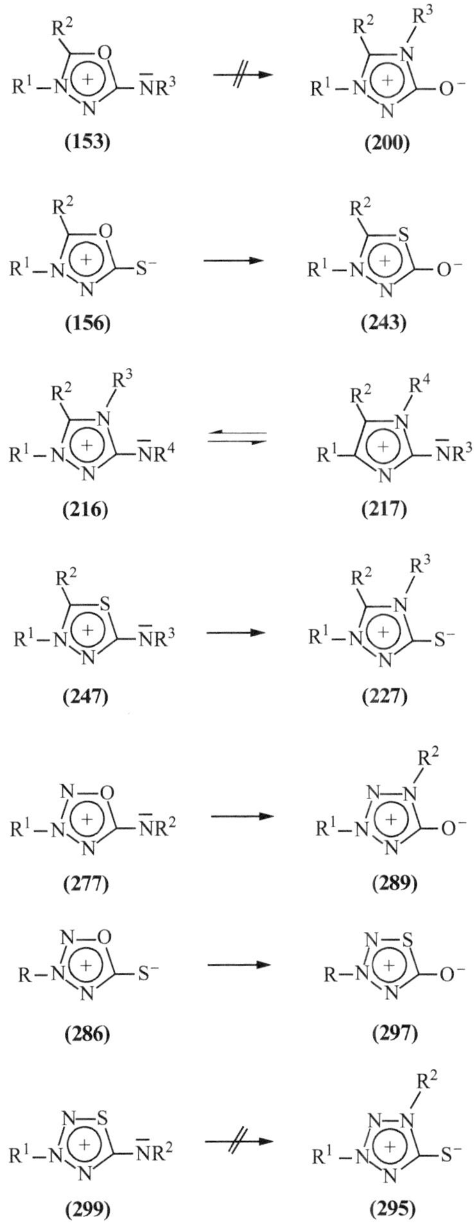

Fig. 2 Interconversion between meso-ionic isomers.

and there is little difference in the relative thermodynamic stability of the two isomers (**216** and **217**). On current knowledge, it is not possible to provide a well-based understanding of the factors that determine the relative thermodynamic stability of meso-ionic isomers.

Various methods and reagents have been used to effect the interconversion between meso-ionic isomers **304** ⇆ **305** including (i) heating in protic solvents such as ethanol or ethyl mercaptan,[103,143,149] (ii) treatment with hot ethanolic ammonia or ethanolic sodium hydroxide,[197,200] (iii) heating,[149] (iv) heating with aryl isocyanates[197] or isothiocyanates.[149] Mechanisms that are probably associated, respectively, with these processes are (i) the formation of betaine intermediates (**306**) (Fig. 3),[103,143,149,197,200] (iii) homolysis or heterolysis of the X—Z bond (**304**) or the X—Z bond (**305**) giving diradical (**307**) or dipolar (**308**) intermediates, (iv) 1,3-dipolar cycloaddition yielding intermediate adducts (e.g., **309**).[149,197] The base-catalyzed rearrangements (ii) present very interesting mechanistic problems suitable for speculation and experimental enquiry.

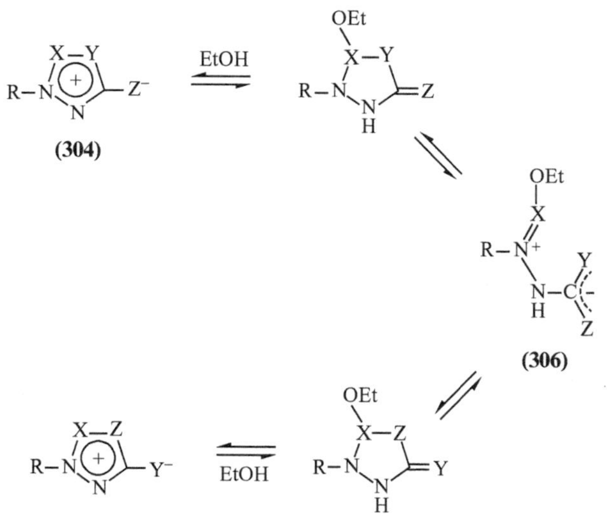

(305)

Fig 3 Equilibration of meso-ionic isomers in protic solvents (e.g., EtOH) involving a betaine intermediate.

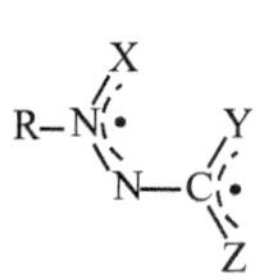

(307)

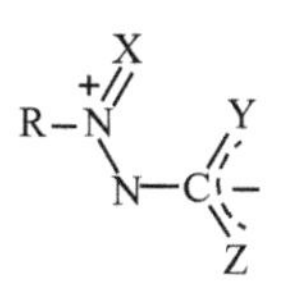

(308)

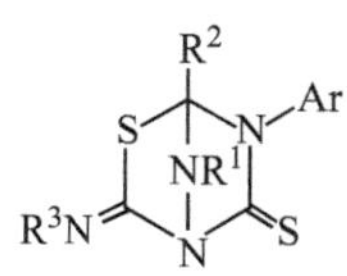

(309)

Two rearrangements may well be regarded as further examples of interconversion between meso-ionic isomers. The 1,3-diazolium chloride (**310**) and warm dilute aqueous potassium hydroxide yield 4-anilino-l-methyl-2-phenylimidazole (**311**), presumably via the indicated meso-ionic intermediates.[60] The transformation **312** → **313** by treatment of the thiazolium chloride with warm aniline may be similarly interpreted.[87]

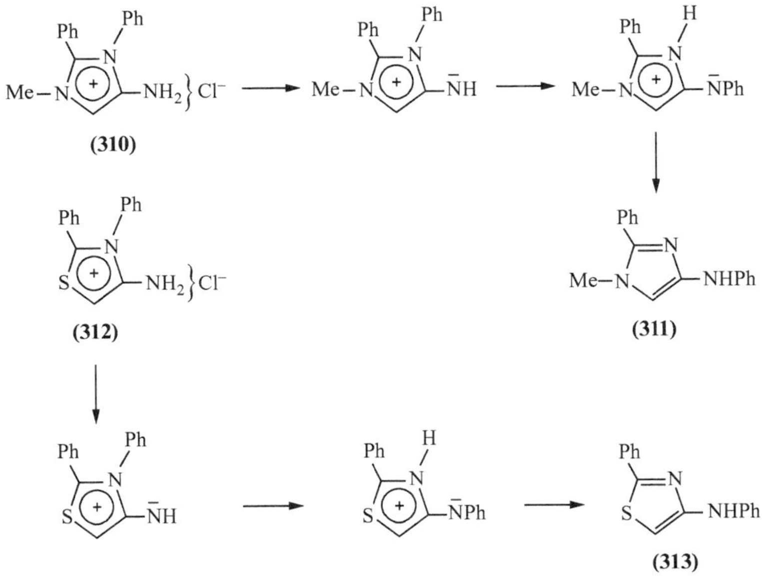

The example[62] claimed for the acid-catalyzed rearrangement **103** → **108** (Section 7.3.2) has not been confirmed[59a,67b]: the meso-ionic 1,3-thiazol-5-imines (**108**) apparently await synthesis.

The most clear-cut distinction between the pairs of meso-ionic isomers given in Fig. 2 is provided by comparison of their mass spectral fragmentation patterns.[201] Comparison of the dipole moments of corresponding isomers is also informative.[108]

9. Photochemistry of type A meso-ionic heterocycles

The investigation of the photochemistry of meso-ionic heterocycles is still rather limited. However, some common mechanistic features can be appreciated in the pathways that have been put forward to account for the products of a number of photoreactions. A summary of the results obtained for various classes of meso-ionic heterocycles follows. Although the decision has been taken not to discuss in detail in this review the general chemistry of sydnones (**1**) and sydnone imines (**2**), it is nevertheless appropriate to include in this section the results obtained in the study of their photochemistry.

9.1 1,2,3-Oxadiazol-5-ones (sydnones) (1)

Irradiation of *N*-phenylsydnone (**1**, R^1=Ph, R^2=H) in benzene has yielded a complex mixture of photolysis products from which 4-phenyl-Δ^2-1,3,4-oxadiazolin-5-one (**317**, R^1=Ph, R^2=H) has been isolated.[213] The mechanism proposed for this transformation[213] is given in Fig. 4. Postulated intermediates include the bicyclic photoisomer (**314**, R^1=Ph, R^2=H), the 1*H*-diazirine (**315**, R^1=Ph, R^2=H), and the 1,3-dipolar nitrilimine (**316**, R^1=Ph, R^2=H). In connection with this proposal, it has been shown[213] that when the irradiation of *N*-phenylsydnone (**1**, R^1=Ph, R^2=H) in dioxan is carried out in the presence of ^{14}C-labeled carbon dioxide, then the radiolabel is incorporated (80% activity of the photolysis products) into the 4-phenyl-Δ^2-1,3,4-oxadiazolin-5-one (**317**, R^1=Ph, R^2=H). Furthermore, the yield of this compound (**317**, R^1=Ph, R^2=H) is related to the pressure of externally supplied carbon dioxide.

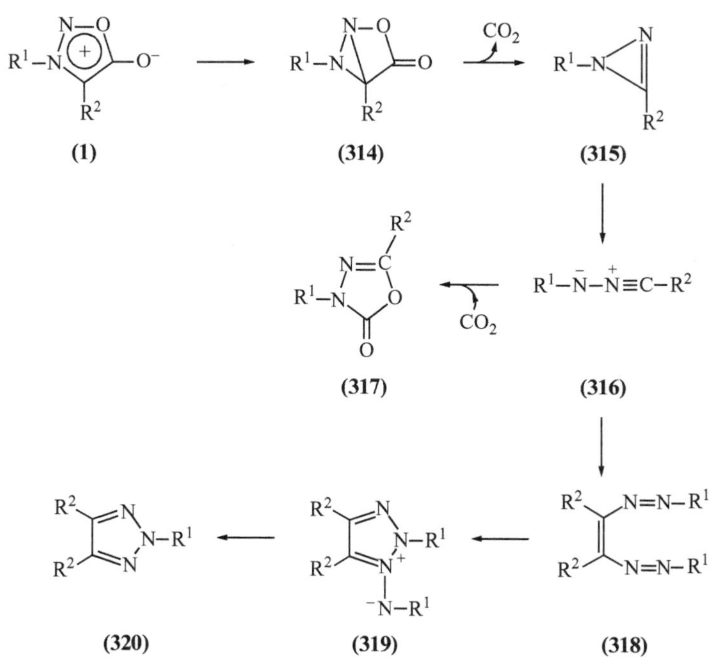

Fig. 4 Photochemistry of sydnones (**1**).

The photochemistry of *NC*-diphenylsydnone (**1**, $R^1 = R^2 = Ph$) has been the subject of almost simultaneous study by a number of groups.[214,215,216,217] Some variations of isolated reaction products occurred, but the formation of 2,4,5-triphenyl-1,2,3-triazole (**320**, $R^1 = R^2 = Ph$) (yield 13–24%) was generally observed: this has been interpreted (Fig. 4) as involving the dimerization of the nitrilimine (**316**). Depending on the reaction conditions, other products obtained from the photolysis of *NC*-diphenylsydnone (**1**, $R^1 = R^2 = Ph$) included benzilosazone[215] and benzanilide.[215] It has been suggested[215] that benzanilide is formed by the hydrolysis of an amidine generated from the nitrilimine (**316**) and aniline.

Irradiation of *N*-cyclohexyl-*C*-phenylsydnone[218] (**1**, R^1 = cyclohexyl, R^2 = phenyl) in benzene under a nitrogen atmosphere gave the 1,3,4-oxadiazolin-5-one (**317**, R^1 = cyclohexyl, R^2 = phenyl; 15%), the bisazoethylene (**318**, R^1 = cyclohexyl, R^2 = phenyl; 15%), and the triazole (**320**, R^1 = cyclohexyl, R^2 = phenyl; 12%). When the nitrogen atmosphere was replaced by carbon dioxide, then the yield of the 1,3,4-oxadiazolin-5-one was increased from 15% to 47%. These results[218] provide good support for the general mechanistic proposals given in Fig. 4.

Photochemical cycloaddition reactions[214–218] between sydnones (**1**) and 1,3-dipolarophiles take place to give products which are different from, but isomeric with, the thermal 1,3-dipolar cycloaddition products.[3] These results are directly interpreted in terms of reactions between the 1,3-dipolarophiles and the nitrilimine (**316**). The photochemical reactions between sydnones and the following 1,3-dipolarophiles have been reported: dicyclopentadiene,[214] dimethyl acetylene dicarboxylate,[215–218] dimethyl maleate,[215] dimethyl fumarate,[215] indene,[216] carbon dioxide,[213] and carbon disulfide.[217]

9.2 1,2,3-Oxadiazol-5-imines (*N*-acylsydnone imines) (2, R^3 = COR)

The photochemical isomerization of *N*-acylsydnone imines (**2**, R^3 = COR) to give meso-ionic 1,2,3-triazol-4-ones (**176**) has been reported.[128b]

Irradiation in benzene–ethanol of the *N*-acylsydnone imine **321**, R = Me, yielded the isomer **322**, R = Me (11%), and the triazole **323** (10%). The *N*-benzoyl derivative **321**, R = Ph, similarly yielded the isomer **322**, R = Ph (2%) and the triazole **323** (35%).

These results are curious in that the photochemistry of *N*-acylsydnone imines is quite different in mechanistic form from that of sydnones (Fig. 4) and other meso-ionic heterocycles.

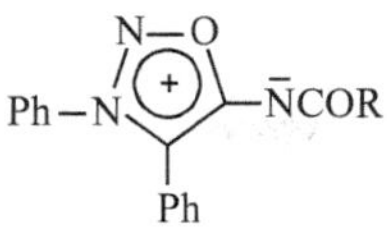

(321)

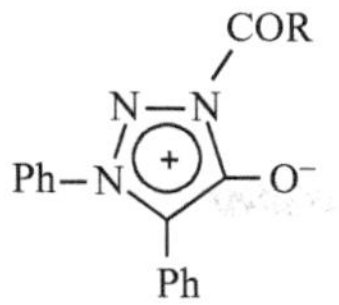

(322)

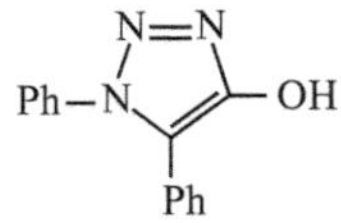

(323)

9.3 1,3-Thiazol-4-ones (114)

Irradiation, in ethanol, of the meso-ionic 1,3,-thiazol-4-one (**324**) yields the isomeric 3-methyl-4-methylthio-5-phenyl-1,3-thiazol-2-one (**325**).[78b] The rather curious mechanism that has been proposed[78b] for the rearrangement **324** → **325** is given below.

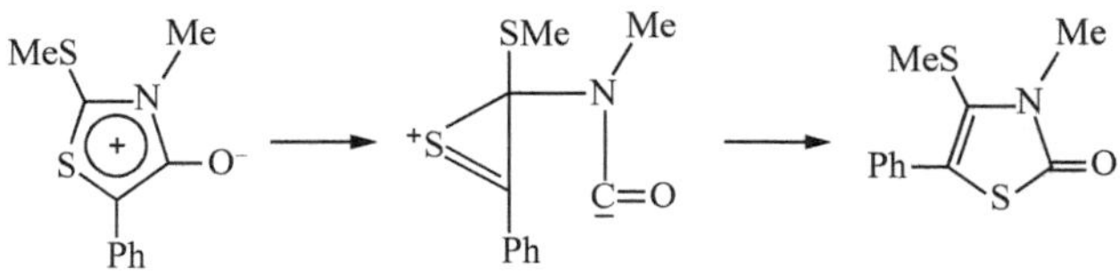

(324) (325)

9.4 1,3-Dithiol-4-ones (134)

Photolysis of 2,5-diphenyl-1,3-dithiol-4-one (**326**) yields tetraphenyl-1,4-dithiin (**331**; 19%), diphenylacetylene (**329**; 16%), and sulfur (15%).[95] These transformations have been interpreted[95] in terms of a diphenylthiiren intermediate (**328**).

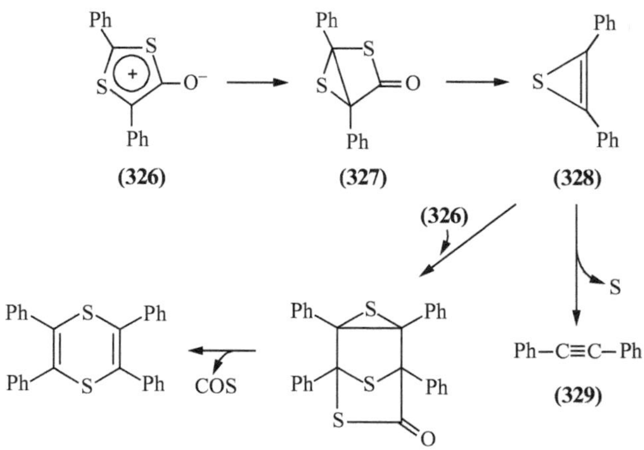

(326) (327) (328) (329) (331) (330)

9.5 1,3-Dithiol-4-imines (141)

Irradiation of a benzene solution of the meso-ionic *N*-benzoyl-2-phenyl-1,3-dithiol-4-imine (**332**) gave the isomer **334** in high yield (80%).[97] The isomerization has been interpreted[97] as involving rearrangement of the bicyclic photoisomer **333**.

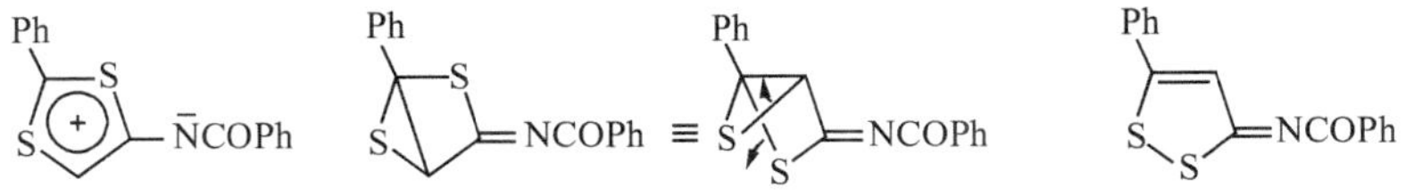

(332) (333) (334)

9.6 1,3,2-Oxathiazol-5-ones (169)

Photolysis, in ether–methylene chloride solution, of the meso-ionic 4-phenyl-1,3,2-oxathiazol-5-one (**335**) gave benzonitrile (77%) and sulfur (91%).[118] This result has been interpreted[118] in terms of elimination of carbon dioxide from the photoisomer (**336**) yielding the thiazirin (**337**) which after isomerization to benzonitrile sulfide disproportionates, giving benzonitrile and sulfur. The intermediate benzonitrile sulfide has been trapped[118a] with dimethyl acetylenedicarboxylate, giving the products **338** and **339**.

An alternative photoinduced pathway has been identified[118c] in the photolysis of the meso-ionic oxathiazolone (**335**). Irradiation in ethanol in the presence of oxygen yields ethyl phenylglyoxylate (**340**; 65%), ethyl benzoate (3%), and benzonitrile (21%). The postulated intermediates (Fig. 5) have been spectroscopically identified by photolysis in ether–isopentane–ethanol–glass at 85 °K.[118c]

9.7 1,2,4-Triazol-3-ones (200)

The results first described[97a] for the photolysis of the meso-ionic 1,4-diphenyl-1,2,4-triazol-3-one (**341**) have been subsequently modified.[97b] Irradiation of this compound (**341**) in methylene chloride solution yields phenylisocyanate (13%)[97a] and diphenyl urea (23%),[97a] whereas in 5% methanol–methylene chloride solution the photolysis products are methyl phenyl carbamate[97b] (25%), azobenzene (**334**; 7%),[97b] and benzimidazole (**348**; 18%).[97b]

Mechanistic proposals (Fig. 6) have been put forward to account for the formation of the significant photolysis products, azobenzene (**344**), phenylisocyanate (**346**), and benzimidazole (**348**). Two possible routes from

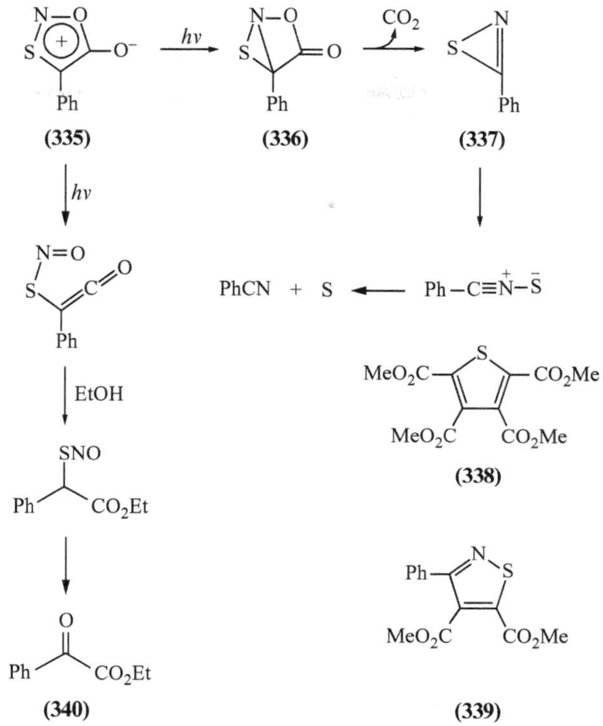

Fig. 5 Photochemistry of 1,3,2-oxathiazol-5-ones.

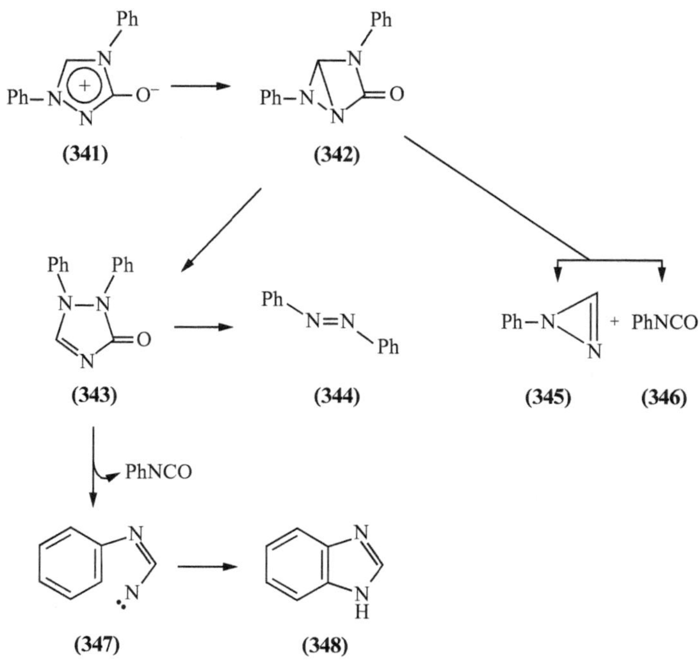

Fig. 6 Photochemical of 1,2,4-triazol-3-ones.

the bicyclic photoisomer (**342**) have been suggested. One route[97] involves its photocleavage to *N*-phenyl-1*H*-diazirine (**345**) and phenylisocyanate. The second route[97b] involves the valence isomerization **342** → **343** which could then fragment yielding azobenzene (**344**), phenylisocyanate, and phenyliminomethyl nitrene (**347**)—a reasonable precursor of benzimidazole (**348**).

9.8 1,3,4-Thiadiazol-2-ones (243)

Irradiation of the meso-ionic 1,3,4-thiadiazol-2-one (**349**) in methyl cyanide solution gives the *N*-isocyanate (**350**) (ν_{max} 2060 cm^{-1}) and *N*-isothiocyanate (**351**) (ν_{max} 2260 cm^{-1}), which were both recognized spectroscopically.[22] Mechanisms for the conversion **350** → **351** have been considered.[22]

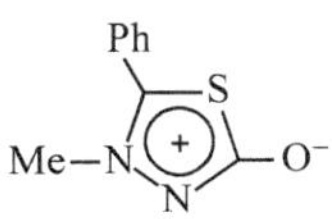

(349)

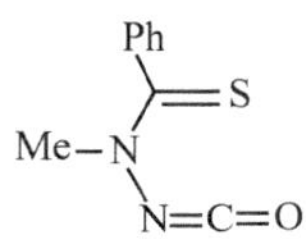

(350)

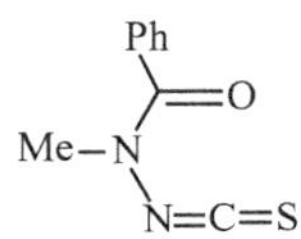

(351)

9.9 1,3,4-Thiadiazole-2-thiones (251)

Irradiation[22,157] in methyl cyanide solution of the compound **352** yields the valence tautomer **353**. This has been recognized spectroscopically[22] and by trapping with *n*-butylamine, which yields the thiourea (**354**).[22] Further photofragmentation of the *N*-isothiocyanate (**353**) yielded the thioamide (**355**) and elemental sulfur.

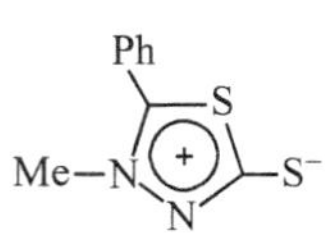

(352)

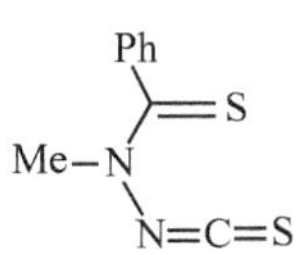

(353)

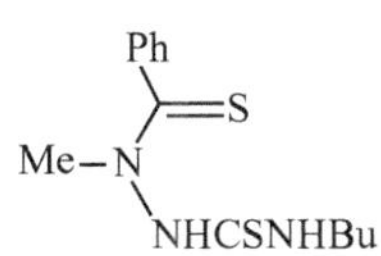

(354)

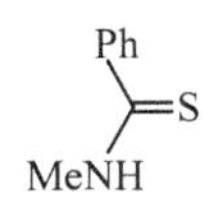

(355)

9.10 1,2,3,4-Oxatriazol-5-ones (271)

Photolysis[97a] of the *N*-cyclohexyl derivative (**356**), followed by hydrolysis, yields cyclohexanone. This has been interpreted in terms of the following sequence: **356** → **357** → **358** → **359**.

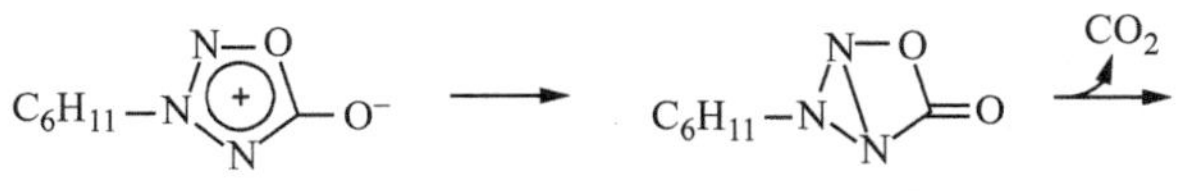

(356) **(357)**

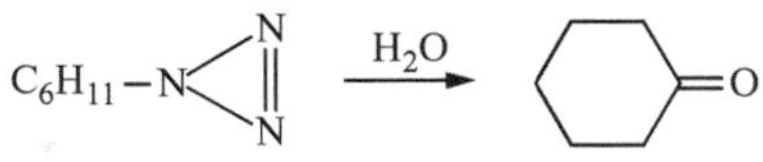

(358) **(359)**

9.11 General comments

Meso-iomic heterocycles show several general types of photochemical transformation. These can be summarized (Fig. 7) as (i) a valence tautomerism giving the heterocumulene **364**, (ii) a valence isomerization giving the isomer **365**, and (iii) a fragmentation reaction giving three-membered heterocycles (**362**) which could be the precursors of 1,3-dipoles (**366**). These transformations have usually been interpreted in terms of a bicyclic photoisomer (**361**).

The suggestion that three-membered heterocycles (**362**) are involved is illustrated by the postulated generation of derivatives of 1*H*-diazirine (**315** and **345**), thiiren (**328**), thiazirin (**337**), and 1*H*-triazirine (**358**). These three-membered heterocycles belong to the class of 4π-antiaromatic heterocycles; their possible role as reaction intermediates[219] is of general interest. There is, however, an interesting difference between the photochemistry of sydnones (Fig. 4) and meso-ionic 1,2,4-triazol-3-ones (Fig. 6). In both cases 1*H*-diazirines (**315** and **345**) have been considered[213–218] as possible intermediates, whereas there is no experimental evidence[97b] that supports the view that 1*H*-diazirines could be involved as intermediates in the photochemical reactions of sydnones (Fig. 4) and meso-ionic 1,2,4-triazol-3-ones (Fig. 6).

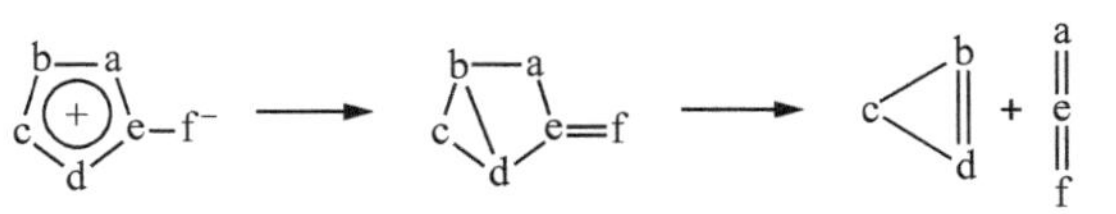

(360) **(361)** **(362)** **(363)**

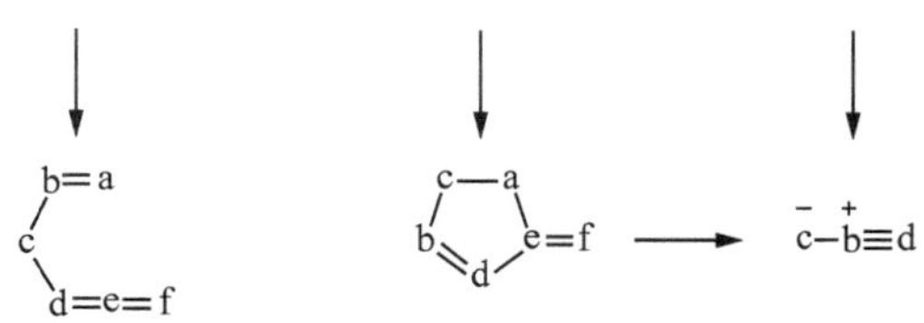

(364) **(365)** **(366)**

Fig. 7 Photochemical transformations of meso-ionic heterocycles.

10. The chemistry of meso-ionic compounds of type B

These are listed under the parent heterocyclic systems in the order given in Table 2 (Section 5).

10.1 Oxazoles

10.1.1 1,2-Oxazol-4-ones (anhydro-4-hydroxy-1,2-oxazolium hydroxides) (367)

Attempts to prepare meso-ionic 1,2-oxazol-4-ones (**367**) have not been successful, although the corresponding salts (**368**) have been isolated. Treatment of the salts (**368**) with basic reagents may have yielded 1,2-oxazol-4-ones (**367**), but these could not be isolated.

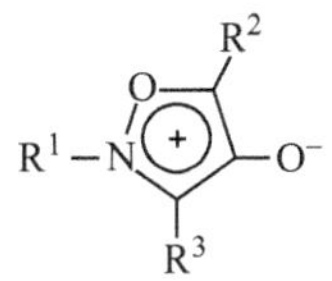

(367)

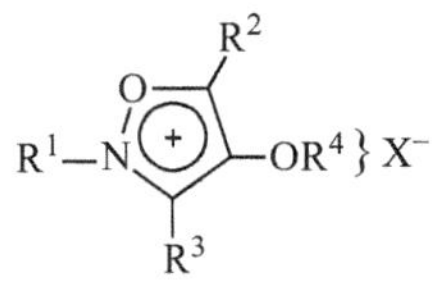

(368)

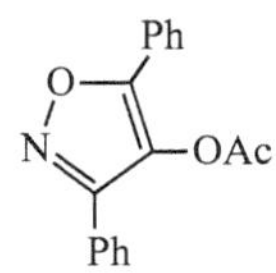

(369)

4-Acetoxy-3,5-diphenylisoxazole (**369**)[220] was *N*-alkylated with triethyloxonium tetrafluoroborate giving the salt **368**, R^1=Et, R^2=R^3=Ph, R^4=Ac, X=BF_4.[221] The salt **368**, R^1=Me, R^2=R^3=Ph, R^4=H, X=ClO_4, was similarly prepared using methyl fluorosulfonate followed by perchloric acid.[222] It is concluded that the spectrum of the salt **368**, R^1=Me, R^2=R^3=Ph, R^4=H, X=ClO_4, in alkaline solution is that of its decomposition products, but a transient yellow color may be due to the meso-ionic 1,2-oxazol-4-one (**367**, R^1=Me, R^2=R^3=Ph).[222]

10.1.2 1,2-Oxazol-4-imines (anhydro-4-amino-1,2-oxazolium hydroxides) (370)

The formation of an *N*-tosyl derivative of a meso-ionic 1,2-oxazol-4-imine (**370**) in solution is possible, but the product was not isolated.

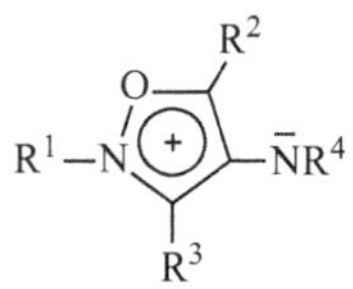

(370)

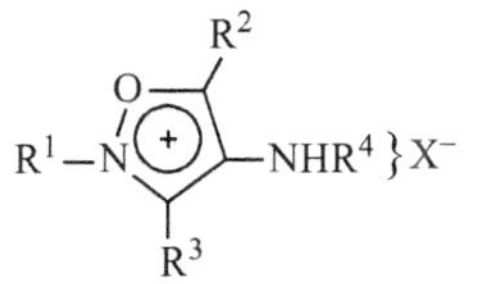

(371)

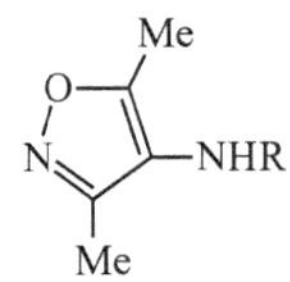

(372)

4-Amino-3,5-dimethylisoxazole (**372**, R=H)[223] gave an *N*-tosyl derivative (**372**, R=tosyl), which by treatment with methyl fluoro-sulfonate followed by perchloric acid gave the salt **371**, $R^1=R^2=R^3=Me$, R^4=tosyl, $X=ClO_4$.[222] Attempts to isolate the meso-ionic sulfonamidate (**370**, $R^1=R^2=R^3=Me$, R^4=tosyl) by treatment of the salt with potassium hydroxide or triethylamine were unsuccessful, but spectroscopic evidence for its formation in solution has been offered.[222]

10.2 Diazoles

10.2.1 1,2-Diazol-4-ones (anhydro-4-hydroxy-1,2-diazolium hydroxides) (373)

Meso-ionic 1,2-diazol-4-ones (**373**) were possibly first prepared in 1900 by Wolff and Fertig[224]; compounds of this type (**373**) were again possibly encountered in 1931.[225] The products obtained in these two studies were not structurally formulated, but the quoted molecular formulae[224,225] match either the monohydrates of the meso-ionic hetero-cycles (**373**) or the onium hydroxides (**374**, X=OH). The synthesis and characterization of the meso-ionic 1,2-diazol-4-ones (**373**) have been recently described.[226,227]

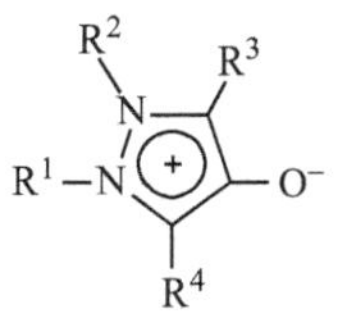

(373)

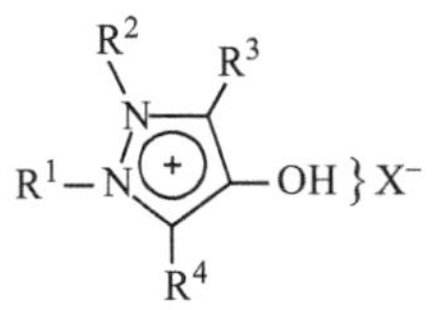

(374)

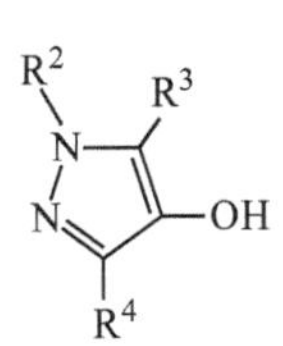

(375)

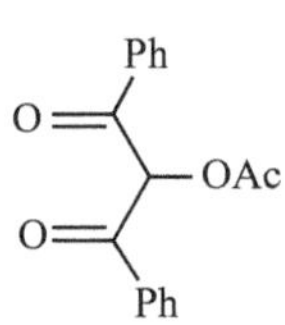

(376)

4-Hydroxy-1-phenylpyrazole (**375**, R^2=Ph, $R^3=R^4$=H) yields a methiodide (**374**, R^1=Me, R^2=Ph, $R^3=R^4$=H, X=I), which with silver oxide gives a product, $C_{10}H_{12}N_2O_2$, m.p. 132°–135°, which may now be formulated either as a monohydrate of the heterocycle **373**, R^1=Me, R^2=Ph, $R^3=R^4$=H, or as the 1,2-diazolium hydroxide **374**, R^1=Me, R^2=Ph, $R^3=R^4$=H, X=OH.[224] 1-Phenyl-4-hydroxy-5-methylpyrazole (**375**, R^2=Ph, R^3=Me, R^4=H) similarly yielded $C_nH_{14}N_2O_2$ (**373** monohydrate, $R^1=R^3$=Me, R^2=Ph, R^4=H, or **374**, $R^1=R^3$=Me, R^2=Ph, R^4=H).[225]

The compound **373**, $R^1=R^2$=Me, $R^3=R^4$=Ph, has been prepared by two routes.[226,227] Reaction of *N*-methylhydrazine with 2-acetoxy-1,

3-diphenylpropane-1,3-dione (**376**) gave 4-hydroxy-l-methyl-3,5-diphenylpyrazole (**375**, R^2=Me, R^3=R^4=Ph), which with dimethyl sulfate gave the methosulfate salt **374**, R^1=R^2=Me, R^3=R^4=Ph, X=$MeSO_4$. This salt and alkali gave the meso-ionic 1,2-diazol-4-one (**373**, R^1=R^2=Me, R^3=R^4=Ph). Alternatively, *N,N'*-dimethylhydrazine and the dione **376**, followed by treatment with alkali, gave the same compound (**373**, R^1=R^2=Me, R^3=R^4=Ph) directly.

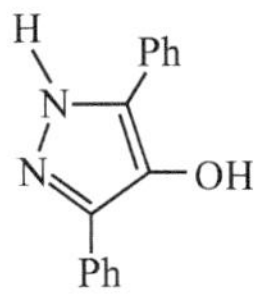
(377)

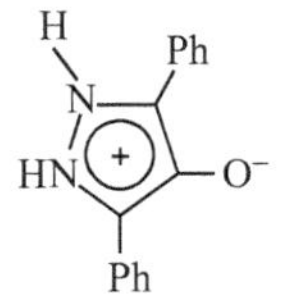
(378)

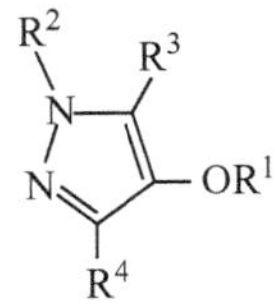
(379)

The equilibrium between the pyrazole **377** and its meso-ionic tautomer (**378**) in 4% methanol–water has been determined.[228] The equilibrium is strongly in favor (K=3.2×10^{-5}; ΔF=6.15 kcal/mol) of the pyrazole (**377**). This result provides an instructive contrast with the approximate 1:1 equilibrium ratio between 3-hydroxypyridine (**380**) and its betaine tautomer (**381**) in aqueous solution.[229]

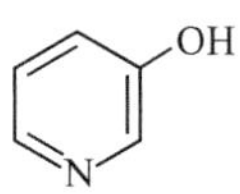
(380)

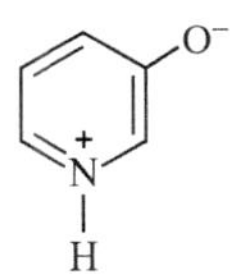
(381)

On heating, the meso-ionic 1,2-diazol-4-one (**373**, R^1=R^2=Me, R^3=R^4=Ph) is transformed into the isomeric ether (**379**, R^1=R^2=Me, R^3=R^4=Ph) and the 4-hydroxypyrazole (**375**, R^2=Me, R^3=R^4=Ph).[222] A mechanistically analogous reaction takes place between the meso-ionic 1,2-diazol-4-one (**373**, R^1=R^2=Me, R^3=R^4=Ph) and dimethyl acetylenedicarboxylate, giving the ether **379**, R^1=C(CO_2Me)=CH(CO_2Me), R^2=Me, R^3=R^4=Ph.[222]

Anhydro-1,2-dimethyl-3,5-diphenyl-4-hydroxy-1,2-diazolium hydroxide (**373**, R^1=R^2=Me, R^3=R^4=Ph) shows a band (**1546** cm^{-1} in dimethyl sulfoxide), which has been assigned as a carbonyl band.[230] This absorption band, which is much lower than, for example, the sydnones (**1**) (ν_{CO} 1750–1770 cm^{-1})[7] has been briefly considered in relation to the electronic characteristics of type A and type B meso-ionic heterocycles.[230] However, the significance of this spectral difference must surely await more extensive comparison between corresponding type A and type B heterocycles. The compound (**373**, $R^1=R^2=Me$, $R^3=R^4=Ph$) also shows a striking solvent dependence of its ultraviolet/visible spectrum (λ_{max} C_6H_6 447; Me_2SO 421; $CHC1_3$ 410; Bu_tOH 370; MeOH 345; H_2O 325 nm).[230]

10.2.2 1,2-Diazol-4-imines (anhydro-4-amino-1,2-diazolium hydroxides) (382)

Meso-ionic 1,2-diazol-4-imines (**382**) have been recently obtained[222] as their *N*-mesyl ($R^4=MeSO_2$) and *N*-tosyl ($R^4=p\text{-}MeC_6H_4SO_2$) derivatives. Mesylation of 4-amino-1,3,5-trimethylpyrazole (**383**, $R^2=R^3=R^5=Me$) followed by (i) methylation with methyl fluorosulfonate and (ii) treatment with perchloric acid gave the salt **384**, $R^1=R^2=R^3=R^5=Me$, $R^4=MeSO_2$, $X=ClO_4$. Deprotonation of this salt with aqueous alkali gave the compound **382**, $R^1=R^2=R^3=R^5=Me$, R^4 $MeSO_2$. The meso-ionic 1,2-diazol-4-imines **382**, $R^1=R^2=R^3=R^5=Me$, $R^4=p\text{-}Me.C_6H_4SO_2$, and **382**, $R^1=R^2=Me$, $R^3=R^5=Ph$, $R^4=p\text{-}Me.C_6H_4SO_2$, were similarly prepared.

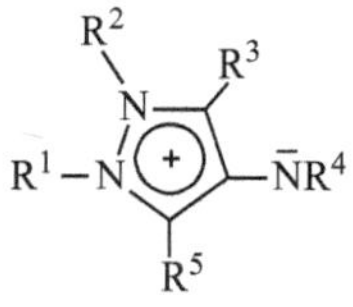

(382)

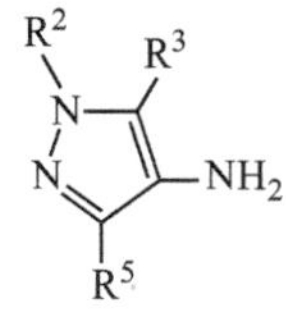

(383)

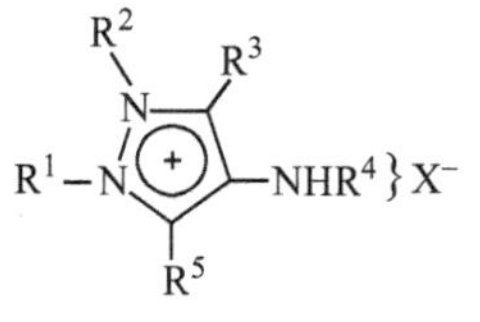

(384)

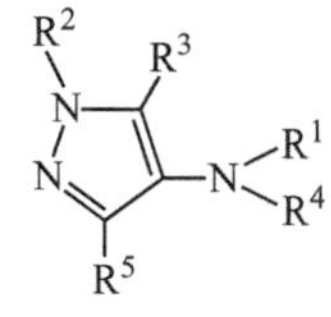

(385)

Thermal isomerization (**382** → **385**, $R^1=Me$ and $R^4=MeSO_2$ or *p*-Me.$C_6H_4SO_2$) has been observed on heating in benzonitrile.[222] A related reaction occurs when the meso-ionic 1,2-diazol-4-imines (**382**, $R^1=Me$, $R^4=MeSO_2$ or *p*-Me.CgH_4SO_2) are heated with dimethyl acetylenedicarboxylate in acetonitrile. This yields the products **385**, $R^1=C(CO_2Me)=CH(CO_2Me)$, $R^4=MeSO_2$ or *p*-Me.$C_6H_4SO_2$.[222]

10.3 Thiazoles

10.3.1 1,2-Thiazol-4-imines (anhydro-4-amino-1,2-thiazolium Hydroxides) (386)

The pale yellow meso-ionic 1,2-thiazol-4-imine (**386**, $R^1=R^4=Me$, $R^2=H$, $R^3=p\text{-}Me.C_6H_4SO_2$) has been prepared from 4-amino-3-methylisothiazole (**387**) by a sequence similar to that used for meso-ionic 1,2-diazol-4-imines (**382**, $R^4=p\text{-}Me.C_6H_4SO_2$).[222]

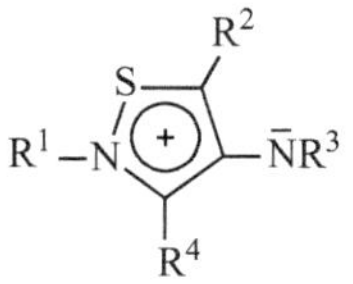

(386)

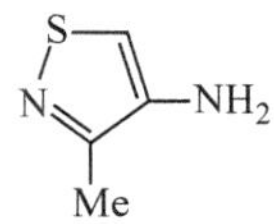

(387)

10.4 Dithioles

10.4.1 1,2-Dithiol-4-ones (anhydro-4-hydroxy-1,2-dithiolium hydroxides) (388)

The synthesis and some reactions of meso-ionic 1,2-dithiol-4-ones (**388**) have been recently reported. The brown compound **388**, $R^1=R^2=Ph$, has been prepared by several methods: (i) the reaction between 1,1,3,3-tetrabromo-1,3-diphenylacetone ($PhCBr_2COCBr_2Ph$) and potassium ethyl xanthate,[24] (ii) the reaction between 1,3-diphenyl-propanetrione hydrate and tetraphosphorus decasulfide,[24] (iii) 1,3-diphenylpropanetrione with hydrogen sulfide–hydrogen chloride in ethanol–chloroform yields the salt **389**, $R^1=R^2=Ph$, X=Cl, which gives the meso-ionic 1,2-dithiol-4-one with triethylamine, pyridine, or aqueous sodium bicarbonate.[23,231]

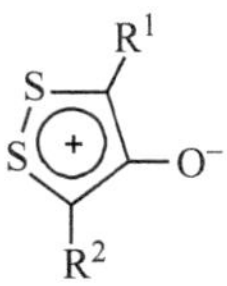

(388)

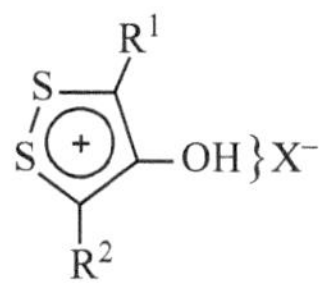

(389)

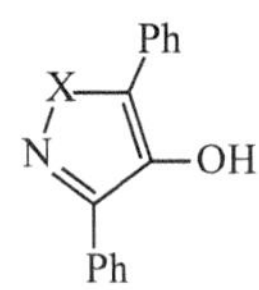

(390) X = NH
(391) X = S

The diphenyl derivative **388**, $R^1=R^2=Ph$, has an absorption band at 1495 cm^{-1}, which has been assigned to the carbonyl group.[24] This assignment could be questioned. The following chemical reactions of the compound **388**, $R^1=R^2=Ph$, may be noted: (a) Raney nickel desulfurization followed by oxidation with manganese dioxide yields dibenzyl ketone,[24] (b) hydrazine yields 3,5-diphenyl-4-hydroxypyrazole (**390**),[24] (c) phenylhydrazine yields $PhCH_2COCPh{=}NNHPh$,[24] (d) heating at 275°

yields sulfur and tetraphenyl-*p*-benzoquinone,[24] (e) methanolic ammonia yields 3,5-diphenyl-4-hydroxyisothiazole(**391**).[23] Reaction with aniline yields a blue compound, $C_{21}H_{15}SN$, which has been formulated as 3,4,5-triphenylisothiazole.[23] This assumption requires further study.

10.5 Tetrazoles

10.5.1 1,2,3,4-Tetrazol-5-ones (anhydro-5-hydroxy-1,2,3,4-tetrazolium hydroxides) (392)

Meso-ionic 1,2,3,4-tetrazol-5-ones (**392**) have been known for many years, although they were not initially recognized as having betaine-type structures. Treatment of 1,5-diphenyl-3-nitroformazan (**393**, X=NO_2) with acetic acid[232,233,234] or isoamyl nitrite[235] yields a white crystalline compound, $C_{13}H_{10}N_4O$, which was initially formulated with the bicyclic structure **394** by Bamberger.[232–235] The same compound, $C_{13}H_{10}N_4O$, was also obtained by the oxidation of diphenyl carbazide (**395**)[235,236] or diphenylcarbazone (**396**)[236] by Cazeneuve, who formulated the compound as having the acyclic structure **397**.[236] Other 2,3-diaryl-1,2,3,4-tetrazol-5-ones (**392**) have been prepared by oxidation with amyl nitrite and hydrogen chloride of 1,5-diaryl-3-chloroformazans (**393**, X=Cl)[237] or 1,5-diaryl-3-nitroformazans (**393**, X=NO_2).[238,239]

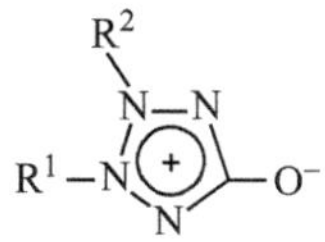

(392)

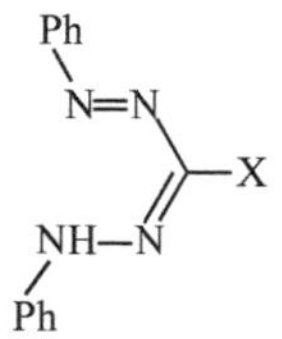

(393)

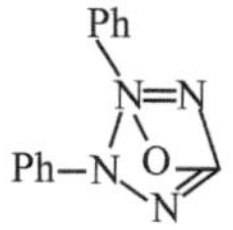

(394)

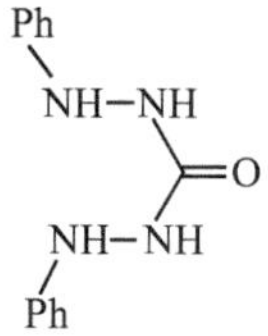

(395)

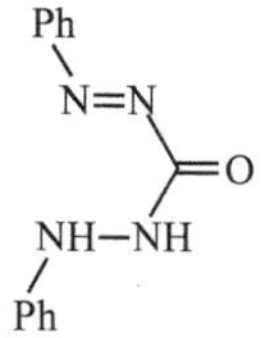

(396)

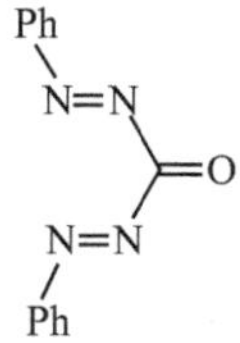

(397)

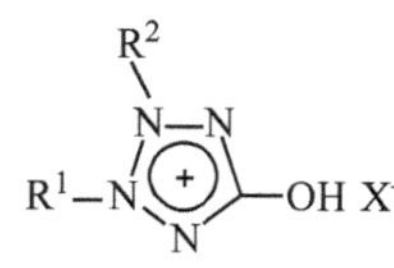

(398)

Clearly the bicyclic structure **394** is unacceptable for the compound $C_{13}H_{10}N_4O$, and its ultraviolet spectrum is regarded[142,240,241] as supporting the meso-ionic formulation **392** and excluding the bisphenylazo structure **397**. The 2,3-diphenyl derivative (**392**, $R^1=R^2=Ph$) forms salts[235] (**398**, $R^1=R^2=Ph$) including a hydrochloride, hydriodide, chloroplatinate, perchlorate, and picrate; it is reduced to diphenylcarbazide (**395**) by ammonium sulfide.[235]

10.5.2 1,2,3,4-Tetrazol-5-imines (anhydro-5-amino-1,2,3,4-tetrazolium hydroxides) (399)

Bamberger, Padova and Ormerod reported in 1926[235] an interesting but incomplete study of several unusual products obtained from 5-amino-2,3-diaryl-1,2,3,4-tetrazolium salts (**400**). Recently, this topic has been the subject of a detailed study by Smith,[242] and the existence of several derivatives of meso-ionic 1,2,3,4-tetrazol-5-imines (**399**) has now been firmly established.[242] Comparison of Smith's recent results[242] with the earlier results of Bamberger et al.[235] is instructive.

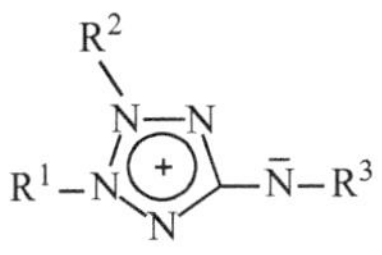

(399)

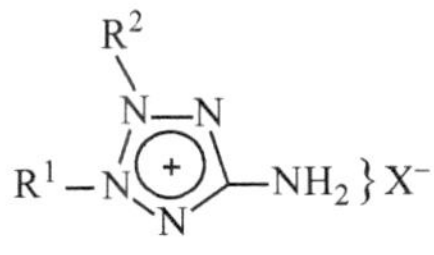

(400)

The molecular formulas and the structures originally proposed by Bamberger[235] for a number of transformation products of 5-amino-2,3-diphenyl-1,2,3,4-tetrazolium chloride (**401**) are given in Fig. 8. The methods used to obtain these products from the salt **401** may be summarized as follows. Schotten-Baumann benzoylation gave a yellow crystalline product, m.p. 232° (**402**), which was reduced by ammonium sulfide in ethanol giving a colorless product, m.p. 222°–223° (**403**). Nitrosation of the salt **401** gave a yellow product, m.p. 176°–177° (**404**), which yielded an orange isomer, m.p. 154°–155° (**405**), on heating in ethanol.

Repetition of the reactions summarized in Fig. 8 produced the results given in Fig. 9. There is a satisfying agreement between the old and new results, with one important exception: the molecular formula $C_{20}H_{15}N_5O$, originally assigned to the benzoylation product (**402**), has had to be replaced by the molecular formula $C_{20}H_{17}N_5O$. Its formulation as 3-benzamido-1,5-diphenylformazan (**406**) is firmly established[242] by (i) its acid hydrolysis, yielding 3-amino-1,5-diphenylformazan (**393**, $X=NH_2$) and benzoic acid, and (ii) its dehydration, yielding 1,5-diphenyl-3-phenylazo-1,2,4-triazole

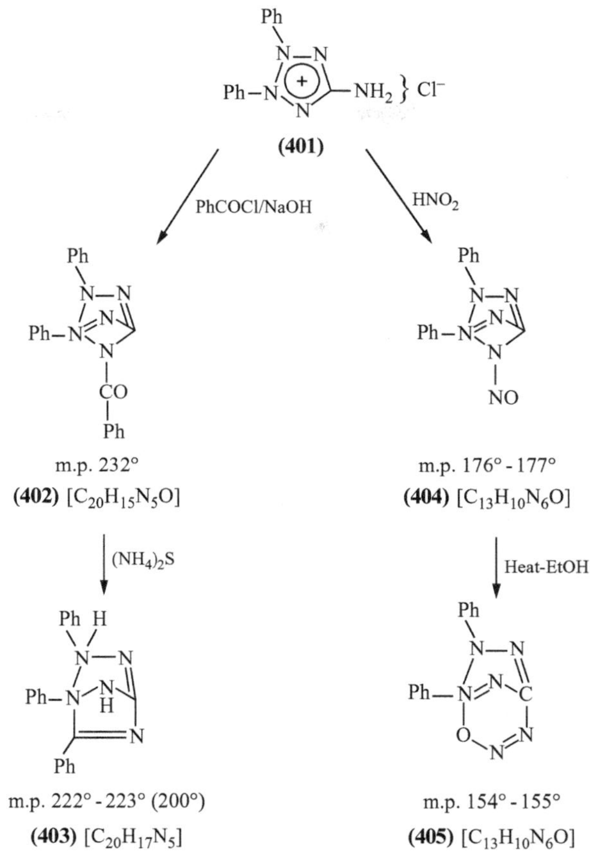

Fig. 8 A summary of the results and structural proposals given by Bamberger, Padova, and Ormerod.[235]

(**410**) by boiling its solution in xylene. This triazole (**410**) is also formed by manganese dioxide oxidation of the hydrazo compound **407**.[242]

The transformation **401** → **406** by Schotten-Baumann benzoylation is a mysterious reaction, the mechanism of which is not obvious. Clearly a change in oxidation level has occurred, but related changes are known, and this is an incompletely understood aspect of the chemistry[243] of formazans (**393**) and the corresponding tetrazolium cations. In contrast, the reaction of the salt with *p*-toluenesulphonyl chloride yields the meso-ionic 1,2,3,4-tetrazol-5-imine (**411**) directly.

The thermal transformation **408** → **409** presumably involves the valence tautomer **412** as an intermediate.[242] As a class, the type B meso-ionic

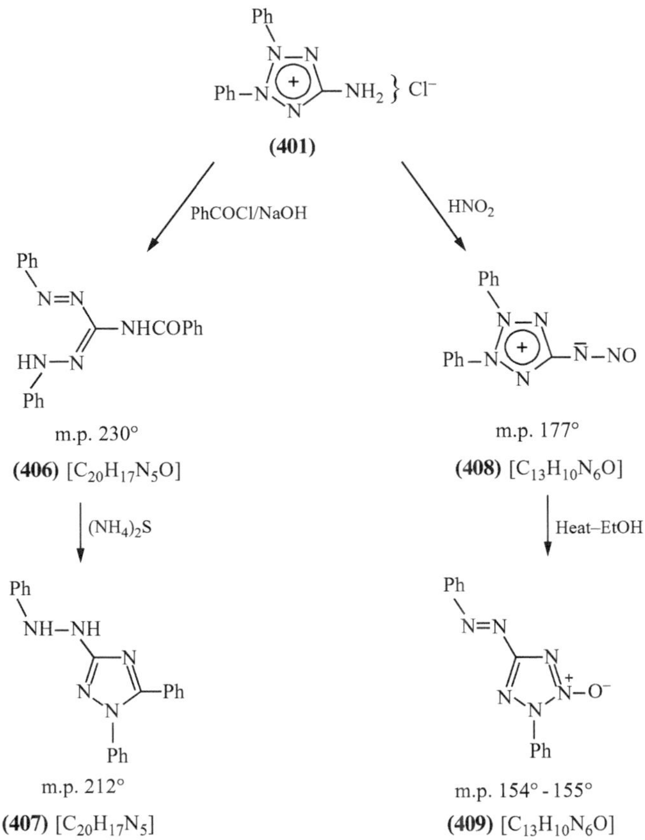

Fig. 9 A reinterpretation[242] of the results summarized in Fig. 8.

1,2,3,4-tetrazol-5-imines are exemplified by the *N*-nitroso and *N*-*p*-tosyl derivatives **408** and **411**. However, our current understanding[242] of the results first reported in 1926 is a clear encouragement for the synthesis and examination of other classes of type B meso-ionic heterocycles.

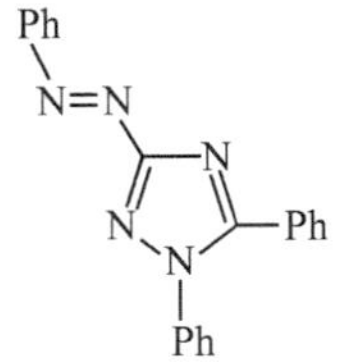

(410)

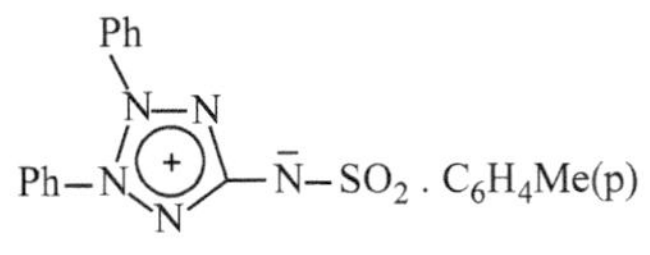

(411)

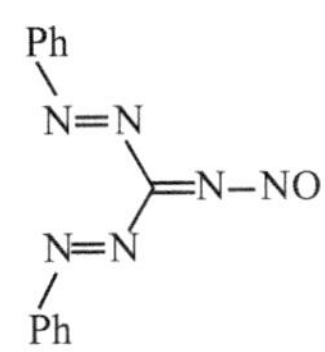

(412)

10.5.3 *1,2,3,4-Tetrazole-5-thiones (anhydro-5-mercapto-1,2,3,4-tetrazolium hydroxides) (413)*

The *last* part of this section is devoted to the *first* meso-ionic heterocycle to be synthesized. It is called dehydrodithizone, and it is the most widely investigated representative of meso-ionic 1,2,3,4-tetrazole-5-thiones (**413**). Dehydrodithizone was first obtained by Emil Fischer in 1882[25] and his collaborator in this investigation was E. Besthorn, who subsequently discovered another meso-ionic compound, Besthorn's Red (**93a**).

Dithizone (**414**) is a well-known reagent for the gravimetric estimation of heavy metal cations. Its oxidation with either manganese dioxide[25] or isoamyl nitrite[235] yields an orange crystalline derivative called dehydrodithizone. This was initially formulated by Besthom and Fischer[25] as a bisphenylazo compound (**415**), whereas Bamberger et al. advocated[235] the equivalent (**416**) of a betaine structure. A cyclic betaine structure was later supported[142b,240,241] on the basis of its spectroscopic properties. Cogent arguments for the representation of dehydrodithizone as a meso-ionic heterocycle (**413**, $R^1=R^2=Ph$) were presented by Ogilvie and Corwin[244] on the basis of its general chemistry and particular properties as a nucleophilic reagent.[245] An X-ray crystallographic study[246] of dehydrodithizone (**413**, $R^1=R^2=Ph$) confirmed its symmetrical structure and the planarity of its five-membered ring. The interatomic distances were in accord with an aromatic meso-ionic structure.

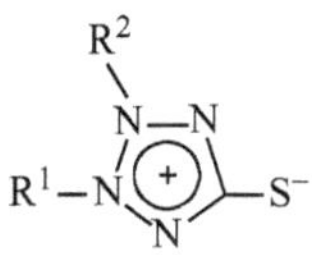

(413)

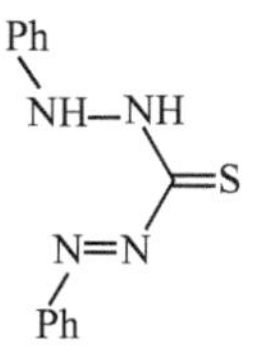

(414)

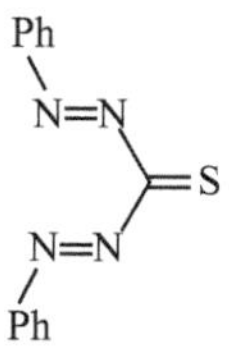

(415)

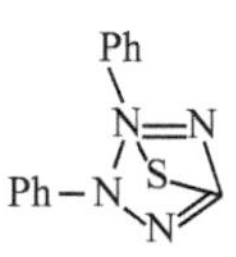

(416)

Oxidation of dithizone (**414**) with selenium dioxide or iodine in water gives the disulfide (**417**) which then disproportionates[26] on standing, yielding dithizone (**414**) and dehydrodithizone (**413**, $R^1=R^2=Ph$). This is closely analogous to the disproportionation **254** → **251** + **253**.

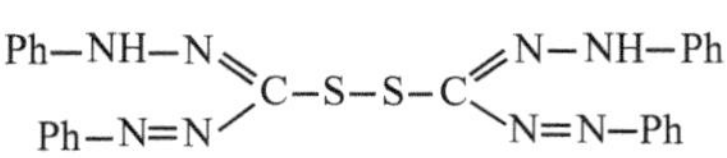

(417)

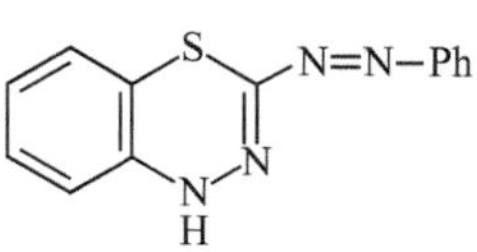

(418)

The purple compound produced[244,247,248] by the acid-catalyzed isomerization of dehydrodithizone (**413**, $R^1 = R^2 = Ph$) has been shown by an X-ray diffraction study[249] to have the constitution **418**.

The formation of various 1:1 adducts from dehydrodithizone **413**, $R^1 = R^2 = Ph$, and various "1,3-dipolarophiles" has been reported.[250] Thus, dimethyl acetylenedicarboxylate yields **419**, X = COOMe, tetracyanoethylene yields **420**, and ethoxycarbonylmethylenetriphenylphosphorane yields the betaine **421**. These transformations have been considered as 1,3-dipolar cycloaddition reactions of a novel type. It seems to us rather unlikely that these transformations are *concerted* 1,3-dipolar cycloadditions: the alternative that they are reactions involving dipolar intermediates (e.g., **422**, **423**, and **424**) should also be considered. The strong nucleophilic character of dithizone[245] could assist in the generation of such dipolar intermediates (**422**, **423**, and **424**), and there is plenty of general analogy for such intermediates.[251] Particular reference may be made to the closely related cycloaddition reactions of meso-ionic 1,2-diazoles **373** and **382**.[222] The transformation **424** → **421** could well involve a pentacovalent phosphorane intermediate as well.

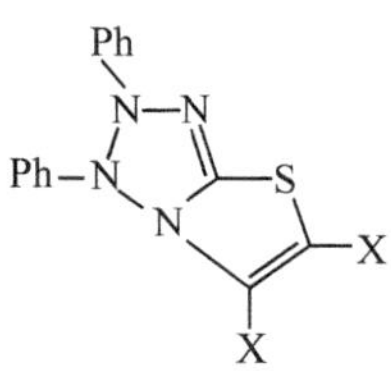

(419)

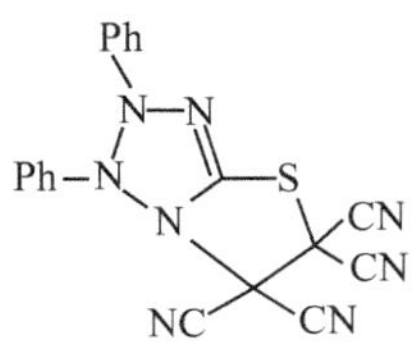

(420)

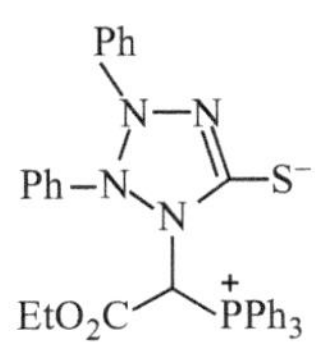

(421)

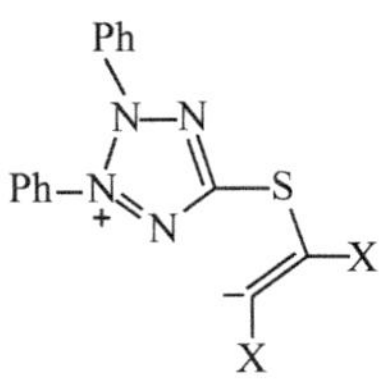

(422)

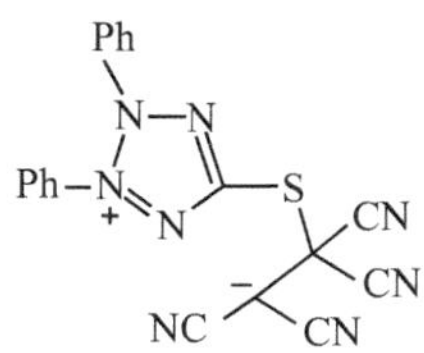

(423)

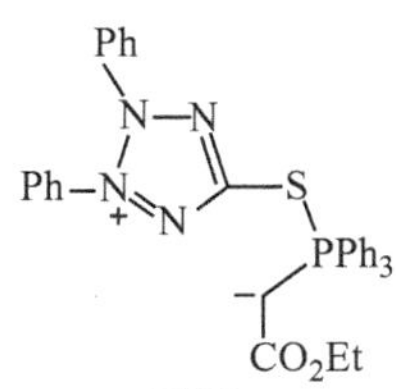

(424)

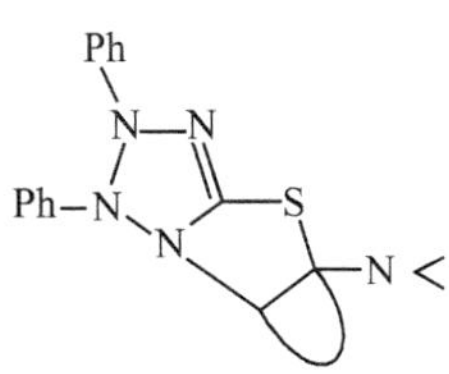

(425)

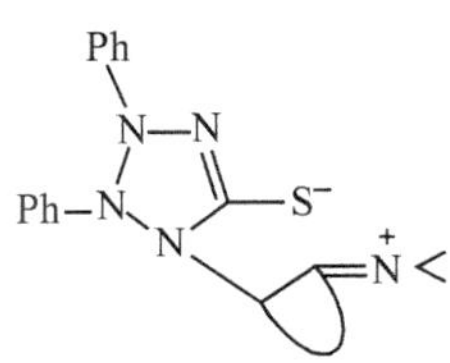

(426)

The reaction between enamines[250] and dehydrodithizone **413**, $R^1 = R^2 = Ph$, may well involve a different type of dipolar intermediate (**426**) whose formation matches the general characteristics of enamine reactions.

The reaction between dehydrodithizone **413**, $R^1=R^2=Ph$, and ynamines has been recently reported.[252] Thus, with diethylaminoprop-1-yne the two products **427** and **428** are obtained. Their formation involves extrusion of sulfur and one mechanistic possibility which has been considered

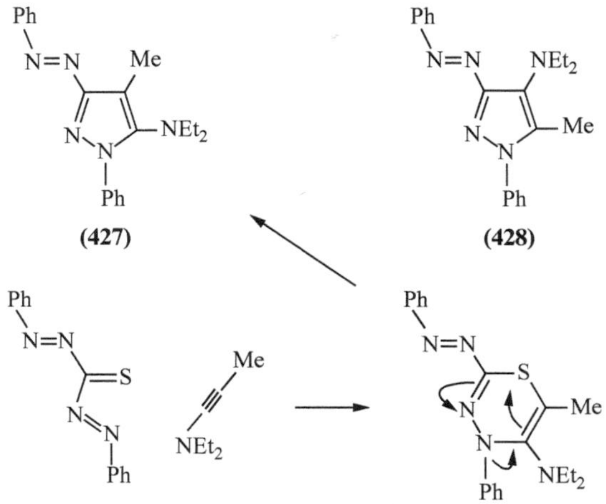

(427) **(428)** **(429)** **(430)**

is that a cycloaddition (**429**) takes place between the valence tautomer of dithizone and the ynamine. This could lead to the intermediate **430**, which after cyclization (**430**; arrows) leads to an azomethine imine, which by extrusion of sulfur could yield the product **427**. The initial cycloaddition (**429**) in the alternative sense could lead to the other product (**428**).

11. The physical study and theoretical treatment of meso-ionic compounds

11.1 Physical methods

The spectroscopic properties of meso-ionic compounds have been discussed in detail elsewhere[4,7,9b] and the reviewers do not feel that it would be useful to include a comprehensive account here. Ultraviolet, infrared, and nuclear magnetic resonance spectra of meso-ionic heterocycles provide general support for the conjugative interaction that would be expected for aromatic heterocycles,[9b] but detailed interpretation of their spectra is not justifiable. Mass spectrometry has been shown to be particularly useful for distinguishing between pairs of meso-ionic isomers (Section 8),[201] and several accounts of the mass spectral fragmentation patterns of meso-ionic compounds have been published.[201,253–258] Measurement of electric dipole moments encouraged the assignment of the original meso-ionic structures to the sydnones.[259–264] More recently, electric dipole moment studies have given powerful support to the formulation of several new classes of heterocycle as meso-ionic compounds.[19a,108,113]

11.2 X-ray crystallography

11.2.1 1,3-Thiazole-5-thiones (109) (Section 7.3.3)

The crystal structure of the 2-*p*-bromophenyl derivative (**431**, Ar=*p*-Br.C_6H_4) has been reported and is consistent with the meso-ionic formulation of the molecule.[72a]

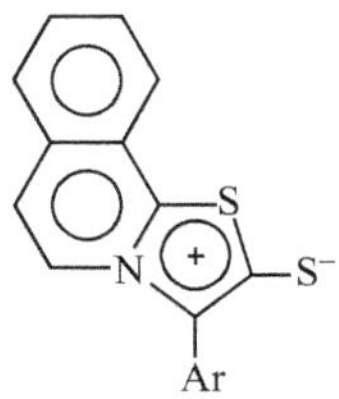

(431)

11.2.2 1,3-Thiazol-4-ones (114) (Section 7.3.4)

The structure of the meso-ionic 1,3-thiazol-4-one (**432**) has been examined, and the observed bond lengths are consistent with a structure resembling the betaine **433**.[78a] Thus, the CC distance (1.31 Å) corresponds to a double bond; the CS distances (1.81 and 1.77 Å) are comparatively long (cf. 1.72 Å in thiophene), and the CO bond length (1.33 Å), which is rather longer than that observed for sydnones (1.2 Å), suggests a structure in which a substantial negative charge is associated with the oxygen atom.

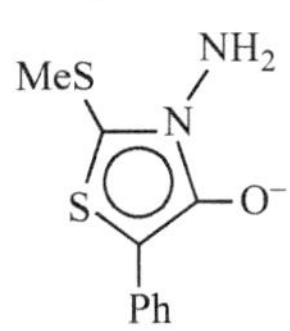

(432)

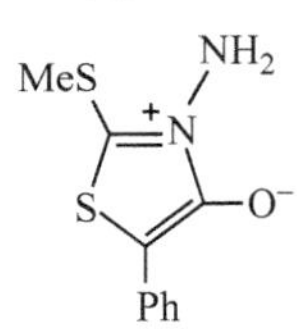

(433)

11.2.3 1,2,3-Oxadiazol-5-ones (sydnones) (1)

The crystal structures of two sydnones have been determined. A study[265] of *N*-*p*-bromophenyl sydnone (**1**, R^1=*p*-Br.C_6H_4, R^2=H) showed that (i) the meso-ionic ring is planar; (ii) the exocyclic CO distance (1.20 Å) is essentially that of a double bond; (iii) the ring CO distance (1.42 Å) is close to that expected for a CO single bond; (iv) the *p*-bromophenyl ring is inclined at a dihedral angle of 27.6° to the plane of the sydnone ring. Because the crystal geometry of the sydnone **1**, R^1=*p*-Br . C_6H_4,

$R^2 = H$, may be influenced by a "charge transfer" interaction between the carbonyl oxygen atom and a neighboring bromine atom, the structure of 4,4′-dichloro-3,3′-ethylene bissydnone (**434**) has also been determined[266]: the structure of this sydnone (**434**) is essentially the same as that observed by the previous workers.

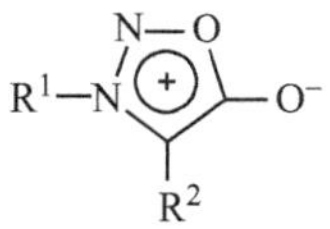

(1)

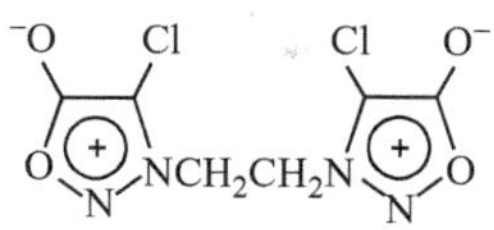

(434)

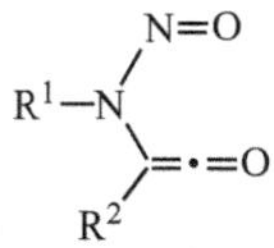

(435)

The sydnone ring geometries obtained in these two studies are not in obvious agreement with the suggestion that the sydnone ring is aromatic. In particular, the ring **CO** distance is longer than expected while the exocyclic **CO** distance is shorter than would be expected for an oxygen atom associated with a large negative charge. These observations, together with the rather large CCO_{exocyc} bond angles (135.7° and 135.5°) have led to the suggestion[266] that the valence tautomeric ketene **435** makes a substantial contribution as a canonical form to the meso-meric sydnone structure (**1**).

11.2.4 1,3,2-Oxathiazol-5-ones (169) (Section 7.6.1)

The crystal structure of 4-phenyl-1,3,2-oxathiazol-5-one (**169**, R = Ph) has been reported.[117] The ring is apparently not planar; the carbonyl carbon atom is out of the plane of the other four ring atoms by 0.026 Å. The dihedral angle between the two rings is 16.6°. The observed bond distances (CO, 1.195 and 1.405 Å; CC, 1.443 Å; CS, 1.659 Å; NO, 1.375 Å; SN, 1.604 Å) support the structure with tetracovalent sulfur (**436**), but this may alternatively be interpreted as indicating that the betaine tautomer (**437**) makes an important contribution to the structure. The C=O bond length (1.195 Å)[117] is almost identical with that observed for sydnones (1.20 Å).[265,266]

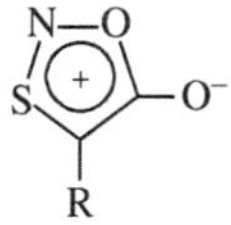

(169)

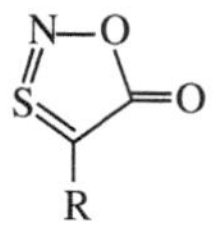

(436)

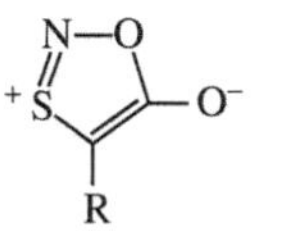

(437)

11.2.5 1,2,4-Triazol-3-imines (216) (Section 7.7.5)

A preliminary account of an X-ray crystallographic study of the 1,2,4-triazol-3-imine (**226**) has been reported.[147]

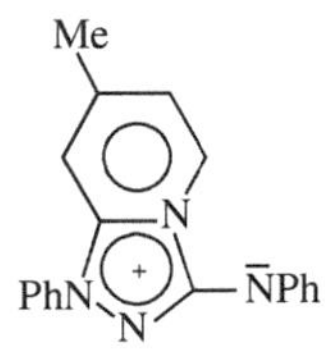

(226)

11.2.6 1,3,4-Thiadiazol-2-enes (261) (Section 7.8.4)

A detailed structural determination of the 1,3,4-thiadiazol-2-ene (**438**) has been reported.[175] The separation of the two sulfur atoms (2.79 Å) indicates some interaction.

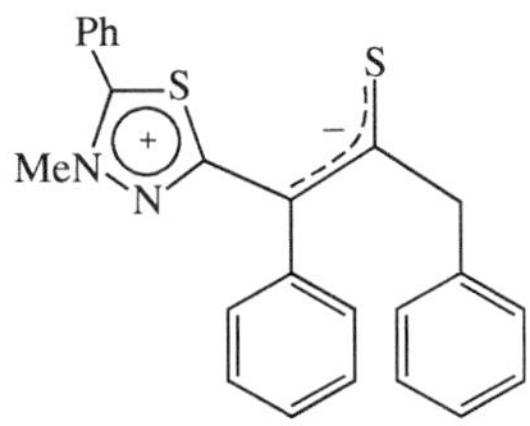

(438)

11.2.7 1,2,3,4-Tetrazole-5-thiones (413) (Section 10.5.3)

A detailed investigation of the crystal structure of dehydrodithizone (**413**, $R^1=R^2=Ph$) has been reported[246] and is of particular interest in that it is the only known crystal structure of a type B meso-ionic heterocycle. The five-membered ring and the sulfur atom are coplanar, and all the bond lengths (NN, 1.313 and 1.318 Å; CN, 1.360 Å; CS, 1.687 Å) are intermediate between single and double bond lengths. These features are all entirely consistent with the meso-ionic structure **413**.

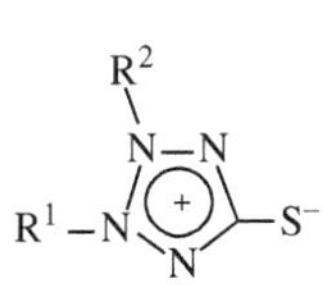

(413)

11.3 ESCA spectroscopy

11.3.1 1,2,3-Oxadiazol-5-ones (sydnones) (1)

The ESCA spectra of *N*-phenylsydnone (**1**, $R^1=Ph$, $R^2=H$)[172,267] and *N*-methylsydnone (**1**, $R^1=Me$, $R^2=H$)[267] have been recorded. The core electron binding energies of the two nitrogen atoms in the sydnone ring (Table 3) differ by 1.4–2.0 eV, suggesting that there is a considerable

Table 3 ESCA spectra of meso-ionic heterocycles.

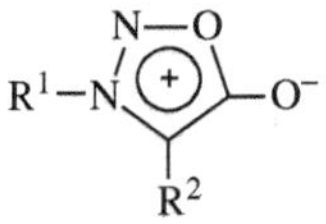
(1)

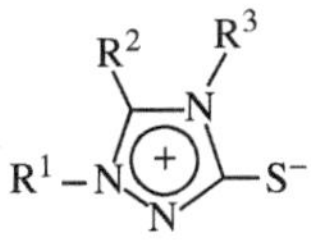
(227)

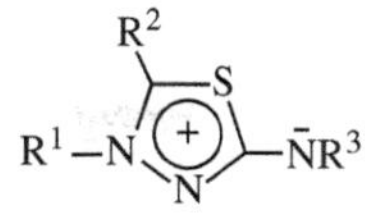
(247)

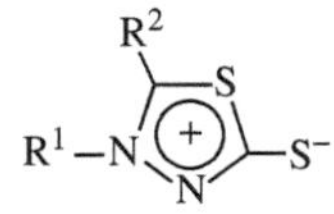
(251)

				Core electron binding energies (eV)			
Compound	**R¹**	**R²**	**R³**	**C(1s)**	**N(1s)**	**O(1s)**	**S(2p)**
Sydnones (**1**)[172,267]					399.4; 401.4;	530.3; 532.7;	
	Ph	H			400.8; 402.2	531.4; 533.8	
	Me	H		285.6; 287.9	401.2; 402.6		
1,2,4-Triazole-3-thiones (**227**)[149b]							
	Ph	m-NO_2.C_6H_4	Allyl		398.2; 399.9; 400.1; 405.0 (NO_2)		160.6
	Ph	t-C_4H,	Allyl		397.8; 399.2; 400.2		160.4
1,3,4-Thiadiazol-2-imines (**247**)[149b]							
	Me	Ph	OEt		397.0; 398.2; 400.7		163.9
	Me	Ph	NMe_2		397.0; 398.2; 398.9 (NMe_2); 401.2		163.5
	Me	Ph	Me		396.6; 398.2; 400.5		163.3
	Me	Ph	t-C_4H_9		397.0; 398.2; 400.4		163.6
	Me	Ph	Ph		396.6; 398.1; 400.1		163.2
	Me	Ph	p-NO_2.C_6H_4		396.8; 398.3; 400.2; 404.6 (NO_2)		163.7
	Me	p-Cl.C_6H_4	OEt		396.9; 398.2; 400.7		163.7
	Me	p-EtO.C_4H_4	OEt		397.1; 398.2; 400.6		163.7
1,3,4-Thiadiazole-2-thiones (**251**)[172]							
	Me	Ph			398.2; 400.6		161.1; 163.7
	Me	p-Cl.C_6H_4			398.3; 400.5		160.7; 163.2

difference in the formal charges associated with these two nitrogen atoms. Similarly, the molecular environment of the two oxygen atoms in the sydnone nucleus are quite different. For *N*-methylsydnone (**1**, $R^1 = Me$, $R^2 = H$), the intramolecular chemical shifts of the core electron binding energies of the carbon, nitrogen, and oxygen atoms have been estimated using *ab initio* calculations and applying Koopmans' theorem.[267] The results are in good agreement with experiment, although the absolute values of the core ionization potentials (Table 3) are calculated to be high (probably owing to the neglect of orbital relaxation on ionization). Attempts to correlate the observed core ionization potentials with formal charges calculated by *ab initio* and semiempirical intermediate neglect of differential overlap (INDO) methods were less successful.[267]

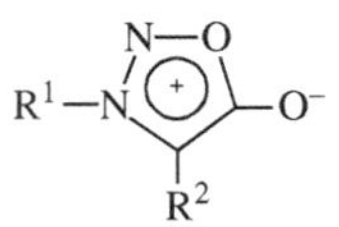

(1)

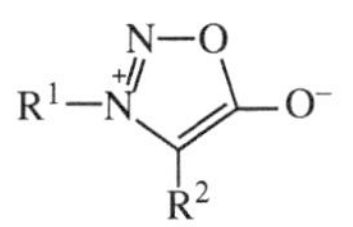

(439)

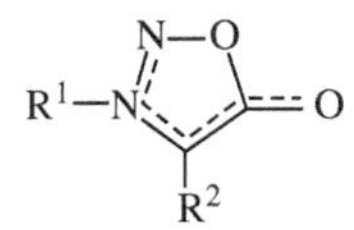

(440)

On the basis of their ESCA spectra, two interpretations of the best representation of the bonding in sydnones have been proposed. Patsch and Thieme[172] concluded that the differences in the binding energies for the two nitrogen atoms and the two oxygen atoms support a betaine structure (**439**). Alternatively, Hillier and his co-workers,[267] who based their conclusion on both spectroscopic evidence and calculations, suggested that the structure **440** best represents 3-alkylsydnones. These two representations, **439** and **440**, are in fact very similar; an important common feature is the nonconjugation of the ring oxygen atom in the cyclic π-electron system.

11.3.2 1,2,4-Triazole-3-thiones (227)

ESCA spectra of two meso-ionic 1,2,4-triazole-3-thiones (**227**) (Table 3) have been measured.[149b] The results which support the meso-ionic formulation **227**, are quite distinct from those obtained for derivatives of the isomeric meso-ionic 1,3,4-thiadiazol-2-imine system (**247**).

11.3.3 1,3,4-Thiadiazol-2-imines (247)

Representatives of the meso-ionic 1,3,4-thiadiazol-2-imines (**247**) have only recently been prepared (Section 7.8.2), although the isomeric meso-ionic 1,2,4-triazole-3-thiones (**227**) have been known for a number of years (Section 7.7.6). ESCA spectra of eight derivatives of the 1,3,4-thiadiazol-2-imines (**247**) have been reported (Table 3).[149b] These spectra are quite distinct from those of the 1,2,4-triazole-3-thiones (**227**) and provide further support for the formulation of these compounds as 1,3,4-thiadiazol-2-imines (**247**). In particular, the sulfur atom in the ring position has a core binding energy (ca. 163.5 eV) (Table 3) quite distinct from that of the exocyclic sulfur atom (ca. 160.5 eV) in the meso-ionic 1,2,4-triazole-3-thiones (**227**) (Table 3).

11.3.4 1,3,4-Thiadiazole-2-thiones (251)

The ESCA spectra of two meso-ionic 1,3,4-thiadiazole-2-thiones (**251**) (Table 3) have been recorded.[172] The difference in the core binding energy of the two nitrogen atoms (ca. 2.3 eV) and the difference between the two sulfur atoms (ca. 2.5 eV) support a betaine structure (**441**) in which similar atoms are associated with quite different electron densities. The core binding energies of the two sulfur atoms have values (Table 3) that correspond to those observed for the ring sulfur atom in the meso-ionic 1,3,4-thiadiazol-2-imines (**247**) and the exocyclic sulfur atom in the meso-ionic 1,2,4-triazole-3-thiones (**227**).

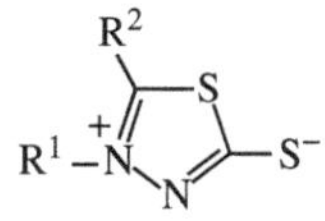

(441)

11.4 Molecular orbital calculations

Ever since the discovery of the sydnones (**1**) and the introduction of the term meso-ionic,[1] the class of meso-ionic heterocycles has aroused much theoretical interest because of the problems associated with their representation. Most theoretical studies have been concerned with molecular orbital (MO) calculations on the sydnones, and increasingly sophisticated calculations have been reported with the development of improved computer facilities. A summary has been recently presented[268] of the various types of MO calculations which are currently of interest to organic chemists.

11.4.1 1,2,3-Oxadiazol-5-ones (sydnones) (1)

(i) *π-Electron calculations*. The first MO calculations carried out for the sydnone ring were Hückel calculations.[269–272] These results were in general agreement with the observed magnitude and direction of the electric dipole moment of the sydnone ring, although the calculated moments were far too large. The use of the ω-Hückel method gave more satisfactory calculated values for dipole moments.[273,274] Later the Pariser–Parr–Pople (PPP) method and modifications of this method were applied to the sydnone ring and used to calculate charge distributions and electronic spectra with modest success.[193,275–279] Since these π-electron methods involve quite serious approximations, particularly when applied to heterocyclic systems, and much more sophisticated and accurate methods are now available, these early calculations are only of historical interest, although they did play an important role in the development of the understanding of bonding in sydnones.

(ii) *All valence electron semiempirical calculations*. Several semiempirical MO calculations of sydnones (**1**) have now been reported. The earliest study was the application of the extended Hückel method (EH) to *N*-phenylsydnone (**1**, R^1 = Ph, R^2 = H).[280] In this calculation, the geometry of the molecule was assumed to be the same as that obtained for it in the X-ray study of *N*-*p*-bromophenylsydnone (**1**, R^1 = *p*-Br. C_6H_4, R^2 = H) (Section 11.2). The calculated electric dipole, moment was rather large (μ_{calc} 12.5 D; μ_{exp} 6.48 D). When the total energy of *N*-phenylsydnone (**1**, R^1 = Ph, R^2 = H) was calculated at several inter-ring torsion angles, an energy minimum was found when the angle was 27°, which is in remarkable agreement with the torsion angle of 27.6° observed for *N*-*p*-bromophenylsydnone (**1**, R^1 = *p*-Br.C_6H_4, R^2 = H).[280] More recently, the charge iterative extended Hückel method (IEH) has been applied to *N*-methylsydnone (**1**, R^1 = Me, R^2 = H).[281] Here, the calculated dipole moments are in rather better agreement with experiment (μ_{calc} 7.04 D; μ_{exp} 7.31 D). A detailed discussion of the charge distribution has also been given.[281]

Complete neglect of differential overlap (CNDO) calculations with inclusion of configuration interaction (CNDO CI) have been reported for the unsubstituted sydnone molecule (**1**, R^1 = R^2 = H) and *N*-methylsydnone (**1**, R^1 = Me, R^2 = H)[282]; geometries were based on *N*-*p*-bromophenylsydnone (**1**, R^1 = *p*-Br.C_6H_4, R^2 = H). The calculated dipole moment of *N*-methylsydnone (**1**, R^1 = Me, R^2 = H) is in very good agreement with experiment (μ_{calc} 7.57 D; μ_{exp} 7.31 D), but the calculated electronic spectrum is less satisfactory.[282] Similar results have been obtained for CNDO/2 calculations on the unsubstituted sydnone (**1**, R^1 = R^2 = H) (μ_{calc} 6.82 D).[283]

INDO calculations for *N*-methylsydnone (**1**, R^1 = Me, R^2 = H) and *N*-phenylsydnone (**1**, R^1 = Ph, R^2 = H) have been reported[267]; again the geometries were based on the X-ray study of *N*-*p*-bromophenylsydnone (1, R^1 = *p*-Br.C_6H_4, R^2 = H). The calculated charge distribution in both these sydnones (1, R^1 = Me, and Ph, R^2 = H) is very similar, and this is consistent with the observation of very similar core binding energies in their ESC A spectra (Table 3) (Section 11.3). The calculations of the charge distribution in A-methylsydnone by CNDO[282] and INDO[267] calculations are very similar. The exocyclic oxygen atom is associated with a substantial negative charge (ca. 0.45 eu), and the adjacent carbon atom carries an almost equal positive charge. The nitrogen atom attached to the methyl group is positively charged (ca. 0.25 eu), and the adjacent carbon and nitrogen atoms carry rather smaller negative charges (ca. 0.1–0.2 eu).

(iii) *Ab initio calculations.* Two *ab initio* studies of *N*-methylsydnone (**1**, R^1 = Me, R^2 = H) have been reported; in each case the molecular geometry was based on that of 3,3′-ethylenebissydnone (**434**) (Section 11.2).[267,284] By applying Koopmans' theorem, the calculations give intramolecular chemical shifts of the core electron binding energies that are in good agreement with the observed ESCA spectrum (Table 3) (Section 11.3).[267] The calculated dipole moment (μ_{calc} 5.98 and 6.28 D; μ_{exp} 7.31 D) and ionization potential (I_{calc} 8.75 eV; I_{exp} 8.9 eV) are in satisfactory agreement with experiment.

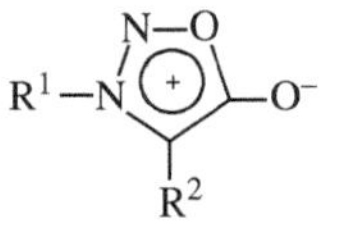

(1)

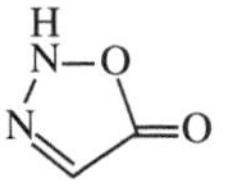

(442)

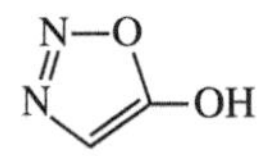

(443)

The charge distributions of the sydnone ring calculated by these *ab initio* methods[267,284] show one significant difference from those given by CNDO and INDO calculations, namely, a reversal of the positive and negative charges on the nitrogen atoms.

In addition, *ab initio* calculations for the unknown unsubstituted sydnone molecule (**1**, R^1 = R^2 = H) and the tautomeric species 2*H*-1,2,3-oxadiazol-5-one (**442**) and 5-hydroxy-1,2,3-oxadiazole (**443**) have been reported.[284] The meso-ionic tautomer (**1**) is calculated to be less stable than the tautomers **442** and **443**. However, since total energy was not minimized with respect to geometry, these results are questionable.

11.4.2 Other meso-ionic systems

A number of π-electron MO calculations on systems other than the sydnones (**1**) have been reported[4]: these include the 1,2,3-oxadiazol-5-imines (sydnone imines) (**444**),[4,271,278] the 1,3,4-oxadiazol-2-ones (isosydnones) (**445**),[193] and the 1,2,3,4-oxatriazol-5-ones (**446**).[193] Since much more sophisticated MO methods are now available, the reviewers do not think that a detailed report of these early calculations is useful.

PPP calculations have been used to predict the properties of the type B meso-ionic 1,2-oxazol-4-one (**447**),[285] although the possibility that the valence tautomer **448** may be more stable was not considered. So far, the three predictions that this as yet unknown compound (**447**) (i) "*should show two absorption maxima at about 460 and 440 mp (approximation ±30 mμ),*" (ii) "*should appear as an orange or red substance,*" and (iii) "*should be handled and kept at low temperatures*" have not been confirmed.[285] We leave to the judgment of others the extent to which their confidence can and should be placed in these three predictions.

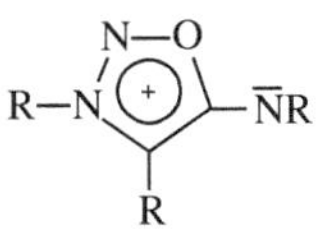

(444)

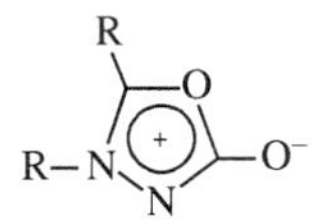

(445)

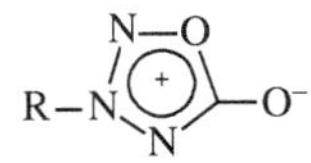

(446)

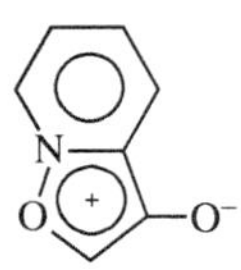

(447)

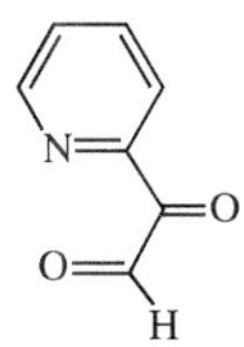

(448)

12. Pharmacological activity of meso-ionic compounds

The wide range of structural variation presented by meso-ionic heterocycles of type A and type B is such that some members have attracted the attention of medicinal chemists. Many different biological activities have been claimed, particularly in the patent literature. The authors of this review are not competent to provide a critical appraisal of the biological activities that have been attributed to meso-ionic compounds. These are therefore only briefly recorded in the following summary, and the original papers or patents should be consulted for details.

For reasons given in Section 1, the chemistry of sydnones and sydnone imines is not discussed in the chemistry section of the review. Examination of their pharmacological activity is, however, given in this section.

The possibility that meso-ionic compounds might have potential value as biologically active substances has been emphasized particularly in an excellent review by Kier and Roche.[4] Kier has also put forward stimulating proposals concerning possible general applications of MO theory to drug research.[286]

12.1 1,2,3-Oxadiazol-5-ones (sydnones) (1)

The earliest reported consideration of sydnones (**1**) as biologically active compounds was made by Brookes and Walker in 1957.[287] The relation between their meso-ionic structure (**1**) and the zwitterionic structure of amino acids suggested that sydnones might exhibit properties as natural amino acid antagonists. However, no antibacterial activity was observed.[287] A wide range of biological properties has been claimed for various sydnone derivatives and these have been well reviewed up to 1966.[4,288] Later references are included in this review.

The following activities have been claimed for sydnone derivatives (**26**): antibacterial,[4,287–300] antitumor,[4] antifungal,[301,302] antimalarial,[303–306] antiparasitic,[307] analgesic,[308–325] anti-inflammatory,[309–311,314–319,321–323] coccidiostatic,[326] hypotensive,[327] hypoglycemic,[4] diuretic,[4] hepatotoxic,[328] and insecticidal.[312,313,329,330] Central nervous system activity—convulsant and anti-convulsant—have been claimed[331–333] as well as the ability to function as inhibitors of monoamine oxidase[334] and other biotransformations.[335–337]

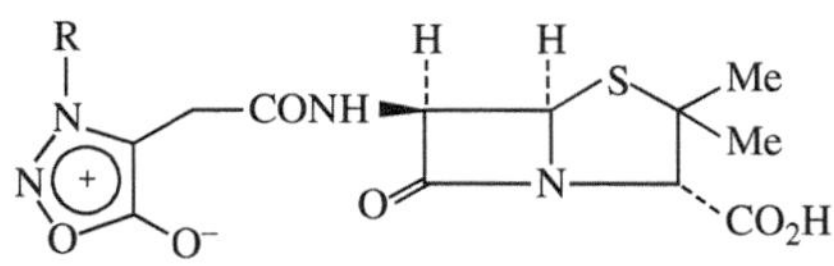

(449)

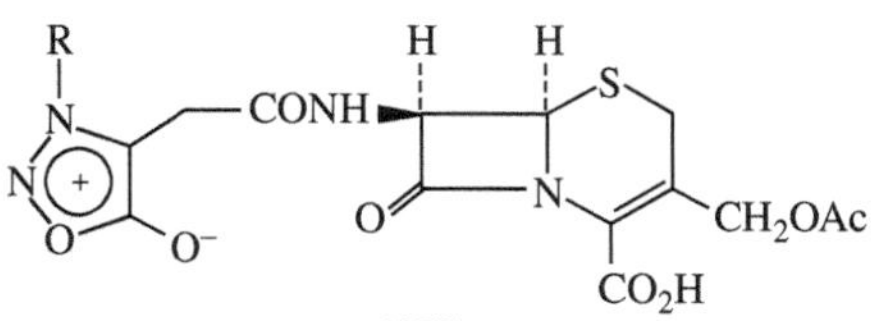

(450)

A large number of derivatives of the penicillin (e.g., **449**)[298] and cephalosporin (e.g., **450**)[300] types have been prepared with *N*-acyl residues containing meso-ionic heterocyclic groupings. Some have shown antistreptococcal and antistaphylococcal activities *in vivo*.

12.2 1,2,3-Oxadiazol-5-imines (sydnone imines) (2)

A number of different biological activities have been claimed for sydnone imines (**2**), their corresponding cations (**3**), and their *N*-acyl derivatives (**4**). The activities that have been listed include analgesics,[338,339] anti-inflammatory agents,[338–340] cardiovascular agents,[341–352] central nervous system stimulants,[340–362] and antidepressants.[363]

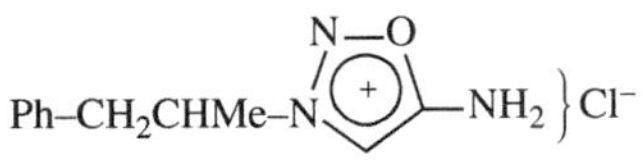

(451) Sydnofen

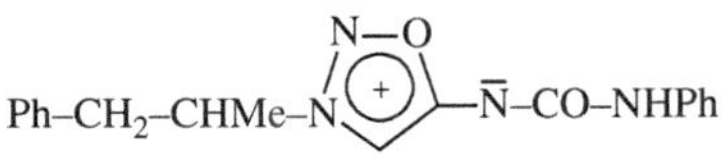

(452) Sydnocarb

Clinical trials have been carried out on sydnofen (**451**)[360] and sydnocarb (**452**)[363] as psychostimulants or antidepressants. These two potential pharmaceuticals are of Russian origin.

12.3 1,3-Diazol-4-ones (91)

The chloroacetylation product of 1-azaphenothiazine has been condensed with various secondary amines giving basic polycyclic meso-ionic 1,3-diazol-4-one derivatives (**453**), some of which showed useful anthelmintic activity.[364]

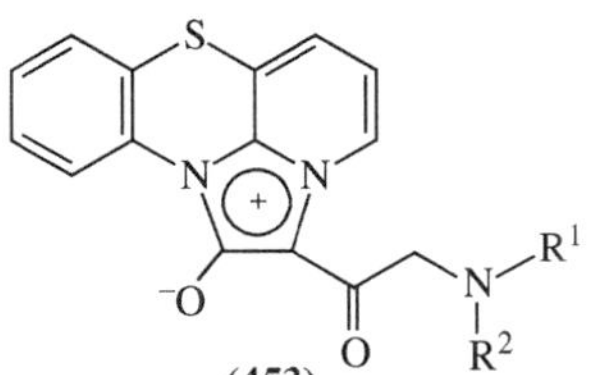

(453)

12.4 1,3-Thiazol-4-ones (114)

Derivatives of this group have been described as useful central nervous system stimulants and anti-inflammatory drugs.[365]

12.5 1,3-Thiazol-4-imines (124)

Various salts (**126**) have been found to act as anti-inflammatory agents,[366–371] sedatives,[366,367] and central nervous system stimulants.[368–371]

12.6 1,2,3-Triazol-4-ones (176)

The activity of these heterocycles as plant growth inhibitors has been described, and their use as pre- or post-emergence herbicides reported.[372]

12.7 1,2,4-Triazol-3-ones (200)

The cycloaddition product (**200**, $R^1 = R^3 = Ph$, $R^2 = H$) obtained from *N*-phenylsydnone and phenyl isocyanate has been reported as a psychostimulant and anti-inflammatory agent.[138] Bicyclic 1,2,4-triazolo-pyridines (**209**) and their tetrahydroderivatives have been claimed as analgesics and anti-inflammatory agents.[373–377]

12.8 1,3,4-Thiadiazol-2-imines (247)

N-Acyl-thiadiazol-2-imines have been claimed as useful sedatives.[378]

12.9 1,3,4-Thiadiazole-2-thiones (251)

The meso-ionic thiadiazoles[4,158,139] show an interesting range of activities against gram-positive and gram-negative micro-organisms. Some derivatives also show cardiovascular effects.

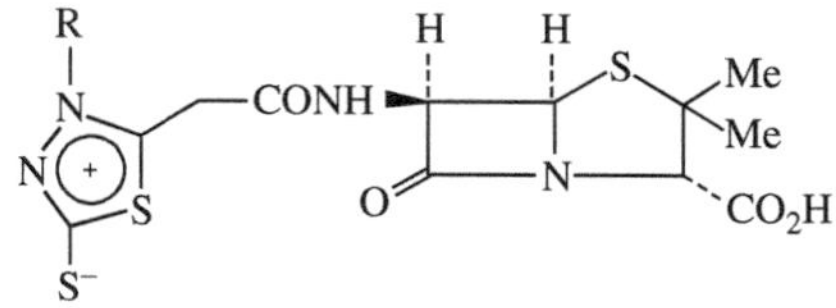

(454)

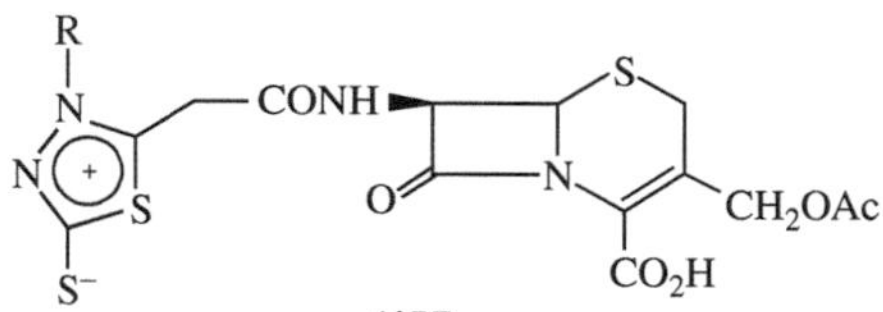

(455)

Various derivatives of 6-aminopenicillanic acid (e.g., **454**) and 7-aminocephalosporanic acid (e.g., **455**) have been prepared and examined as possible antibacterials.[379]

12.10 1,2,3-Thiadiazol-4-ones (265)

Recently, the inhibition of monoamine oxidase by a series of meso-ionic 1,2,3-thiadiazol-4-ones (**265**) has been described.[177,178] A model of enzyme–inhibitor interaction and an analysis of structural features controlling the mode of enzyme inhibition have been presented.[178]

12.11 1,2,3,4-Oxatriazol-5-ones (271)

Their activity as hypotensive agents[192] has been reviewed.[4]

12.12 1,2,3,4-Oxatriazol-5-imines (277)

The *N*-acyl derivatives (**277**, $R^2{=}Ac$)[380] and the salts (**285**)[381] have been claimed as hypotensive agents.

13. Meso-ionic: Definition and delineation

In the definition of the term meso-ionic as originally proposed by Baker and Ollis,[2] it was stated that "a compound may be appropriately called meso-ionic if it is a five- or possibly a six-membered heterocyclic compound which cannot be represented *satisfactorily* by any one covalent or polar structure and possesses a sextet of electrons in association with the atoms comprising the ring." This definition has formed the basis of this review, but we now propose a modification.

One hundred and forty-four meso-ionic heterocycles of type A (**13**, **19**) and 84 meso-ionic heterocycles of type B (**14**, **20**) are possible. The numbers of these two types which are now known (Table 1: type A, 44 representatives) and (Table 2: type B, 8 representatives) encourage us to put forward the proposal that the term meso-ionic should in future be restricted to five-membered heterocycles belonging to type A (**13**, **19**) and type B (**14**, **20**). This clear restriction upon the use of the term meso- ionic has obvious advantages. It still embraces 228 different classes of heterocycles with a common structural characteristic, and the many types of meso-ionic compounds included in the authoritative review by Ohta and Kato[9b] are included. Needless to say, the restriction upon the definition of the term meso-ionic to five-membered heterocycles of type A and type B still includes, for example, benz derivatives such as the compounds **67**, **71**, **110**, **123**, **133**, **151**, **206**, **209**, **226**, **255**, and **258**.

It is very important to emphasize that the terms meso-ionic compound and sydnone are not interchangeable. The term sydnone should be used only for the heterocyclic system (**1**). It just causes confusion if the word sydnone is used to describe other heterocyclic systems, although they are meso-ionic. The term sydnone should not be used to describe derivatives of 1,3,2-oxathiazol-5-one (**169**)[117] and 1,2,3,4-tetrazole-5-thione (**413**).[246]

If heterocyclic chemists can be persuaded to restrict the use of the term "meso-ionic" only to five-membered heterocycles of type A and type B, then this will necessarily exclude some compounds that have previously been described as meso-ionic. Acceptance of this suggestion would then mean that the term meso-ionic should no longer be used to describe various six-membered heterocyclics (**456–469**)[382–411] and a number of polyazapentalenes.[412] These compounds (**470–477**)[413–432] have been described as meso-ionic, but they are surely best regarded as mesomeric betaines and should be formulated and described as such.

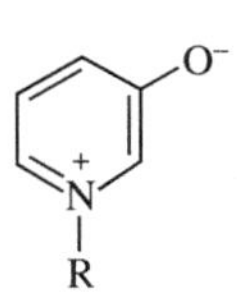

(456)[382–385]

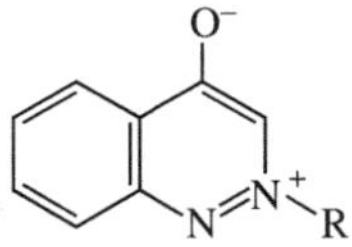

(457)[386–388]

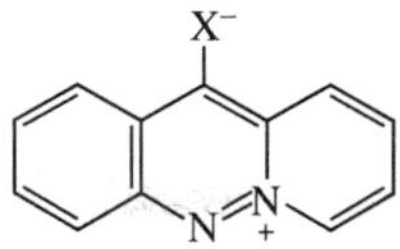

X = NR, O, S

(458)[389]

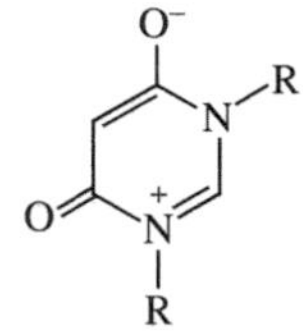

(459)[390–394]

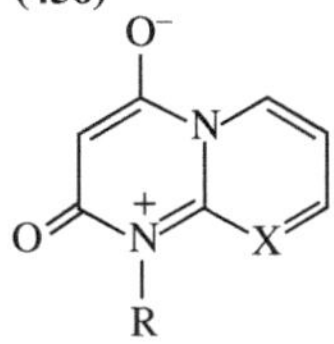

X = CH, N

(460)[395,396]

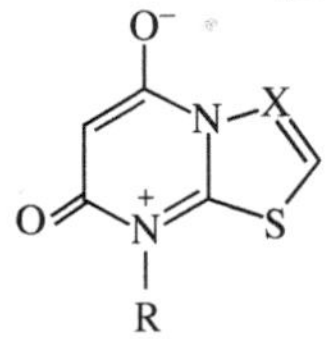

X = CH, N

(461)[397,398]

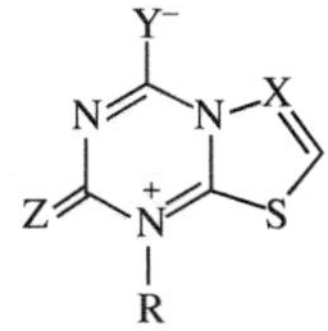

X = CH, N
Y, Z = O, S

(462)[399]

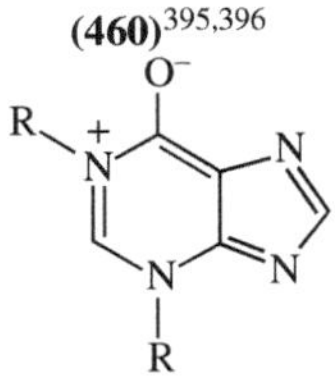

(463)[400]

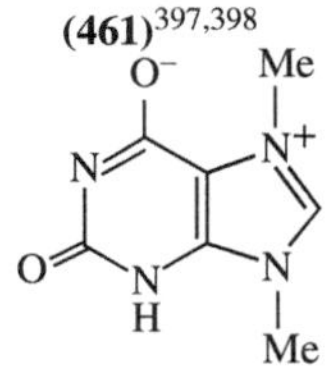

(464)[401]

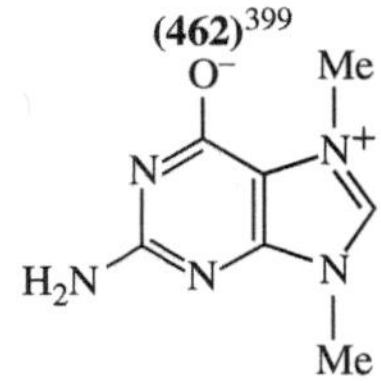

(465)[402]

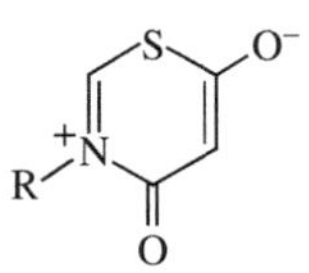

(466)[403]

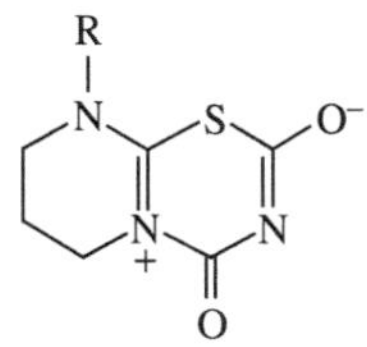

(467)[404,405]

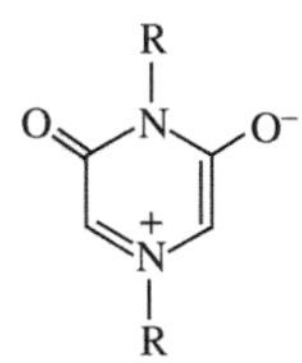

(468)[406–410]

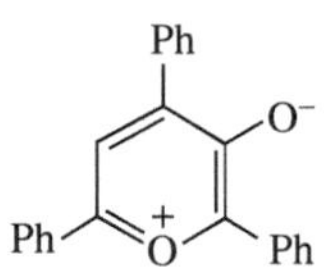

(469)[411]

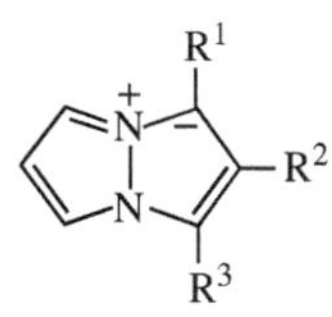

(470)[413–417]

4,8-Diazapentalenes

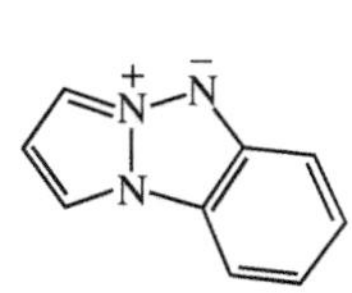

(471)[415]

1,4,8-Triazapentalenes

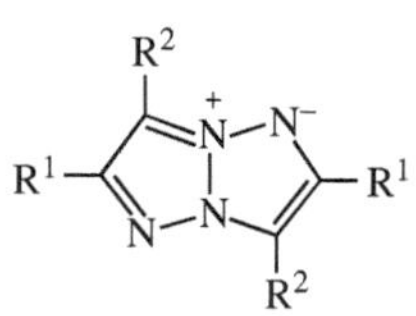

(472)[418–423]

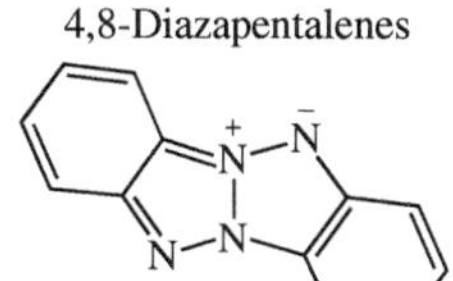

(473)[424–427]

1,4,5,8-Tetrazapentalenes

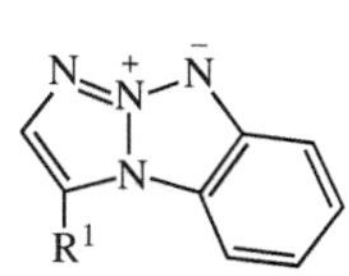

(474)[426–429]

1,4,7,8-Tetrazapentalenes

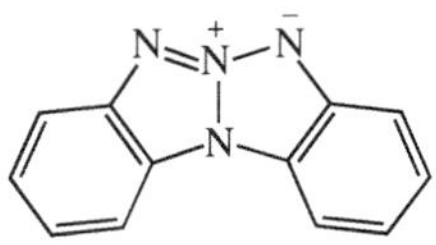

(**475**)[426,429]
1,4,7,8-Tetrazapentalenes

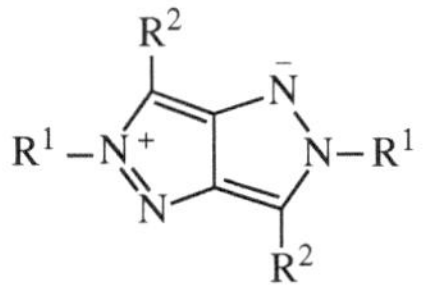

(**476**)[430,431]
1,2,5,6-Tetrazapentalenes

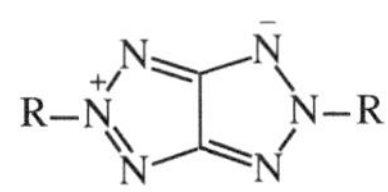

(**477**)[432]
1,2,3,5,6,7-Hexazapentalenes

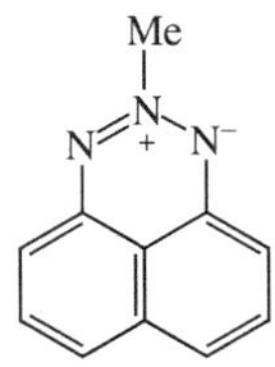

(**478**)[433]

Other polyaza derivatives of aromatic systems, such as the blue triazaphentalene derivative (**478**),[433] are also adequately described as mesomeric betaines or inner salts.

Ohta and Kato, in a section entitled "Bridged Heterocyclic Compounds" (Ref. 9b, page 241), have referred to seven types of compounds and the stereochemically unacceptable structures (**479–485**) which had been allocated to them. The possibility was considered by Ohta and Kato that some of these structures might have to be replaced by their meso-ionic equivalents. However, we have now established[434] that this is not the case, and the relation between the original structures (**479–485**) and the correct structures (**486–492**) is given in Fig. 10.

If our proposal for the restriction of the term "meso-ionic" finds general support and it is agreed that compounds exemplified by the structures (**456–478**) should not be called meso-ionic, then the definition we would wish to propose is as follows:

A compound may be appropriately called meso-ionic if it is a five-membered heterocycle which cannot be represented satisfactorily by any one covalent or polar structure and possesses a sextet of electrons in association with the five atoms comprising the ring.

Ohta and Kato, in 1969, in their useful and comprehensive review[9b] on sydnones and other meso-ionic compounds, expressed their "*strong desire to see the unfortunately chaotic present state of meso-ionic chemistry promptly rectified.*" We hope that this review covering the literature available to us up to September 1974 goes some way toward meeting their desire.

Original structural proposal[9b]

Correct structure[434]

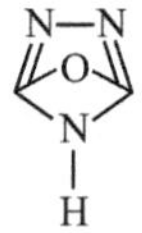

(479)

$$H_2N-\overset{O(S)}{\overset{\|}{C}}-NH-NH-CO-NH_2$$

(486)

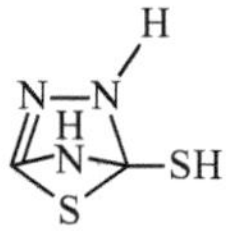

(480)

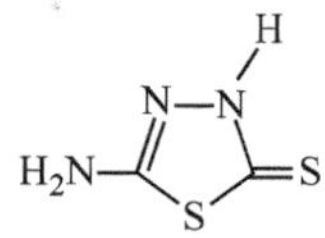

(487)

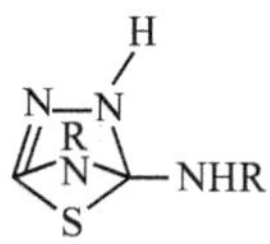

(481)

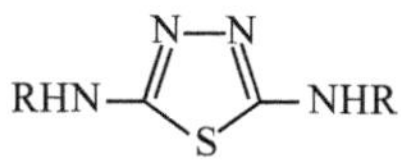

(488)

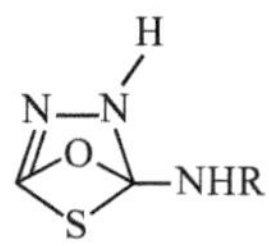

(482)

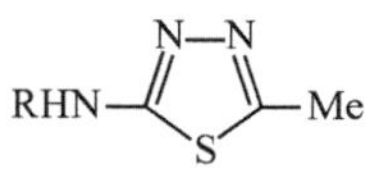

or

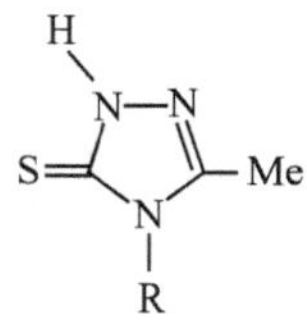

(489)

Original structural proposal[9b]

Correct structure[434]

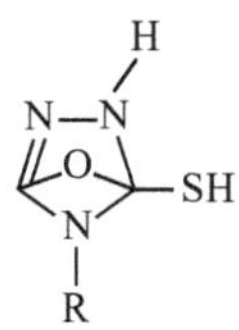

(483)

$$R-NH-CS-NH-\overset{+}{N}H_3 \} \; Cl^-$$

(490)

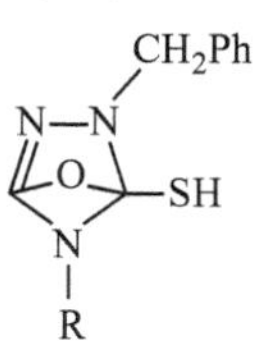

(484)

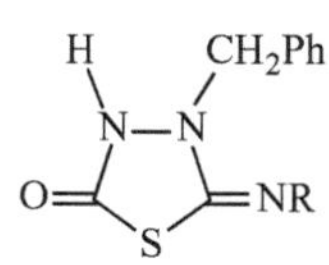

(491)

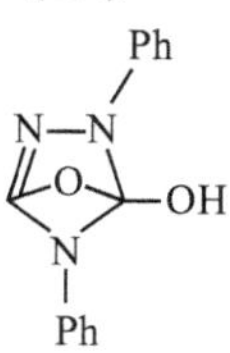

(485)

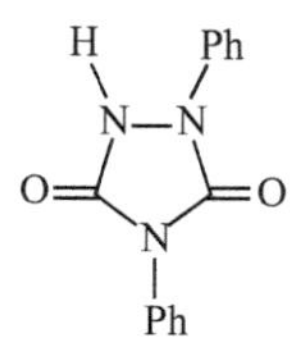

(492)

Fig. 10 Correlation of original and correct structures of some "bridged heterocyclic compounds" reviewed by Ohta and Kato (Ref. 9b, page 241).

Appendix added in proof

The extent of current interest in meso-ionic chemistry is illustrated by the fact that since the completion of the manuscript of this review a number of significant papers have been published. These include descriptions of four new representatives of the type A meso-ionic heterocycles. Three of these systems belong to the 144 structural possibilities described in Section 6. This brings the number of known type A meso-ionic systems up to 47. The fourth new system is the first example of a type A meso-ionic heterocycle having a selenium atom in the meso- ionic ring. The extension of the general formula **19** to include selenium provides the possibility of a further 176 type A meso-ionic systems of which one is now known. The structures of these four new meso-ionic systems are given in Table A1 (Appendix to Table A1).

Table A1

		Atom or group					
Parent system	**Hetero-cycle**	**a**	**b**	**c**	**d**	**e**	**f**
Oxazoles							
1,3-Oxazol-4-ones	**481**	NR	CR	O	CR	C	O
Oxathioles							
1,3-Oxathiol-4-ones	**498**	S	CR	O	CR	C	O
1,3-Oxathiol-5-ones	**502**	O	CR	S	CR	C	O
Selenazoles							
1,3-Selenazol-4-ones	**505**	NR	CR	Se	CR	C	O

This appendix covers the literature available to the reviewers up to September 1975. Section headings are used for easy reference to the main text. New headings are used for new classes of compounds.

A.1 Section 7.1

7.1.1. *1,3-Oxazol-5-ones* (**66**) (Section 7.1.1)

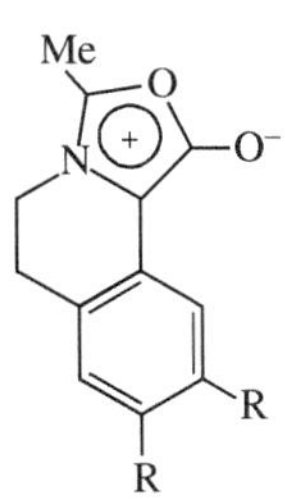

(479)

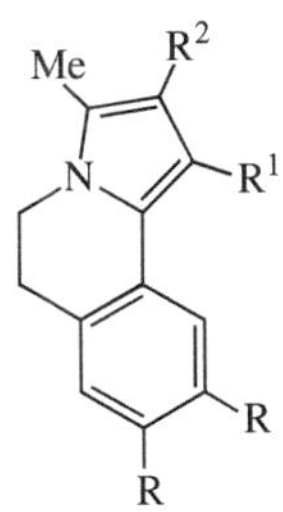

(480)

The ring fused pyrroles **480** have been prepared by *in situ* trapping of the meso-ionic 1,3-oxazol-5-ones (**479**) with alkynes. This 1,3-dipolar cycloaddition was found to be regiospecific when phenyl acetylene was used as 1,3-dipolarophile, the only products being **480**, R = H or MeO, R^1 = Ph, R^2 = H.[435]

Polymers containing pyrrole units have been prepared by the *in situ* condensation of bis-1,3-oxazol-5-ones with diacetylenes.[436]

7.1.4. *1,3-Oxazol-4-ones (Anhydro-4-hydroxy-1,3-oxazolium Hydroxides)* (**481**)

Representatives of this new class of meso-ionic heterocycles have been prepared by catalytic decomposition of the diazo compounds (**482**) using cupric acetylacetonate in hot benzene.[437,438] For example, compound **481**, R^1 = Ph, R^2 = Me, R^3 = $C_6H_4.NO_2$, m.p. 187°–189°, was obtained as red crystals in 85% yield. These compounds readily participate in 1,3-dipolar cycloaddition reactions. The adduct **483** was formed in quantitative yield when compound **481**, R^1 = Ph, R^2 = Me, R^3 = $C_6H_4.NO_2$, reacted with dimethyl fumarate in hot benzene. With acetylenic 1,3-dipolarophiles ($R^4C{=}CR^5$), adducts of the general type **484** are readily formed together with furan derivatives, generated by loss of an isocyanate (R^2NCO) from the primary adduct **484**.[437,438]

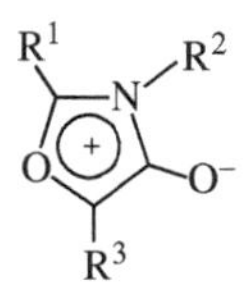

(481) (See Table A-I)

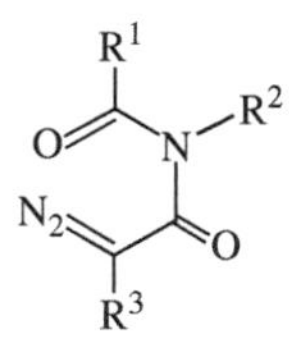

(482)

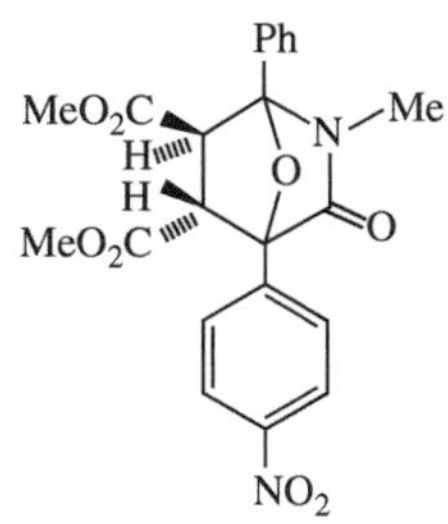

(483)

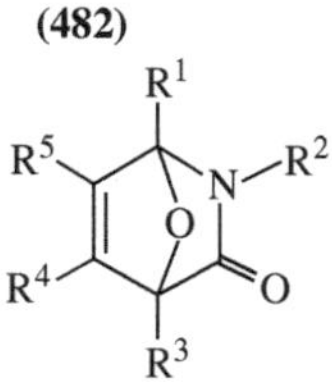

(484)

A.2 Section 7.2

7.2.1. *1,3-Diazol-4-ones* (91) (Section 7.2.1)

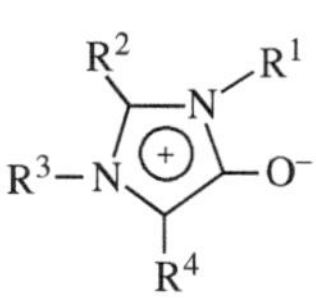

(91)

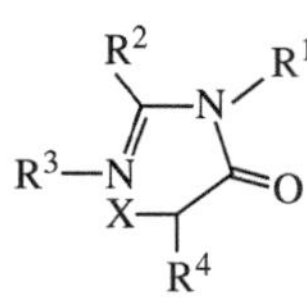

(485)

A new method of preparing meso-ionic 1,3-diazol-4-ones (91) *in situ* has been described.[439] Condensation of α-haloacyl chlorides ($R^4CHX.COCl$) with benzamidines ($R^2.C{=}NR^1.NHR^3$) in benzene solution gives α-haloaeylamidines (**485**) which in the presence of triethylamine cyclize to the meso-ionic 1,3-diazol-4-ones (**91**). Using this procedure, the 1,3-diazol-4-ones (**91**) were not isolated but evidence for their formation was provided by (*i*) the development of a yellow coloration and (*ii*) the formation of pyrroles in good yield by 1,3-dipolar cycloaddition of alkynes.[439]

A.3 Section 7.3

7.3.1. *1,3-Thiazol-5-ones* (105) (Section 7.3.1)

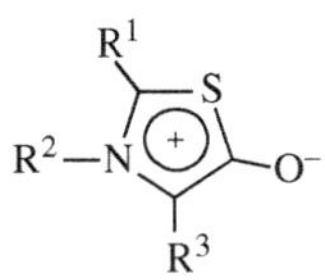

(105)

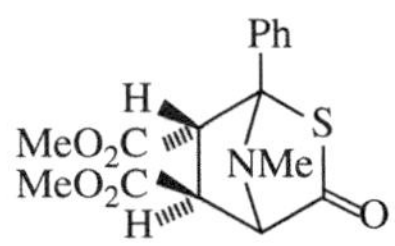

(486)

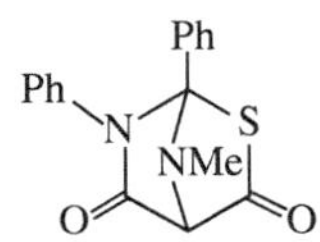

(487)

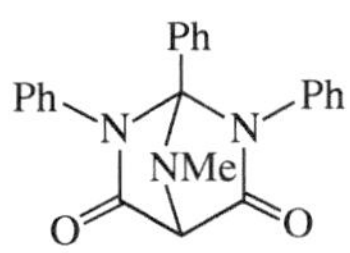

(488)

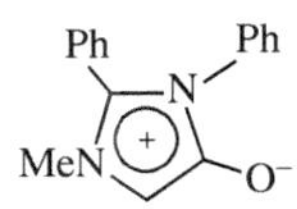

(489)

A full account of the preparation and 1,3-dipolar cycloaddition reactions of this class of meso-ionic heterocycle (**105**) has now been published.[440] Using olefinic dipolarophiles or heterocumulenes, a number of structurally interesting 1:1 bicyclic adducts have been prepared. For example, with dimethyl fumarate in benzene compound **105**, $R^1{=}Ph$, $R^2{=}Me$, $R^3{=}H$, gave the adduct **486**. The same derivative with phenyl isocyanate at 100° gave the primary adduct **487** (11%) and also the adduct **488** (48%). Compound **488** is formed by addition to the meso-ionic 1,3-diazol-4-one (**489**) (Section 7.2.1) which is almost certainly generated by loss of carbonyl sulfide from the adduct **487**.[440]

7.3.4. *1,3-Thiazol-4-ones* (114) (Section 7.3.4)

A full account of the preparation and 1,3-dipolar cycloadditions of the meso-ionic 1,3-thiazol-4-ones (**114**) has now been published.[441]

A detailed chemical and spectroscopic study of the stereochemistry of the 1:1 adducts of the 1,3-thiazol-4-ones (**114**) with olefinic 1,3-dipolarophiles has been reported.[442]

A.4 Section 7.5

7.5.1. *1,3,4-Oxadiazol-2-ones* (146) (Section 7.5.1)

The kinetics of the acid hydrolysis of isosydnones (**146**) to *N*-acylhydrazines (**147**, X = O) has been examined in detail and the mechanism has been discussed.[443,444]

A.5 Section 7.7

7.7.1. *1,2,3-Triazol-4-ones* (176) (Section 7.7.1)

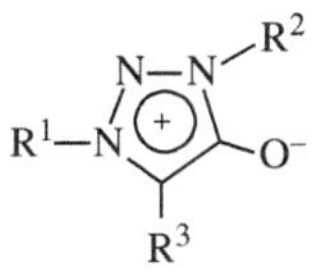

(176)

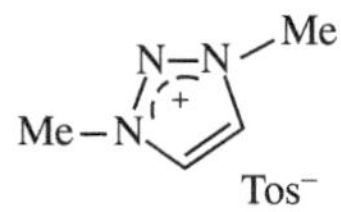

(490)

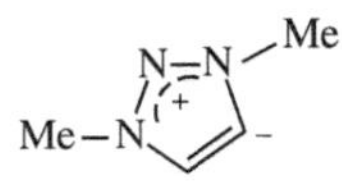

(491)

Treatment of 1,3-dimethyl-1,2,3-triazolium tosylate (**490**) with sodium hydride in dimethylformamide gives the ylide (**491**). Oxidation of this ylide (**491**) using oxygen and cupric chloride catalyst gave the meso-ionic 1,2,3-triazol-4-one **176**, $R^1 = R^2 = Me$, $R^3 = H$.[445]

7.7.3. *1,2,3-Triazole-4-thiones* (**196**) (Section 7.7.3)

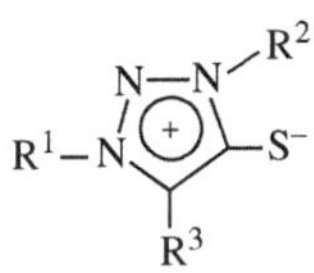

(196)

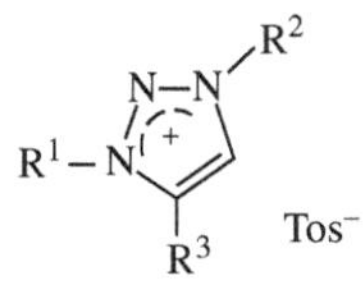

(492)

A new and convenient method of preparing the meso-ionic 1,2,3-triazole-4-thiones (**196**) has been reported. This involves the direct thiation of 1,3-disubstituted 1,2,3-triazolium tosylates (**492**) using sulfur and sodium hydride in dimethylformamide.[446]

7.7.6. *1,2,4-Triazole-3-thiones* (**227**) (Section 7.7.6)

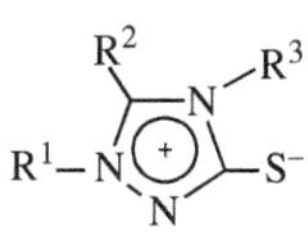

(227)

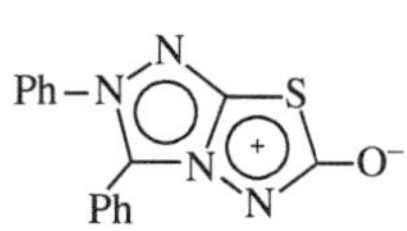

(493)

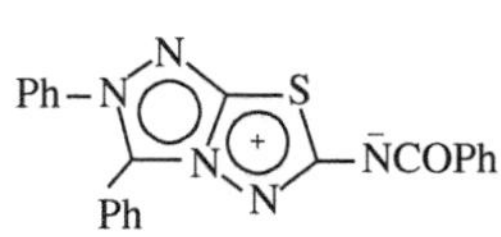

(494)

Treatment of the meso-ionic 1,2,4-triazole-3-thione, **227**, $R^1 = R^2 = Ph$, $R^3 = NH_2$, with phosgene gives a novel bicyclic heterocyclic system which can be represented as a bicyclic meso-ionic 1,3,4-thiadiazol-2-one (**493**) (Section 7.8.1). A similar reaction using *N*-benzoyl isocyanide dichloride ($PhCO.N{\equiv}CCl_2$) gives the novel meso-ionic 1,3,4-thiadiazol-2-imine (**494**) (Section 7.8.2).[447]

A.6 Section 7.8

7.8.3. *1,3,4-Thiadiazole-2-thiones* (**251**) (Section 7.8.3)

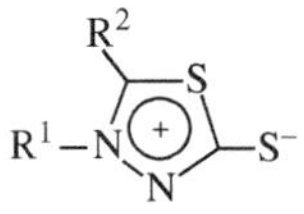

(251)

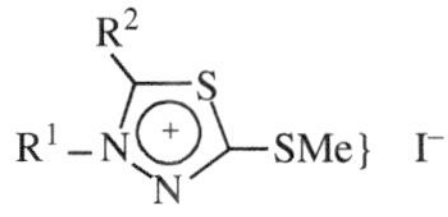

(495)

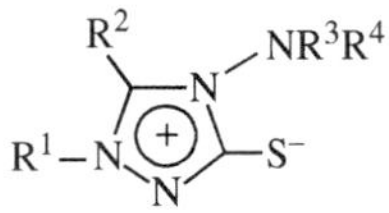

(496)

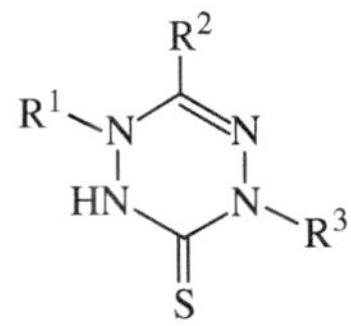

(497)

Predictably, treatment of the methiodides (**495**) with 1,1-disubstituted hydrazines ($R^3R^4N.NH_2$) gives the meso-ionic 1,2,4-triazole-3-thiones (**496**). An interesting variation is observed when the iodides (**495**) are treated with monosubstituted hydrazines ($R^3NH.NH_2$). In this case the product is a tetrazine (**497**).[448]

A.7 Section 7.11. Oxathioles

7.11.1. *1,3-Oxathiol-4-ones* (*Anhydro-4-hydroxy-1,3-oxathiolium Hydroxides*) (**498**)

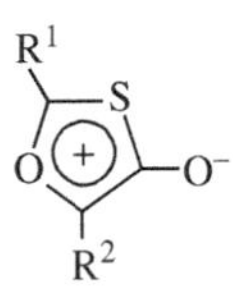

(498) (See Table A-I)

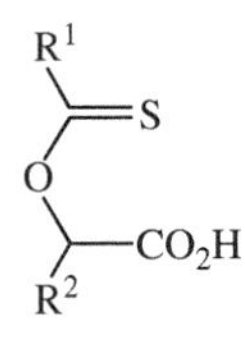

(499)

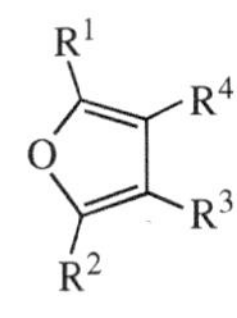

(500)

The meso-ionic 1,3-oxathiol-4-ones (**498**) are a new class of meso- ionic heterocycle which appear to be too unstable to be isolated at room temperature. They can be generated by cyclodehydration of the acids **499**, $R^1{=}NR_2$ $R^2{=}Ph$, using acetic anhydride, and trapped *in situ* by 1,3-dipolar cycloaddition with acetylenes; the product is a furan (**500**) formed via the bicyclic adduct (**501**).[449]

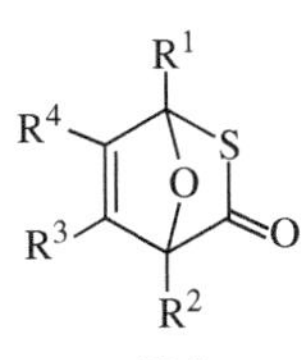

(501)

7.11.2. *1,3-Oxathiol-5-ones (Anhydro-5-hydroxy-1,3-oxathiolium Hydroxides)* (**502**)

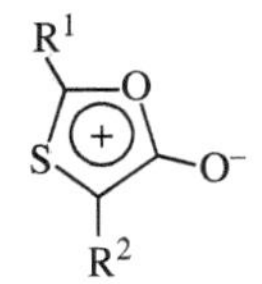

(502) (See Table A-I)

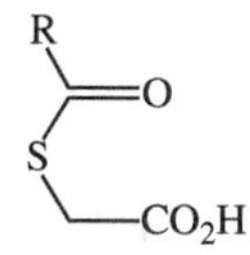

(503)

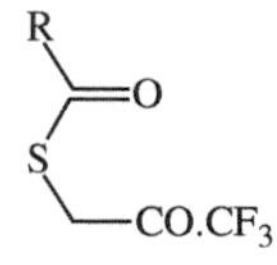

(504)

Representatives of this new type of meso-ionic system (**502**) have been prepared by cyclodehydration of the acid derivatives **503** using trifluoroacetic anhydride.[450] For example, the acid **503**, R=*p*.ClC_6H_4, gave the red crystalline meso-ionic 1,3-oxathiol-5-one (**502**, R^1 =p.Cl.C_6H_4, R^2=CF_3.CO) which is readily hydrolysed to the ketone (**504**, R=*p*.Cl.C_6H_4). The trifluoroacetyl group in the 4-position appears to be essential for the stability of the meso-ionic system (**502**).

A.8 Section 7.13. Selenazoles

7.13.1. *1,3-Selenazol-4-ones (Anhydro-4-hydroxy-1,3-selenazolium Hydroxides)* (**505**)

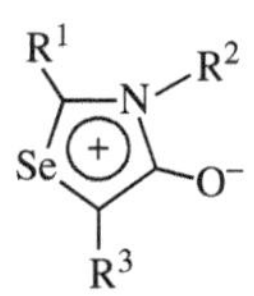

(505) (See Table A-I)

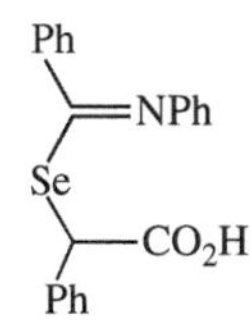

(506)

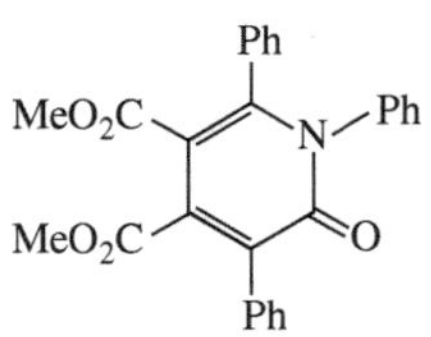

(507)

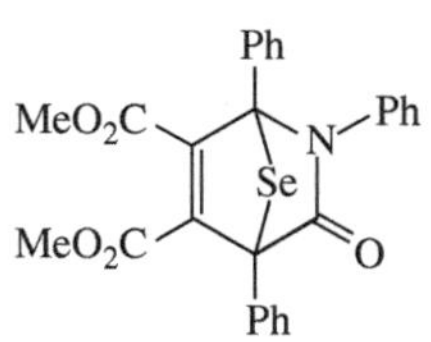

(508)

The α-selenoacid (**506**) with acetic anhydride-triethylamine at room temperature gave the meso-ionic 1,3-selenazol-4-one (**505**, R^1=R^2=R^3=Ph) as magenta needles in 80% yield.[451] This new meso-ionic system (**505**) appears to be only weakly reactive as a 1,3-dipole. When a benzene solution was heated under reflux with dimethyl acetylenedicarboxylate for one week, the pyridone diester (**507**) was formed. This product (**507**) is presumably formed by loss of selenium from the 1,3-dipolar cycloadduct (**508**).

It is interesting to compare the mode of reaction of **505** with alkynes with that of the sulfur analog (**114**) whose mode of reaction depends upon the nature of the substituent in position 5 (see Section 7.3.4).

A.9 Section 10.5

10.5.3. *1,2,3,4-Tetrazole-5-thiones* (**413**) (Section 10.5.3)

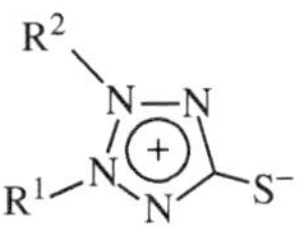

(413)

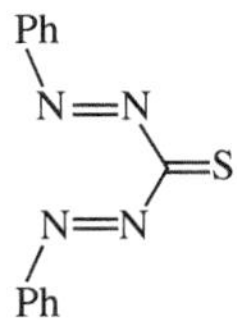

(415)

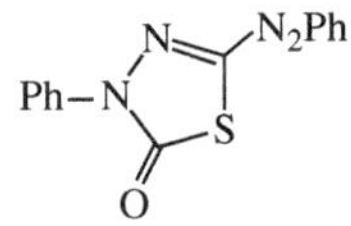

(509)

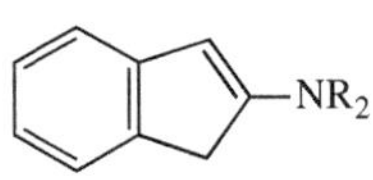

(510)

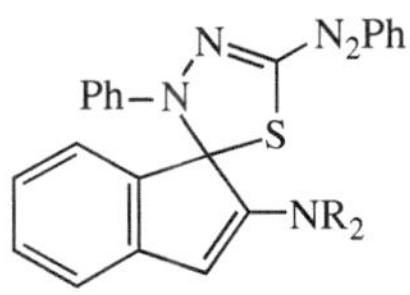

(511)

Two unusual transformations of dehydrodithizone (**413**, $R^1=R^2=Ph$) have been reported,[452,453] and the structures of the products determined by X-ray crystallography. Both these reactions suggest that the dehydrodithizone (**413**, $R^1=R^2=Ph$) may well be reacting as its valence tautomer (**415**).

The first study[452] has shown that dehydrodithizone (**413**, $R^1=R^2=Ph$) reacts with pentacarbonyliron to give the 1,3,4-thiadiazol-2-one (**509**).

The second[453] has shown that ynamines of the general type **510** (NR_2=pyrrolidino, piperidino, morpholino) combine with dehydrodithizone (**413**, $R^1=R^2=Ph$) in chloroform at room temperature giving the red crystalline *spiro* 1*H*-indene-1,2′(3′*H*)-1,3,4-thiadiazoles (**511**). The mode of formation of **511** is not clear but a reasonable mechanism would appear to require participation by the acyclic valence tautomer (**415**).

A.10 Section 11.1

The ionization potentials of four sydnones (**1**) have been compared with those of glycines and *N*-nitrosoglycines.[454] The results suggest that in the gas phase the sydnones (**1**) exist as the nitrosoketen isomer (**47**).

A.11 Section 11.4

11.4.1. *1,2,3-Oxadiazol-5-ones (Sydnones)* (**1**) (Section 11.4.1)

A carbon-13 NMR study of four sydnones (**1**) has been reported.[455] In 4-unsubstituted sydnones (**1**, R^1 = Me, Ph or $PhCH_2$, R^2 = H), the chemical shift of C-4 is substantially higher than has been measured for carbon in other unsubstituted heterocycles. This shielding of C-4 in the sydnone molecule is consistent with the results of *ab initio* and semi-empirical calculations which predict a high electron density at C-4.

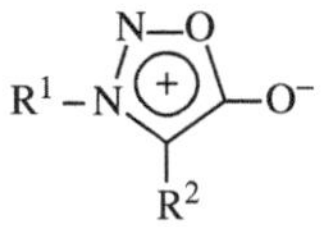

(1)

A further CNDO study of 3-methylsydnone (**1**, R^1 = Me, R^2 = H) has been reported.[456] The position of protonation and the mechanism of alkaline cleavage of 3-methylsydnone (**1**, R^1 = Me, R^2 = H) has been discussed in terms of CNDO/2 calculations.[457]

11.4.2. *Other meso-ionic systems* Section 11.4.2)

PPP and CNDO/2 calculations have been reported for the 1,3-dithiol-4-ones (**134**) (Section 7.4.1) and the 1,2-dithiol-4-ones (**338**) (Section 10.4.1).[458]

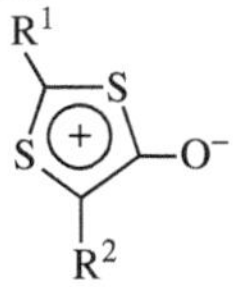

(134)

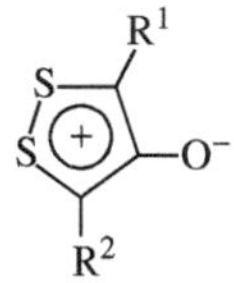

(388)

A.12 Section 12

12.1. *1,2,3-Oxadiazol-5-ones (Sydnones)* (**1**) (Section 12.1)

New cephalosporin derivatives of the sydnones (**1**) have been prepared[459] and some sydnones (**1**) useful as anti-inflammatory drugs have been described.[460,461]

For a series of biologically active substituted 3-phenylsydnones, quantitative structure–activity relationships have been described.[462]

12.2. *1,2,3-Oxadiazol-5-imines (Sydnone Imines)* (**2**) (Section 12.2)

Further studies on the preparation and activity of sydnofen (**451**)[463] and sydnocarb (**452**)[464,465] have been reported. A number of sydnone imine derivatives have also been described as useful hypotensive agents,[466] vasodilators,[466] muscle relaxants,[466] and monoamine oxidase inhibitors.[467]

12.7. *1,2,4-Triazol-3-ones* (**200**) (Section 12.7)

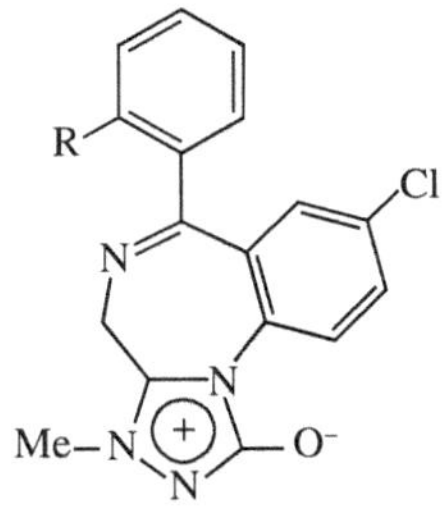

(512)

The derivatives **512** have been described as tranquilizers.[468]

A.13 Section 13

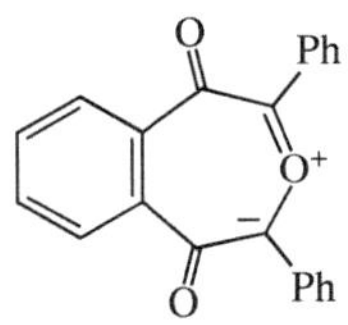

(513)[469]

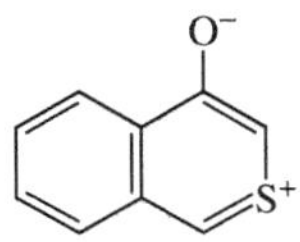

(514)[470]

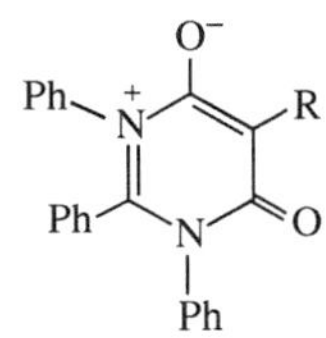

(515)[471]

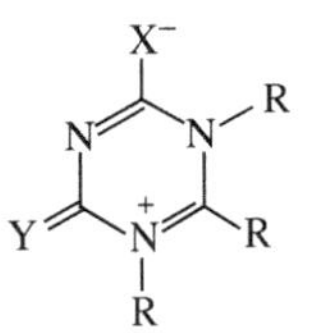

(516a) X = Y = O[472]
(516b) **X** = O, Y = S
(516c) X = Y = S

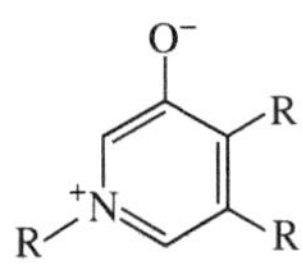

(517)[473]

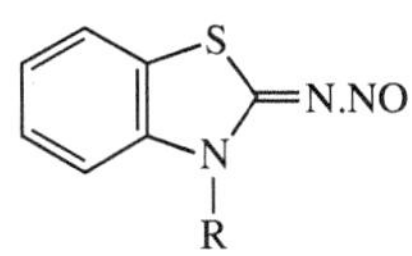

(518)[474]

The compounds **513–517** have recently been described as meso-ionic.[469–473] According to our modified definition of the term "meso-ionic" (Section 13) these interesting compounds (**513–517**) should now be described as mesomeric betaines.

The heterocyclic nitrosoimines (**518**) have been described as meso-ionic.[474] These compounds (**518**) are represented *satisfactorily* by a covalent structure (**518**) and are *not* therefore meso-ionic.

A.14 References added in proof (January 1976)

The following references[475–489] extend the literature coverage up to December 31, 1975. The preparation[475] of derivatives of one previously unknown type A meso-ionic system has come to the attention of the authors since the completion of the Appendix. These new meso-ionic compounds[475] are 1,2,3-thiadiazole-5-ones [Table A1 (page 111): a=S, b=N, c=NR, d=CR, e=C, f=O]. They are the first examples of sulfur analogs of the sydnones (**1**).

References

1. Baker, W.; Ollis, W. D.; Poole, V. D. *J. Chem. Soc.* **1950**, *1542* (1949), 307.
2. (a) Baker, W.; Ollis, W. D. *Chem Ind. (London)* **1955**, 910; (b) Baker, W.; Ollis, W. D. *Quart. Rev.* **1957**, *11*, 15.
3. (a) Huisgen, R. *Angew. Chem., Int. Ed. Engl.* **1963**, ***2***, 565; (b) Huisgen, R. *Chem. Soc. Spec. Publ.* **1967**, *21*, 51; (c) Huisgen, R. *Rev. Real Acad. Cienc. Exactas, Fis. Natur. Madrid* **1971**, ***65***, 293 [*CA* **75**, 109421n (1971)].
4. Kier, L. B.; Roche, E. B. *J. Pharm. Sci.* **1967**, *56*, 149. For discussion of more recent papers and additional references, see later section.
5. Zuman, P. *Chem. Listy* **1959**, *53*, 1029.
6. (a) Noël, Y. *Bull. Soc. Chim. Fr.* **1964**, 173; (b) Kubbinga, H. H. *Chem. Tech. (Amsterdam)* **1973**, ***28***, 4 [*CA* **78**, 111165n (1973)].
7. Stewart, F. H. C. *Chem. Rev.* **1964**, *64*, 129.
8. Suciu, N. *Stud. Cercet. Chim.* **1968**, *16*, 117.
9. (a) Ohta, M.; Kato, H. *Nippon Kagaku Zasshi* **1965**, ***86***, 661 [*CA* **65**, 2247h (1966)]; (b) Ohta, M.; Kato, H. In *Nonbenzenoid Aromatics*; Snyder, J. P., Ed.; Academic Press: New York, 1969; pp. 117–248; (c) Ohta, M. *Yuki Gosei Kagaku Kyokai Shi* **1970**, ***28***, 281 [*CA* **72**, 131633t (1970)].
10. Baker, W. *Proc. Chem. Soc.* **1959**, 75. in "Perspectives in Organic Chemistry" (Sir Alexander Todd, ed.), pp. 28–67. Interscience, New York, 1956.
11. Boulton, A. J.; Katritzky, A. R. *Tetrahedron* **1961**, *12*, 41.
12. Hetzheim, A.; Möckel, K. *Advan. Heterocyc. Chem.* **1966**, 7, 193.
13. Katritzky, A. R.; Lagowski, J. M. *Chemistry of the Heterocyclic N-Oxides*; Academic Press: New York, 1971; Ochiai, E. *Aromatic Amine Oxides*; Elsevier: Amsterdam, 1967.
14. Volodarsky, L. B.; Lisack, A. N.; Koptyug, V. A. *Tetrahedron Lett.* **1965**, 1565.
15. Petersen, S.; Heitzer, H. *Angew. Chem., Int. Ed. Engl.* **1970**, *9*, 67.
16. Bauer, H.; Boulton, A. J.; Fedeli, W.; Katritzky, A. R.; Majid-Hamid, A.; Mazza, F.; Vaciago, A. *Angew. Chem., Int. Ed. Engl.* **1971**, *10*, 129.
17. (a) Angadiyavar, C. S.; Sukumaran, K. B.; George, M. V. *Tetrahedron Lett.* **1971**, 633; (b) Middleton, W. J.; Engelhardt, V. A. *J. Amer. Chem. Soc.* **1938**, *80*, 2788; (c) Hayman, D. F.; Jackman, G. B.; Petrow, V.; Stephenson, O.; Wild, A. M. *J. Pharm. Pharmacol.* **1964**, *16*, 677; (d) Zimmerman, H. E. *Angew. Chem., Int. Ed. Engl.* **1969**, *8*, 1; (e) Cameron, A. F.; Hair, N. J. Morris, D. G.; Hawley, D. M. *Chem. Commun.* **1971**, 725; (f) CA 75, 81352q (1971);; (g) CA 75, 54K (1971).
18. Dewar, M. J. S. *The Molecular Orbital Theory of Organic Chemistry*; McGraw-Hill: New York, 1969; p. 194.
19. (a) McCarthy, A. R.; Ollis, W. D.; Barnes, A. N. M.; Sutton, L. E.; Ainsworth, C. *J. Chem. Soc. B* **1969**, 1185; (b) Talukdar, P. B.; Sengupta, S. K.; Datta, A. K. *J. Indian Chem. Soc.* **1971**, *48*, 591.
20. McCarthy, A. R.; Ollis, W. D.; Ramsden, C. A. *J. Chem. Soc., Perkin Trans.I* **1974**, 624.
21. Huisgen, R.; Funke, E.; Schaefer, F. C.; Knorr, R. *Angew. Chem., Ed. Engl.* **1967**, *6*, 367; Funke, E.; Huisgen, R. *Chem. Ber.* **1971**, *104*, 3222.

22. Moriarty, R. M.; Mukherjee, R.; Chapman, O. L.; Eckroth, D. R. *Tetrahedron Lett.* **1971**, 397.
23. Inouye, K.; Sato, S.; Ohta, M. *Bull. Chem. Soc. Jap.* **1970**, *43*, 1911.
24. Schönberg, A.; Frese, E. *Tetrahedron Lett.* **1969**, 4063. Chem. Ber. 103, 3885 (1970).
25. Fischer, E.; Besthorn, E. *Ann.* **1882**, *212*, 316.
26. Ramakrishna, R. S.; Irving, H. M. N. H. *Chem. Commun.* **1969**, 1356.
27. Lawson, A.; Miles, D. H. *Chem. Ind. (London)* **1958**, 461.
28. Lawson, A.; Miles, D. H. *J. Chem. Soc.* **1960**, (1959), 2865, 1945.
29. Huisgen, R.; Gotthardt, H.; Bayer, H. O.; Schaefer, F. C. *Angew. Chem., Ed. End.* **1964**, *3*, 136.
30. Bayer, H. O.; Huisgen, R.; Knorr, R.; Schaefer, F. C. *Chem. Ber.* **1970**, *103*, 2581.
31. Knorr, R.; Huisgen, R. *Chem. Ber.* **1970**, *103*, 2598.
32. Singh, G.; Singh, S. *Tetrahedron Lett.* **1964**, 3789.
33. Burrows, W. D. *J. Org. Chem.* **1966**, *31*, 3435.
34. Ohta, M.; Mase, T. *Nippon Kagaku Zasshi* **1968**, *89*, 714 [CA 70, 11603v (1969)].
35. Greco, C. V.; Gray, R. P.; Grosso, V. G. *J. Org. Chem.* **1967**, *32*, 4101.
36. Boyd, G. V. *Chem. Commun.* **1968**, 1410; Boyd, G. V.; Wright, P. H. *J. Chem. Soc., Perkin Trans. I* **1972**, 909–914.
37. Boyd, G. V.; Wright, P. H. *Chem. Commun.* **1969**, 182.
38. Boyd, G. V.; Wright, P. H. *J. Chem. Soc. C* **1970**, 1485; Boyd, G. V. In *Aromaticity, Pseudo-aromaticity, Anti-aromaticity*; Bergmann, E. D., Pullman, B., Eds.; Israel Acad. Sci. Humanities: Jerusalem, 1971; p. 166.
39. Gotthardt, H.; Huisgen, R.; Schaefer, F. C. *Tetrahedron Lett.* **1964**, 487.
40. Huisgen, R.; Gotthardt, H.; Bayer, H. O.; Schaefer, F. C. *Chem. Ber.* **1970**, *103*, 2611.
41. Gotthardt, H.; Huisgen, R. *Chem. Ber.* **1970**, *103*, 2625.
42. Knorr, R.; Huisgen, R.; Staudinger, G. K. *Chem. Ber.* **1970**, *103*, 2639.
43. Huisgen, R.; Funke, E.; Gotthardt, H.; Panke, H.-L. *Chem. Ber.* **1970**, *104*, 1532.
44. Funke, E.; Huisgen, R.; Schaefer, F. C. *Chem. Ber.* **1971**, *104*, 1550.
45. Brunn, E.; Funke, E.; Gotthardt, H.; Huisgen, R. *Chem. Ber.* **1971**, *104*, 1562.
46. (a) Potts, K. T.; Singh, U. P. *Chem. Commun.* **1969**, 66; (b) Potts, K. T.; McKeough, D. *J. Amer. Chem. Soc.* **1973**, *95*, 2749; (c) Potts, K. T.; Baum, J. *J. Chem. Soc., Chem. Commun.* **1973**, 833; (d) Matsukubo, H.; Kato, H. *J. Chem. Soc., Chem. Commun.* **1974**, 412. *J. Chem. Soc., Perkin Trans. I*, 632 (1975); T. Eicher and V. Schäfer, *Tetrahedron* **30**, 4025 (1974); (e) Martin, H.-D.; Hekman, M. *Angew. Chem., Ed. Engl.* **1972**, *11*, 926.
47. Kille, G.; Fleury, J.-P. *Bull. Soc. Chim. Fr.* **1968**, 4636.
48. Gotthardt, H.; Huisgen, R.; Bayer, H. O. *J. Amer. Chem. Soc.* **1970**, *92*, 4340.
49. Bayer, H. O.; Gotthardt, H.; Huisgen, R. *Chem. Ber.* **1970**, *103*, 2356.
50. Huisgen, R.; Gotthardt, H.; Bayer, H. O. *Chem. Ber.* **1970**, *103*, 2368.
51. (a) Mohan, S.; Kumar, B.; Sandhu, J. S. *Chem. Ind. (London)* **1971**, 671; (b) Potts, K. T.; Marshall, J. *J. Chem. Soc. Chem. Commun.* **1972**, 1000.
52. Cobb, R. L.; McEwen, W. E. *J. Amer. Chem. Soc.* **1955**, 77, 5042.
53. McEwen, W. E.; Mineo, I. C.; Shen, Y. H.; Han, G. Y. *Tetrahedron Lett.* **1968**, 5157; McEwen, W. E.; Stott, P. E.; Zepp, C. M. *J. Amer. Chem. Soc.* **1973**, ***95***, 8452.
54. Sato, S.; Mase, T.; Ohta, M. *Bull. Chem. Soc. Jap.* **1968**, *41*, 2218.
55. Roesler, P.; Fleury, J.-P. *Bull. Soc. Chim. Fr.* **1968**, 631.
56. Götz, M.; Zeile, K. *Tetrahedron* **1970**, *26*, 3185.
57. Chinone, A.; Sato, S.; Mase, T.; Ohta, M. *Bull. Chem. Soc. Jap.* **1969**, *42*, 2310.
58. (a) Roche, E. B.; Stansloski, D. W. *J. Heterocyc. Chem.* **1970**, 7, 139; (b) Stele, J. W.; Anderson, W. K.; Trager, W. F. *Can. J. Chem.* **1972**, *50*, 1026.
59. (a) Shiba, T.; Kato, H. *Bull. Chem. Soc. Jap.* **1970**, *43*, 3941; (b) Grashey, R.; Jänchen, E.; Litzke, J. *Chem. Ztg.* **1973**, *97*, 657; (c) Singh, G.; Pande, P. S. *Tetrahedron Lett.* **1974**, 2169.

60. Potts, K. T.; Husain, S.; Husain, S. *Chem. Commun.* **1970**, 1360; Potts, K. T.; Husain, S. *J. Org. Chem.* **1971**, *36*, 3368.
61. Chinone, A.; Sato, S.; Ohta, M. *Bull. Chem. Soc. Jap.* **1971**, *44*, 826.
62. Huisgen, R.; Funke, E.; Schaefer, F. C.; Gotthardt, H.; Brunn, E. *Tetrahedron Lett.* **1967**, 1809.
63. Potts, K. T.; Roy, D. N. *Chem. Commun.* **1968**, 1062.
64. Lawson, A.; Searle, C. E. *J. Chem. Soc.* **1957**, 1556.
65. Ohta, M.; Chosho, H.; Shin, C.-G.; Ichimura, K. *Nippn Kagaku Zasshi* **1964**, ***85***, 440 [*CA* **61**, 14657h (1964)]; Ohta, M.; Shin, C. *Bull. Chem. Soc. Jap.* **1965**, ***38***, 704.
66. Potts, K. T.; Houghton, E.; Husain, S. *Chem. Commun.* **1970**, 1025.
67. (a) Potts, K. T.; Roy, D. N. *Chem. Commun.* **1968**, 1061; (b) Shiba, T.; Kato, H. *Bull. Chem. Soc. Jap.* **1971**, *44*, 1864; (c) 46, 964 (1973).
68. Barrett, G. C.; Khokhar, A. R.; Chapman, J. R. *Chem. Commun.* **1969**, 818.
69. Kröhnke, F.; Steuemagel, H. H. *Angew. Chem.* **1961**, *73*, 26.
70. Kröhnke, F.; Steuemagel, H. H. *Chem. Ber.* **1964**, *97*, 1118.
71. Huisgen, R.; Grashey, R.; Steingruber, E. *Tetrahedron Lett.* **1963**, 1441.
72. (a) Baldwin, J. E.; McDaniel, M. C.; Newton, M. G.; Paul, I. C. *Tetrahedron Lett.* **1966**, 4239; (b) Talukdar, P. B.; Sengupta, S. K.; Datta, A. K. *J. Chem. Soc., Chem. Commun.* **1972**, 696. *Indian J. Chem.* 11, 1257 (1973); [*CA* **80**, 133344p (1974)].
73. Ohta, M.; Chosho, H.; Shin, C.-G.; Ichimura, K. *CA* 61, 14657h (1964). *Nippon Kagaku Zasshi* **1964**, *85*, 440.
74. Potts, K. T.; Singh, U. P.; Houghton, E. *Chem. Commun.* **1969**, 1128.
75. (a) Ohta, M.; Sato, S. *J. Chem. Soc. Jap.* **1968**, *89*, 199; (b) Duffin, G. F.; Kendall, J. D. *J. Chem. Soc.* **1956**, 361. E. B. Knott, *ibid.*, 916, 937 (1955); (c) Hashimoto, M.; Ohta, M. *Bull. Chem. Soc. Jap.* **1960**, ***33***, 1394 [*CA* **55**, 15470c (1961)]; (d) Kochergin, P. M. *Zh. Obshch. Khim.* **1961**, ***31***, 3257. [*CA* **57**, 2208c (1962)]; M. Ohta and K. Kishimoto, *Bull. Chem. Soc. Jap.* **34**, 1402 (1961); (f) Salmond, W. G. *Quart. Rev., Chem. Soc.* **1968**, *22*, 272.
76. Potts, K. T.; Houghton, E.; Singh, U. P. *Chem. Commun.* **1969**, 1129; Potts, K. T.; McKeough, D. *J. Amer. Chem. Soc.* **1973**, ***95***, 2750.
77. (a) Nakazawa, S.; Kiyosawa, T.; Hirakawa, K.; Kato, H. *J. Chem. Soc., Chem. Commun.* **1974**, 621; (b) Takayanagi, Z.; Kato, H.; Ohta, M. *Bull. Chem. Soc. Jap.* **1967**, *40*, 2930.
78. (a) Abrahamsson, S.; Westerdahl, A.; Isaksson, G.; Sandström, J. *Acta Chem. Scand.* **1967**, *21*, 442; (b) Buchardt, O.; Domanus, J.; Harrit, N.; Holm, A.; Isaksson, G.; Sandström, J. *J. Chem. Soc., Chem. Commun.* **1974**, 376.
79. Koenigs, E.; Geisler, H. *Ber.* **1924**, *57*, 2076.
80. Tschitschibabin, A. E.; Woroshtzow, N. N. *Ber.* **1933**, *66*, 364.
81. Duffin, G. F.; Kendall, J. D. *J. Chem. Soc.* **1951**, 734.
82. Kato, H.; Tanaka, K.; Ohta, M. *Bull. Chem. Soc. Jap.* **1962**, *35*, 1901.
83. Kato, H.; Tanaka, K.; Ohta, M. *Bull. Chem. Soc. Jap.* **1966**, *39*, 1248.
84. Ohta, M.; Yoshida, K.; Sato, S. *Bull. Chem. Soc. Japan.* **1966**, *39*, 1269.
85. Chosho, H.; Ichimura, K.; Ohta, M. *Bull. Chem. Soc. Jap.* **1964**, *37*, 1670.
86. Ichimura, K.; Ohta, M. *Bull. Chem. Soc. Jap.* **1965**, *38*, 707.
87. Sato, S.; Ohta, M. *Bull. Chem. Soc. Jup.* **1968**, *41*, 2801.
88. Chinone, A.; Huseya, Y.; Ohta, M. *Bull. Chem. Soc. Jap.* **1970**, *43*, 2650.
89. Kato, H.; Ohta, M. *Bull. Chem. Soc. Jap.* **1966**, *39*, 1253.
90. Ohta, M.; Sugiyama, M. *Bull. Chem. Soc. Jap.* **1965**, *38*, 596.
91. Potts, K. T.; Singh, U. P. *Chem. Commun.* **1969**, 569.
92. Gotthardt, H.; Christl, B. *Tetrahedron Lett.* **1968**, 4743.
93. Gotthardt, H.; Christl, B. *Tetrahedron Lett.* **1968**, 4747.
94. Gotthardt, H.; Christl, B. *Tetrahedron Lett.* **1968**, 4751.
95. Kato, H.; Kawamura, M.; Shiba, T.; Ohta, M. *Chem. Commun.* **1970**, 959.
96. Ohta, M.; Sugiyama, M. *Bull. Chem. Soc. Jap.* **1963**, *36*, 1437.

97. (a) Kato, H.; Shiba, T.; Yoshida, H.; Fujimori, S. *Chem. Commun.* **1970**, 1591; (b) Kato, H.; Shiba, T.; Miki, Y. *J. Chem. Soc., Chem. Commun.* **1972**, 498.
98. Hashimoto, M.; Ohta, M. *Bull. Chem. Soc. Jap.* **1961**, *34*, 668.
99. Roche, E. B.; Kier, L. B. *J. Pharm. Sci.* **1965**, *54*, 1700.
100. Ainsworth, C. *Can. J. Chem.* **1965**, *43*, 1607.
101. McCaustland, D. J.; Burton, W. H.; Cheng, C. C. *J. Heterocycl. Chem.* **1971**, *8*, 89.
102. Talukdar, P. B.; Banerjee, S.; Chakraborty, A. *Indian J. Chem.* **1972**, *10*, 610.
103. McCarthy, A. R.; Ollis, W. D.; Ramsden, C. A. *Chem. Commun.* **1968**, 499. *J. Chem. Soc. Perkin Trans. I*, 627 (1974).
104. Huisgen, R.; Gotthardt, H.; Grashey, R. *Chem. Ber.* **1968**, *101*, 536.
105. (a) Hoegerle, K. *Helv. Chim. Acta* **1958**, *41*, 548; (b) Balasubramanian, A.; McIntosh, J. M.; Snieckus, V. *J. Org. Chem.* **1970**, *35*, 433.
106. Mackay, D.; Wong, L. L. *J. Chem. Soc., Chem. Commun.* **1974**, *621*.
107. (a) Ollis, W. D.; Ramsden, C. A. *Chem. Commun.* **1971**, 1223. *J. Chem. Soc., Perkin Trans. I*, 642 (1974); (b) Grashey, R.; Hamprecht, R.; Keramaris, N.; Baumann, M. *Tetrahedron Lett.* **1972**, 2943; (c) Lazaris, A. Y.; Shmuilovich, S. M.; Egorochkin, A. N. *Khim. Geterotsikl. Soedin.* **1973**, 1345 [CA **80**, 271752 1974)1; A. Ya. Lazaris and A. N. Egorochkin, ibid., 648 (1975) [CA **83**, 79164b (1975)].
108. Hanley, R. N.; Ollis, W. D.; Ramsden, C. A.; Rowlands, G. R.; Sutton, L. E. *J. Chem. Soc., Perkin Trans. I* **1978**, 600.
109. Sandstròm, J.; Wennerbeck, I. *Acta Chem. Scand.* **1966**, *20*, 57.
110. Lazaris, A. Y. *J. Org. Chem. USSR* **3** **1967**, 1856 [CA 68, 12910p (1968)].
111. Lazaris, A. Y. *J. Org. Chem. USSR* **1968**, *4*, 1786 [CA 70, 19992s (1969)].
112. (a) Grashey, R.; Keramaris, N.; Baumann, M. *Tetrahedron Lett.* **1970**, 5087; (b) Lazaris, A. Y.; Egorochkin, A. N. *J. Org. Chem. USSR* **1972**, *8*, 1569 [CA 77, 139907j (1972)].
113. Atkin, C. W.; Barnes, A. N. M.; Edgerley, P. G.; Sutton, L. E. *J. Chem. Soc. B* **1969**, 1194.
114. Grashey, R.; Baumann, M.; Hamprecht, R. *Tetrahedron Lett.* **1970**, 5083.
115. Boyd, G. V.; Harms, M. D. *J. Chem. Soc. C* **1970**, 807.
116. Bacchetti, T.; Alemagna, A. *Atti Accad. Naz. Lincei, Rend. Cl. Sci. Fis. Mat. Nat.* **1960**, ***28***, 646 [CA **55**, 9382c (1961)].
117. Andreetti, G. D.; Bocelli, G.; Cavalca, L.; Sgarabotto, P. *Gazz. Chim. Ital.* **1972**, ***102***, 23 [CA 77, 10803t (1972)].
118. (a) Gotthardt, H. *Tetrahedron Lett.* **1971**, *1277*, 1281. *Chem. Ber.* 105, 188, 196 (1972); (b) Alemagna, A.; Bacchetti, T. *Chim. Ind. (Milan)* **1972**, *54*, 1105 [CA 78, 97561e (1973)]; (c) Holm, A.; Harrit, N.; Bechgaard, K.; Buchardt, O.; Harnung, S. E. *J. Chem. Soc., Chem. Commun.* **1972**, 1125.
119. Scarpati, R.; Sica, D.; Lionetti, A. *Gazz. Chim. Ital.* **1963**, *93*, 90.
120. Begtrup, M.; Pedersen, C. *Acta Chem. Scand.* **1965**, *19*, 2022.
121. Begtrup, M.; Pedersen, C. *Acta Chem. Scand.* **1966**, *20*, 1555.
122. Begtrup, M.; Pedersen, C. *Acta Chem. Scand.* **1967**, *21*, 633.
123. Begtrup, M.; Hansen, K.; Pedersen, C. *Acta Chem. Scand.* **1967**, *21*, 1234.
124. Begtrup, M.; Pedersen, C. *Acta Chem. Scand.* **1969**, *23*, 1091.
125. Begtrup, M.; Kristensen, P. A. *Acta Chem. Scand.* **1969**, *23*, 2733. M. Begtrup and K. V. Poulsen, *ibid.* **25**, 2087 (1971).
126. Begtrup, M. *Acta Chem. Scand.* **1971**, ***25***. 249, 795, 803, 3500.
127. Potts, K. T.; Husain, S. *J. Org. Chem.* **1970**, *35*, 3451. **37**, 2049 (1972).
128. (a) Daeniker, H. U.; Druey, J. *Helv. Chim. Acta* **1962**, ***45***, 2441. H. U. Daeniker, *ibid.* **47**, 33 (1964); (b) Chinone, A.; Ohta, M. *Chem. Lett.* **1972**, 969.
129. Saito, Y.; Teraji, T.; Kamiya, T. *Tetrahedron Lett.* **1971**, 2893.
130. Begtrup, M. *Tetrahedron Lett.* **1971**, 1577. Acta Chem. Scand. **26**, 1243 (1972).
131. Potts, K. T.; Roy, S. K.; Jones, D. P. *J. Heterocycl. Chem.* **1965**, *2*, 105.

132. (a) Potts, K. T.; Roy, S. K.; Jones, D. P. *J. Org. Chem.* **1967**, *32*, 2245; (b) Grashey, R.; Baumann, M. *Tetrahedron Lett.* **1972**, 2947.
133. Palazzo, G.; Baiocchi, L. *Ann. Chim. (Rome)* **1966**, ***56***, 190 [*CA* **64**, 19613 (1966)]; Shimizu, B.; Saito, A. *Japan Kokai* **1974**, 7362, 796 [*CA* **80**, 3533p (1974)].
134. Boots, S. G.; Cheng, C. C. *J. Heterocycl. Chem.* **1967**, *4*, 272.
135. Grashey, R.; Baumann, M.; Lubos, W.-D. *Tetrahedron Lett.* **1968**, 5877.
136. Palazzo, G.; Baiocchi, L. *Ann. Chim. (Rome)* **1965**, ***55***, 935 [*CA* 63, 16335g (1965)].
137. Kato, H.; Sato, S.; Ohta, M. *Tetrahedron Lett.* **1967**, 4261.
138. Ota, M. *(Takeda Chem. Ind., Ltd.), Japan 70 07,741*; 1970 [CA 73, 14857d (1970)].
139. Potts, K. T.; Roy, S. K.; Schneller, S. W.; Huseby, R. M. *J. Org. Chem.* **1968**, *33*, 2559.
140. Palazzo, G.; Baiocchi, L. *Gazz. Chim. Ital.* **1966**, ***96***, 1020 [CA **66**, 115652u (1967)].
141. Evans, G. W.; Milligan, B. *Aust. J. Chem.* **1967**, *20*, 1779; Smith, R. F.; Deutsch, J. L.; Almeter, P. A.; Johnson, D. S.; Roblyer, S. M.; Rosenthal, T. C. *J. Heterocycl. Chem.* **1970**, 7, 671.
142. (a) Talukdar, P. B.; Sengupta, S. K.; Datta, A. K. *Indian J. Chem.* **1971**, ***9***, 179 [CA **75**, 27561u (1971)]; (b) **9**, 1018 (1971).
143. Ollis, W. D.; Ramsden, C. A. *Chem. Commun.* **1971**, 1224. *J. Chem. Soc., Perkin Trans. I*, 638 (1974).
144. Kühle, E.; Anders, B.; Zumach, G. *Angew. Chem., Int. Ed. Engl.* **1967**, *6*, 649.
145. Scott, F. L.; Cronin, D. A. *Chem. Ind. (London)* **1964**, 1757; Scott, F. L.; Cronin, J. A.; Donovan, J. *Tetrahedron Lett.* **1969**, 4615.
146. E. Cawkill, W. D. Ollis, C. A. Ramsden, and G. P. Rowson, *J. Chem. Soc., Perkin Trans. I*, 1979, 724.
147. Grout, R. J.; King, T. J.; Partridge, M. W. *Chem. Commun.* **1971**, 898.
148. Ohta, M.; Kato, H.; Kaneko, T. *Bull. Chem. Soc. Jap.* **1967**, *40*, 579.
149. (a) Ollis, W. D.; Ramsden, C. A. *Chem. Commun.* **1971**, 1222. *J. Chem. Soc., Perkin Trans. I*, 633 (1974); (b) Thieme, P.; Patsch, M.; Konig, H. *Ann.* **1972**, ***764***, 94. H. König, P. Thieme, and A. Amann (Badische Anilin-und Soda-Fabrik A.-G.) Ger. Offen. 2,147,025 [*CA* 78, 159618q (1973)].
150. Duffin, G. F.; Kendall, J. D.; Waddington, H. R. J. *J. Chem. Soc.* **1959**, 3799; Talukdar, P. B.; Sengupta, S. K.; Datta, A. K. *J. Indian Chem. Soc.* **1973**, *50*, 154.
151. Hinman, R. L.; Fulton, D. *J. Amer. Chem. Soc.* **1958**, *80*, 1895.
152. (a) Lazaris, A. Y.; Egorochkin, A. N. *Zh. Org. Khim.* **1970**, *6*, 2342 [CA 74, 42317s (1971)]; (b) Grashey, R.; Baumann, M.; Hamprecht, R. *Tetrahedron Lett.* **1972**, 2939.
153. Landquist, J. K. *J. Chem. Soc. C* **1970**, 323.
154. Lazaris, A. Y.; Shmuilovich, S. M.; Egorochkin, A. N. *Zh. Org. Khim.* **1972**, *8*, 2621 [CA **78**, 72015s (1973)].
155. Grashey, R.; Baumann, M. *Angew. Chem., Int. Ed. Engl.* **1969**, *8*, 133.
156. Potts, K. T.; Sapino, C. *Chem. Commun.* **1968**, 672.
157. Moriarty, R. M.; Mukherjee, R. *Tetrahedron Lett.* **1969**, 4627.
158. Kier, L. B.; Dodd, M. C.; Sapko, P.; Stewart, T. G. *Nature (London)* **1964**, *204*, 697.
159. (a) Stewart, T. G.; Kier, L. B. *J. Pharm. Sci.* **1965**, *54*, 731; (b) Kuftinec, J.; Kolbah, D. *Croat. Chem. Acta* **1971**, *43*, 73.
160. Weintraub, P. M.; Highman, F. E. *J. Org. Chem.* **1969**, *34*, 254.
161. Baker, W.; Ollis, W. D.; Phillips, A.; Strawford, T. *J. Chem. Soc.* **1951**, 289.
162. Talukdar, P. B.; Sengupta, S. K. *J. Indian Chem. Soc.* **1968**, *45*, 356.
163. Kier, L. B.; Scott, M. K. *J. Heterocycl. Chem.* **1968**, *5*, 277.
164. Grashey, R.; Baumann, M.; Lubos, W.-D. *Tetrahedron Lett.* **1968**, 5881.
165. Huisgen, R.; Grashey, R.; Krischke, R. *Tetrahedron Lett.* **1962**, 387.
166. Kiwan, A. M.; Irving, H. M. N. H. *Chem. Commun.* **1970**, 928. *J. Chem. Soc. B*, 901 (1971).

167. Talukdar, P. B.; Banerjee, S.; Chakraborty, A. C. *Bull. Chem. Soc. Jap.* **1970**, *43*, 125. *Indian J. Chem.* **9**, 827 (1971); P. B. Talukdar, S. K. Sengupta, and A. K. Datta, *ibid.* **9**, 1417 (1971); P. B. Talukdar, S. Banerjee, and A. C. Chakraborty, *ibid.* **10**, 929 (1972); P. B. Talukdar, S. K. Sengupta, and A. K. Datta, *ibid.* **10**, 1070 (1972); **11**, 753 (1973); *J. Indian Chem. Soc.* **50**, 154 (1973).
168. Moriarty, R. M.; Kliegman, J. M.; Desai, R. B. *Chem. Commun.* **1967**, 1255.
169. Talukdar, P. B.; Sengupta, S. K. *J. Indian Chem. Soc.* **1970**, *47*, 49.
170. Kiwan, A. M.; Irving, H. M. N. H. *J. Chem. Soc. B* **1971**, 898.
171. Worley, S. D. *Chem. Rev.* **1971**, *71*, 295.
172. Patsch, M.; Thieme, P. *Angew. Chem., Ini. Ed. Engl.* **1971**, *10*, 569.
173. Moriarty, R. M.; Kliegman, J. M.; Desai, R. B. *Chem. Commun.* **1967**, 1045.
174. Mosby, W. L.; Vega, M. L. *Chem. Commun.* **1971**, 837; Moriarty, R. M.; Chin, A. *J. Chem. Soc., Chem. Commun.* **1972**, 1300.
175. Moriarty, R. M.; Mukherjee, R.; Flippen, J. L.; Karle, J. *Chem. Commun.* **1971**, 1436; Flippen, J. L. *Acta Crystallogr. B* **1972**, ***28***, 2749.
176. Duffin, G. F.; Kendall, J. D. *J. Chem. Soc.* **1956**, 3189.
177. Cameron, D. P. *(Pfizer Inc.), U.S. Patent 3,580,921*; 1971 [CA **75**, 49095x (1971)].
178. Wiseman, E. H.; Cameron, D. P. *J. Med. Chem.* **1969**, *12*, 586.
179. Pacha, W.; Prijs, B. *Helv. Chim. Acta* **1958**, *41*, 521.
180. Talukdar, P. B.; Sengupta, S. K.; Datta, A. K. *Indian J. Chem.* **1971**, *9*, 1018.
181. Goerdeler, J.; Roth, W. *Chem. Ber.* **1963**, *96*, 534.
182. Boyer, J. H.; Canter, F. C. *J. Amer. Chem. Soc.* **1955**, *77*, 1280.
183. Thomas, T. L.; Fedorchuk, M.; Shetty, B. V.; Anderson, F. E. *J. Med. Chem.* **1970**, *13*, 196.
184. Boyer, J. H.; Hernandez, J. A. *J. Amer. Chem. Soc.* **1956**, *78*, 5124.
185. Hashimoto, M.; Ohta, M. *Bull. Chem. Soc. Jap.* **1962**, *35*, 766.
186. Widman, O. *Ber.* **1895**, *28*, 1925.
187. von Pechmann, H. *Ber.* **1896**, *29*, 2161.
188. Farrar, W. V. *J. Chem. Soc.* **1964**, 906.
189. Quilico, A. *Guzz. Chim. Itul.* **1932**, ***62***. 503, 912 (1932); **63**, 269 (1933); A. Quilico and R. Juston, *ibid.* **63**, 862 (1933); **65**,201 (1935); A. Quilko and M. Simonetta, *ibid.* **76**, 259 (1946).
190. Ponzio, G. *Guzz. Chim. Ztul.* **1915**, ***45*** (2), 12. **46**(2), 56 (1916); **63**, 471 (1933).
191. Hünig, S.; Boes, O. *Ann.* **1953**, *579*, 28.
192. Kier, L. B.; Al-Shamma, A.; Hahn, R.; Tye, A. *J. Pharm. Sci.* **1966**, *55*, 1467; Tubaro, E. *Boll. Chim. Farm.* **1966**, ***105***, 641 [CA **66**, 17790t (1967)].
193. Sundaram, K.; Purcell, W. P. *Int. J. Quant. Chem.* **1968**, *2*, 145.
194. Busch, M.; Becker, J. *Ber.* **1896**, *29*, 1686.
195. Busch, M.; Schmidt, W. *Ber.* **1929**, *62*, 1449.
196. Christopherson, C.; Treppendahl, S. *Acta Chem. Scand.* **1971**, *25*, 625.
197. Hanley, R. N.; Ollis, W. D.; Ramsden, C. A. *J. Chem. Soc., Perkin Trans. I* **1979**, 736.
198. Masuda, K.; Kamiya, T.; Kashiwa, K. *Chem. Pharm. Bull.* **1971**, *19*, 559.
199. Finnegan, W. G.; Henry, R. A. *J. Org. Chem.* **1965**, *30*, 567.
200. Hanley, R. N.; Ollis, W. D.; Ramsden, C. A. *J. Chem. Soc., Chem. Commun.* **1976**, 306; *J. Chem. Soc., Perkin Trans. I*, 732 (1979).
201. Ollis, W. D.; Ramsden, C. A. *J. Chem. Soc., Perkin Trans. I* **1974**, 645. R. N. Hanley, W. D. Ollis, and C. A. Ramsden, *ibid.*, 747 (1979).
202. Backer, H. J. *Rec. Trav. Chim. Pays-Bas* **1951**, *70*, 733.
203. Dubenko, R. G.; Neplyuev, V. M.; Pel'kis, P. S. *J. Org. Chem. USSR* **1966**, **2**, 506 [CA **65**, 7085b (1966)].
204. Phillips, R. R. *Org. React.* **1959**, *10*, 143.
205. Hatton, K.; Lieber, E.; Horwitz, J. P. *J. Amer. Chem. Soc.* **1956**, *78*, 411.

206. Tamura, Z.; Iitaka, Y.; Tanabe, H.; Uchiyama, S. *Chem. Pharm. Bull.* **1970**, ***18***, 2359 [ICA **74**, 58177g (1971)].
207. Hanley, R. N.; Ollis, W. D.; Ramsden, C. A. *J. Chem. Soc., Chem. Commun.* **1976**, 308; *J. Chem. Soc., Perkin Trans. I*, 741 (1979).
208. Bryden, J. H.; Henry, R. A.; Finnegan, W. G.; Boschan, R. H.; McEwan, W. S.; Van Dolah, R. W. *J. Amer. Chem. Soc.* **1953**, *75*, 4863. R. A. Henry, W. G. Finnegan, and E. Lieber, *ibid.* 76, 2894 (1954).
209. Shchipanov, V. P.; Sheinker, Y. N.; Postovskii, I. Y. *J. Org. Chem. USSR* **1966**, ***2***, 342 [CA **65**, 2248a (1966)].
210. Shchipanov, V. P. *J. Org. Chem. USSR* **1966**, ***2***, 347 [*CA* **65**, 2248c (1966)].
211. Shchipanov, V. P. *J. Org. Chem. USSR* **1966**, *2*, 1471 [CA 66, 55441n (1967)].
212. Hanley, R. N.; Ollis, W. D.; Ramsden, C. A.; Smith, I. S. *J. Chem. Soc., Perkin Trans. I* **1979**, *744*.
213. Krauch, C. H.; Kuhls, J.; Piek, H.-J. *Tetrahedron Lett.* **1966**, 4043.
214. Huseya, Y.; Chinone, A.; Ohta, M. *Bull. Chem. Soc. Jap.* **1971**, *44*, 1667.
215. Angadiyavar, C. S.; George, M. V. *J. Org. Chem.* **1971**, *36*, 1589.
216. Marky, M.; Hansen, H.-J.; Schmid, H. *Helv. Chim. Acta* **1971**, *54*, 1275.
217. Gotthardt, H.; Reiter, F. *Tetrahedron Lett.* **1971**, 2749.
218. Huseya, Y.; Chinone, A.; Ohta, M. *Bull. Chem. Soc. Jap.* **1972**, *45*, 3202.
219. Dewar, M. J. S.; Ramsden, C. A. *J. Chem. Soc., Chem. Commun.* **1973**, 688.
220. Blatt, A. H.; Hawkins, W. L. *J. Amer. Chem. Soc.* **1934**, *56*, 2190.
221. Bianchi, G.; Cook, M. J.; Katritzky, A. R. *Tetrahedron* **1971**, *27*, 6133.
222. Boyd, G. V.; Noms, T. *J. Chem. Soc., Perkin Trans. I* **1974**, 1028.
223. Morgan, G. T.; Burgess, H. *J. Chem. Soc.* **1921**, *119*, 697.
224. Wolff, L.; Fertig, E. *Ann.* **1900**, *313*, 12.
225. Chattaway, F. D.; Irving, H. *J. Chem. Soc.* **1931**, 786.
226. Nye, M. J.; O'Hare, M. J.; Tang, W.-P. *Abstr. 3rd Int. Congr. Heterocycl. Chem., B* **1971**, 514.
227. Nye, M. J.; Tang, W.-P. *Tetrahedron* **1972**, *28*, 455.
228. Nye, M. J.; Tang, W.-P. *Tetrahedron* **1972**, *28*, 463.
229. Metzler, D. E.; Snell, E. E. *J. Amer. Chem. Soc.* **1955**, 77, 2431.
230. Nye, M. J.; O'Hare, M. J.; Tang, W.-P. *J. Chem. Soc., Chem. Commun.* **1973**, 402.
231. Chinone, A.; Inouye, K.; Ohta, M. *Bull. Chem. Soc. Jap.* **1972**, *45*, 213.
232. Bamberger, E. *Genfer Arch. Sci. Phys. Nat.* **1898**, *6*, 384.
233. Bamberger, E. *Chem. Zentralbl.* **1898**, *2*, 1050.
234. Bamberger, E. *Ber.* **1911**, *44*, 3743.
235. Bamberger, E.; Padova, R.; Ormerod, E. *Ann.* **1926**, *446*, 260.
236. Cazeneuve, P. *Bull. Soc. Chim. Fr.* **1901**, *25*, 315.
237. Lozinskii, M. O.; Pelkis, P. S. *J. Gen. Chem. USSR* **1963**, ***33***, 106 [*CA* **59**, 612g (1963)].
238. Dubenko, R. G.; Pel'kis, P. S. *J. Gen. Chem. USSR* **1959**, ***29***, 200 [*CA* **53**, 21906f (1959)].
239. Pupko, L. S.; Dychenko, A. I.; Pelkis, P. S. *Ukr. Khim. Zh.* **1965**, ***31***, 1306 [*CA* **64**, 12661f (1966)].
240. Grammaticakis, P.; Acad, C. R. *Sci.* **1952**, *234*, 528.
241. Talukdar, P. B.; Sengupta, S. K.; Datta, A. K.; Chakravorty, A. *Indian J. Chem.* **1973**, *11*, 611.
242. I. S. Smith, M.Sc. Thesis, University of Sheffield (1972).
243. Nineham, A. W. *Chem. Rev.* **1955**, *55*, 355; Cheronis, N. D.; Stein, H. *J. Chem. Educ.* **1956**, *33*, 120; Hooper, W. D. *Rev. Pure Appl. Chem.* **1969**, *19*, 221; Laćan, M.; Tabaković, I.; Čeković, Ž. *Tetrahedron* **1974**, *30*, 2911.
244. Ogilvie, J. W.; Corwin, A. H. *J. Amer. Chem. Soc.* **1961**, *83*, 5023.

245. Ogilvie, J.; Miyamoto, V. K.; Bruice, T. C. *J. Amer. Chem. Soc.* **1961**, *83*, 2493.
246. Kushi, Y.; Fernando, Q. *Chem. Commun.* **1969**, 1240. *J. Amer. Chem. Soc.* **92**, 1965 (1970).
247. Irving, H. M. N. H.; Mahnot, U. S. *Talanta* **1968**, *15*, 811.
248. C. H. Carlin and A. H. Corwin, *Abstr. 157th Nat. Meeting, Amer. Chem. Soc., Minneapolis, ORGN 123* (1969).
249. McDonald, W. S.; Irving, H. M. N. H.; Raper, G.; Rupainwar, D. C. *Chem. Commun.* **1969**, 392.
250. Rajagopalan, P.; Penev, P. *Chem. Commun.* **1971**, 490.
251. Gompper, R. *Angew. Chem., Int. Ed. Engl.* **1969**, ***8***, 312. "Cyclo-addition Reactions" (R. Gompper, ed.). Butterworths, London, 1972; C. K. Bradsher, *Advan. Heterocycl. Chem.* **16**, 289 (1974).
252. Boyd, G. V.; Norris, T.; Lindley, P. F. *J. Chem. Soc., Chem. Commun.* **1974**, 639.
253. Bowie, J. H.; Eade, R. A.; Earl, J. C. *Aust. J. Chem.* **1968**, *21*, 1665.
254. Goudie, R. S.; Preston, P. N.; Palmer, M. H. *Org. Mass Spectrom.* **1969**, *2*, 953.
255. Shima, T.; Ouchida, A.; Asahi, Y. *Shitsuryo Bunseki* **1969**, ***17***, 661 [*CA* 73, 87149j (1970)].
256. Dougherty, R. C.; Foltz, R. L.; Kier, L. B. *Tetrahedron* **1970**, *26*, 1989.
257. Anderson, W. K.; Friedman, A. E. *Org. Mass Spectrom.* **1972**, *6*, 797.
258. Potts, K. T.; Armbruster, R.; Houghton, E.; Kane, J. *Org. Mass Spectrom.* **1973**, 7, 203.
259. Baker, W.; Ollis, W. D.; Poole, V. D.; Barltrop, J. A.; Hill, R. A. W.; Sutton, L. E. *Nature (London)* **1947**, *160*, 366.
260. Earl, J. C.; Leake, E. M. W.; Le Fèvre, R. J. W. *Nature (London)* **1947**, *160*, 366.
261. Earl, J. C.; Leake, E. M. W.; Le Fèvre, R. J. W. *J. Chem. Soc.* **1948**, 2269.
262. Hill, R. A. W.; Sutton, L. E. *J. Chim. Phys.* **1949**, *46*, 244.
263. Hill, R. A. W.; Sutton, L. E. *J. Chem. Soc.* **1949**, 746. 1482 (1953).
264. Borod'ko, Y. G.; Syrkin, Y. K. *Dokl. Akad. Nauk SSSR* **1960**, ***134***, 1127 [*ICA* **55**, 12039h (1961)].
265. Bämighausen, H.; Jellinek, F.; Munnik, J.; Vos, A. *Acta Crystallogr.* **1963**, *16*, 471.
266. Thiessen, W. E.; Hope, H. *J. Amer. Chem. Soc.* **1967**, *89*, 5977.
267. Barber, M.; Broadbent, S. J.; Connor, J. A.; Guest, M. F.; Hillier, I. H.; Puxley, H. J. *J. Chem. Soc., Perkin Trans. II* **1972**, 1517; Aarons, L. J.; Guest, M. F.; Hillier, I. H. *J. Chem. Soc., Faraday Trans. II* **1972**, ***68***, 1866.
268. Ramsden, C. A. *Chem. Brit* **1978**, 396.
269. Longuet-Higgins, H. C. *J. Phys. Chim.* **1949**, *46*, 246.
270. Orgel, L. E.; Cottrell, T. L.; Dick, W.; Sutton, L. E. *Trans. Faraday Soc.* **1951**, *47*, 113.
271. Bochvar, D. A.; Bagaturyants, A. A. *Zh, Fiz. Khim.* **1965**, *39*, 1631 [CA 64, 3329a (1966)].
272. Singer, J. A.; Purcell, W. P. *J. Med. Chem.* **1967**, *10*, 754.
273. Kier, L. B.; Roche, E. B. *J. Pharm. Sci.* **1966**, *55*, 807.
274. Roche, E. B.; Kier, L. B. *Tetrahedron* **1968**, *24*, 1673.
275. Bochvar, D. A.; Bagatur'yants, A. A. *Zh. Fiz. Khim.* **1965**, ***39***, 1631 [CA **64**, 3329a (1966)].
276. Bochvar, D. A.; Bagaturyants, A. A. *Teor. Eksp. Khim.* **1969**, ***5***, 19 [CA **71**, 33577n (1969)].
277. Borisov, E. V.; Kholodov, L. E.; Bagatur'yants, A. A.; Yashunskii, V. G. *Khim. Geterotsikl. Soedin.* **1971**, **7**, 1407 [CA **76**, 98939e (1972)].
278. Bochvar, D. A.; Bagaturyants, A. A.; Borisov, E. V. *Zh. Fiz. Khim.* **1972**, ***46***, 523 [*CA* **76**, 146892h (1972)].
279. Sauer, J.; Jung, C. *Z. Chem.* **1973**, *13*, 434.
280. Kier, L. B. *Tetrahedron Lett.* **1967**, 1233.
281. Martensson, O. *J. Mol. Struct.* **1971**, *9*, 321.

282. Schmid, G. H. *J. Mol. Struct.* **1970**, *5*, 236.
283. Bloor, J. E.; Gilson, B. R.; Billingsley, F. P., II. *Theor. Chim. Acta* **1968**, *12*, 360.
284. Palmer, M. H.; Gaskell, A. J.; Barber, M. S. *J. Mol. Struct.* **1972**, *12*, 197.
285. Paoloni, L.; Ciampi, A. *J. Heterocycl. Chem.* **1968**, *5*, 7.
286. Kier, L. B. *Molecular Orbital Theory in Drug Research*; Academic Press: New York, 1971.
287. Brookes, P.; Walker, J. *J. Chem. Soc.* **1957**, 4409.
288. Ackermann, E. *Pharmazie* **1967**, ***22***, 537 [CA **68**, 28171j (1968)].
289. Zotova, S. A.; Padeiskaya, E. N.; Yashunskii, V. G.; Pershin, G. N. *Zh. Org. Khim.* **1966**, ***2***, 728 [CA **65**, 10657f (1966)].
290. Naito, T.; Nakagawa, S.; Takahashi, K.; Masuko, K.; Fujisawa, K.; Kawaguchi, H. *J. Antibiot. (Tokyo)* **1968**, ***21***, 290 [CA **70**, 28862b (1969)].
291. Naito, T.; Nakagawa, S.; Takahashi, K.; Fujisawa, K.; Kawaguchi, H. *J. Antibiot. (Tokyo)* **1968**, ***21***, 300 *[*CA **70**, 37698e (1969)].
292. T. Takano (1969) (Fujisawa Pharmaceutical Co., Ltd.), S. African 68 02,695 [CA 71, 124458r (1969)].
293. Pala, G.; Coppi, G.; Mantegani, A.; Crescanzi, E. *Chim. Ther.* **1969**, ***4***, 26 [CA **71**, 3329r (1969)].
294. K. Hattori and S. Horibe, (1971) Japan 70 37,975 [CA **74**, 100034k (1971)].
295. Fujisawa Pharm. Co., Ltd., (1970) Brit 1,187,323 [CA 73, 3923w (1970)].
296. (a) Fujisawa Pharm. Co., Ltd. Japan. 71 02,341; 71 02,342; 71 02,343; 71 02,344 [CA 74, 141838u; 141837t; 141831m; 141832n] (1971); (b) Fujisawa Pharm. Co., Ltd. Japan. 71 02,340; 71 02,345; 71 02,346 *[*CA 75,5924n; 5922k; 5921j] (1971).
297. Roussel-Uclaf. BE 757792-Q.
298. G. R. Forker and V. P. Francois (1972) (Beecham Group Ltd.), S. African 71 00,171 [CA **76**, 140793e (1972)].
299. M. Kurita, Y. Saito, and T. Teraji (1971) (Fujisawa Chem. Co., Ltd.), Japan. 71 24,037 [CA **75**, 118330) (1971)].
300. T. Takano (1972) (Fujisawa Pharm. Co., Ltd.), Japan. 71 21,715 [CA **76**, 14557t (1972)].
301. Davis, D.; Becker, H. J.; Rogers, E. F. *Phytopathology* **1959**, *49*, 821.
302. E. F. Roger and D. Davis (1965) (Merck and Co., Inc.), U.S. 3,189,520 [CA **63**, 7605d (1965)].
303. Nyberg, W. H.; Cheng, C. C. *J. Med. Chem.* **1965**, *8*, 531.
304. Boots, S. G.; Cheng, C. C. *J. Heterocycl. Chem.* **1967**, *4*, 272.
305. Popoff, I. C.; Singhal, G. H. *J. Med. Chem.* **1968**, *11*, 631.
306. McCaustland, D. J.; Burton, W. H.; Cheng, C. C. *J. Heterocycl. Chem.* **1971**, *8*, 89.
307. Pala, G.; Mantegani, A.; Coppi, G.; Genova, R. *Chim. Ther.* **1969**, ***4***, 31 [*CA* **71**, 3328q (1969)].
308. Y. Imashiro and K. Masuda (1970) (Takeda Chem. Ind., Ltd.) Ger. Offen. 1,802,568 [*CA* **73**, 35379q (1970)].
309. Y. Saito and T. Kamitani (1970) (Takeda Chem. Ind., Ltd.), Japan. 70 10,509 [*CA* **73**, 14852y (1970)].
310. Y. Saito and T. Kamitani (1970) (Takeda Chem. Ind., Ltd.), Japan. 70 10,510 [*CA* **73**, 14853z (1970)].
311. T. Kamitani, Y. Saito, and T. Teraji (1970) (Fujisawa Pharm. Co., Ltd), Japan. 70 17,188 [*CA* **73**, 66586j (1970)].
312. K. Masuda and T. Okutani (1970) (Takeda Chem. Ind., Ltd.), Japan. 70 20,902 [*CA* **73**, 87927m (1970)].
313. K. Masuda and T. Okutani (1970) (Takeda Chem. Ind., Ltd.), Japan. 70 20,903 [*CA* **73**, 87928n (1970)].
314. Y. Saito and T. Kamitani (1970) (Fujisawa Pharm. Co., Ltd.), Japan. 70 21,507 [*CA* **73**, 87923g (1970)].

315. Y. Saito and T. Kamitani (1970) (Fujisawa Pharm. Co., Ltd.), Japan. 70 21,508 [*CA* **73**, 87924h (1970)].
316. Y. Saito and T. Kamitani (1970) (Fujisawa Pharm. Co., Ltd.), Japan. 70 21,509 [*CA* **73**, 87925j (1970)].
317. Y. Saito and T. Kamitani (1970) (Fujisawa Pharm. Co., Ltd.), Japan. 70 21,710 [*CA* **73**, 87926k (1970)].
318. T. Kamitani, Y. Saito, and T. Teraji (1970) (Fujisawa Pharm. Co., Ltd.), Japan. 70 26,087 [*CA* **73**, 109786u (1970)].
319. T. Kamitani and Y. Saito (1970) (Fujisawa Pharm. Co., Ltd.), Japan. 70 26,091 [*CA* **73**, 109787v (1970)].
320. T. Kamitani, Y. Saito, and T. Teraji (1971) (Fujisawa Pharm. Co., Ltd.), Japan. 70 26,092 [*CA* **74**, 3633x (1971)].
321. T. Kamiya, Y. Saito, and T. Teraji (1973) (Fujisawa Pharm. Co., Ltd.), Japan. 72 32,072 [*CA* **78**, 4257x (1973)].
322. T. Kamiya, Y. Saito, and T. Teraji (1973) (Fujisawa Pharm. Co., Ltd.), Japan. 72 32,073 [*CA* **78**, 4259z (1973)].
323. T. Kamitani, Y. Saito, and T. Terachi (1972) (Fujisawa Pharm. Co., Ltd.), Japan. 72 07,376 [G4 **76**, 153748u (1972)].
324. Y. Imashiro and K. Masuda (1972) (Takeda Chem. Ind., Ltd.), U.S. 3,642,793 [*CA* **76**, 140838y (1972)].
325. Takeda Chem. Ind., Ltd., (1973) Fr. M. 7421 [*CA* **78**, 58430s (1973)].
326. P. M. Weintraub, C. O. Baughn, and R. E. Bambury (1971) (Richardson-Merrell Inc.), U.S. 3,574,856 [*CA* **75**, 67496h (1971)].
327. K. Masuda, K. Kikuchi, and Y. Imajo (1973) (Takeda Chem. Ind., Ltd.), Japan. 72 29,511 [*CA* **78**, 124597e (1973)].
328. Lange, P.; Jung, F. *Acta Biol. Med. Ger.* **1971**, *27*, 425 [*CA* **76**, 136508y (1972)].
329. Farbenfabriken Bayer Akt.-Ges, Brit. 823,001 [*CA* **54**, 8854b (1960)].
330. P. F. Wiley (1966) (Eli Lilly and Co.), U.S. 3,224,937 [*CA* **64**, 14194e (1966)].
331. Goeres, E.; Faehndrich, A. *Acta Biol. Med. Ger.* **1968**, *20*, 641 [CA 69, 50740m (1968)].
332. Y. Imashiro and K. Masuda (1970) (Takeda Chem. Ind. Ltd.), Japan. 69 32,411 [CA 72, 111482q (1970)].
333. K. Masuda and Y. Imashiro (1970) (Takeda Chem. Ind. Ltd.), Japan. 70 06,016 [CA 72, 132745t (1970)].
334. Cameron, D. P.; Wiseman, E. H. *J. Med. Chem.* **1968**, *11*, 820.
335. Ackermann, E. *Acta Biol. Med. Ger.* **1966**, *17*, 498 [CA 66, 17865w (1967)].
336. Faulhaber, H. D.; Spinner, H.; Oehme, P.; Schwarz, K. *Folia Haematol.* **1965**, *83*, 267 [CA 67, 1969t (1967)].
337. H. A. Wagner, (1970) (G. D. Searle and Co.), U.S. 3,524,859 [CA 73, 131006m (1970)].
338. T. Kamitani and Y. Saito (1971) (Fujisawa Pharm. Co., Ltd.), Japan. 70 26,085 [CA 74, 13165k (1971)]; T. Kamitani, Y. Saito, and T. Teraji (Fujisawa Pharm. Co., Ltd.), Japan. 70 26,086; 70 26,088; 70 26,089; 70 26,090 [CA 74, 13164j; 13161f; 13166m; 13160e (1971)].
339. T. Kamitani, Y. Saito, and T. Terachi (1972) (Fujisawa Pharm. Co. Ltd.), Japan. 72 07,375 [CA 76, 153751q (1972)].
340. M. Ota (1970) (Takeda Chem. Ind. Ltd.), Japan. 70 06,015 [*CA* **73**, 3918y (1970)].
341. Y. Aramaki, M. Hirata, K. Masuda, and Y. Tanimoto (1968) (Takeda Chemical Industries, Ltd.), Japan. 6342 (1967) [*CA* **68**, 29702v (1968)].
342. Kikuchi, K.; Hirata, M.; Nagaoka, A.; Aramaki, Y. *Jap. J. Pharmacol.* **1970**, ***20***, 23 [*CA* **73**, 2425y (1970)].
343. Hirata, M.; Kikuchi, K. *Jap. J. Pharmacol.* **1970**, ***20***, 187 [*CA* **73**, 64778z (1970)].

344. K. Masuda and Y. Imashiro (1976) (Takeda Chem. Ind., Ltd.), U.S. 3,312,690 [*CA* **67**, 108662q (1967)].
345. Masuda, K.; Imashiro, Y.; Kaneko, T. *Chem. Pharm. Bull.* **1970**, *18*, 128.
346. Kikuchi, K.; Hirata, M.; Nagaoka, A.; Aramaki, Y. *Jap. J. Pharmacol.* **1970**, ***20***, 23 [*CA* **73**, 2425y (1970)].
347. C. H. Boehringer, (1970) Sohn, Fr. Demande 2,004,770 [*CA* **72**, 111479u (1970)].
348. M. Goetz, K. Freter, and K. Zeile (1970) (C. H. Boehringer, Sohn), Ger. Offen. 1,813,752 [*CA* **73**, 66584g (1970)].
349. G. Wehlmann, K. Zeile, M. Goetz, and K. Freier, (1970) (C. H. Boehringer, Sohn), Ger. Offen. 1,942,854 [*CA* **73**, 131005k (1970)].
350. Hashimoto, K.; Taira, N.; Hirata, M.; Kokubun, M. *Arzneim.-Forsch.* **1972**, ***21***, 1329 [*CA* **76**, 41952z (1972)].
351. G. Regnier, R. Canevari, and M. Laubie (1973) (Science-Union et Cie.—Sociéte Française de Recherche Médicale), Ger. Offen. 2,241,991 [*CA* **78**, 159672c (1973)].
352. Götz, M.; Grozinger, K.; Oliver, J. T. *J. Med. Chem.* **1973**, *16*, 671.
353. Kholodov, L. E.; Tishchenkova, I. F.; Al'tshuler, R. A.; Yashunskii, V. G.; Mashkovskii, M. D. *Khim.-Farm. Zh.* **1968**, ***2***, 3 [*CA* **69**, 96595w (1968)].
354. K. Masuda, Y. Imashiro, and T. Kamitani (1970) (Takeda Chem. Ind., Ltd.), Japan. 69 18,302 [*CA* **72**, 3502n (1970)].
355. K. Masuda and Y. Imashiro (1970) (Takeda Chem. Ind., Ltd.), Japan. 70 06,265 [*CA* **73**, 25485g (1970)].
356. Yashunskii, V. G.; Mashkovskii, M. D.; Gorkin, V. Z.; Kolodov, L. E.; Al'tshuler, R. A.; Polezhaeva, A. I.; Verevkina, I. V. *Farmakol. Toksikol. (Moscow)* **1970**, ***33***, 297 [*CA* **73**, 54388h (1970)].
357. Gorkin, V. Z.; Yashunskii, V. G.; Mashkovskii, M. D.; Al'tshuler, R. A.; Verevkina, I. V.; Kholodov, L. E. *J. Med. Chem.* **1971**, *14*, 1013.
358. M. D. Mashkovskii, V. G. Yashunskii, R. Al'tshuler, L. E. Kholodov, G. Y. Avrutskii, Y. A. Aleksandrovskii, and A. B. Shmulevich (1972) (Ordzhonikidze, S., All-Union Sei. Res. Chem.-Pharm. Inst.), Ger. Offen. 2,028,880 [*CA* **76**, 72521n (1972)].
359. Mashkovskii, M. D.; Al'tshuler, R. A.; Avrutskii, G. Y.; Aleksandrovskii, Y. A.; Smulevich, R. L.; Nevropatol, Z. *Psikhiat. im. S.S. Korsakova* **1971**, ***71***, 1704 [*CA* **76**, 81319e (1972)].
360. Ordzhonikidze, S. *All-Union Sci.-Res. Chem.-Pharm. Inst., Fr. M. 8171*; 1973 [*CA* **78**, 62182e (1973)].
361. Al'tshuler, R. A.; Mashkovskii, M. D.; Roshchina, L. F. *Farmakol. Toksikol. (Moscow)* **1973**, ***36***, 18 [*CA* **78**, 92685p (1973)].
362. Kholodov, L. E.; Yashunskii, V. G.; Al'tshuler, R. A.; Mashkovskii, M. D.; Roshchina, L. F.; Shershneva, S. I.; Leibel'man, F. Y.; Volzhina, O. N.; Gorodetskii, L. S.; Petrova, N. A. *Khim.-Farm. Zh.* **1973**, 7, 50 [*CA* **78**, 119341n (1973)].
363. M. D. Mashkovskii, V. G. Yashunskii, R. A. Al'tshuler, L. E. Kholodov, G. Y. Avrutskii, Y. A. Aleksandrovskii, and A. B. Shmulevich (1972) (Ordzhonikidze, S., All-Union Sci.-Res. Chem. Pharm. Inst.), British Patent 1,262,830 [*CA* **76**, 140836w (1972)].
364. Steele, J. W.; Anderson, W. K.; Trager, W. F. *Can. J. Chem.* **50**, 1026 (1972). *Can. J. Pharm. Sci.* **1972**, 7, 103.
365. Ota, M. (Takeda Chem. Ind., Ltd.), Japan. 70 14,069; 1970 [*CA* **73**, 77241u (1970)].
366. M. Ota (1967) (Takeda Chem. Ind., Ltd.), Japan. 15,581 (1966) [*CA* **66**, 10948v (1967)].
367. M. Ota (1967) (Takeda Chem. Ind., Ltd.), Japan. 15,582 (1966) [*CA* **66**, 10949y (1967)].

368. M. Ota (1970) (Takeda Chem. Ind., Ltd.), Japan. 70 06,014 [*CA* **73**, 25444t (1970)].
369. M. Ota (1970) (Takeda Chem. Ind., Ltd.), Japan. 70 11,505 [*CA* **73**, 45501n (1970)].
370. Takeda Chemical Industries Ltd., Japan. 73 11,711.
371. Takeda Chemical Industries, Ltd., Japan. 73 11,712.
372. W. Beil, H.-J. Wolff, A. Hoeppener, and H. C. Beil (Pfizer Inc.), DT 22239400.
373. T. Kamitani and T. Teraike (1971) (Fujisawa Pharm. Co. Ltd.), Japan. 71 21,033 [*CA* **75**, 118324k (1971)].
374. T. Kamitani and T. Teraike (1971) (Fujisawa Pharm. Co. Ltd.), Japan. 71 21,034 [*CA* **75**, 129814w (1971)].
375. T. Teraji and T. Kamitani (1971) (Fujisawa Pharm. Co. Ltd.), Japan. 71 27,466 [*CA* **75**, 151806s (1971)].
376. T. Teraji and T. Kamitani (1971) (Fujisawa Pharm. Co. Ltd.), Japan. 71 27,467 [*CA* **75**, 118325m (1971)].
377. T. Teraji and T. Kamitani (1971) (Fujisawa Pharm. Co. Ltd.), Japan. 71 27,468 [*CA* **75**, 151807t (1971)].
378. H. Koenig, P. Thieme, and A. Amann (1973) (Badische Anilin- und Soda-Fabrik A.-G.), Ger. Offen. 2,147,025 [*CA* **78**, 159618q (1973)].
379. T. Takano and T. Kamitani (1972) (Fujisawa Pharm. Co. Ltd.), Japan. 71 25,747 [*CA* **76**, 3876v (1972)].
380. K. Masuda and T. Kamitani (1970) (Takeda Chem. Ind. Ltd.), Japan. 70 21,102 [*CA* **73**, 87922f (1970)].
381. K. Masuda and T. Kamitani (1970) (Takeda Chem. Ind. Ltd.), Japan. 70 20,904 [*CA* **73**, 87930g (1970)].
382. Harris, S. A.; Webb, T. J.; Folkers, K. *J. Amer. Chem. Soc.* **1940**, *62*, 3198.
383. Katritzky, A. R.; Takeuchi, Y. *J. Amer. Chem. Soc.* **1970**, *92*, 4134.
384. Katritzky, A. R.; Takeuchi, Y. *J. Chem. Soc. C* **1971**, 874.
385. Dennis, N.; Katritzky, A. R.; Matsuo, T.; Parton, S. K.; Takeuchi, Y. *J. Chem. Soc., Perkin Trans. I* **1974**, 746.
386. Ames, D. E.; Kucharska, H. Z. *J. Chem. Soc.* **1963**, 4924.
387. Lunt, E.; Threlfall, T. L. *Chem. Ind. (London)* **1964**, 1805.
388. Ames, D. E.; Novitt, B. *J. Chem. Soc. C* **1969**, 2355.
389. Ning, R. Y.; Chen, W. Y.; Sternbach, L. H. *J. Heterocycl. Chem.* **1974**, *11*, 125.
390. Prystaš, M. *Collect. Czech. Chem. Commun.* **1967**, *32*, 4241.
391. Kappe, T.; Lube, W. *Angew. Chem., Int. Ed. Engl.* **1971**, *10*, 925.
392. Kappe, T.; Lube, W. *Monatsh. Chem.* **1971**, *102*, 781.
393. Maki, Y.; Sako, M.; Suzuki, M. *J. Chem. Soc., Chem. Commun.* **1972**, 999.
394. Potts, K. T.; Šorm, M. *J. Org. Chem.* **1972**, *37*, 1422.
395. Potts, K. T.; Šorm, M. *J. Org. Chem.* **1971**, *36*, 8.
396. Potts, K. T.; Hsia, R. K. C. *J. Org. Chem.* **1973**, *38*, 3485.
397. Coburn, R. A.; Glennon, R. A. *J. Heterocycl. Chem.* **1973**, *10*, 487.
398. Coburn, R. A.; Glennon, R. A. *J. Pharm. Sci.* **1973**, *62*, 1785.
399. Cobum, R. A.; Bhooshan, B. *J. Org. Chem.* **1973**, *38*, 3868.
400. Coburn, R. A.; Carapellotti, R. A. *Tetrahedron Lett.* **1974**, 663.
401. Bredereck, H.; Kupsch, G.; Wieland, H. *Chem. Ber.* **1959**, *92*, 566.
402. Bredereck, H.; Christmann, O.; Koser, W. *Chem. Ber.* **1960**, *93*, 1206.
403. Kappe, T.; Golser, W. *Synthesis* **1972**, 312.
404. Hagemann, H.; Ley, K. *Angew. Chem., Int. Ed. Engl.* **1972**, *11*, 1012.
405. Hagemann, H.; Ley, K. *Angew. Chem.* **1972**, *84*, 1063.
406. Ohta, M.; Kato, H. *Nippon Kagaku Zasshi* **1957**, *78*, 1400 [*CA* **54**, 511h (1960)].
407. Huisgen, R.; Mäder, H. *Angew. Chem., Int. Ed. Engl.* **1969**, *8*, 604.
408. Honzl, J.; Šorm, M. *Tetrahedron Lett.* **1969**, 3339.
409. Honzl, J.; Šorm, M.; Hanuš, V. *Tetrahedron* **1970**, *26*, 2305.

410. Šorm, M.; Honzl, J. *Tetrahedron* **1972**, *28*, 603.
411. Potts, K. T.; Elliott, A. J.; Šorm, M. *J. Org. Chem.* **1972**, *37*, 3838.
412. Matsumoto, A.; Lee, J. H.; Yoshida, M. *Yuki Gosei Kagaku Kyokai Shi* **1970**, ***28***, 1097 [*CA* **74**, 87705h (1971)].
413. Solomons, T. W. G.; Fowler, F. W. *Chem. Ind. (London)* **1963**, 1462.
414. Solomons, T. W. G.; Fowler, F. W.; Calderazzo, J. *J. Amer. Chem. Soc.* **1965**, *87*, 528.
415. Trofimenko, S. *J. Amer. Chem. Soc.* **1965**, *87*, 4393. **88**, 5588 (1966).
416. Solomons, T. W. G.; Voigt, C. F. *J. Amer. Chem. Soc.* **1965**, *87* (5256), 88. 1992 (1966).
417. Boekelheide, V.; Fedoruk, N. A. *Proc. Nat. Acad. Sci. U.S.A.* **1966**, ***55***, 1385 [*CA* **65**, 13683d (1966)].
418. Metze, R. *Angew. Chem.* **1956**, *68*, 580.
419. Pfleger, R.; Hahn, H.-G. *Chem. Ber.* **1957**, *90*, 2411.
420. Pfleger, R.; Garthe, E.; Rauer, K. *Chem. Ber.* **1963**, *96*, 1827.
421. R. Pfleger, E. Garthe, and K. Rauer, (1971) Ger. Offen. 1,620,103 [*CA* **75**, 49090s (1971)].
422. Brafani, M.; Fedeli, W.; Giacomello, G.; Vaciago, A. *Chem. Ber.* **1963**, *96*, 1840.
423. Carboni, R. A.; Kauer, J. C.; Hatchard, W. R.; Harder, R. J. *J. Amer. Chem. Soc.* **1967**, *89*, 2626.
424. Carboni, R. A.; Castle, J. E. *J. Amer. Chem. Soc.* **1962**, *84*, 2453.
425. Burke, M. E.; Sparks, R. A.; Trueblood, K. N. *Acta Crystallogr.* **1963**, *16*, A64.
426. Carboni, R. A.; Kauer, J. C.; Castle, J. E.; Simmons, H. E. *J. Amer. Chem. Soc.* **1967**, *89*, 2618.
427. Harder, R. J.; Carboni, R. A.; Castle, J. E. *J. Amer. Chem. Soc.* **1967**, *89*, 2643.
428. Kauer, J. C.; Carboni, R. A. *J. Amer. Chem. Soc.* **1967**, *89*, 2633.
429. Chia, Y. T.; Simmons, H. E. *J. Amer. Chem. Soc.* **1967**, *89*, 2638.
430. Lee, J. H.; Matsumoto, A.; Simamura, O.; Yoshida, M. *Chem. Commun.* **1969**, 1393.
431. Matsumoto, A.; Lee, J. H.; Yoshida, M.; Simamura, O. *Bull. Chem. Soc. Jap.* **1974**, *47*, 946.
432. Yoshida, M.; Matsumoto, A.; Simamura, O. *Bull. Chem. Soc. Jap.* **1970**, *43*, 3587.
433. Perkins, M. J. *J. Chem. Soc.* **1964**, 3005.
434. Smith, I. S. *Ph.D. Thesis*; University of Sheffield, 1975.
435. Hershenson, F. M. *J. Org. Chem.* **1975**, *40*, 740.
436. Manecke, G.; Klawitter, J. *Makromol. Chem.* **1974**, *175*, 3383.
437. Hamaguchi, M.; Ibata, T. *Tetrahedron Lett.* **1974**, 4475.
438. Ibata, T.; Hamaguchi, M.; Kiyohara, H. *Chem. Lett.* **1975**, 21.
439. Hamaguchi, M.; Ibata, T. *Chem. Lett.* **1975**, 169.
440. Potts, K. T.; Baum, J.; Houghton, E.; Roy, D. N.; Singh, U. P. *J. Org. Chem.* **1974**, *39*, 3619.
441. Potts, K. T.; Houghton, E.; Singh, U. P. *J. Org. Chem.* **1974**, *39*, 3627.
442. Potts, K. T.; Baum, J.; Houghton, E. *J. Org. Chem.* **1974**, *39*, 3631.
443. Buglass, A. J.; Tillett, J. G. *J. Chem. Soc., Perkin Trans. II* **1973**, 1687.
444. Isukul, E. A.; Tillett, J. G. *J. Chem. Soc., Perkin Trans. II* **1975**, 230.
445. Begtrup, M. *J. Chem. Soc., Chem. Commun.* **1975**, 334.
446. Begtrup, M. *Acta Chem. Scand.* **1975**, *B29*, 141.
447. Lazaris, A. Y.; Shmuilovich, S. M.; Egorochkin, A. N. *Zh. Org. Khim.* **1974**, *10*, 2236 [CA 82, 43270y (1975)].
448. Grashey, R.; Knom, C.; Weidner, M. *Chem. Ztg., Chem. App.* **1973**, *97*, 565 [CA 80, 27223p (1974)].
449. Gotthardt, H.; Weisshuhn, M. C.; Dörhöfer, K. *Angew. Chem. Int. Ed. Engl.* **1975**, *14*, 422.
450. Potts, K. T.; Kane, J.; Carnahan, E.; Singh, U. P. *J. Chem. Soc., Chem. Commun.* **1975**, 417.

451. Cava, M. P.; Saris, L. E. *J. Chem. Soc., Chem. Commun.* **1975**, 617.
452. Preston, P. N.; Robinson, N. J.; Turnbull, K.; King, T. J. *J. Chem. Soc., Chem. Commun.* **1974**, 998.
453. Boyd, G. V.; Norris, T.; Lindley, P. F. *J. Chem. Soc., Chem. Commun.* **1975**, 100.
454. Undheim, K.; El-Gendy, M. A. F.; Hurum, T. *Org. Mass. Spectrom.* **1974**, *9*, 1242 [CA 83, 27104n (1975)].
455. Hearn, M. T. W.; Potts, K. T. *J. Chem. Soc., Perkin Trans. II* **1974**, 875.
456. Sauer, J.; Jung, C.; Phys, Z. *Chem. (Leipzig)* **1974**, *255*, 412.
457. Borisov, E. V.; Lebedev, V. L.; Yashunskii, V. G. *Zh. Fiz. Khim.* **1975**, *49*, 769 [CA 83, 57860y (1975)].
458. Fabian, J. *J. Prakt. Chem.* **1973**, *315*, 690.
459. M. Ochiai, O. Aki, A. Morimoto, and T. Okada (1975) (Takeda Chem. Ind., Ltd.), Ger. Offen. 2,429,135 [*CA* **82**, 156342s (1975)].
460. Wagner, R. H.; Hill, J. B. *J. Med. Chem.* **1974**, *17*, 1337.
461. Hill, J. B.; Ray, R. E.; Wagner, R. H.; Aspinall, R. L. *J. Med. Chem.* **1975**, *18*, 50.
462. Franke, R.; Gäbler, E.; Oehme, P. *Acta Biol. Med. Germ.* **1974**, *32*, 545.
463. Bobritskaya, Z. M. *Fiziol. Zh. (Kiev)* **1974**, ***20***, 163 [*CA* **81**, 33442w (1974)].
464. Roshchina, L. F.; Al'tshuler, R. A.; Mashkovskii, M. D. *Farmakol. Toksikol. (Moscow)* **1975**, ***38***, 263 [*CA* **83**, 71997e (1975)].
465. Mekhedova, A. Y.; Luk'yanova, S. N. *Farmakol. Toksikol. (Moscow)* **1975**, ***38***, 267 [*CA* **83**, 71998f (1975)].
466. K. Masuda and Y. Imashiro (1975) (Takeda Chem. Ind., Ltd.), U.S. 3,812,128 [*CA* **82**, 170958d (1975)].
467. Slyusarenko, I. S.; Verevkina, I. V.; Bryantsev, B. I.; Gorkin, V. Z.; Yashunskii, V. G. *Khim. Farm. Zh.* **1975**, ***9***, 19 [*CA* **83**, 9926c (1975)].
468. J. B. Hester (1975) (Upjohn Co.) Ger. Offen. 2,441,436 [**CA 83**, 28296p (1975)].
469. Kato, H.; Yamaguchi, K.; Tezuka, H. *Chem. Lett.* **1974**, 1089.
470. Undheim, K.; Baklien, S. *J. Chem. Soc., Perkin Trans. I* **1975**, 1366.
471. Kappe, T.; Zadeh, R. K. *Synthesis* **1975**, 247.
472. Coburn, R. A.; Bhooshan, B. *J. Heterocyc. Chem.* **1975**, *12*, 187.
473. Maki, Y.; Suzuki, M.; Furuta, T.; Hiramitsu, T.; Kuzuya, M. *Tetrahedron Lett.* **1974**, 4107.
474. Akiba, K.; Hisoaka, M.; Inamoto, N.; Ohta, T.; Kuroda, H. *Chem. Lett.* **1975**, 347.
475. Burmistrov, S. I.; Kozinskii, V. A. *Zh. Org. Khitn.* **1974**, ***10***, 891 [*CA* **81**, 13444v (1974)].
476. Section 7.1.1 Matsukubo, H.; Kato, H. *J. Chem. Soc., Chem. Commun.* **1975**, 840. R. Verbruggen and H. G. Viehe, *Chimia* **29**, 350 (1975).
477. Section 7.3.3 Tominaga, Y.; Matsuda, Y.; Kobayashi, G. *J. Pharm. Soc. Jap.* **1975**, ***95***, 980; Talukdar, P. B.; Chakraborty, A. *Indian J. Chem.* **1975**, ***13***, 661.
478. Section 7.3.4 Baudy, M.; Robert, A. *J. Chem. Soc., Chem. Commun.* **1976**, 23.
479. Section 7.6.1 Alemagna, A.; Bacchetti, T. *Chim. Ind.* **1972**, *54*, 1105.
480. Section 7.8.2 Gannon, M.; McCormick, J. E.; McElhinney, R. S. *Proc. Roy. Irish Acad. Sect. B* **1974**, ***74***, 331; Grashey, R.; Weidner, M. *Chem.-Zt., Chem. App.* **1973**, ***97***, 623.
481. Section 7.8.3 Grashey, R.; Knorn, C.; Weidner, M. *Chem.-Zt., Chem. App.* **1973**, ***97***, 565. R. Grashey and C. Knorn, *ibid* **97**, 566 (1973).
482. Section 11.2 Boyd, G. V.; Davies, C. G.; Donaldson, J. D.; Silver, J.; Wright, P. H. *J. Chem. Soc. Perkin Trans. II* **1975**, 1280; Tominaga, Y.; Tamura, C.; Sato, S.; Hata, T.; Natsuki, R.; Matsuda, Y.; Kobayashi, G. *Chem. Pharm. Bull.* **1973**, ***21***, 1651.
483. Section 11.2.3 Schmidt, G. M. J. *Bull. Res. Counc. Isr.* **1951**, ***1***, 121 [*CA* **46**, 2871 (1952)].
484. Section 11.4.1 Borisov, E. V.; Gorskaya, N. V.; Yashunskii, V. G.; Vasyanina, L. K. *Khim. Geterotsikl. Soedin.* **1975**, 1049 [*CA* **83**, 205381k (1975)].

485. Section 11.4.2. M. J. S. Dewar and I. J. Turchi, *J. Chem. Soc. Perkin Trans. 2*, 1976, 548.
486. Section 12.5 Litzinger, E. F. *(Allied Chem. Corp.), U.S. 3,617,248*; 1972 [*CA* **76**, 25284c (1972)].
487. Section 12.8 Amann, A.; Koenig, H.; Thieme, P. C.; Giertz, H.; Kretzschmar, R. *(BASF A.-G), Ger. Offen. 2,408,288*; 1975 [*CA* **83**, 193341g (1975)].
488. Section 13 Coburn, R. A.; Glennon, R. A.; Chmielewicz, Z. F. *J. Med. Chem.* **1974**, *17*, 1025.
489. *Historical Note.* Much of the early work on meso-ionic compounds was pioneered by Max Busch. It seems appropriate that the final reference in this review should recognize this important contribution. Max Busch (1865–1941), *Ber.* **74**, (A)225 (1941); *Sitzungsber. Phys.-Med. Soz. Erlangen* **72**, xxxiii (1941) [*CA* **44**, 3314h (1950)].

APPENDIX 2 (REPRINTED FROM NEWTON, C.G.; RAMSDEN, C.A. TETRAHEDRON 1982, *38*, 2965.)

CHAPTER SIX

Meso-ionic heterocycles (1976–1980)

Christopher G. Newton[a] and Christopher A. Ramsden[b,*]
[a]The Research Laboratories, May & Baker Limited, Essex, United Kingdom
[b]Lennard-Jones Laboratories, School of Physical and Geographical Sciences, Keele University, Staffordshire, United Kingdom
*Corresponding author: e-mail address: c.a.ramsden@keele.ac.uk

Contents

Advances in Heterocyclic Chemistry, Volume 137
ISSN 0065-2725
https://doi.org/10.1016/bs.aihch.2021.11.002

Abstract

This review updates a previous review of mesoionic compounds by Ollis and Ramsden (*Adv. Heterocycl. Chem.* **1976**, *19*, 1–122) and extends the literature coverage to the end of 1980. The review adopts the same systematic approach and identifies examples of sixty known mesoionic rings of type A and thirteen known mesoionic rings of type B. Developments in the preparation and reactions of these known systems during the period 1976–1980 are covered. During this period considerable attention was directed to 1,3-dipolar cycloadditions of type A mesoionic rings, particularly cycloadditions of 1,3-oxazolium-5-olates (münchnones). Developments in the chemistry of sydnones and sydnone imines are also reviewed. The advantage of restricting the use of the term mesoionic to type A and type B five-membered rings is emphasized.

Keywords: Mesoionic, Heterocycle, Type A, Type B, 1,3-Dipolar cycloadditions, Munchnones, Sydnones, Mesomeric betaines

1. Introduction

It is one hundred years since Emil Fischer and Emil Besthorn described the preparation of dehydrodithizone (**1**) and thus provided the first example of a meso-ionic compound.[1] This initial contribution was soon followed by the pioneering work of Max Busch who, during the period 1895–1905, described the preparation and chemical properties of a number of meso-ionic heterocycles (e.g., **2**–**5**).[2] Later, three important further developments paved the way for extensive investigations which were to follows: (i) Earl and Mackney (1935) prepared *N*-phenylsydnone (**6**; R = Ph) by cyclodehydration of *N*-nitroso *N*-phenylglycine[3]; (ii) Schonberg (1938) recognized that the derivatives (**2**) and (**5**), which had been formulated as bicyclic compounds by Busch, were resonance hybrids of several dipolar canonical forms[4]; (iii) Baker and Ollis (1946) demonstrated that the sydnones (**6**) belong to a large family of heterocycles which includes the derivatives (**2**–**5**) and which they described as meso-ionic (*meso*meric + *ionic*).[2] The possible existence of other meso-ionic compounds was predicted by Baker and Ollis and subsequently many new systems have been prepared. Pre-eminent among these are the munchnones (**7**) whose preparation and cycloaddition reactions have been described by Huisgen (1964).[5] The recognition that the munchnones (**7**) react as 1,3-dipoles soon led to the investigation of other meso-ionic systems and was largely responsible for the extensive studies of meso-ionic compounds which have been reported over the last two decades.

The chemistry of meso-ionic compounds was comprehensively reviewed by Ollis and Ramsden in 1976.[6] In the period 1976–1981 considerable progress has been made in this area of chemistry. The purpose of this Report is primarily to up-date the earlier review while the number of references is still manageable. The literature up to the end of 1980 has been covered.

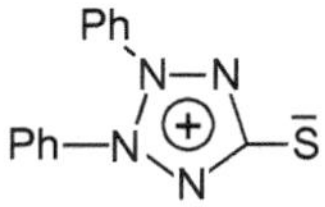

1

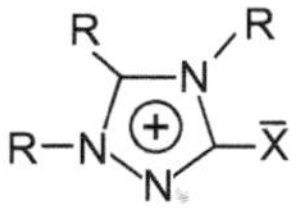

2 X=NR
3 X=O
4 X=S

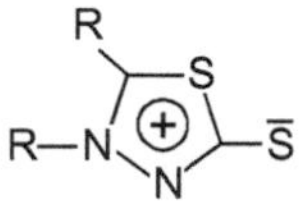

5

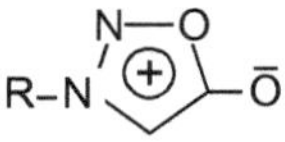

6

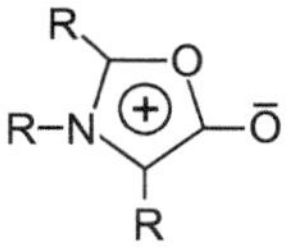

7

2. Meso-ionic compounds of types A and B

The definition of the term meso-ionic and the possible types of meso-ionic compounds have been discussed in detail elsewhere.[6] Although the original definition of the term meso-ionic recognized that some 6-membered rings could be described as meso-ionic, more recently it has been recommended that the name should be restricted to 5-membered rings. The following modified definition has been proposed:

"a compound may be appropriately called meso-ionic if it is a five-membered heterocycle which cannot be represented satisfactorily by any one covalent or polar structure and possesses a sextet of electrons in association with the five atoms comprising the ring."[6]

In accord with this definition meso-ionic compounds are represented by formulae of the general type[8] where a–f are atoms or groups derived from carbon or heteroatoms.

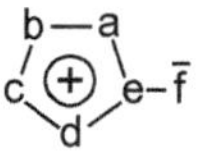

8

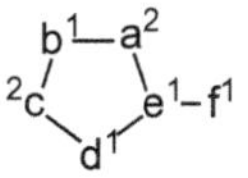

9 Type A

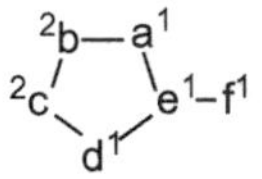

10 Type B

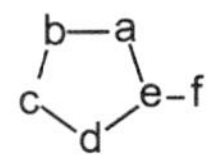

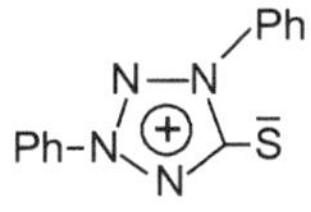

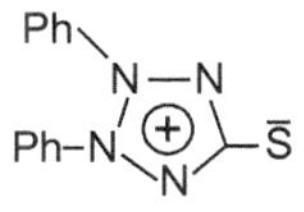

11 **12** **13**

The different types of heterocycle having the σ-skeleton (**11**) have been discussed by Ollis and Ramsden.[6] Analysis of these possible systems shows that there are two types of meso-ionic heterocycle having structures (**9**) or (**10**), where the superscripts indicate the origin of the eight π-electrons. These meso-ionic systems are referred to as type A (**9**) or type B (**10**). The 1,2,3,4-tetrazolium-5-thiolate (**12**) is an example of a type A meso-ionic system whereas the isomeric dehydrodithizone (**13**) belongs to type B. The two general types of meso-ionic compound show quite distinct chemical properties.

A major aspect of the chemistry of type A meso-ionic compounds is their participation in 1,3-dipolar cycloaddition reactions. Scheme 1 shows two examples of typical cycloaddition reactions of 2,4-diphenyl-3-methyl-1,3-oxazolium-5-olate (**14**).[5]

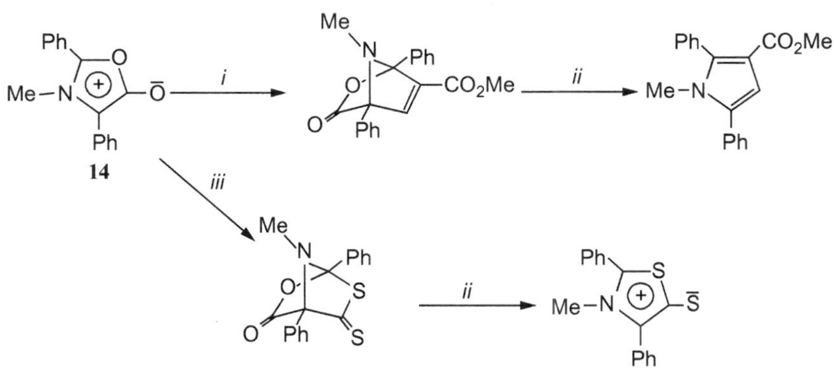

Scheme 1 Reagents: i, HC≡C, CO_2Me; ii, spontaneous elimination of CO_2; iii, CS_2.

The chemical properties of type B meso-ionic compounds are quite different to those of type A. An important feature of type B meso-ionic heterocycles is their facile ring opening to give acyclic valence tautomers. For example, many of the reactions of dehydrodithizone (**13**) are consistent with a mechanism in which valence tautomerism (**13** → **15**; Scheme 2) precedes reaction.[6]

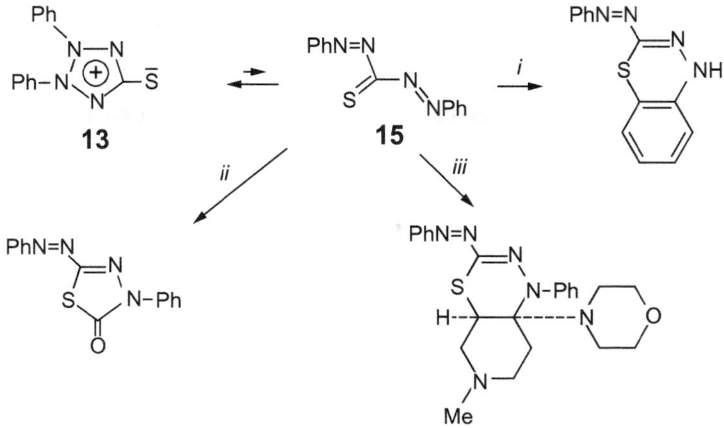

Scheme 2 Reagents: i, hot $MeCO_2H$; ii, $Fe_2(CO)_{10}$; iii, 1,2,5,6-tetrahydro-l-methyl-4-morpholinopyridine.

3. The chemistry of meso-ionic compounds of type A (1976–1980)

These systems are discussed in the order given in Table 1. Since the review by Ollis and Ramsden in 1976,[6] derivatives of 12 new type A meso-ionic systems have been reported bringing the total number of known systems up to 60.

3.1 Dioxoles

3.1.1 1,3-Dioxolium-4-olates (16)

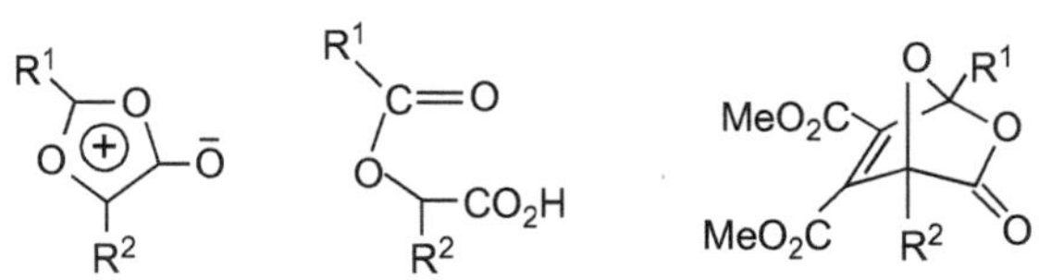

16 **17** **18**

Cyclodehydration of *O*-acetylmandelic acid (**17**; R^1=Me, R^2=Ph) in the presence of dimethyl acetylenedicarboxylate gives 2-methyl-5-phenyl-3,4-dimethoxycarbonyl furan (32%). This reaction probably proceeds *via* the meso-ionic 1,3-dioxolium-4-olate (**16**; R^1=Me, R^2=Ph) and its 1,3-dipolar cycloadduct (**18**; R^1=Me, R^2=Ph). This is the first report of a meso-ionic 1,3-dioxolium-4-olate (**16**).[7]

Table 1 Known meso-ionic heterocycles of type A (**9**).

		Atom or group[a]					
Parent system	**Heterocycle**	**a**	**b**	**c**	**d**	**e**	**f**
Dioxoles							
1,3-Dioxolium-4-olates	(**16**)	O	CR	O	CR	C	O
Oxazoles							
1,3-Oxazolium-4-olates	(**19**)	NR	CR	O	CR	C	O
1,3-Oxazolium-4-aminides	(**28**)	NR	CR	O	CR	C	NR
1,3-Oxazolium-5-olates	(**29**)	O	CR	NR	CR	C	O
1,3-Oxazolium-5-aminides	(**53**)	O	CR	NR	CR	C	NR
Oxathioles							
1,3-Oxathiolium-4-olates	(**62**)	S	CR	O	CR	C	O
1,3-Oxathiolium-5-olates	(**67**)	O	CR	S	CR	C	O
Diazoles							
1,3-Diazolium-4-olates	(**70**)	NR	CR	NR	CR	C	O
1,3-Diazolium-4-aminides	(**78**)	NR	CR	NR	CR	C	NR
1,3-Diazolium-4-thiolates	(**80**)	NR	CR	NR	CR	C	S
1,3-Diazolium-4-methylides	(**88**)	NR	CR	NR	CR	C	CXY
Thiazoles							
1,3-Thiazolium-4-olates	(**90**)	NR	CR	S	CR	C	O
1,3-Thiazolium-4-aminides	(**114**)	NR	CR	S	CR	C	NR
1,3-Thiazolium-5-olates	**(115)**	S	CR	NR	CR	C	O
1,3-Thiazolium-5-aminides	(**127**)	S	CR	NR	CR	C	NR
1,3-Thiazolium-5-thiolates	(**128**)	S	CR	NR	CR	C	S
Selenazoles							
1,3-Selenazolium-4-olates	(**134**)	NR	CR	Se	CR	C	O
Dithioles							
1,3-Dithiolium-4-olates	(**140**)	S	CR	S	CR	C	O
1,3-Dithiolium-4-aminides	(**146**)	S	CR	S	CR	C	NR

Table 1 Known meso-ionic heterocycles of type A (**9**).—cont'd

Parent system	Heterocycle	Atom or group[a] a	b	c	d	e	f
Oxadiazoles							
1,2,3-Oxadiazolium-5-olates	(**149**)	O	N	NR	CR	C	O
1,2,3-Oxadiazolium-5-aminides	(**175**)	O	N	NR	CR	C	NR
1,2,3-Oxadiazolium-5-thiolates	(**178**)	O	N	NR	CR	C	S
1,3,4-Oxadiazolium-2-olates	(**181**)	O	CR	NR	N	C	O
1,3,4-Oxadiazolium-2-aminides	(**187**)	O	CR	NR	N	C	NR
1,3,4-Oxadiazolium-2-thiolates	(**190**)	O	CR	NR	N	C	S
1,3,4-Oxadiazolium-2-methylides	(**193**)	O	CR	NR	N	C	CXY
Oxathiazoles							
1,3,2-Oxathiazolium-5-olates	(**195**)	O	N	S	CR	C	O
Triazoles							
1,2,3-Triazolium-4-olates	(**202**)	NR	N	NR	CR	C	O
12,3-Triazolium-4-aminides	(**205**)	NR	N	NR	CR	C	NR
1,2,3-Triazolium-4-thiolates	(**206**)	NR	N	NR	CR	C	S
1,2,4-Triazolium-3-olates	(**207**)	NR	CR	NR	N	C	O
1,2,4-Triazolium-3-aminides	(**219**)	NR	CR	NR	N	C	NR
1,2,4-Triazolium-3-thiolates	(**220**)	NR	CR	NR	N	C	S
1,2,4-Triazolium-3-methylides	(**226**)	NR	CR	NR	N	C	CXY
1,2,4-Triazolium-3-selenolates	(**227**)	NR	CR	NR	N	C	Se
Thiadiazoles							
1,2,3-Thiadiazolium-4-olates	(**230**)	NR	N	S	CR	C	O
1,2,3-Thiadiazolium-5-olates	(**231**)	S	N	NR	CR	C	O
1,2,3-Thiadiazolium-5-aminides	(**233**)	S	N	NR	CR	C	NR
1,2,3-Thiadiazolium-5-thiolates	(**237**)	S	N	BR	CR	C	S
1,2,3-Thiadiazolium-5-methylides	(**240**)	S	N	NR	CR	C	CXY
1,2,4-Thiadiazolium-3-olates	(**241**)	NR	CR	S	N	C	O

Continued

Table 1 Known meso-ionic heterocycles of type A (**9**).—cont'd

		Atom or group[a]					
Parent system	**Heterocycle**	**a**	**b**	**c**	**d**	**e**	**f**
1,2,4-Thiadiazolium-3-aminides	(**245**)	NR	CR	S	N	C	NR
1,3,4-Thiadiazolium-2-olates	(**247**)	S	CR	NR	N	C	O
1,3,4-Thiadiazolium-2-aminides	(**252**)	S	CR	NR	N	C	NR
1,3,4-Thiadiazolium-2-thiolates	(**254**)	S	CR	NR	N	C	S
1,3,4-Thiadiazolium-2-methylides	(**257**)	S	CR	NR	N	C	CXY
Oxatriazoles							
1,2,3,4-Oxatriazolium-5-olates	(**258**)	O	N	NR	N	C	O
1,2,3,4-Oxatriazolium-5-aminides	(**259**)	O	N	NR	N	C	NR
1,2,3,4-Oxatriazolium-5-thiolates	(**261**)	O	N	NR	N	C	S
Tetrazoles							
1,2,3,4-Tetrazolium-5-olates	(**263**)	NR	N	NR	N	C	O
1,2,3,4-Tetrazolium-5-aminides	(**264**)	NR	N	NR	N	C	NR
1,2,3,4-Tetrazolium-5-thiolates	(**265**)	NR	N	NR	N	C	S
1,2,3,4-Tetrazolium-5-methylides	(**267**)	NR	N	NR	N	C	CXY
1,2,3,4-Tetrazolium-5-selenolates	(**269**)	NR	N	NR	N	C	Se
Thiatriazoles							
1,2,3,4-Thiatriazolium-5-olates	(**271**)	S	N	NR	N	C	O
1,2,3,4-Thiatriazolium-5-aminides	(**273**)	S	N	NR	N	C	NR
1,2,3,4-Thiatriazolium-5-thiolates	(**274**)	S	N	NR	N	C	S
1,2,3,4-Thiatriazolium-5-methylides	(**275**)	S	N	NR	N	C	CXY
Dithiadiazoles							
1,3,2,4-Dithiadiazolium-5-olate	(**276**)	S	N	S	N	C	O
1,3,2,4-Dithiadiazolium-5-aminides	(**280**)	S	N	S	N	C	NR

[a]The grouping a and c each contribute 2 electrons to the π-electron system of the heterocycle; b, d, e, and f each contribute 1 electron.

3.2 Oxazoles

3.2.1 1,3-Oxazolium-4-olates (19)

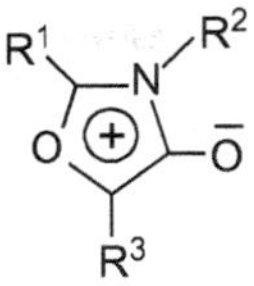
19

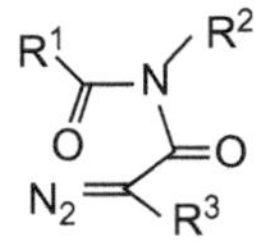
20

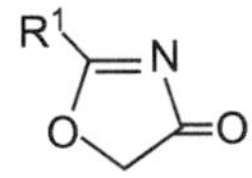
21

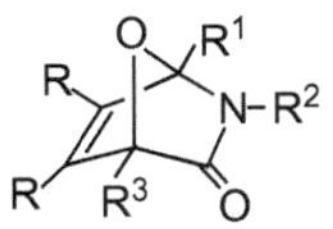
22

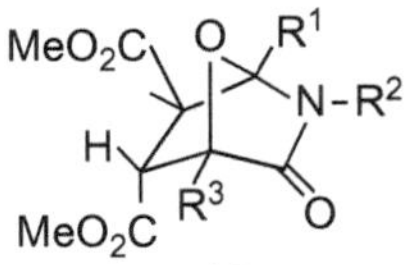
23

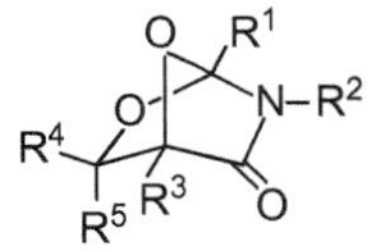
24

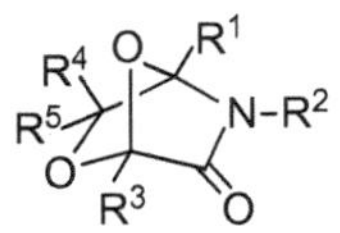
25

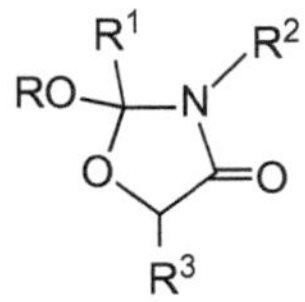
26

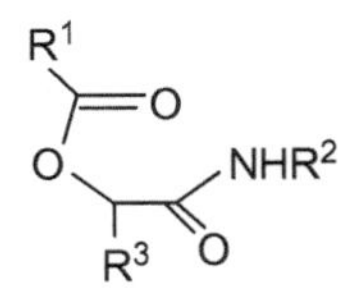
27

Examples of this meso-ionic system (**19**) were first described in 1974 and were obtained by cupric acetylacetonate catalyzed thermal cyclization of the diazo-ketones (**20**).[8,9] The 1,3-oxazolium-4-olates (**19**) are reactive 1,3-dipolarophiles. With acetylenes they form the primary adducts (**22**) which undergo thermal elimination of an isocyanate molecule (R^2NCO) to give the corresponding furan. 4(5*H*)-Oxazolones (**21**) also react with acetylenes giving furans and evidence that these reactions proceed *via* the meso-ionic valence tautomers (**19**; $R^2=R^3=H$) has been presented.[10] 1,3-Dipolar cycloadducts are also formed with olefinic dipolarophiles: compound (**19**; $R^1=Ph$, $R^2=Me$, $R^3=p.\ C_6H_4{\cdot}NO_2$) and dimethyl fumarate, in hot benzene, gave a quantitative yield of the adduct (**23**, $R^1=Ph$, $R^2=Me$, $R^3=p{\cdot}C_6H_4{\cdot}NO_2$).[8,11]

An interesting feature of the chemistry of the 1,3-oxazolium-4-olates (**19**) is their formation of 1,3-dipolar cycloadducts with carbonyl compounds. The 2,3-diphenyl derivative (**19**; $R^1=R^2=Ph$, $R^3=H$) reacts with aromatic ketones and aldehydes giving the adducts (**24**). Surprisingly, when benzyl is used as 1,3-dipolarophile, the regiospecificity is reversed giving the adduct (**25**; $R^1=R^2=R^4=Ph$, $R^3=H$; $R^5=COPh$).[12] Reaction of the *p*-nitrophenyl derivative (**19**; $R^1=Ph$, $R^2=Me$, $R^3=p \cdot NO_2 \cdot C_6H_4$) with tetrachloro-*o*-benzoquinone gives a [4 + 4] cycloadduct.[13]

In contrast to the isomeric 1,3-oxazolium-5-olates (**29**) (Section 3.2.3) which undergo nucleophilic attack at the carbon atom associated with the exocyclic oxygen atom, the 1,3-oxazolium-4-olate (**19**; R^1=Ph, R^2=Me, R^3=*p* · NO_2·C_6H_4) reacts with alcohols giving the adducts (**26**; R^1=Ph, R^2=Me, R^3=*p*·NO_2·C_6H_4). When the nucleophile is water, cleavage of the initial adduct (**26**) gives a quantitative yield of *N*-methyl-*α*-benzoyloxy-*α*-*p*-nitrophenyl acetamide (**27**; R^1=Ph, R^2=Me, R^3=*p*·NO_2·C_6H_4).[14]

3.2.2 1,3-Oxazolium-4-aminides (28)

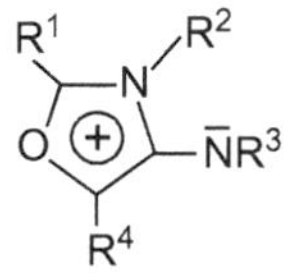

28

No new work on this system (**28**) has been reported since the subject was last reviewed.[6]

3.2.3 1,3-Oxazolium-5-olates (Munchnones) (29)

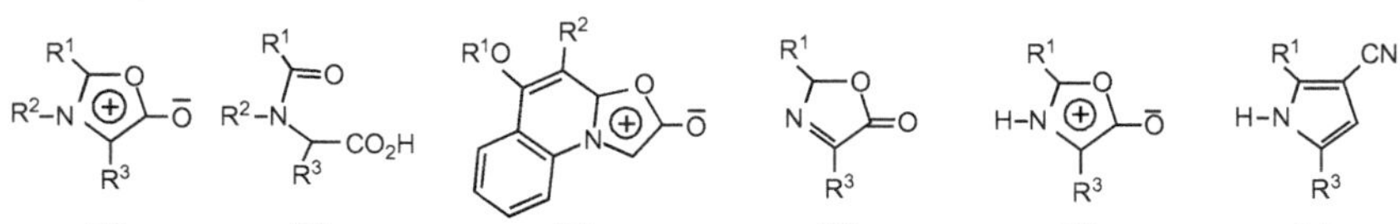

29 **30** **31** **32** **33** **34**

1,3-Oxazolium-5-olates (**29**) are prepared by cyclodehydration of *N*-acylamino acids (**30**) and have been the subject of considerable study. Recently an alternative synthetic route has been described. Treatment of acyltetracarbonyl ferrate salts [ArCOFe$(CO)_4^-$ M^+]with imidoyl chlorides (R^2N=CC1 · Ar) gives the meso-ionic 1,3-oxazolium-5-olates (**29**; R^1=R^3=Ar, R^2=Ar or alkyl) in good yield. This reaction fails to give alkyl derivatives (**29**, R^1 or R^3=alkyl).[15]

4-Unsubstituted munchnones (**29**; R^3=H) are usually too unstable to be isolated and the preparation of the unsubstituted polycyclic derivatives (**31**; R^1=COMe, p · Me · $C_6H_4SO_2$; R^2=H, Me) using the cyclodehydration route is therefore of interest: these compounds undergo electrophilic substitution at position 4.[16,17]

1,3-Dipolar cycloadditions of meso-ionic 1,3-oxazolium-5-olates (**29**) are valuable synthetic reactions. Often it is unnecessary to isolate the meso-ionic reagent. Either the oxazolone (**32**) may be used as a source of its meso-ionic tautomer (**33**) or the meso-ionic species can be generated *in situ* by cyclodehydration of *N*-acylamino acids (**30**). It is interesting to note that both the oxazolone (**32**; $R^1=R^3=Ph$), m.p. 103.5–104.5°, and the meso-ionic tautomer (**33**; $R^1=R^3=Ph$), m.p. 113–115°, have been isolated and characterized.[18]

The most important source of the 1,3-oxazolium-5-olate synthon (**29**) is still the cyclodehydration of *N*-acylamino acids.[19–33] Some recent examples of the use of munchnones (**29**) as *in situ* intermediates in organic synthesis are given in Scheme 3. These reactions include syntheses of (i) natural products; (ii) compounds of potential value in chemotherapy and (iii) novel heterocyclic systems. It is notable that the use of dibenzoylacetylene as the 1,3-dipolarophile led to the first isolable 1:1 cycloadduct of a munchnone and an acetylene (Scheme 3; reaction g).[28]

Reactions with olefines have also led to novel products. Cyclodehydration of *N*-acylaminoacids (**30**; $R^2=H$) in the presence of 2-chloroacrylonitrile affords, after elimination of carbon dioxide and hydrogen chloride from the initial adduct, 3-cyanopyrroles (**34**) in good yield. This pyrrole synthesis is particularly useful if the substituents (**34**; R^1 and R^3) are the same or when the cycloaddition is regiospecific.[34]

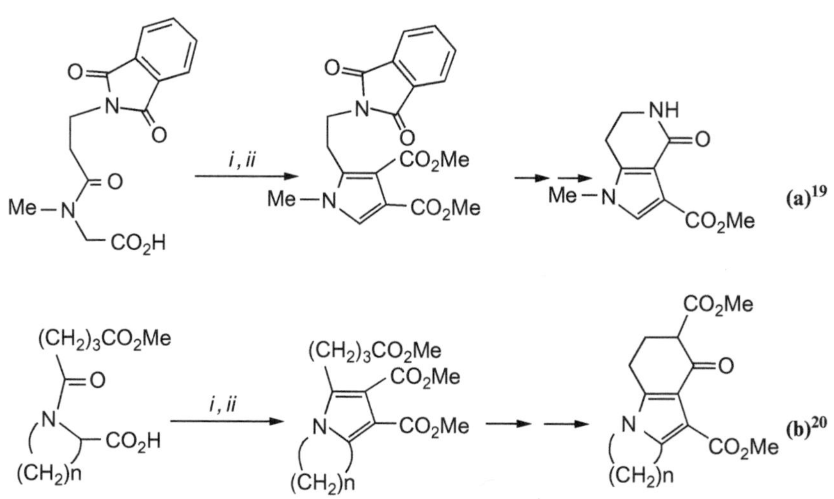

Scheme 3 Reagents: i, Ac_2O; ii, dimethyl acetylenedicarboxylate; iii, $HC{\equiv}C.CO_2Et$; iv, $PhCO.C{\equiv}C.COPh$, v, 1,2-dicyanocyclobutene.

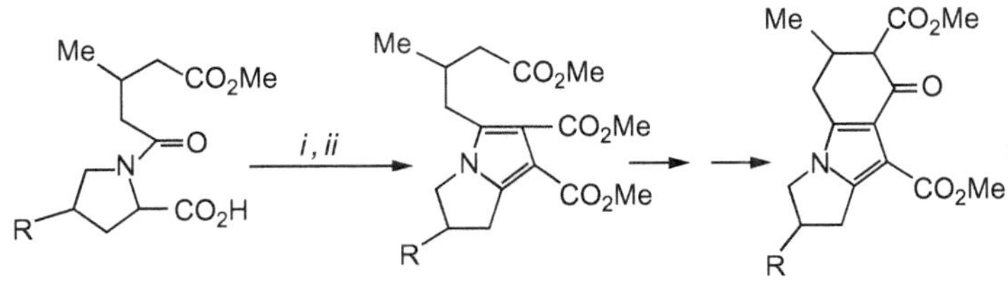

(c)[21]

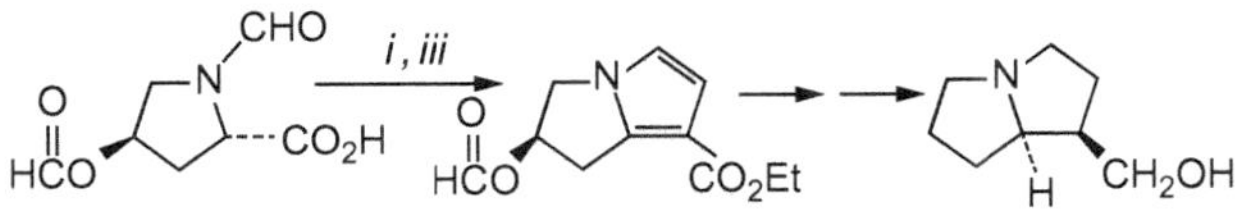

(d)[25]

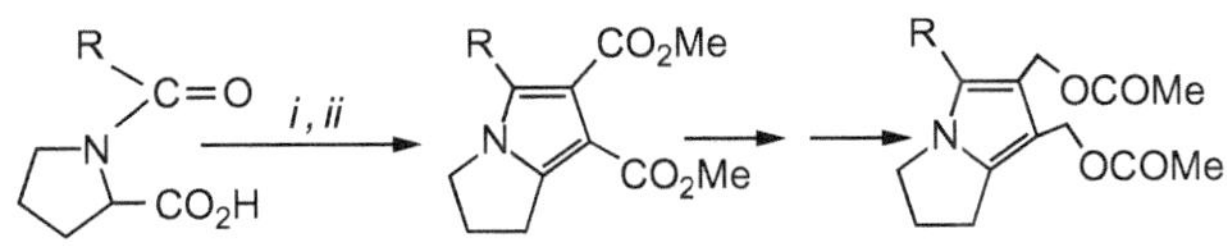

(e)[26]

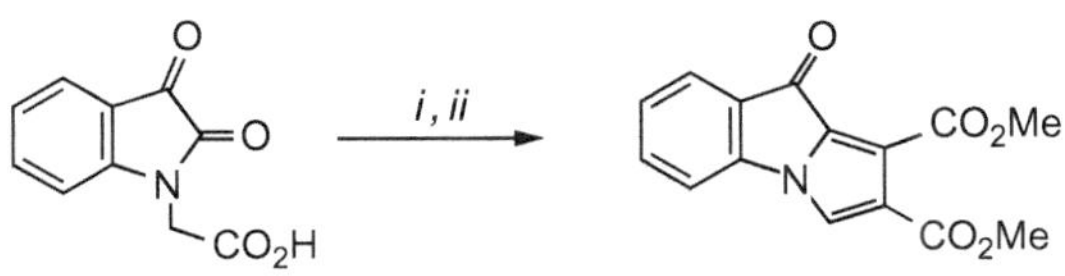

(f)[27]

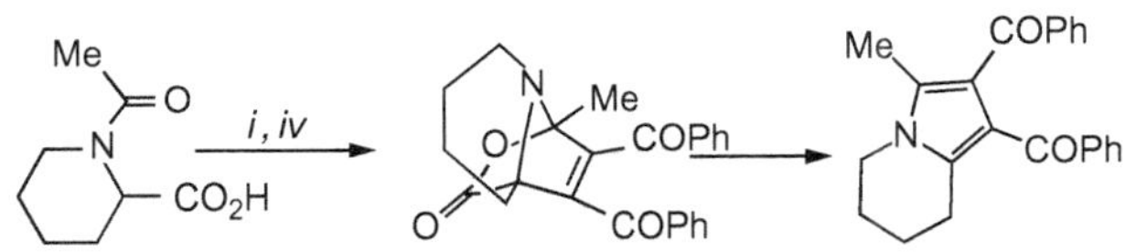

(g)[28]

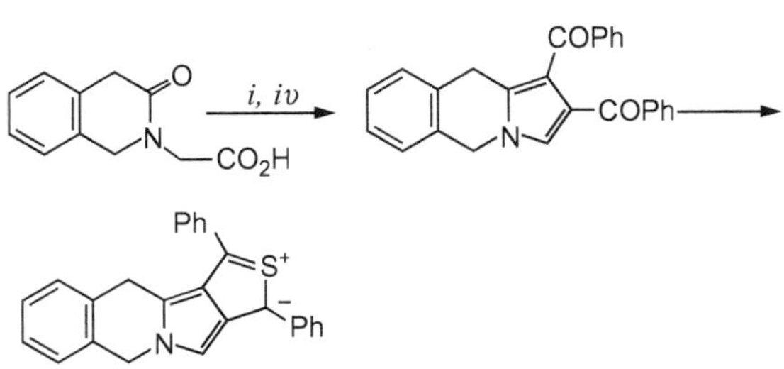

(h)[29]

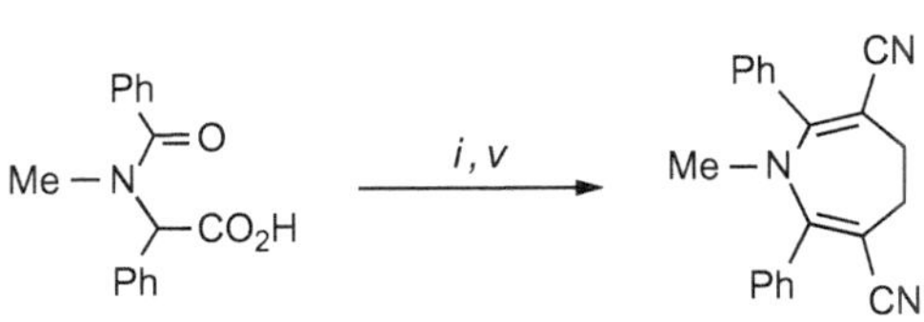

(i)[33]

Scheme 3—Cont'd

With the oxabicycloheptadiene (**35**; X=O), the 1,3-oxazolium-5-olate (**29**; $R^1=R^3=Ph$, $R^2=Me$) undergoes non-selective addition to the double bonds: spontaneous fragmentation of the resulting adducts leads to pyrrole derivatives and the olefin (**35**; X=O) is thus behaving as a masked acetylene.[35] With the azabicycloheptadene (**35**; X=N $SO_2 \cdot C_6H_4 \cdot p \cdot Me$), compound (**29**; $R^1=R^3=Ph$, $R^2=Me$) adds specifically to the unsubstituted double bond.[36] In a similar reaction, compound (**29**; $R^1=R^3=Ph$, $R^2=Me$) and cyclopentadienequinone (**36**) give the acid (**38**) formed by cleavage of the initial adduct (**37**). Treatment of compound (**38**) with aqueous base results in oxidative decarboxylation giving the pyrrole derivative (**39**) and pyrolysis in hot benzene gives compound (**40**).[37–40] Reactions with several other quinones have also been described.[41]

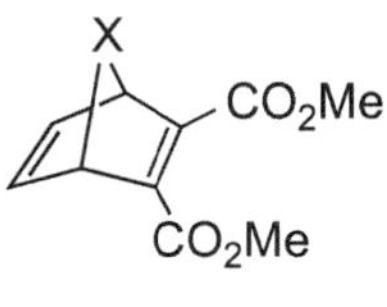

35

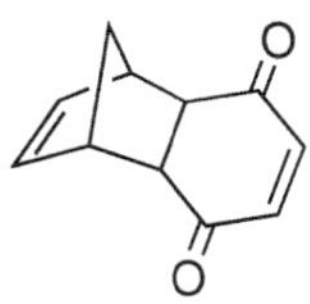

36

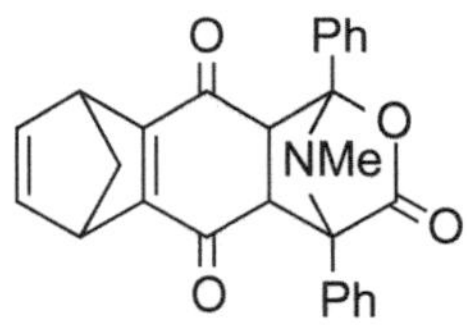

37

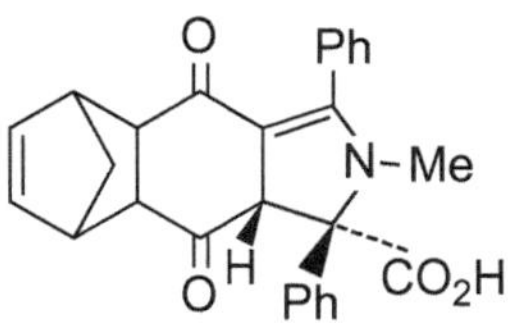

38

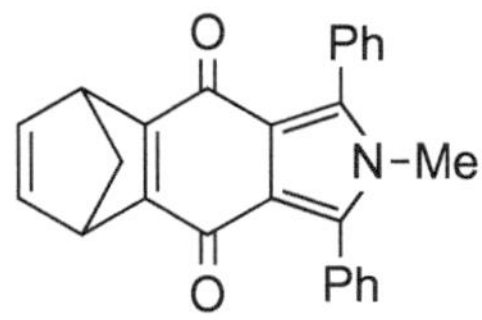

39

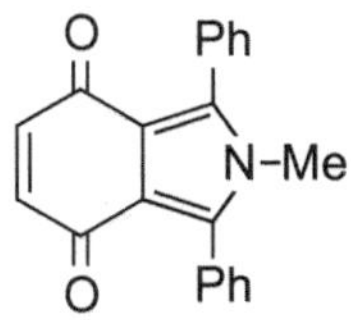

40

Intramolecular additions of olefines to munchnones have been reported. For derivatives of the type (**41**), the regioselectivity of the addition (i.e., **42** or **43**) depends upon the substituents on the meso-ionic ring (**41**; R^1 and R^3).[42] The use of diphenylcyclopropenone and its thione as dipolarophile affords 4-pyridones and 4-thiopyridones, respectively.[43,44] When the munchnones (**29**) are reacted with carbon disulfide, the initial cycloadduct (**45**) eliminates carbon dioxide providing a useful synthesis of meso-ionic 1,3-thiazolium-5-thiolates (**44**) (see Section 3.5.5).[45,46] If isocyanates are used as 1,3-dipolarophile, the primary adduct (**46**) is stable and can be isolated.[47]

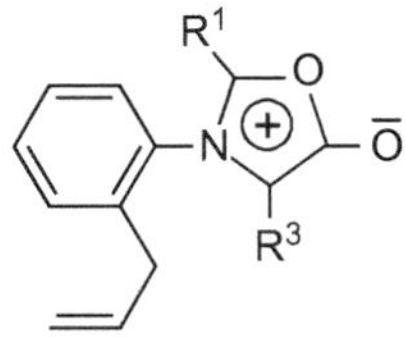
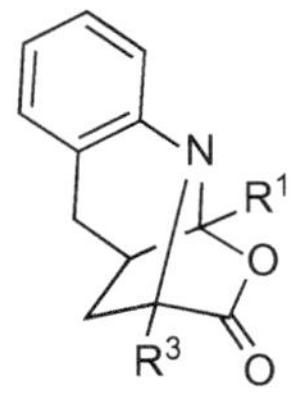
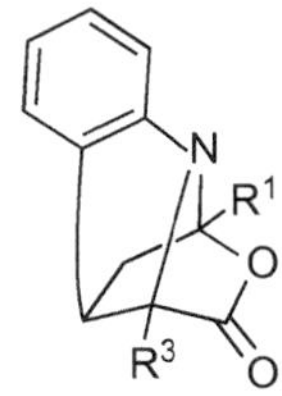

41 **42** **43**

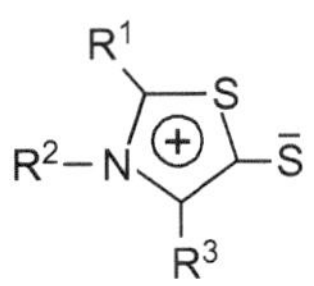
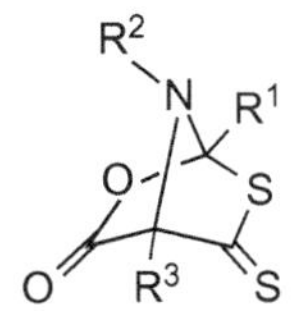
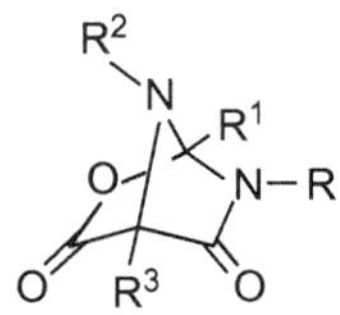

44 **45** **46**

An interesting aspect of the munchnones (**29**) is that in addition to their reactivity as 1,3-dipoles, they also have the capacity to react in terms of a [4 + 4] cycloaddition.[48,49] In this context, *o*-quinones react with 3-methyl-l, 3-oxazolium-5-olates (**29**; $R^2 = Me$) giving the adducts (**48**; $X = O$).[48,49] Quinone imines give similar adducts (**48**; $X = NSO_2Ph$)[48] and cinnamaldehyde anils give *α*-pyridones.[50] Compound (**29**; $R^1 = R^3 = Ph$, $R^2 = Me$) reacts with 2,2-dimethyl-3-dimethylamino azirine yielding the oxazoline derivative (**49**) whose structure has been confirmed by X-ray crystallography.[51,52] Hydrolysis of munchnones (**29**) is comparable to that of a ketene: compound (**29**; $R^1 = EtS$, $R^2 = Me$, $R^3 = COCF_3$) undergoes hydrolysis and decarboxylation to give compound (**50**; $R^1 = EtS$, $R^2 = CF_3$).[53]

The photochemistry of the 1,3-oxazolium-5-olates (**29**) has only recently received attention. Irradiation of compound (**29**; $R^1 = Ph$, $R^2 = Me$, $R^3 = H$) in chloroform-ethanol affords the ester (**50**; $R^1 = Ph$, $R^2 = OEt$), possibly by photo-generation of the ketene tautomer (**47**).[54] Photo-oxidation of the derivative (**29**; $R^1 = R^3 = Ph$, $R^2 = Me$) gives *N*-methyldibenzamide (**51**; $R^1 = R^3$-Ph, $R^2 = Me$) as the major product. This reaction (29 → 51) is rationalized by assuming initial formation of the endoperoxide (**52**).[55] Interestingly, oxazolones (**32**) also undergo photo-oxidation to dibenzamides (**51**) in solvents in which the meso-ionic tautomer (**33**) might be expected to predominate,[55,56]

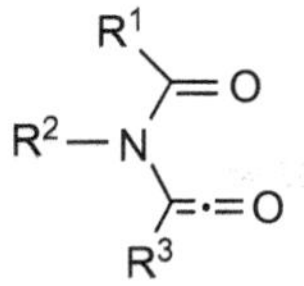
47

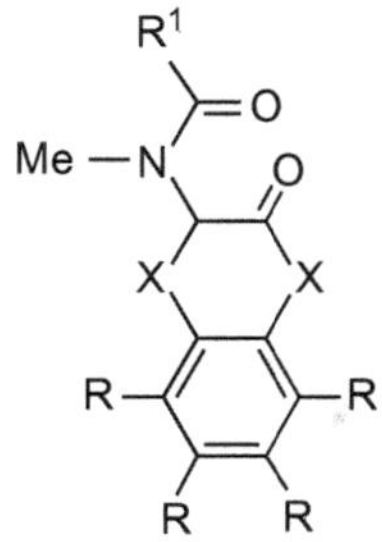
48

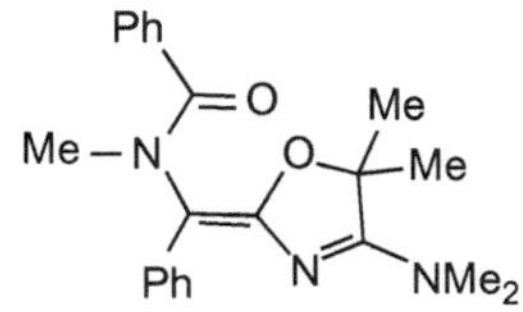
49

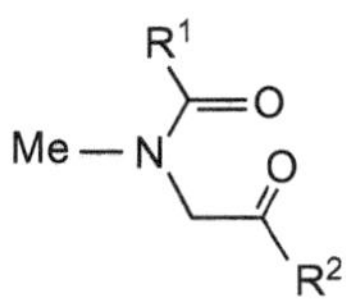
50

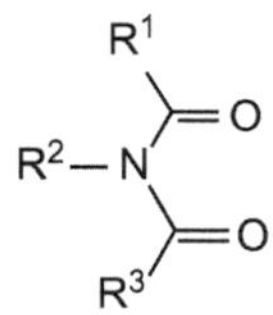
51

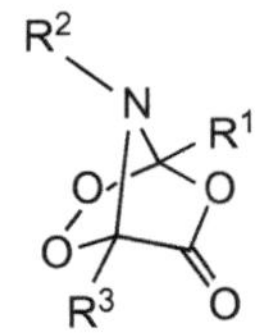
52

3.2.4 1,3-Oxazolium-5-aminides (53)

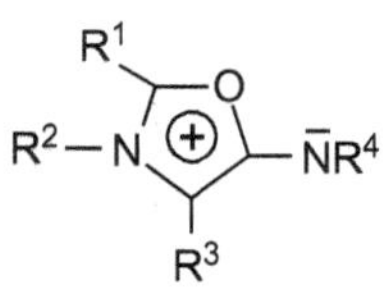
53

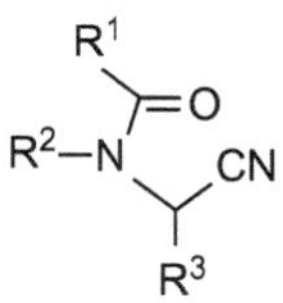
54

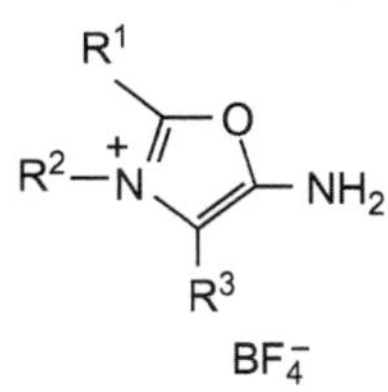
55

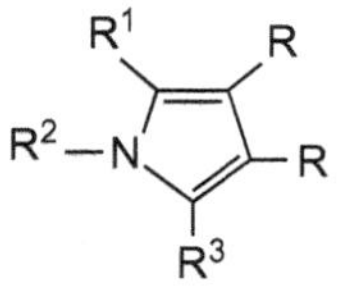
56

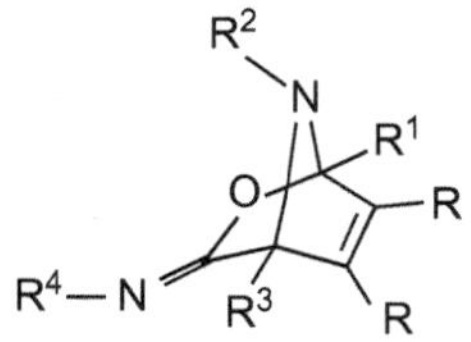
57

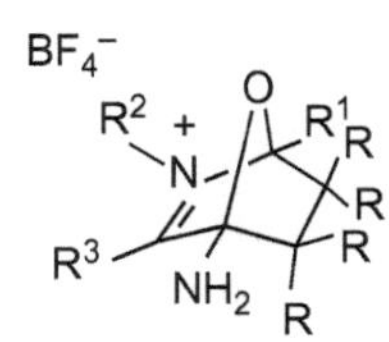
58

Cyclodehydration of the nitriles (**54**; $R^3 = H$) using trifluoroacetic anhydride gives the *N*-trifluoroacetyl-1,3-oxazolium-5-aminides (**53**; $R^3 = H$,

$R^4=COCF_3$): the *N*-acylated structure of these products has been confirmed by ^{13}C-NMR spectroscopy.[57] Treatment of these meso-ionic derivatives (**53**; $R^3=H$, $R^4=COCF_3$) with dimethyl acetylenedicarboxylate gives 3,4-dimethoxycarbonyl-pyrroles (**56**; $R=CO_2Me$) in good yield.[57] Cyclization of the nitriles (**54**; $R^3 \neq H$) with tetrafluoroboric acid similarly gives the 5-amino oxazolium tetrafluoroborates (**55**). The free meso-ionic systems (**53**; $R^4=H$) have not been isolated but in solution a small equilibrium concentration can be trapped by acetylenes to give good yields of the pyrroles (**56**) *via* the 1,3-dipolar cycloadducts (**57**). The meso-ionic system is not trapped, however, by alkenes. In this case, a Diels-Alder cyclo-addition to the oxazolium salt (**55**) takes place giving 2-aroylpyrroles (**56**; $R^1=ArCO$, $R^2=H$) *via* the Diels-Alder cyclo-adducts (**58**).[58]

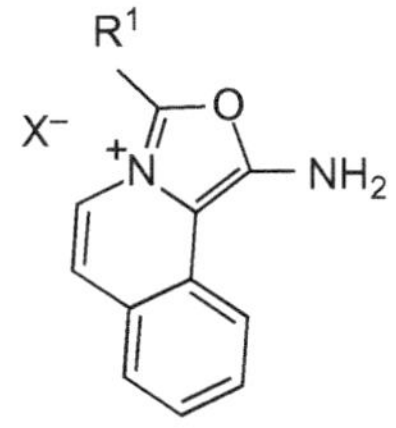

59

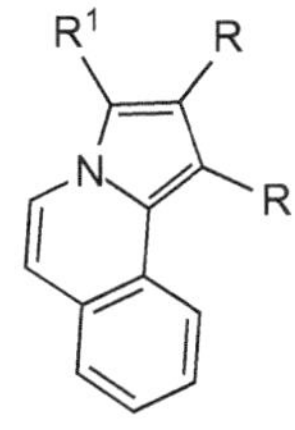

60

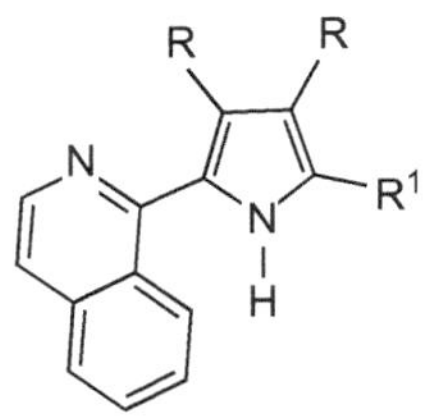

61

A similar behavior toward alkynes and alkenes has been observed for Reissert salts. These have been shown to have the 5-amino-1,3-oxazolium structure (**59**).[59] Reaction with alkynes leads to pyrrolo[1,2-*a*] isoquinolines (**60**) whereas treatment with alkenes gives the l-(2-pyrrolo) isoquinolines (**61**).[60] In a similar manner, addition of acetylenes to phthalazine Reissert salts gives pyrrolo [2,1-*a*] phthalazines.[61]

3.3 Oxathioles

3.3.1 1,3-Oxathiolium-4-olates (62)

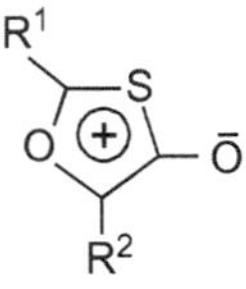

62

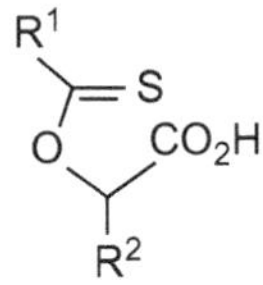

63

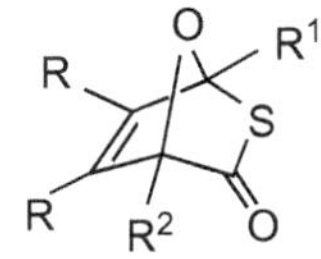

64

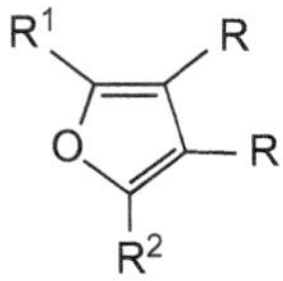

65

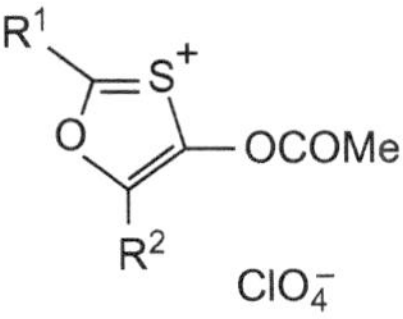

66

The 1,3-oxathiolium-4-olates (**62**) have not been isolated but they can be generated *in situ* by cyclodehydration of the acids (**63**; R^1 = NR_2 or SR) using acetic anhydride or dicyclohexylcarbodiimide. The meso-ionic product (**62**) can be trapped by a variety of acetylenic 1,3-dipolarophiles to give furans (**65**) *via* the primary adducts (**64**). Treatment of the acid (**63**; R^1 = SMe, R^2 = Ph) with acetic anhydride and perchloric acid affords the isolable oxathiolium perchlorate (**66**; R^1 = SMe, R^2 = Ph).[62,63]

3.3.2 1,3-Oxathiolium-5-olates (67)

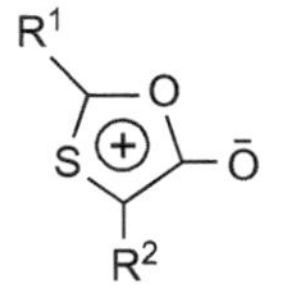

67

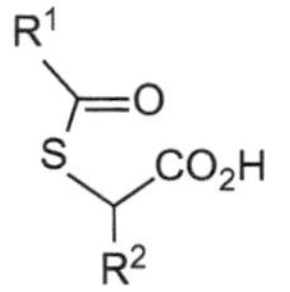

68

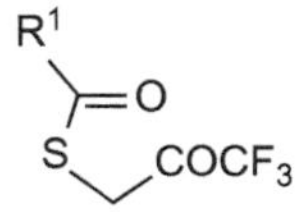

69

Cyclodehydration of aroylthioglycollic acids (**68**; R^1 = Ar, R^2 = H) using trifluoroacetic anhydride yields the red crystalline 1,3-oxathiolium-5-olates (**67**; R^1 = Ar, R^2 = $COCF_3$).[64,65] These novel meso-ionic derivatives (**67**) are readily hydrolyzed to the thioesters (**69**). The diphenyl derivative (**67**; R^1 = R^2 = Ph) has been generated in solution either by cyclodehydration of benzoylphenylthioglycollic acid (**68**; R^1 = R^2 = Ph)[65] or by condensation of thiobenzoic acid (PhCO·SH) with α-bromophenylacetyl chloride (PhCHBr · COCl).[66] This derivative (**67**; R^1 = R^2 = Ph) can be trapped by dimethyl acetylenedicarboxylate to give 3,4-dimethoxycarbonyl-2,5-diphenylthiophene.[65,66]

3.4 Diazoles

3.4.1 1,3-Diazolium-4-olates (70)

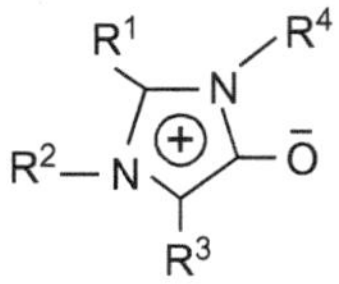

70

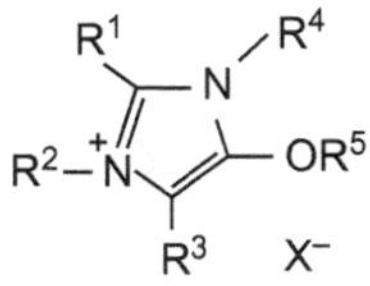

71

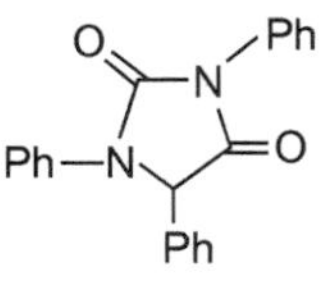

72

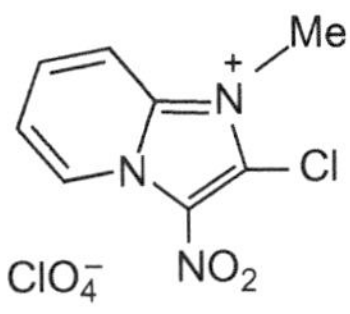

73

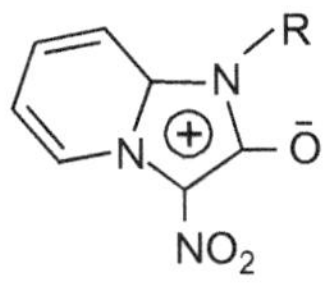

74

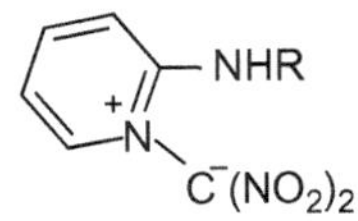

75

Recent studies on the preparation of this ring system (**70**) have utilized the base catalyzed condensation of *N*,*N*′-disubstituted amidines ($R^2NH{\cdot}CR^1{=}NR^4$) with α-bromoacylchlorides ($R^3CHBr{\cdot}COCl$) which gives good yields of the meso-ionic derivatives (**70**). In accord with their meso-ionic structure, reaction of 1,3-diazolium-4-olates (**70**) with electron deficient alkynes gives pyrroles. Hydrolysis of the 2-methylmercapto derivative (**70**; $R^1{=}SMe$, $R^2{=}R^3{=}R^4{=}Ph$) gives the imidazoline dione (**72**). Meerwein alkylation gives the 4-ethoxydiazolium salts (**71**; $R^5{=}Et$).[67,68]

Demethylation of 4-methoxy-l,2,3-trimethyl-5-nitro-diazolium perchlorate (**71**; $R^1{=}R^2{=}R^4{=}R^5{=}Me$, $R^3{=}NO_2$, $X{=}ClO_4$) affords the unstable 5-nitro derivative (**70**; $R^1{=}R^2{=}R^4{=}Me$, $R^3{=}NO_2$).[69] In contrast, the benz-derivative (**74**; R=Me), obtained by treatment of the salt (**73**) with aqueous potassium carbonate, is quite stable. The meso-ionic tautomer (**74**; R=H) appears to be more stable than the corresponding hydroxy tautomer: methylation of compound (**74**; R=H) gives the *N*-methyl derivative (**74**; R=Me). Nitration (HNO_3/H_2SO_4) of the bicyclic meso-ionic derivatives (**74**) gives the dinitromethylides (**75**).[70] Meso-ionic 1,3-diazolium-4-olates react with tetrachloro-*o*-benzoquinone to give [4+4] cycloadducts, the structure of one of which has been verified by X-ray crystallography.[13]

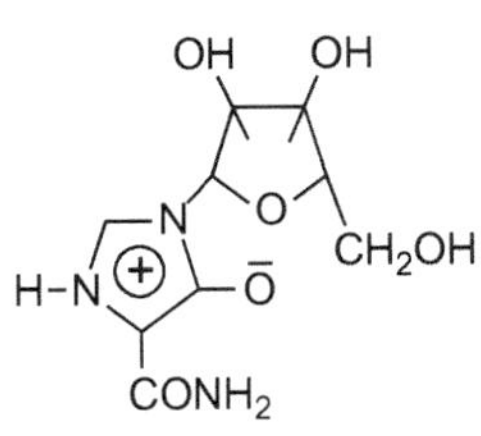

76

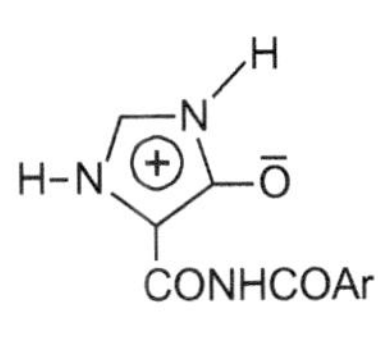

77

The structure of the derivative (**76**) has been investigated by X-ray analysis.[71] This compound (**76**) is known as Bredinin and is a potential immuno-suppressant.[72–74] Acylation of the meso-ionic derivatives (**77**) also leads to compounds of pharmaceutical interest.[75,76]

3.4.2 1,3-Diazolium-4-aminides (*78*)

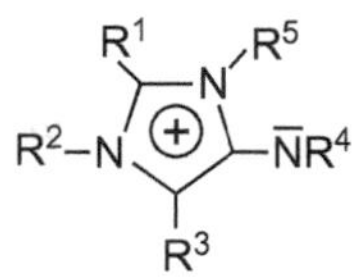

78

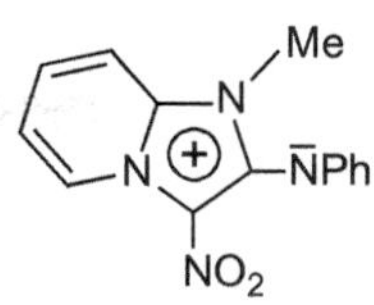

79

There have been no new studies on derivatives of the monocyclic system (**78**) but the orange bicyclic derivative (**79**) has been prepared by treatment of the perchlorate (**73**) with aniline in the presence of triethylamine.[70]

3.4.3 1,3-Diazolium-4-thiolates (*80*)

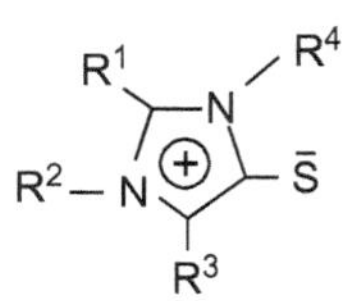

80

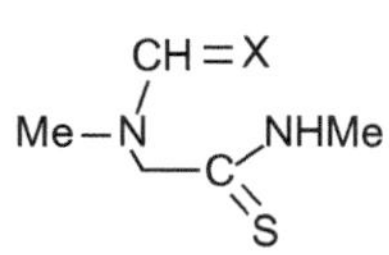

81

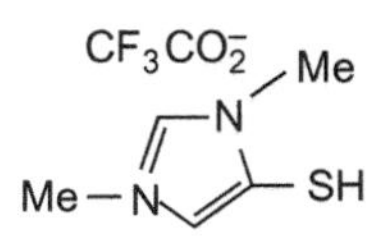

82

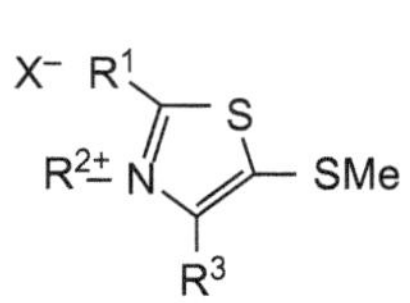

83

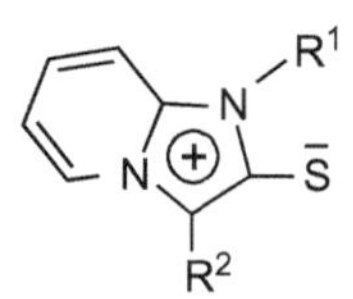

84

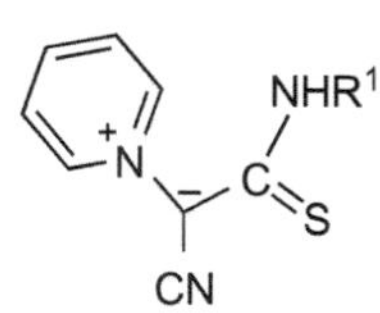

85

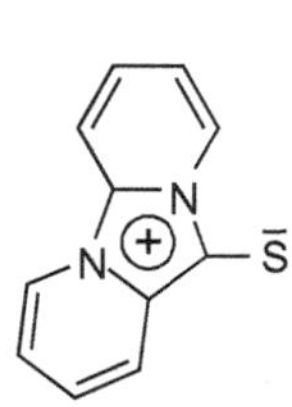

86

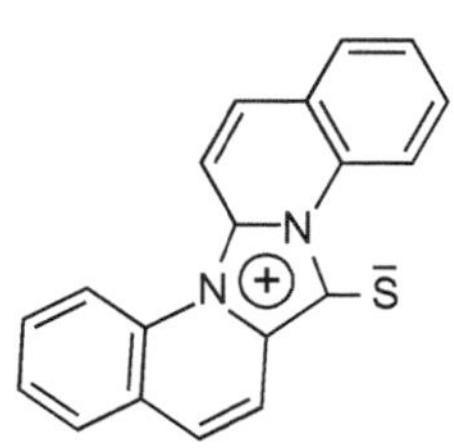

87

The formyl (**81**; X=O) and thioformyl derivatives (**81**; X=S) cyclize in aqueous trifluoroacetic acid yielding the 1,3-diazolium trifluoroacetate (**82**) which upon deprotonation gives the meso-ionic derivative (**80**; $R^1=R^3=H$, $R^2=R^4=Me$). The thioformyl derivative (**81**; X=S) also yields this product (**80**; $R^1=R^3=H$, $R^2=R^4=Me$) by treatment with Amberlite IRA 400 ion exchange resin in the OH form.[77,78]

The mechanism of formation of meso-ionic 1,3-diazolium-4-thiolates (**80**) from 5-methylthio-l,3-thiazolium salts (**83**) and primary amines has been investigated. This reaction proceeds better with aliphatic amines than with aromatic amines: a mechanism has been proposed.[78,79]

The yellow, bicyclic system (**84**; $R^1=Me$, $R^2=NO_2$) has been prepared by treatment of the perchlorate (**73**) with sodium hydrosulfide and triethylamine in water.[70] Treatment of the pyridinium methylides (**85**) with mercuric oxide gives the corresponding cyano derivatives (**84**; $R^2=CN$).[80]

An amusing reaction is the formation of the red tricyclic derivative (**86**) by reaction of 2-picoline with sulfur dioxide and benzoyl chloride in pyridine solution. This reaction was first reported in 1907 but the structure of the product was not recognized. A similar reaction using quinaldine affords compound (**87**) which is the thio-derivative of the dye Besthorn's red.[81]

3.4.4 1,3-Diazolium-4-methylides (88)

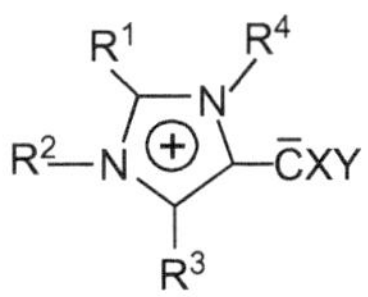

88

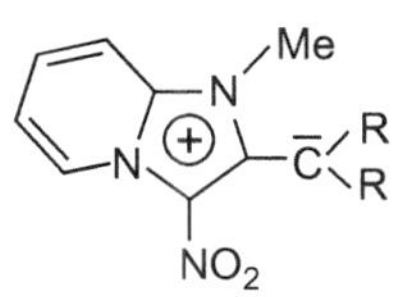

89

Treatment of the 2-chloro-3-nitroimidazo[1,2-*a*]pyridinium perchlorate (**73**) with malononitrile or diethyl malonate in the presence of triethylamine gives the orange-red meso-ionic derivatives (**89**; $R=CN$) and (**89**; $R=CO_2Et$) respectively. These are the first reported examples of this class of meso-ionic system (**88**).[70]

3.5 Thiazoles

3.5.1 1,3-Thiazolium-4-olates (90)

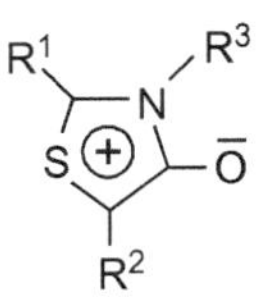

90

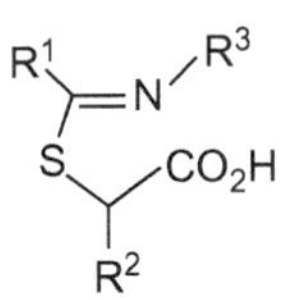

91

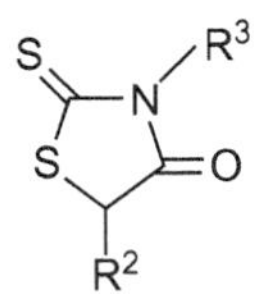

92

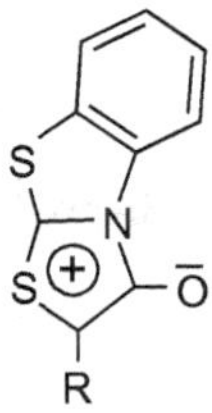

93

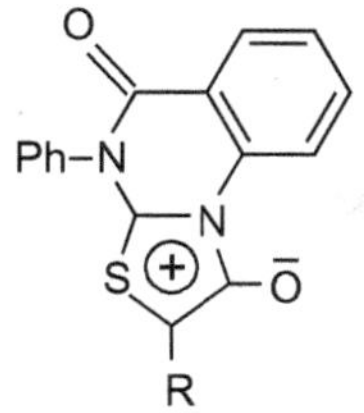

94

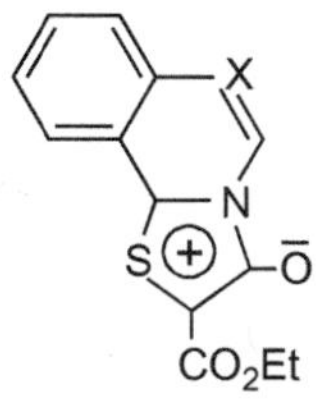

95

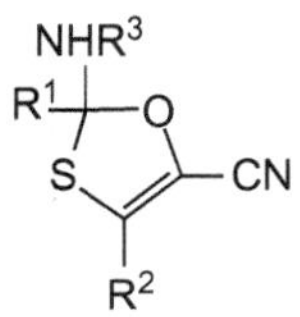

96

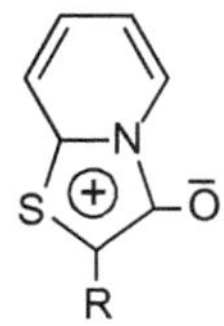

97

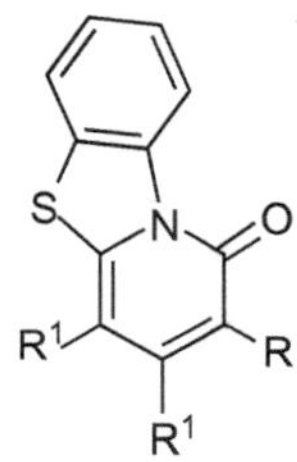

98

The meso-ionic 1,3-thiazolium-4-olates (**90**) have received considerable attention since 1975. Dehydration of the acids (**91**) continues to be an important method for preparing polycyclic derivatives. Recent examples include the preparation of the systems (**93**) and (**94**).[82,83] A second preparative route involves alkylation of rhodanines (**92**) followed by treatment with base: several 2-thioalkyl derivatives have recently been prepared by this method.[84]

Two new routes to this ring system have been described. Complications often encountered using the cyclodehydration synthesis (**91** → **90**) have led to the introduction of an alternative preparation which involves reaction of α-haloacyl chlorides ($R^2CHX{\cdot}COCl$) with *N*-monosubstituted thioamides ($R^1CS{\cdot}NHR^3$). This route enables a wide variety of substituents to be incorporated into the heterocyclic ring. Both monocyclic and polycyclic derivatives have been prepared by this method.[61,82,85–87] In a similar approach, reaction of isoquinolinyl-1-thiol or quinazolinyl-4-thiol with ethylbroinomalonate yields the thiazolo[2,3-*a*] isoquinolinium-3-olate (**95**, X=CH) and the thiazolo [3,2-*c*] quinazolinium-3-olate (**95**; X=N), respectively.[88,89]

An alternative route utilizes the reaction of *gem*-dicyano epoxides with thioamides and leads to the formation of the red meso-ionic 1,3-thiazolium-4-olates (**90**; R^1=aryl, alkyl, CN, S-aryl, NR_2; R^2=aryl; R^3-alkyl or aryl) in good yield. The mechanism of the multi-step process has been discussed and is believed to involve a tetrahedral intermediate (**96**) which rearranges to the meso-ionic product (**90**) with elimination of hydrogen cyanide.[90–93]

Several examples of electrophilic substitution at the 5-position of meso-ionic 1,3-thiazolium-4-olates have been described. Reaction of 2-pyridinylthioglycollic acid and derivatives with acetic anhydride affords the acylated products, e.g., (**97**; R = $COCH_2S{\cdot}C_5H_4N$).[85,94,95] The unsubstituted system (**97**; R = H) has been shown to readily undergo electrophilic substitution with diazonium salts and aldehydes.[96,97] Derivatives of this system may be useful as dyes.[98] Similar results have been obtained in the thiazolo [3,2-*a*] quinolinium-l-olate system.[99,100]

Cycloaddition reactions of meso-ionic 1,3-thiazolium-4-olates (**90**) have attracted considerable attention. With acetylenes two possible modes of fragmentation of the primary adduct (**100**) can occur (Scheme 4): Loss of a molecule of isocyanate generates a thiophene (**99**) whereas loss of sulfur affords a pyridone derivative (**101**). In general, monocyclic 1,3-thiazolium-4-olates (**90**) react with alkynes to give thiophenes (**99**)[66] but this mode of reaction is suppressed by ring annélation and polycyclic 1,3-thiazolium-4-olates give pyridone derivatives. For example, the polycyclic system (**93**) reacts with acetylenes to give the derivatives (**98**)[82] and similar products have been obtained using other polycyclic derivatives.[86,87] Surprisingly, the thiazolo[3,2-*a*]quinolinium-l-olate system yields pyrrolo[l, 2-*a*] quinolines on reaction with acetylenes: a mechanism has been proposed for this unusual reaction.[85]

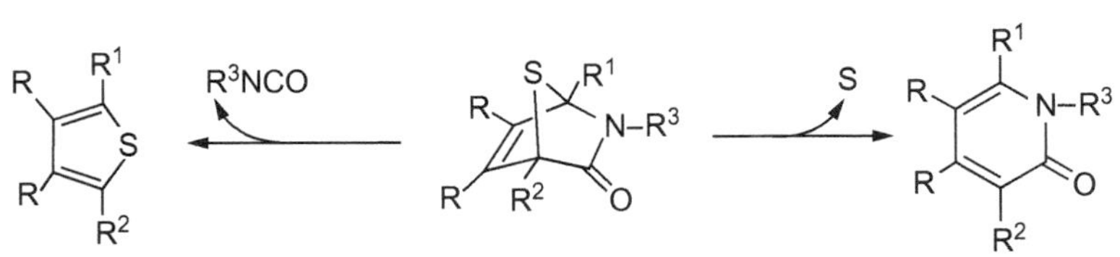

99 **100** **101**

Scheme 4

4-Hydroxythiazoles react with diethyl acetylenedicarboxylate giving the adducts (**100**; $R^3 = C(CO_2Et) = CH \cdot CO_2Et$) which fragment either to a thiophene (**99**) or a pyridone (**101**) depending upon the nature of the substituents. These reactions involve initial formation of the meso-ionic intermediates (**90**; $R^3 = C(CO_2Et) = CH \cdot CO_2Et$).[101,102] The mode of fragmentation of the benzyne adducts (**102**) can be controlled by the reaction conditions:

thermal fragmentation gives isobenzothiophenes (**103**) whereas extrusion of sulfur occurs on photochemical irradiation giving isoquinolones (**104**).[30]

Several studies of the cycloaddition of olefines to meso-ionic 1,3-thiazolium-4-olates (**90**) have been reported. The “masked acetylene” (**35**; X=O) gives a mixture of cycloadducts which upon heating undergo double fragmentation affording thiophenes.[35] Similar reactions have been observed using azabicycloheptadienes (**35**; X=N · Tosyl)[36] and a tricyclodecatriene.[103] An X-ray study of the cycloadduct formed between the meso-ionic 1,3-thiazolium-4-olate (**90**; $R^1=R^3=Ph$, $R^2=C_6H_4 \cdot p \cdot Cl$) and dimethyl maleate has shown it to possess the constitution (**105**) in which the ester groups have exostereo-chemistry.[104] Dimethyl maleate and dimethyl fumarate also react with the meso-ionic tautomers of 4-hydroxythiazoles generating 1:1 cycloadducts.[101,104] A number of polycyclic meso-ionic 1,3-thiazolium-4-olates have been treated with a variety of olefinic dipolarophiles and the thermal decomposition of the primary adducts has been described.[82,85–87] Diphenylcyclopropenone and diphenylcyclopropenethione afford the cycloadducts (**106**): thermolysis of these adducts (**106**) leads to a variety of products.[43] Reaction of 1,3-thiazolium-4-olates (**90**) with isocyanates or isothiocyanates gives adducts of the general structure (**107**). Attempts to prepare the 6-membered cross-conjugated mesomeric betaines (**108**) by removal of the sulfur bridge were unsuccessful.[47]

Some oxidation reactions of this system (**90**) have recently been reported.[105]

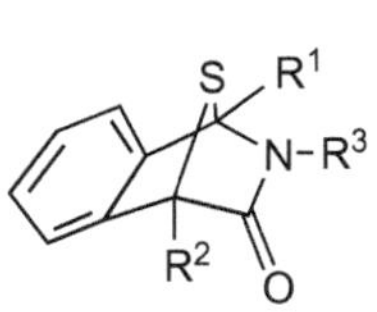

102

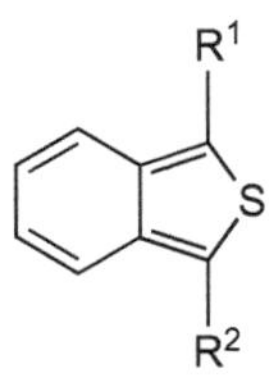

103

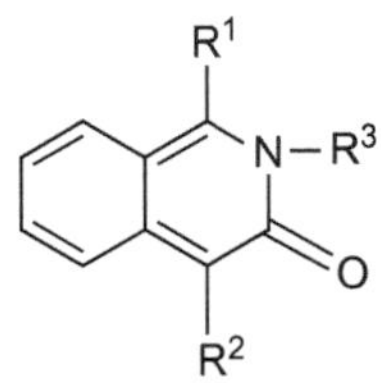

104

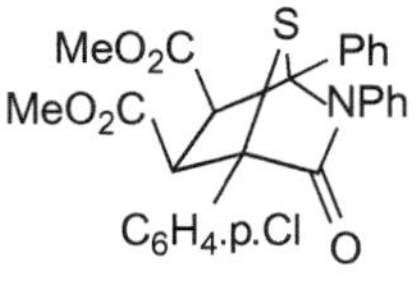

105

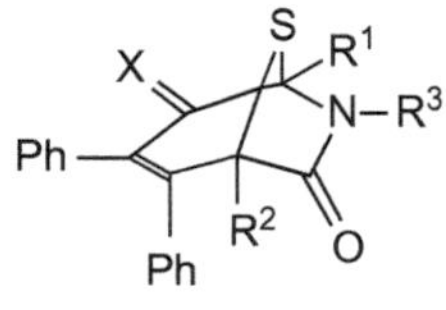

106

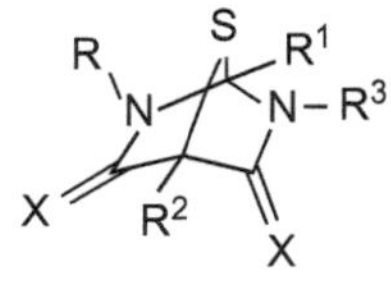

107

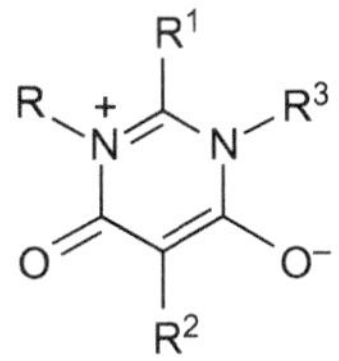

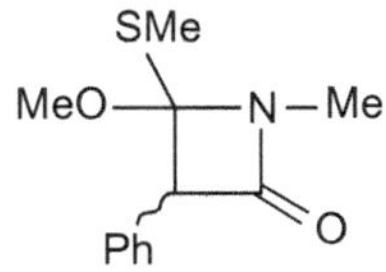

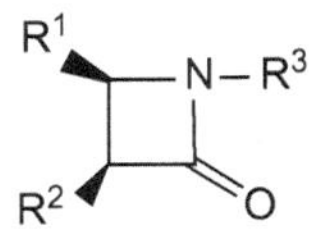

108 **109** **110**

Several studies of the photochemistry of meso-ionic 1,3-thiazolium-4-olates (**90**) have been reported.[54,55] Photolysis of compound (**90**; R^1 = SMe, R^2 = Ph, R^3 = Me) in methanol solution generates the *β*-lactam (**109**) as a mixture of diasterioisomers.[84] A better method of converting meso-ionic 1,3-thiazolium-4-olates (**90**) to *β*-lactams involves treatment with Raney nickel. Using this reagent, desulfurization of the derivative (**90**; R^1 = R^2 = R^3 = Ph) in a variety of solvents afforded the *cis*-*β*-lactam (**110**; R^1 = R^2 = R^3 = Ph) in excellent yield. This transformation (**90** → **110**) has been carried out on other meso-ionic 1,3-thiazolium-4-olates (**90**) and the mechanism has been discussed.[106,107]

The effect of substituents on the color of the polycyclic derivatives (**111**) has been discussed.[108] With alcohols these systems give the adducts (**112**).[109] A kinetic study of the acid catalyzed hydrolysis of the derivatives (**113**) has been reported.[110]

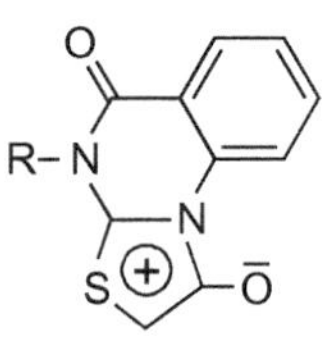

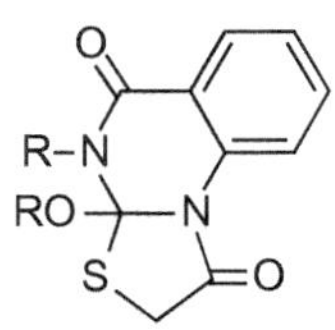

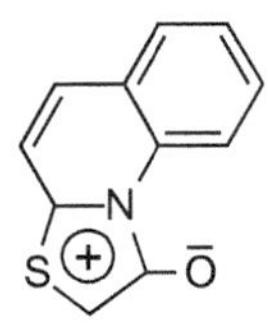

111 **112** **113**

3.5.2 1,3-Thiazolium-4-aminides (114)

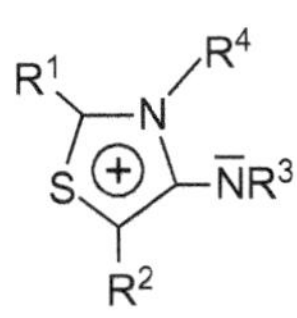

114

No new work on this system (**114**) has been reported since the subject was last reviewed.[6]

3.5.3 1,3-Thiazolium-5-olates (115)

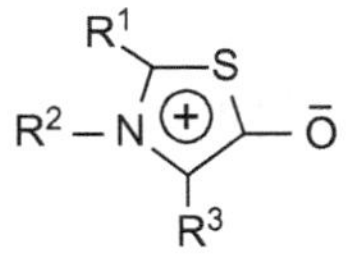

115

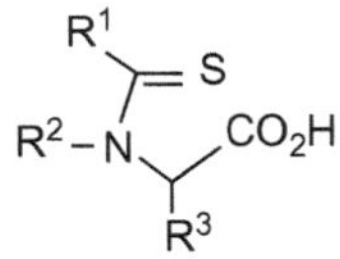

116

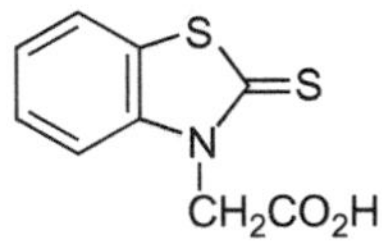

117

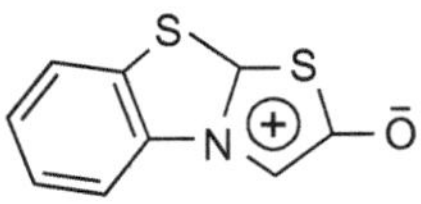

118

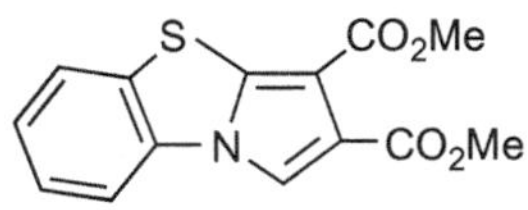

119

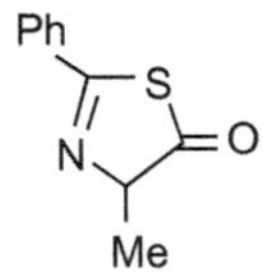

120

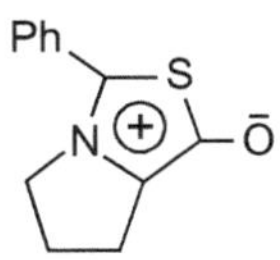

121

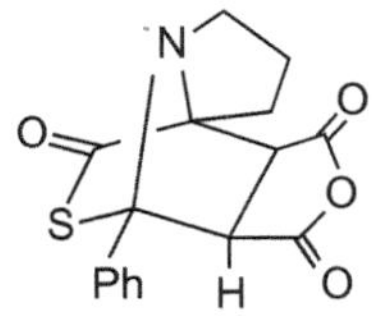

122

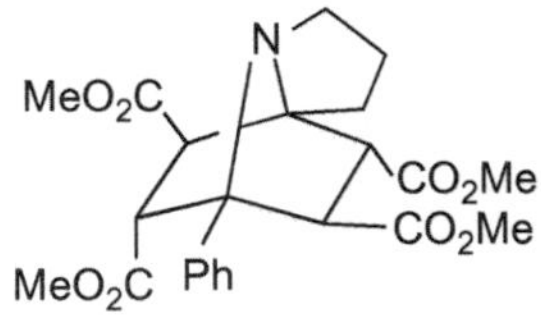

123

The 1,3-thiazolium-5-olates (**115**) are well known and have continued to receive attention. Cyclodehydration of *N*-thioacylglycines (**116**) is the best route to this system and this method has been used to prepare a series of monocyclic derivatives unsubstituted at position 4 (**115**; $R^1=R^2=Ar$, $R^3=H$).[111] Electrophilic substitution gives 4-acyl derivatives (**115**; $R^1=R^2=Ar$, $R^3=COMe$ or COPh).[111] A similar cyclodehydration of the acid (**117**) using acetic anhydride generates the unstable polycyclic meso-ionic 1,3-thiazolium-5-olate (**118**) which is trapped by dimethyl acetylenedicarboxylate yielding the pyrrolo [2,1-*b*] benzothiazole (**119**).[82]

It has been demonstrated that the 5(4*H*) thiazolone (**120**) is in equilibrium with the meso-ionic tautomer (**115**; $R^1=Ph$, $R^2=H$, $R^3=Me$) which can be trapped by dimethyl acetylenedicarboxylate to give 3,4-dimethoxycarbonyl-2-methyl-5-phenylpyrrole (66%).[112]

Addition reactions with other dipolarophiles have been investigated. Compound (**115**; $R^1=R^3=Ph$, $R^2=Me$) and benzyne give 1,3-diphenyl-2-methyl isoindole in low yield.[30]

Oxabicyclo [2.2.1] heptane (**35**; X=O) and compound (**115**; $R^1=R^3=Ph$, $R^2=Me$) give mainly 2,5-diphenyl-1-methyl pyrrole.[35]

With maleic anhydride, compound (**121**) gives the primary adduct (**122**) whereas with dimethyl fumarate the same compound gave the adduct (**123**).[113]

Two groups have investigated the reaction between the meso-ionic compounds (**115**; R^1=Ar, R^2=Me, R^3=H, Ph) and the diphenylcyclopropene derivatives (**124**; X=O, S, $NSO_2{\cdot}C_6H_4{\cdot}p{\cdot}CH_3$, $C(CN)_2$, C(CN) · CO_2Et).[43,44] With the thione (**124**; X=S), pyrid-4-thiones (**125**) were obtained. Additions with heterocumulenes and *o*-quinones have also been described.[113,48]

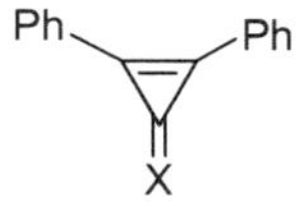

124

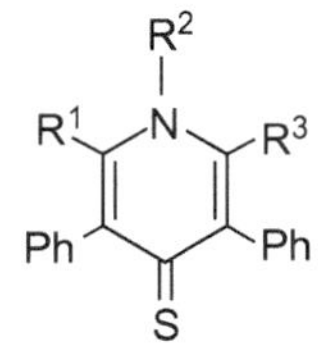

125

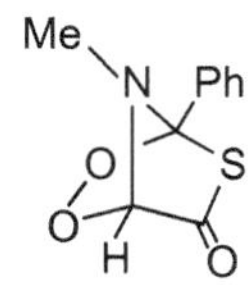

126

Irradiation of the derivative (**115**; R^1=Ph, R^2=Me, R^3=H) in methanol solution yields the methyl ester of the acid (**116**, R^1=Ph, R^2=Me, R^3=H) and similar results have been reported for the derivative (**115**; R^1=SMe, R^2=Me, R^3=$COCF_3$). Interestingly, in the presence of oxygen, partial formation of the oxygen adduct (**126**) occurs: this adduct (**126**) then undergoes fragmentation giving carbonoxysulfide and *N*-formyl-*N*-methylbenzamide.[53,54]

3.5.4 1,3-Thiazolium-5-aminides (127)

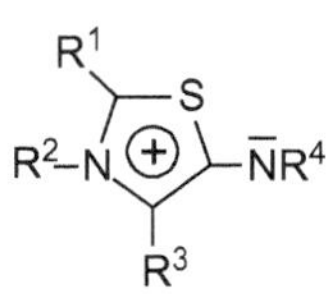

127

No new work on this system (**127**) has been reported since the subject was last reviewed.[6]

3.5.5 1,3-Thiazolium-5-thiolates (128)

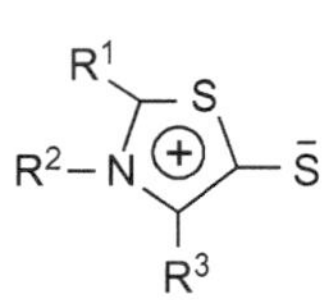

128

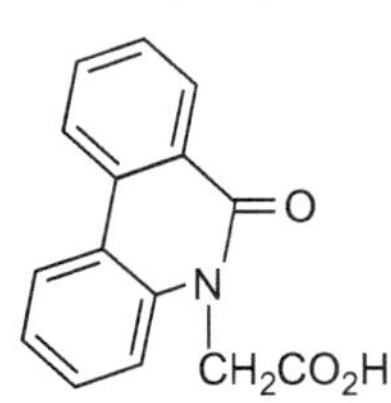

129

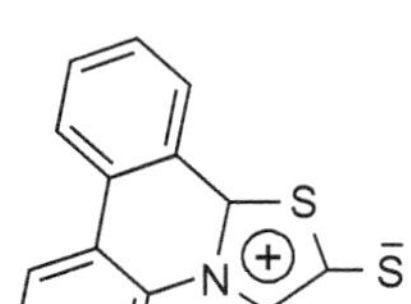

130

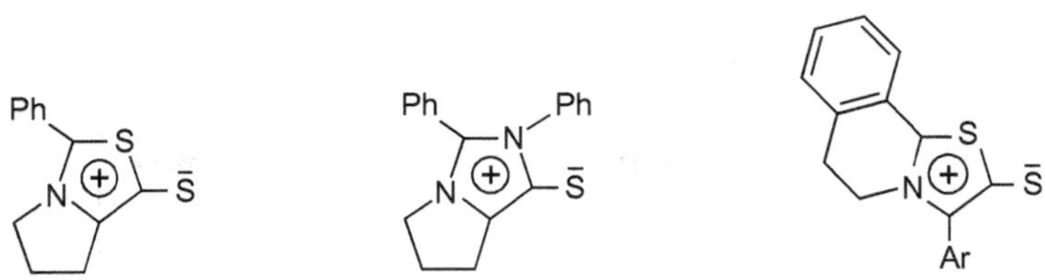

131 **132** **133**

Treatment of the acid (**129**) with phosphorus pentasulfide in boiling pyridine yields the polycyclic 1,3-thiazolium-5-thiolate (**130**). However, attempts to extend this method to other meso-ionic 1,3-thiazolium-5-thiolates failed.[46] Treatment of meso-ionic 1,3-oxazolium-5-olates[29] (Section 3.2.3) or 1,3-thiazolium-5-olates (**115**) (Section 3.5.3) with carbon disulfide remains the best route to this system (**128**).[46,113] Reactions of 3,4-dihydro isoquinolinium methylide salts with carbon disulfide in the presence of base generates meso-ionic 1,3-thiazolium-5-thiolates (**133**). The mechanism of this transformation has been discussed in detail.[114]

In general, meso-ionic 1,3-thiazolium-5-thiolates (**128**) are regarded as unreactive toward 1,3-dipolarophiles. Compound (**131**) did react with phenylisothiocyanate to give the 1,3-diazolium-4-thiolate (**132**) and with dimethylfumarate to give adduct (**123**).[115]

3.6 Selenazoles

3.6.1 1,3-Selenazolium-4-olates (134)

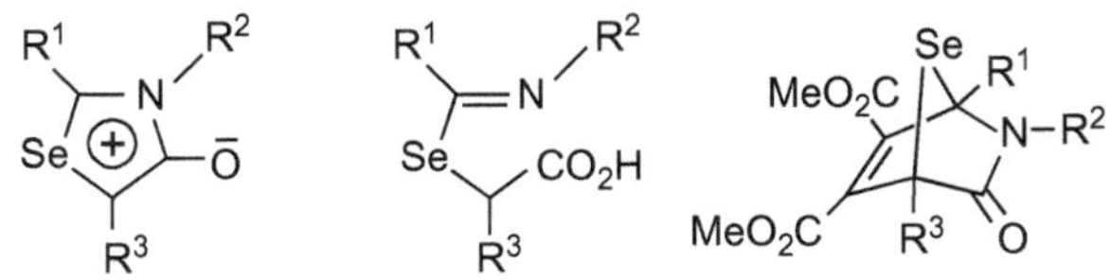

134 **135** **136**

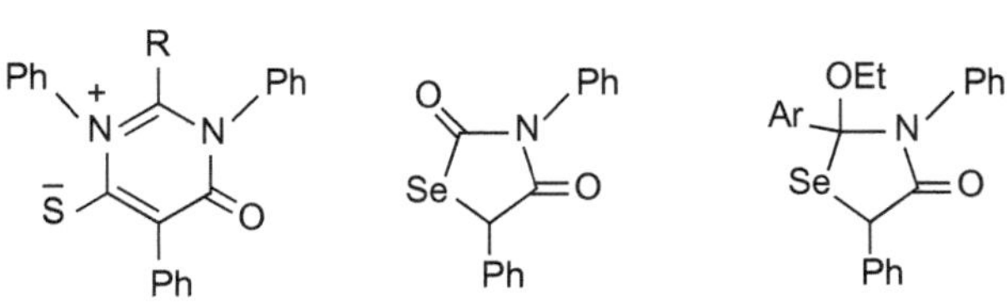

137 **138** **139**

The first example of a selenium containing meso-ionic compound (**134**; $R^1=R^2=R^3=Ph$) was described in 1975 and was obtained by cyclodehydration of the *α*-selenoacid (**135**; $R^1=R^2=R^3=Ph$).[116] More recently, a variety of meso-ionic 1,3-selenazolium-4-olates (**134**) have been prepared by condensation of selenoamides ($R^1C=Se \cdot NHR^2$) with *α*-bromoacyl chlorides ($R^3CHBr \cdot COCl$).[117] With dimethyl acetylenedicarboxylate, the 1,3-selenazolium-4-olates (**134**) yield 2-pyridones by elimination of selenium from the initial adducts (**136**). The derivative (**134**; $R^1=p \cdot MeO \cdot C_6H_4$, $R^2=R^3=Ph$) and phenylisothiocyanate undergo a similar reaction giving the cross-conjugated mesomeric betaine (**137**; $R=p \cdot MeO \cdot C_6H_4$). Compound (**134**; $R^1=SMe$, $R^2=R^3=Ph$) undergoes hydrolysis to the selenazolidine dione (**138**) and reaction of compound (**134**; $R^1=p \cdot MeO \cdot C_6H_4$; $R^2=R^3=Ph$) with alcohol gives the adduct (**139**; $Ar=p \cdot MeO \cdot C_6H_4$).[117]

3.7 Dithioles

3.7.1 *1,3-Dithiolium-4-olates (140)*

The meso-ionic 1,3-dithiolium-4-olates (**140**) can be generated either by condensation of dithiobenzoic acids ($ArCS_2H$) with *α*-haloacyl halides ($R^2CHX \cdot COX$)[66] or by cyclodehydration of the dithioester derivatives (**141**).[118,119] A compound previously assigned the 4-unsubstituted structure (**140**; $R^1=Ph$, $R^2=H$)[120] has been shown to be incorrectly formulated.[119,121] The product, m.p. 185–86°, obtained by treatment of thiobenzoylthioglycollic acid (**141**; $R^1=Ph$, $R^2=H$) with boron trifluoride etherate and acetic anhydride has been reformulated as the intramolecular cyclo-adduct (**142**)[121] and is not the meso-ionic derivative (**140**; $R^1=Ph$, $R^2=COCH_2S_2CPh$) as previously described.[119]

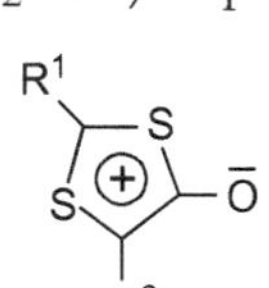

140

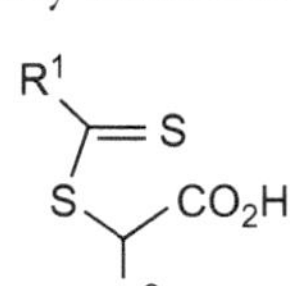

141

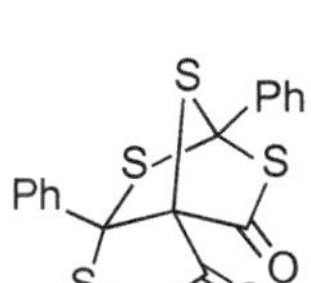

142

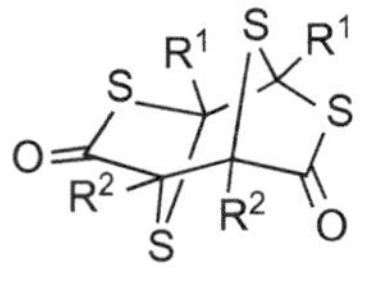

143

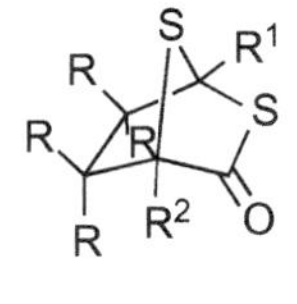

144

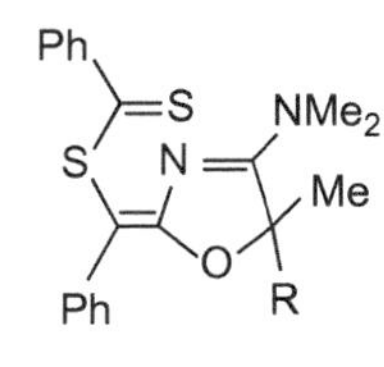

145

An interesting aspect of some 1,3-dithiolium-4-olates (**140**) is their reversible dimerization (**140** → **143**). The dimeric structures (**143**) have been confirmed by X-ray crystallography. In solution, equilibrium between the colorless dimers (**143**) and the red monomers (**140**) is established and the meso-ionic monomer can be trapped by 1,3-dipolarophiles.[122,123]

The meso-ionic derivatives (**140**) are very reactive 1,3-dipoles. Reaction with acetylenes furnishes substituted thiophenes.[119,124–126] Regioselective additions of unsymmetrical alkynes and alkenes have been rationalized by frontier orbital theory.[126,127] Addition of benzyne generates isobenzothiophenes.[30] The 1:1 cycloadducts (**144**) have been obtained by reaction with a wide variety of olefinic 1,3-dipolarophiles, including *N*-phenylmaleimide,[119,123,128] acrylonitrile,[119] ethylacrylate,[119] maleic anhydride,[123,128] dimethyl maleate,[128] dimethyl fumarate,[128] 1,2,3-triphenylcyclopropene,[129] *p*-quinone,[129] 7-oxa-bicycloheptadienes,[35] 7-aza-bicycloheptadienes,[36] diphenylcyclopropenone derivatives (**124**),[44] and tricyclodecatriene.[103] Primary adducts are also formed with hetero double bonds such as formaldehyde[130] and 2,3-diphenylazirine.[119] 2-Dimethylamino-azirines, however, have been found to give the adducts (**145**) which may be formed by initial valence tautomerism of the meso-ionic precursor (**140**).[52]

Photo-oxygenation of the diphenyl derivative (**140**; $R^1 = R^2 = Ph$) gives a low yield of the dimer (**143**; $R^1 = R^2 = Ph$) together with dibenzoyldisulfide.[55] The same derivative (**140**; R^1–$R^2 = Ph$) with aniline yields the meso-ionic 1,3-thiazolium-4-olate (**90**; $R^1 = R^2 = R^3 = Ph$) (see Section 3.5.1).[131]

3.7.2 1,3-Dithiolium-4-aminides (146)

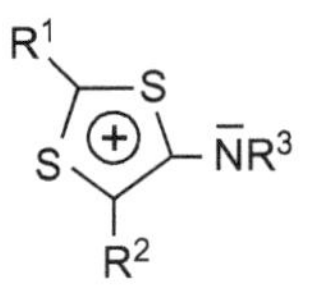

146

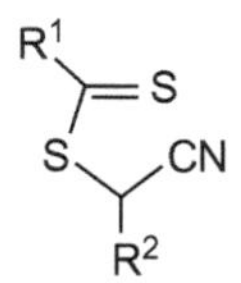

147

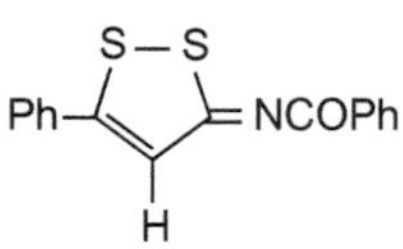

148

Treatment of the nitrile (**147**; R^1 = morpholino, $R^2 = Ph$) with acetyl chloride followed by deprotonation gives the meso-ionic derivative (**146**; R^1 = morpholino, $R^2 = Ph$, $R^3 = COMe$) in good yield.[125] Mechanistic aspects of the photochemical rearrangement (**146**; $R^1 = Ph$, $R^2 = H$, $R^3 = COPh$ → **148**) have been described.[132]

3.8 Oxadiazoles

3.8.1 1,2,3-Oxadiazolium-5-olates (Sydnones) (149)

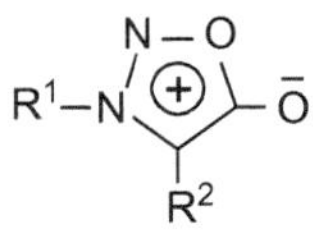

149

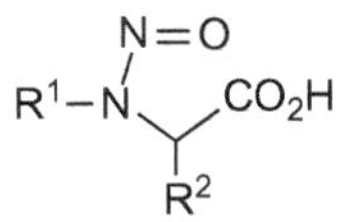

150

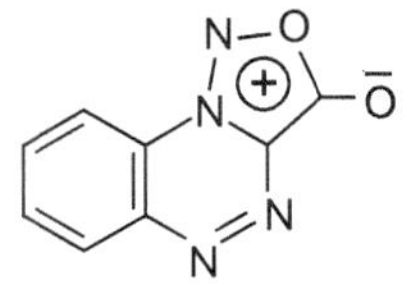

151

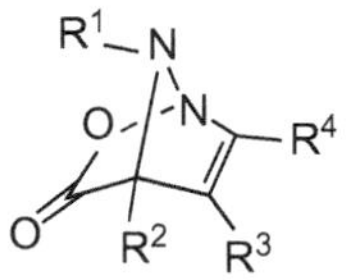

152

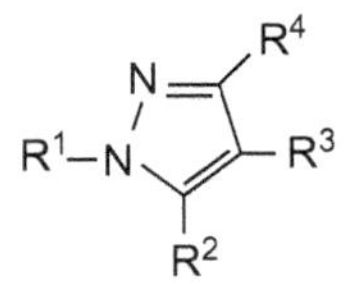

153

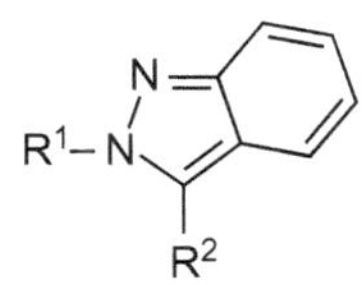

154

Cyclodehydration of *N*-nitroso-*α*-amino-acids (**150**) remains the sole synthetic route available for the preparation of sydnones (**149**).[3] Recent examples include the synthesis of 3-methyl sydnone (**149**; R^1 = Me, R^2 = H) by cyclization of the acid (**150**; R^1 = Me, R^3 = H) using trifluoroacetic anhydride[133] and similar preparations of several 3-benzylsydnones (**149**; R^1 = $ArCH_2$, R^2 = H, Ph).[134] For 4-unsubstituted derivatives (**149**; R^2 = H), modification of the ring substituents can be achieved by electrophilic substitution and a number of novel sydnones (**149**; R^1 = Ar, Me_2N, piperidino and morpholino, R^2 = COMe, Br, SMe, SPh, SO · Ph, CN and CH=NOH) have been made in this way.[135] Anodic halogenation of 4-unsubstituted sydnones gives 3-aryl-4-bromo and 4-chlorosydnones (**149**; R^1 = Ar, R^2 = Br, Cl) but fluoridation and iodination were not successful.[136] 4-Iodo-3-(3′-pyridyl)sydnone (**149**; R^1 = 3-pyridyl, R^2 = I) has been prepared from the unsubstituted sydnone and *N*-iodo succinimide.[137]

Earlier work has demonstrated that nitration of 3-phenylsydnone (**149**; R^1 = Ph, R^2 = H) yields the 4-nitro derivative (**149**; R^1 = Ph, R^2 = NO_2)[138] whereas nitration of 3-(4′-tolyl) sydnone (**149**; R^1 = *p* · Me · C_6H_4, R^2 = H) occurs on the tolyl substituent.[139] Recently, nitration of 3-benzylsydnone has been shown to give 3-(3’-nitrobenzyl) sydnone (**149**; R^1 = *m* · NO_2 · $C_6H_4CH_2$, R^2 = H) and nitration of 3,4-diphenylsydnone (**149**; R^1 = R^2 = Ph) gave 3-phenyl-4(4’-nitrophenyl)syndnone (**149**; R^1 = Ph. R^2 = *p* · NO_2 · C_6H_4).[140] Further examples of electrophilic substitution of 4-unsubstituted sydnones (**149**; R^2 = H) include introduction of diazo substituents: intramolecular addition gives the tricyclic derivative (**151**).[141] 4-Acetylsydnones (**149**; R^1 = Ar, $ArCH_2$, R^2 = COMe) are formed by acetylation using a mixture of acetic anhydride and 85% phosphoric acid[142] and can be converted into 4-diazo acetylsydnones (**149**; R^1 = Ar, R^2 = $COCHN_2$).[143,144] 4-Chlorosulfonation readily yields 4-chlorosulfonylsydnones (e.g., **149**; R' = Ph, R^2 = SO_2Cl)[145] and 4-formylation is achieved under Vilsmeier–Haack conditions.[146] These products (**149**; R^1 = Ar, R^2 = CHO) have been converted to olefines using the Wittig reaction.[146]

An alternative route to 4-olefinic sydnones involves metalation, reaction with a ketone and subsequent dehydration of the intermediate alcohol.[147–150] In another method, 4-hydroxymethylation is accomplished by treating the 4-unsubstituted sydnones with formaldehyde/NaBr affording some of the derivatives (**149**; R^1 = Ar, R^2 = CH_2OH)).[151] Use of excess formaldehyde or formaldehyde in the presence of sulfuric acid gives *bis*-sydnones linked by a methylene group[152] *Bis*-sydnones linked at the 4-position by a sulfur atom have been prepared by the action of thionyl chloride on 4-unsubstituted sydnones[153] and 3,3′-diphenyl-4,4′-*bis*-sydnone has been prepared by treating 4-bromo-3-phenylsydnone (**149**; R^1 = Ph, R^2 = Br) with copper bronze.[154] Synthesis of 3,4-*bis*(4-sydnonylcarbonyl) furoxans is achieved by treating the sydnones (**149**; R^1 = Me, Ph, R^2 = $CO{\cdot}CH_2{\cdot}CO{\cdot}CO_2{\cdot}Me$) with fuming nitric acid.[155]

1,3-Dipolar cycloaddition reactions of sydnones continue to receive attention. Reaction with acetylenes is a well established route to pyrazoles (**153**). Recent papers describe (i) preparation of 3-benzoylpyrazoles (**153**; R^1 = $PhCH_2$, *iso*-Bu, R^2 = CH_2CO_2Me, R^3 = H, R^4 = PhCO) using benzoylacetylene[156]; (ii) 5-chloropyrazoles (**153**; R^1 = $PhCH_2$, R^2 = C1, R^3 = R^4 = CO_2Me) using4-chlorosydnones[157] and (iii) *N*-amino pyrazoles (**153**; R^1 = NMe_2, morpholino, piperidino) using *N*-amino-sydnones (**149**; R^1 = NMe_2, morpholino, piperidino).[158] Gotthardt and Reiter have investigated both thermally[159] and photochemically[160] induced reactions of a wide variety of sydnones with acetylenes. Thermal cycloaddition generates the adducts (**152**) which undergo loss of carbon dioxide generating the pyrazoles (**153**).[159] Photochemical reaction of sydnones with acetylenes also gives pyrazoles but by a different mechanism (see Scheme 5 and related discussion). Photochemical addition is a useful adjunct to the thermal reaction since the pyrazoles formed are isomeric to those formed thermally.

3-Benzyl-4-phenylsydnone (**149**; $R^1 = PhCH_2$, $R^2 = Ph$) and 4-methyl-3-phenylsydnone (**149**; R^1 = Ph $R^2 = Me$) react thermally with benzyne giving the benzo[*c*]pyrazoles (**154**; $R^1 = PhCH_2$, $R^2 = Ph$) (68%) and (**154**; $R^1 = Ph$, $R^2 = Me$) (64%), respectively.[30]

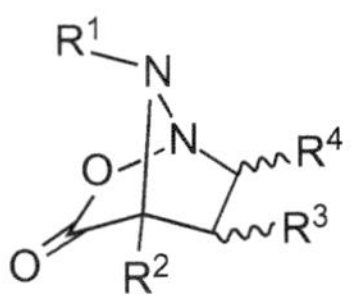

155

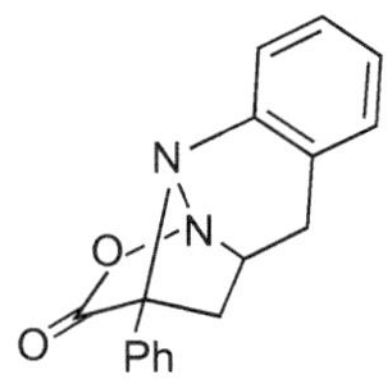

156

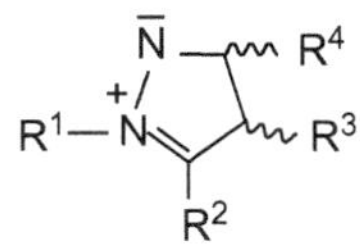

157

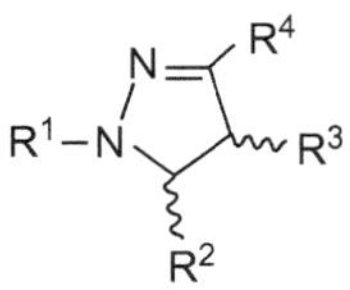

158

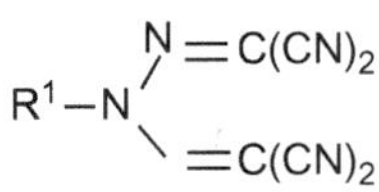

159

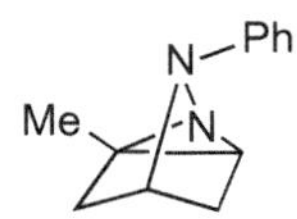

160

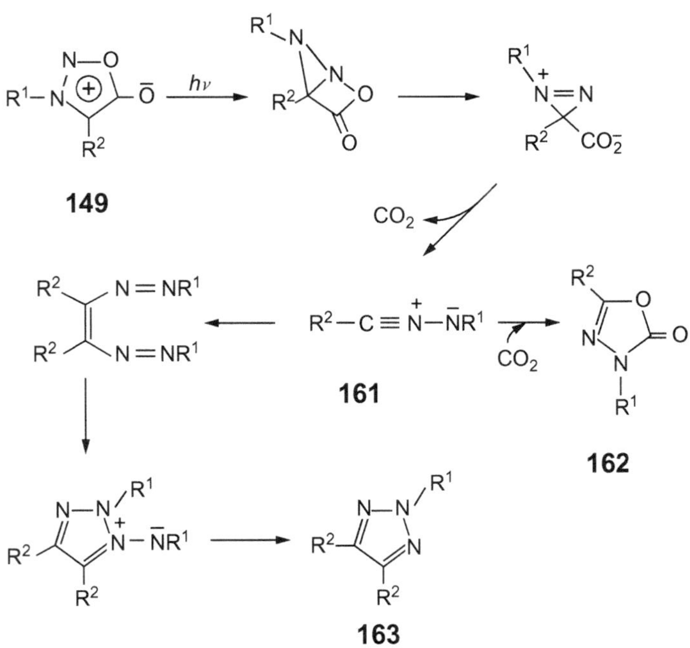

Scheme 5

With olefinic dipolarophiles, sydnones (**149**) yield unstable 1:1 adducts (**155**) which can only be isolated in special circumstances: intramolecular 1,3-dipolar cycloaddition of the sydnone (**149**; $R^1=2'$-allylphenyl, $R^2=Ph$) yields the bridged adduct (**156**).[161,162] More usually, cycloreversion involving loss of carbon dioxide giving the dipolar species (**157**) occurs and proton transfer then yields pyrazolines (**158**).[163,164] In the presence of dipolarophiles the intermediate (**157**) may be trapped: reaction of 3-phenylsydnone with isoprene generates some of the caged product (**160**) together with 3-isopropenyl-1-phenyl pyrazole.[165] An alternative method of fragmentation of the dipolar intermediates (**157**) has been observed when the dipolarophile is tetracyanoethylene[166–168]; the species [**157**; $R^3=R^4=(CN)_2$] undergoes CC-bond cleavage yielding [(dicyanovinyl)hydrazono] malononitriles (**159**).

Photolysis of sydnones continues to attract interest[169–171] In the absence of external trapping agents, irradiation of sydnone derivatives usually yields either 1,2,3-triazoles (**163**) and/or 1,3,4-oxadiazol-2-ones (**162**) as major products. Recent mechanistic proposals for the formation of these products are shown in Scheme 5. This mechanism avoids high energy antiaromatic diazirene intermediates which have previously been proposed. The formation of various minor photolysis products can also be rationalized using this mechanism as a basis (Scheme 5).[170,171] Evidence that formation of the nitrileimines (**161**) involves creation of a bond between the 2 and 4-positions in photoexcited sydnones (**149**) has been provided by the observation that syndones labelled with ^{15}N at position 2 give 1,2,3-triazoles (**163**) in which the label is incorporated at both the 1 and 3 positions.[170]

The intermediacy of the nitrileimines (**161**) has been demonstrated by trapping with dipolarophiles. Acetylenes yield pyrazoles[160] which are isomeric to those formed thermally from the sydnone and the acetylene. With alkenes, 2-pyrazolines are generated[160] For example, 3,4-diphenylsydnone (**149**; $R^1=R^2=Ph$) on irradiation in the presence of cyclopentadiene affords the 2-pyrazoline (**164**)[160] and 2-vinylpyridine and 3-methyl-4-phenylsydnone (**149**; $R^1=Me$, $R^2=Ph$) yield the 2-pyrazoline.[165,172] For unsymmetrical alkenes, the observed regioselectivity of addition has been interpreted qualitatively in terms of M.O. perturbation theory.[160] In the presence of carbon disulfide, irradiation of diphenylsydnone (**149**; $R^1=R^2=Ph$) yields the adducts (**166**; R=Ph) and (**167**).[160] 3-Methyl-4-phenylsydnone (**149**; $R^1=Me$, $R^2=Ph$) similarly yields compound (**166**; R=Me) with carbon disulfide and the triazole (**168**) with phenyl isocyanate.[172] Irradiation of diphenylsydnone (**149**; $R^1=R^2=Ph$) in the presence of tetracyclone gives the oxadiazospiro compound (**169**).[173] The mechanism of formation of the unusual photoproduct (**170**), formed from 3-t-butyl-4-phenylsydnone (**149**; $R^1 = t\text{-Bu}$, $R^2 = Ph$), has been unravelled.[170]

Miscellaneous studies of the photochemistry of sydnones include the Z/E photoisomerism of 4-styrylsydnones[174,175] and 4-azomethine sydnones.[176] Irradiation of 3-morpholino-4-thiophenyl sydnone (**149**; R^1 = morpholino, R^2 = SPh) results in a remarkable rearrangement yielding the tetrapolar product (**171**).[177,178] In some cases, photooxygenation of sydnones has been shown to yield *N,N′*-diaroylhydrazines in addition to triazoles.[179]

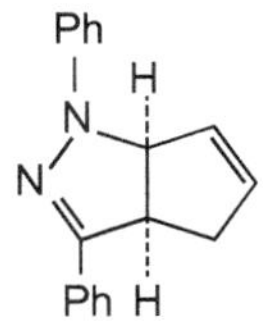
164

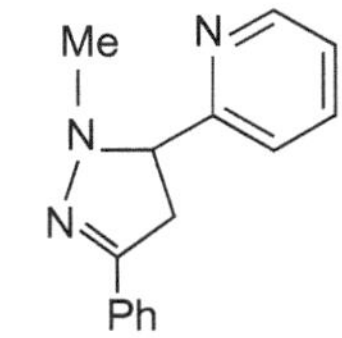
165

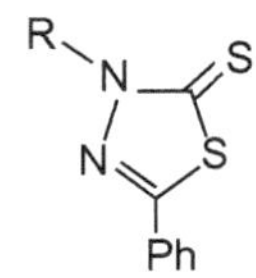
166

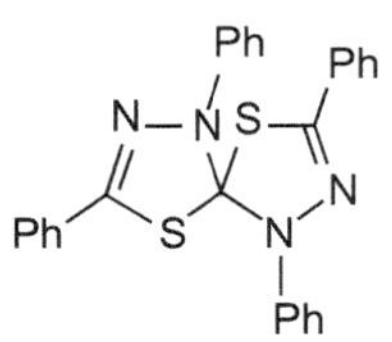
167

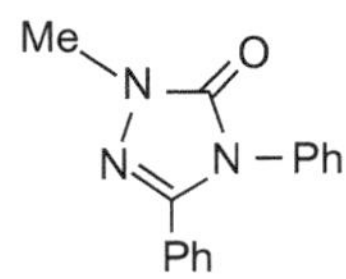
168

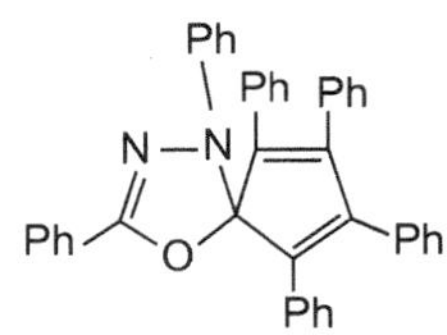
169

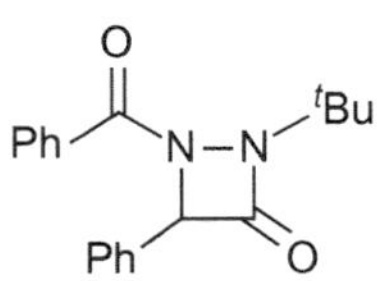
170

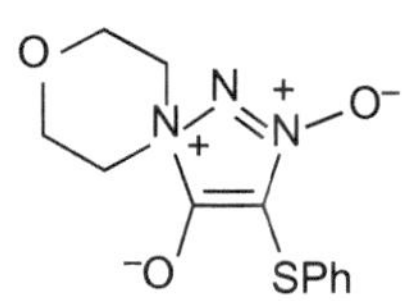
171

Acid catalyzed hydrolysis of sydnones (**149**) yielding hydrazines provides a useful method of converting amines to hydrazines and has been used to synthesize the optically active (−)-(S) hydrazine (**173**) from the (+)-(S) amine (**172**).[180] Alkylhydrazines for oxidative studies have also been prepared by this method.[181]

The kinetics of base catalyzed decomposition of 3-carboxyalkyl sydnones (**149**; $R^1 = (CH_2),CO_2H$, $R^2 = H$) have been studied.[182] Other chemical studies on sydnones include: (i) reaction with 1,3,2-dioxaborolane-4-ones forming 1:1 adducts[183]; (ii) ring opening with triethylborane giving compounds of the type (**174**)[184]; (iii) silylation of 4-1ithiosydnones (**149**; $R^2 = Li$) to give 4-silasydnones.[185]

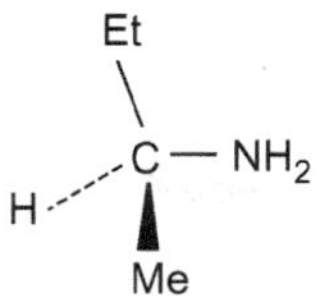

172

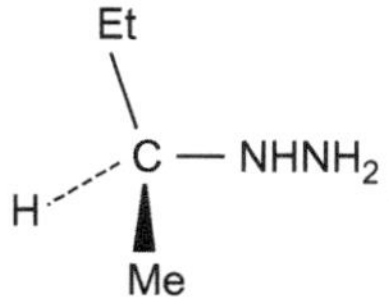

173

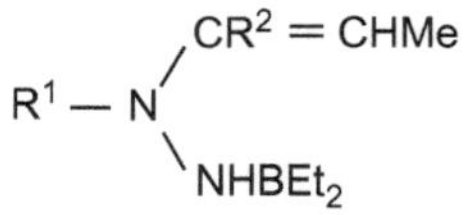

174

Physical and spectral properties of the sydnones have been the subject of several recent papers. X-ray studies have been reported for 3-phenyl-4-(3-methyl-1-buten-2-yl) sydnone (**149**; R^1=Ph, R^2=CH_2=C-[CH$(CH_3)_2$]),[186] 4-bromo-3-(3-pyridyl)sydnone (**149**; R^1=3-pyridyl, R^2=Br),[187] 4(1-cyclohexenyl)-3-phenylsydnone (**149**; R^1=Ph, R^2=1-cyclohexenyl),[188] 3(2-aminophenyl)sydnone (**149**; R^1=*o*-NH_2·C_6H_4, R^2=H),[205] and the tricyclic derivative (**151**).[205] ^{14}N- and ^{15}N-NMR studies are in agreement with the meso-ionic formulation of sydnones.[189,190] Mass spectra[135,191] and UV spectra[192] have been recorded. Quantum yields and fluorescence decay times have been measured for several sydnones undergoing photoreactions.[193] The photochromic properties of sydnones have been the subject of several papers.[194–201] Dielectric constants, viscosities, densities and refractive indices have been recorded for several liquid 3-alkylsydnones[202]: 3-methylsydnone has been described as a good non-aqueous solvent which promotes ionization of solutes.[203] The polarographic behavior of several sydnones has also been studied[204] and theoretical investigations of the structure of sydnones have been described.[206–208]

The biological activity of sydnones was reviewed in 1967[209] and notable activity up to 1976 has been summarized.[6] Recent investigations of sydnones have claimed anti-inflammatory,[210–213] antimicrobial,[212–214] antimalarial,[215] and CNS activity,[216] or a combination of activities.[217–219] A number of cephalosporin derivatives of sydnones have been prepared.[220–230]

3.8.2 1,2,3-Oxadiazolium-5-aminides (Sydnone imines) (175)

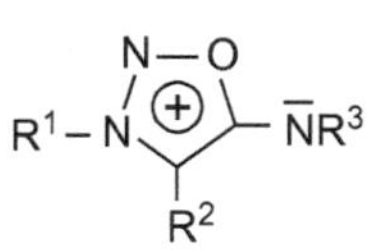

175

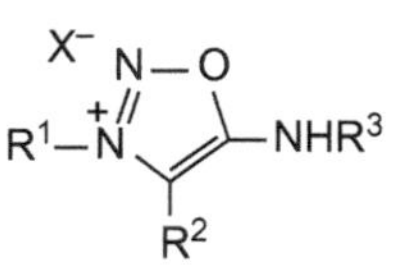

176

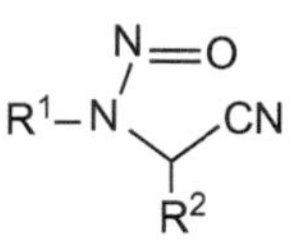

177

A comprehensive review of the chemistry and pharmacological properties of the sydnone imines (**175**) has been published recently.[231] The *N*-unsubstituted sydnone imines (**175**; R^3=H) are unstable and isolable only as the salts (**176**; R^3 = H) which are prepared by acid-catalyzed cyclization of the appropriately substituted *N*-nitroso-α-amino acetonitriles (**177**)[232–234]: the mechanism of this cyclization has been examined in detail.[235–236] Reaction of the 5-amino-oxadiazolium salts (**176**; R^3 = H) with isocyanates[234,237] or acetic anhydride[238] yields *N*-carbamoyl and *N*-acetyl sydnone imines (**175**; R^3 = CONHR or COMe) respectively. Cyclodehydration of *N*-nitroso-α-amino acetonitriles (**177**) with an acylating agent in the presence of triethylamine is a good method of preparing the *N*-acyl derivatives (**175**; R^3 = COR).[239,240]

Electrophilic halogenation of 4-unsubstituted sydnone imines affords 4-halo derivatives but halogenation of aryl substituents may also occur.[241] Alkaline hydrolysis of sydnone imines[242,243] and the mechanism of polarographic reduction[244] of *N*-acyl derivatives have been investigated. Various spectroscopic properties have been studied including ^{14}N-NMR,[189] IR[238] and UV spectra.[245]

Much work on sydnone imines (**175**) has been prompted by the pharmacological properties of Sydnophen (**176**; R^1 = CHMe.CH_2Ph, R^2 = R^3 = H, X = Cl) and Sydnocarb (**175**; R^1 = CHMe.CH_2Ph, R^2 = H, R^3 = CONHPh) which have significant psychotropic properties. Extensive studies of their properties have been reported.[246–277] Another sydnone imine of pharmacological interest is Molsidomine (**175**; R^1 = *N*-morpholino, R^2 = H, R^3 = CO_2Et) which possesses hemodynamic properties.[278–289]

Work continues on the preparation of new sydnone imines and the investigation of their pharmacological activity.[290,291] *N*-Nitrososydnone imines (**175**; R^1 = NO) have been used as cross-linking agents in epoxy resins.[292–294]

3.8.3 1,2,3-Oxadiazolium-5-thiolates (178)

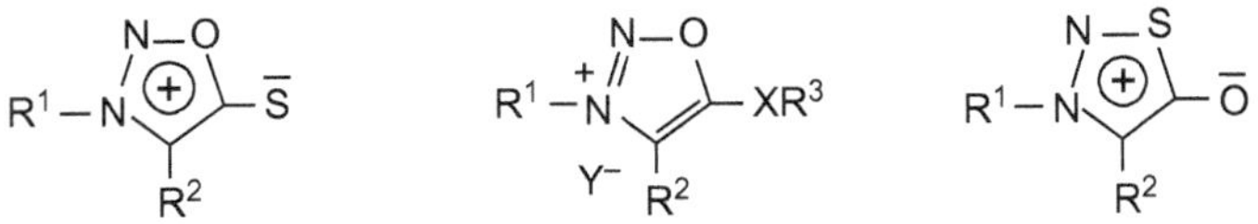

178 **179** **180**

O-Alkylation of sydnones (**149**; R^1 = Ph, Me; R^2 = H, Me, Ph) with triethyloxonium tetrafluoroborate yields the corresponding salts (**179**; X = O, R^3 = Et, Y = BF_4) which on treatment with sodium hydrosulfide at low temperature afford the yellow meso-ionic 1,2,3-oxadiazolium-5-thiolates (**178**; R^1 = Ph, Me; R^2 = H, Me, Ph). This novel meso-ionic system (**178**); rearranges to the isomeric 1,2,3-thiadiazolium-5-olates (**180**) in ethanolic ammonia. With methyl iodide the meso-ionic derivatives (**178**) give the oxadiazolium iodides (**179**; X=S, R^3=Me, Y=I).[295]

3.8.4 1,3,4-Oxadiazolium-2-olates (Isosydnones) (181)

The isosydnones (**181**) are usually prepared by condensation of *N*-acyl-*N*-aryl (or alkyl) hydrazines [$R^1CO \cdot N(R^2)NH_2$] with phosgene. In a similar approach the bicyclic system (**182**) has been prepared by reaction of 4,6-diphenyl-*N*-amino-*α*-pyridone (**183**) with *N*,*N*′-carbonyl diimidazole.[296] The same system (**182**) is also formed by treatment of compound (**183**) with arylisocyanates and pyrolysis of the resulting semicarbazide (**184**).[296]

In a different approach, isosydnones (**181**) have been prepared in excellent yield by treatment of 1,3,4-oxadiazolium-2-thiolates (**190**) (Section 3.8.6) with 2,4-6-trimethylphenyl cyanate.[197]

The mechanism and kinetics of the acid hydrolysis of isosydnones (**181**) to *N*-aroyl-*N*-arylhydrazines have been discussed.[298] Reaction of the derivatives (**181**; R^1=Ph, R^2=Me, Ph) with the dihydroisoquinoline-*N*-imines (**185**) yields the triazoloisoquinolines (**186**).[299] An X-ray spectroscopic study of 4,5-diphenylisosydnone (**181**; R^1=R^2=Ph) has been reported.[205]

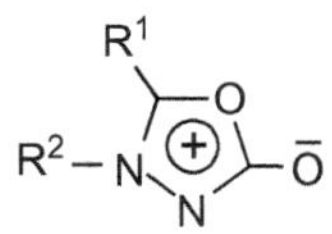

181

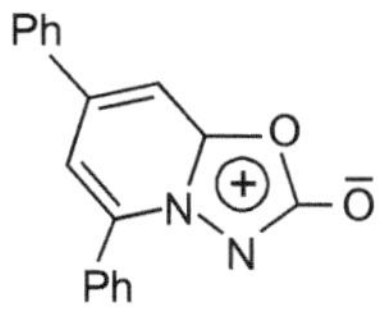

182

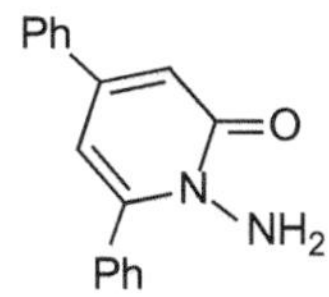

183

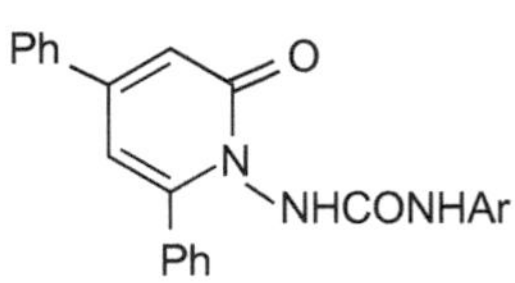

184

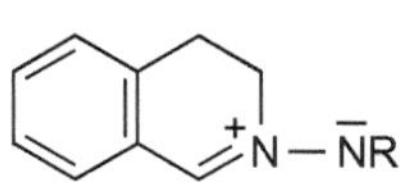

185

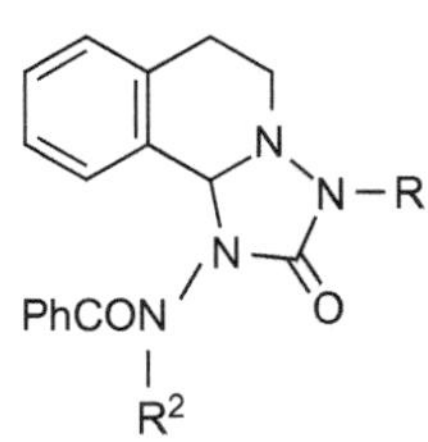

186

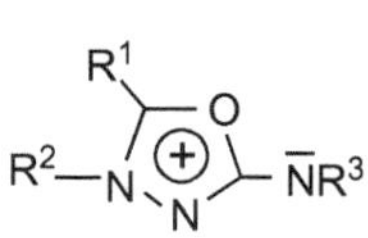

187

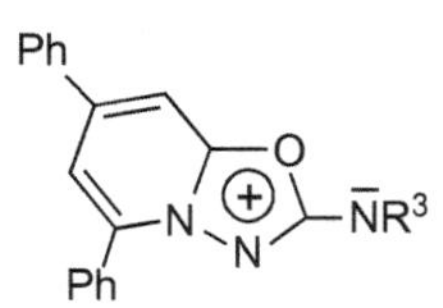

188

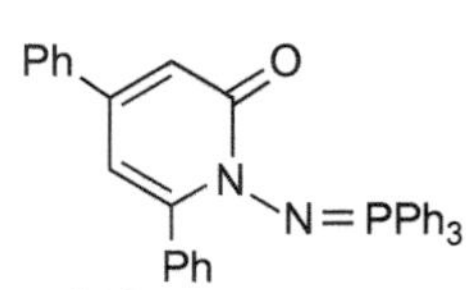

189

3.8.5 1,3,4-Oxadiazolium-2-aminides (187)

Treatment of the *N*-aminopyridone (**183**) with triphenylphosphine dibromide gives the phosphine imine (**189**) which with aryl isocyanates or aryl isothiocyanate yields the bicyclic 1,3,4-oxadiazolium-2-aminides (**188**). Alternatively, de-oxygenation of the semicarbazides (**184**) using triphenylphosphine also yields the derivatives (**188**).[296]

3.8.6 1,3,4-Oxadiazolium-2-thiolates (190)

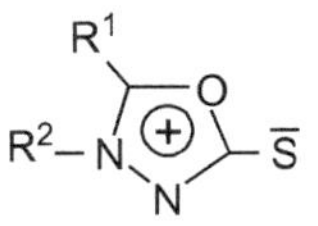

190

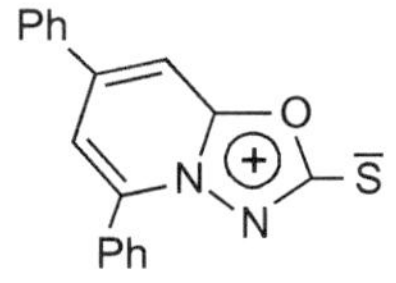

191

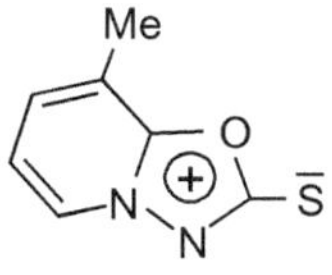

192

The bicyclic system (**191**) has been prepared by reaction of the phosphine imine (**189**) with carbon disulfide[296] Compound (**192**) has been made by condensation of 1-amino-3-methyl pyrid-2-one with thiophosgene.[300]

The monocyclic derivatives (**190**) have been converted to 1,3,4-oxadiazolium-2-olates (**181**) by treatment with 2,4,6-trimethylphenyl cyanate.[297] Reaction of the compound (**190**; $R^1 = Ph$, $R^2 = Me$) with 3,4-dihydroisoquinolin-*N*-imine (**185**; R = H) yields the 1,2,4-triazolium-3-thiolate (**194**) (59%).[299]

3.8.7 1,3,4-Oxadiazolium-2-methylides (193)

No new work on this meso-ionic system (**193**) has been reported since the subject was last reviewed.[6]

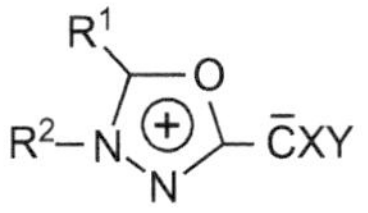

193

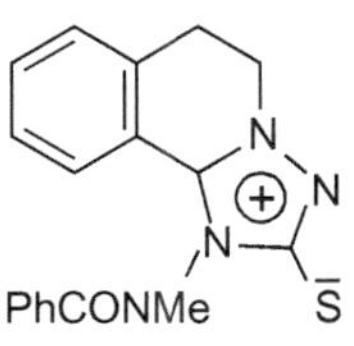

194

3.9 Oxathiazoles

3.9.1 1,3,2-Oxathiazolium-5-olates (195)

The photochemistry of the phenyl derivative (**195**; R = Ph) has attracted special attention. Two photochemical processes can be distinguished. In the first process (Scheme 6) photolysis of the meso-ionic ring (**195**) reversibly generates the acyclic valence tautomer (**196**) which can be identified spectroscopically at low temperature.[301] Photochemical N-S bond cleavage then generates the radical (**197**) which is photooxidized and trapped by ethanol yielding the ester (**198**), The plausible intermediate (**199**) is not involved in the mechanism of formation of the ester (**198**).[302] A second photochemical pathway (Scheme 7) competes with the first pathway under some conditions and yields benzonitrile *via* the nitrile sulfide (**200**).[301,303–305] Evidence for the intermediacy of the nitrile sulfide (**200**) has been provided by UV spectroscopy[303] and by trapping with dimethyl acetylenedicarboxylate yielding 3-phenyl-4,5-dimethoxycarbonyl isothiazole (**201**; R = Ph).[303]

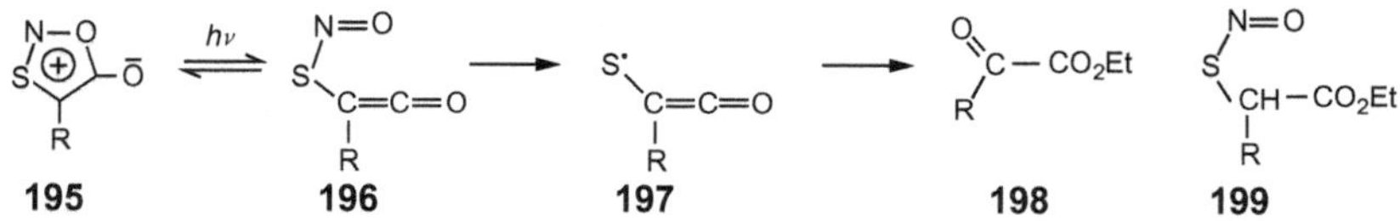

Scheme 6

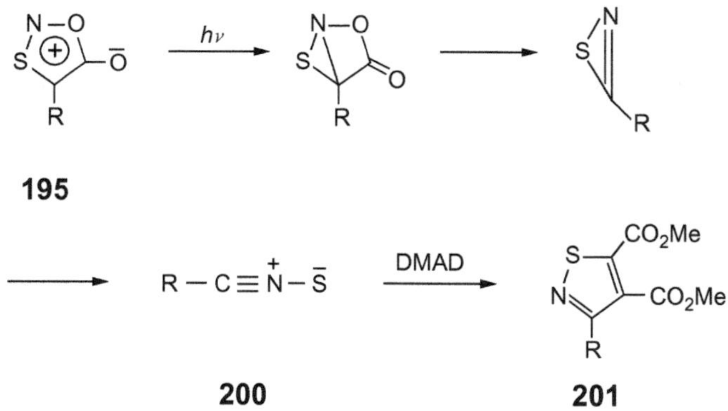

Scheme 7

3.10 Triazoles

3.10.1 1,2,3-Triazolium-4-olates (202)

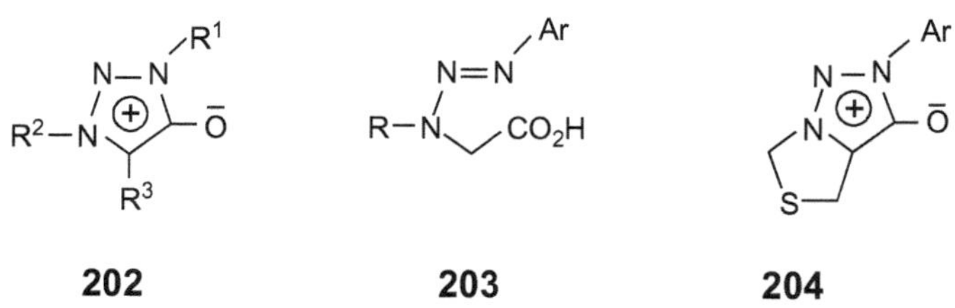

Treatment of aryldiazonium salts with α-aminoacids yields the arylazoaminoacetic acids (**203**) which are cyclodehydrated (Ac_2O/pyridine) giving meso-ionic 1,2,3-triazolium-4-olates (**202**; $R^1 = Ar$, $R^2 = alkyl$, $R^3 = H$) in good yield. The compounds (**202**; $R^1 = Ar$, $R^2 = alkyl$, $R^3 = H$) and some bicyclic derivatives (e.g., **204**) have been described as potent herbicides.[306,307]

3.10.2 1,2,3-Triazolium-4-aminides (205)

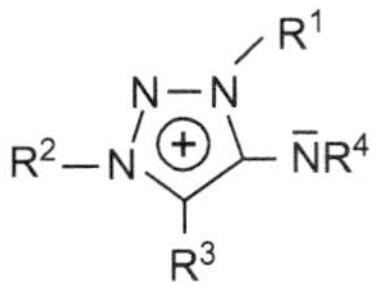

205

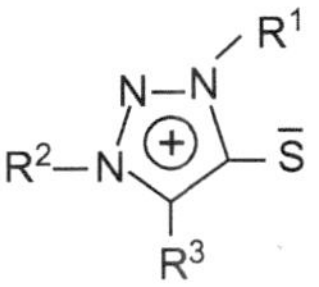

206

No new studies on this meso-ionic system (**205**) have been reported since the subject was last reviewed.[6]

3.10.3 1,2,3-Triazolium-4-thiolates (206)

An X-ray structure determination of compound (**206**; $R^1 = R^2 = Me$, $R^3 = H$) has been reported.[308]

3.10.4 1,2,4-Triazolium-3-olates (207)

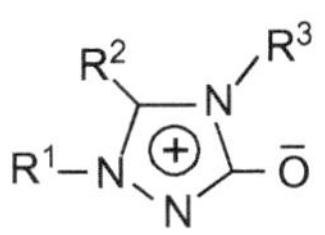

207

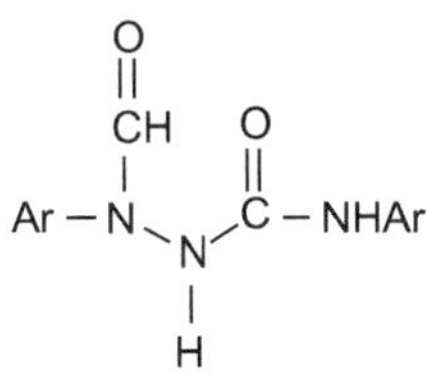

208

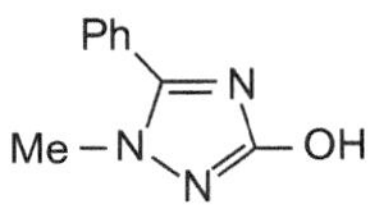

209

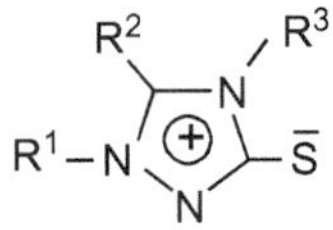

210

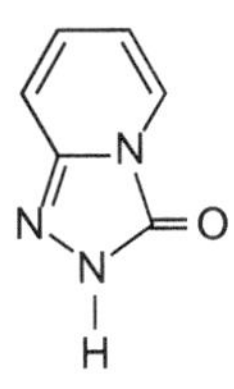

211

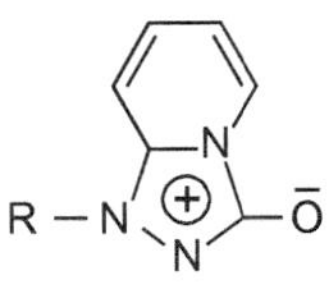

212

Meso-ionic 1,2,4-triazolium-3-olate derivatives (**207**) have been prepared by a wide variety of methods. Thermal ring closure of 1,4-diaryl-l-formyl semicarbazides (**208**) yields 5-unsubstituted systems (**207**; $R^1 = R^3 = Ar$, $R^2 = H$).[169] An alternative synthesis is exemplified by the cycloaddition of phenylisocyanate to 3-phenylsydnone (**149**; $R^1=Ph$, $R^2=H$) giving compound (**207**; $R^1=R^3=Ph$, $R^2=H$).[169] Methylation of the hydroxytriazole (**209**) with methyl iodide in aqueous sodium hydroxide gives primarily the meso-ionic product (**207**; $R^1=R^3=Me$, $R^2=Ph$).[309] Treatment of meso-ionic 1,2,4-triazolium-3-thiolates (**210**) (Section 3.10.6) with 2,4,6-trimethylphenyl cyanate also yields 1,2,4-triazolium-3-olates (**207**).[297]

Bicyclic systems (**212**) have also been prepared. Benzylation of 1,2, 4-triazolo [4,3-*a*] pyridin-3(2*H*)-one (**211**) yields some of the meso-ionic product (**212**; $R=PhCH_2$). A better method of preparing these compounds (**212**; R = alkyl) from (**211**) involves the following procedure: (i) trimethylsilylation; (ii) alkylation in the presence of $HgBr_2$ and (iii) decomposition of the $HgBr_2$ complex with H_2S or ion-exchange resin.[310] The system (**212**; R = Me) has been made by reaction of 1-methyl-l-(2-pyridyl) hydrazine with phosgene.[311]

The benzyloxytriazolopyridine (**213**) undergoes thermal rearrangement to the meso-ionic isomer (**212**; $R=PhCH_2$).[310] In contrast, the meso-ionic derivatives (**207**; $R^1=Me$, $R^2=SMe$, $R^3=Ar$) rearrange to the 5-thioxo-1, 2,4-triazolin-3-ones (**214**) on heating,[312] Pyrolysis of 1,4,5-triaryl-l,2,4-triazolium-3-olates (**207**; $R^1=R^2=R^3=Ar$) yields *N*-(diarylmethylene) anilines (**215**).[169]

Photolysis of meso-ionic 1,2,4-triazolium-3-olates (**207**) has been investigated in some detail. The mixture obtained from photolysis of the 1,4-diphenyl derivative (**207**; $R^1=R^3=Ph$, $R^2=H$) consisted of benzimidazole (18%), azobenzene (4%) and methyl phenylcarbamate (25%).[169] The mechanisms of these transformations have been discussed. Irradiation of the triphenylderivative (**207**; $R^1=R^2=R^3=Ph$) is believed to yield a polycyclic product having structure (**216**) or (**217**).[169]

A variety of polycyclic meso-ionic 1,2,4-triazolium-3-olates (**218**; X = O) have been prepared as anti-inflammatory agents.[313] 1,2,4-Triazolium-3-olate derivatives of cephalosporins have been patented as potential antibacterial agents.[314]

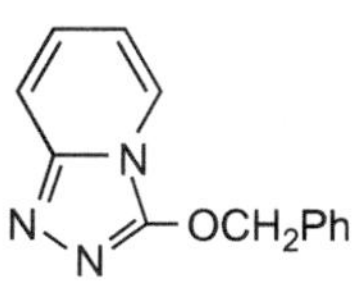

213

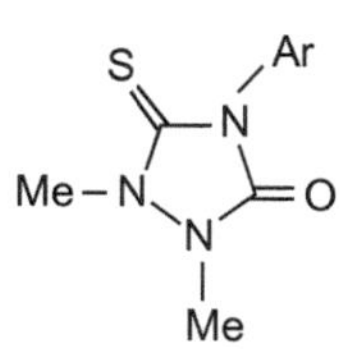

214

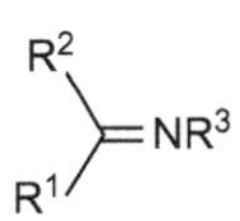

215

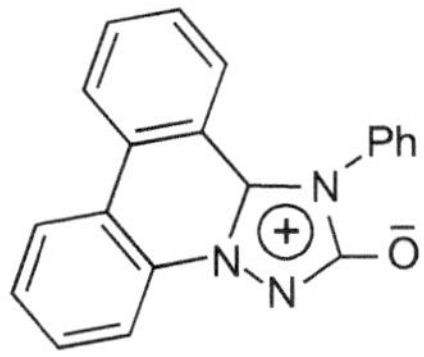

216

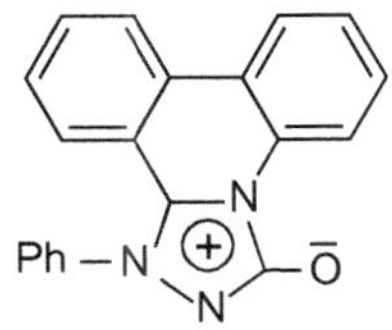

217

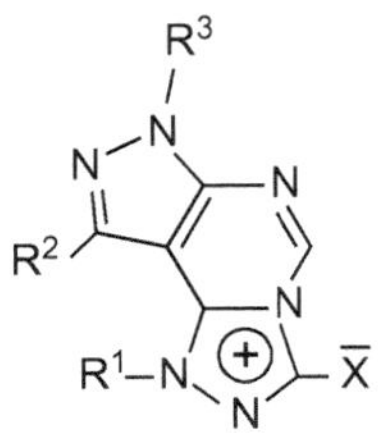

218

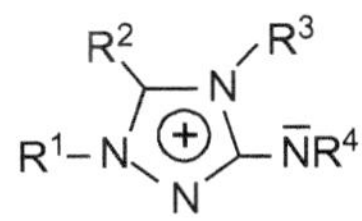

219

3.10.5 1,2,4-Triazolium-3-aminides (219)

The structure of nitron (**219**; $R^1=R^3=R^4=Ph$; $R^2=H$) has been determined by X-ray crystallography. The results are consistent with its meso-ionic formulation.[315]

Meso-ionic 1,2,4-triazolium-3-benzylidenehydrazinides (**219**; $R^4=N{=}CHAr$) have been prepared by reaction of 4-aryl-1,1-dibromo-2,3-diazabuta-1,3-dienes ($Br_2C{=}NN{=}CHAr$) with *N*-amino-*N*-methyl-*N'*-phenylbenzamidine.[316]

3.10.6 1,2,4-Triazolium-3-thiolates (220)

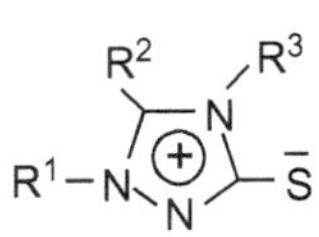

220

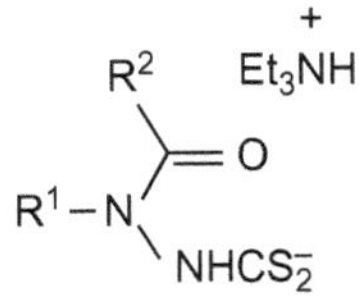

221

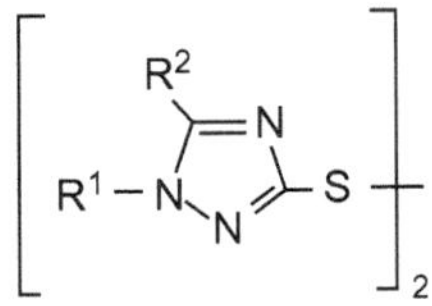

222

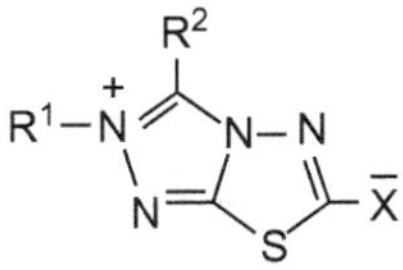
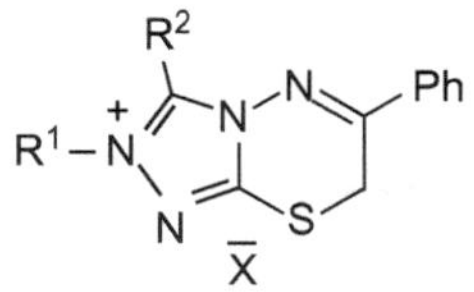
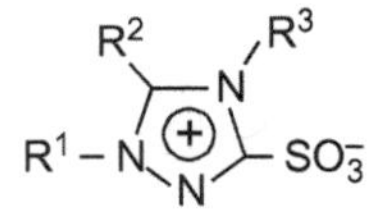

223 **224** **225**

Rearrangement of the meso-ionic 1,3,4-thiadiazolium-2-benzylidenehydrazinides (**252**; R^1=Ph. R^2=Me or Ph, R^3=N=CH·*p*·C1C_6H_4 (Section 3.11.9) in ethanolic ammonia solution gives the hydrazone derivatives (**220**; R^1=Me or Ph, R^2=Ph, R^3=N=CH·*p*·CI·C_6H_4) which are readily hydrolyzed to the 4-amino derivatives (**220**; R^3=NH_2).[316] 4-Amino-1,2,4-triazolium-3-thiolates (**220**; R^3=NH_2) have also been prepared by reaction of the dithiocarbazates (**221**) with chloroacetic acid followed by cyclization with hydrazine.[317,318] Treatment of the derivative (**220**; R^1=R^2=Ph, R^3=NH_2) with nitrous acid yields the disulfide (**222**; R^1=R^2=Ph).[319] Alkylations and acylations of the *N*-amino derivatives (**220**; R_3=NH_2) have been reported.[318] Condensation of these derivatives, (**220**; R^3=NH_2) with phosgene yields the mesomeric betaines (**223**; X=O) and the derivatives (**223**; X=NCOPh, NSO_2Ph) have been similarly prepared.[320,321] Treatment of the 4-amino derivatives (**220**; R^3=NH_2) with phenacyl bromide affords the salts (**224**).[322]

Syntheses of several bicyclic and polycyclic 1,2,4-triazolium-3-thiolates have been described[311,323–326] and antiinflammatory activity has been claimed.[313] Cephalosporin derivatives have been synthesized as anti-bacterial agents.[314]

Oxidation of the derivatives (**220**; R^1=Me or $PhCH_2$, R^2=H, R^3=Ph) gives the sulfonates (**225**).[327]

3.10.7 1,2,4-Triazolium-3-methylides (226)

No new work on this meso-ionic system (**226**) has been reported since the subject was last reviewed.[6]

3.10.8 1,2,4-Triazolium-3-selenolates (227)

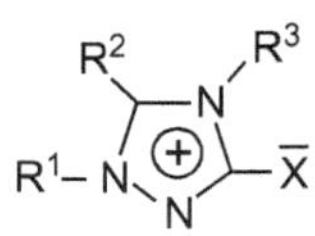
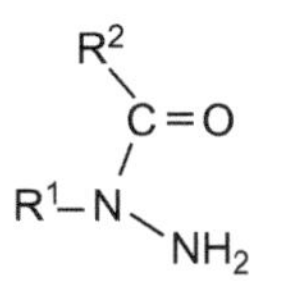
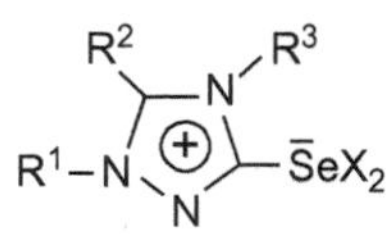

226 X = CR_2 **228** **229**

227 X = Se

Reaction of the hydrazide derivatives (**228**) with phenylisoselenocyanate (PhNCSe) yields the meso-ionic 1,2,4-triazolium-3-selenolates (**227**).[328] Oxidation of compound (**227**; R^1=Me, R^2=H, R^3=Ph) in the presence of magnesium perchlorate yields 1-methyl-4-phenyl-1,2,4-triazolium perchlorate.[329] Bromination of these compounds results in expansion of the selenium co-ordination sphere giving the derivatives (**229**; X=Br).[329] Chloro (**229**; X=C1) and thiocyano (**229**; X=SCN) derivatives have been similarly prepared.[327]

3.11 Thiadiazoles

3.11.1 1,2,3-Thiadiazolium-4-olates (230)

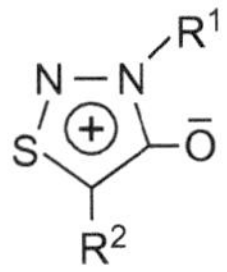

230

The 5-unsubstituted derivative (**230**; R^1=*p*·MeO·C_6H_4, R^2=H) undergoes electrophilic substitution giving 5-formyl, 5-chloro, 5-dimethylaminomethyl and 5-methythio derivatives (**230**; R^1=*p*·MeO·C_6H_4, R^2=CHO, C1, Me_2NCH_2 or SMe). The 5-formyl derivative (**230**; R^1=*p*·MeO·C_6H_4, R^2=CHO) has been converted into the novel meso-ionic 1,2,3-thiadiazolium-5-aminide (**233**; R^1=*p*·MeO·C_6H_4, R^2=OMe, R^3=CO_2Et) (Section 3.11.3).[330]

3.11.2 1,2,3-Thiadiazolium-5-olates (231)

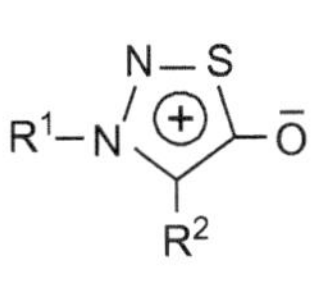

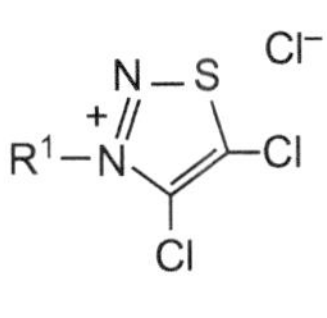

231 **232**

A general route to this new meso-ionic system (**231**) involves rearrangement of the isomeric mesoionic 1,2,3-oxadiazolium-5-thiolates (**178**) (Section 3.8.3) in hot ethanolic ammonia.[295] This general type of rearrangement of type A meso-ionic systems has been reviewed.[6] The system (**231**) is readily alkylated at the exocyclic oxygen atom by Meerwein's reagent.[295]

The first examples of this system to be reported were 3-aryl-4-chloro-derivatives (**231**; R^1=Ar, R^2=CI) prepared by hydrogen peroxide oxidation of the dichlorothiadiazolium salts (**232**; R^1=Ar).[331]

The 4-chloro-3-phenyl derivative (**231**; R^1=Ph, R^2=C1) prepared by this method is identical with a sample prepared by chlorination of the 4-unsubstituted compound (**231**; R^1=Ph, R^2=H).[295]

3.11.3 1,2,3-Thiadiazolium-5-aminides (233)

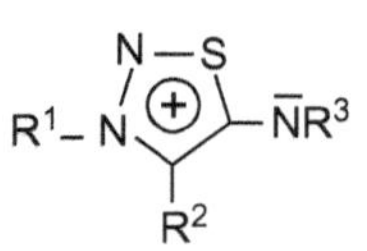

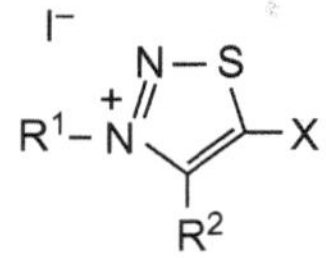

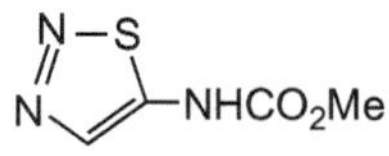

233 **234** X = SMe **236**

235 X = NHMe

The methiodide (**234**; R^1=Ph, R^2=H) prepared from the meso-ionic 1,2,3-thiadiazolium-5-thiolate (**237**; R^1=Ph, R^2=H) (Section 3.11.4) reacts with methylamine giving compound (**235**; R^1=Ph, R^2=H). Deprotonation of this salt gives the yellow meso-ionic 1,2,3-thiadiazolium-5-aminide (**233**; R^1=Ph, R^2=H, R^3=Me).[295]

Alkylation of the meso-ionic 1,2,3-thiadiazolium-4-olate (**230**; R^1=*p*·MeO·C_6H_4, R^2=$NHCO_2Et$) (Section 3.11.1) with excess methyl fluorosulfonate and subsequent deprotonation yields the derivative (**233**; R^1=*p*·MeO·C_6H_4, R^2=OMe, R^3=CO_2Et).[330] A third example of this system (**233**) has been prepared by alkylation of the 1,2,3-thiadiazole (**236**). The structure of this product (**233**; R^1=Me, R^2=H, R^3=CO_2Me) has been confirmed by X-ray crystallography.[332]

3.11.4 1,2,3-Thiadiazolium-5-thiolates (237)

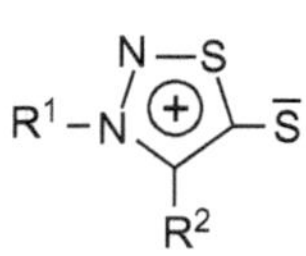

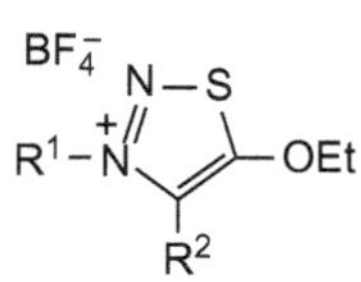

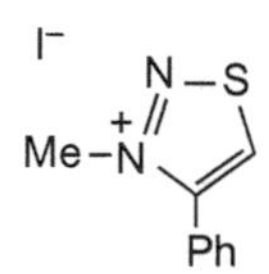

237 **238** **239**

The first examples of this meso-ionic system (**237**) have been prepared by treatment of the salts (**238**) with sodium hydrosulfide at −70°C.[295] Alternatively, treatment of the iodide (**239**) with cold methanolic sodium methoxide yields compound (**237**; R^1=Me, R^2=Ph) whose structure has been confirmed by X-ray crystallography. A mechanism has been proposed for this unexpected reaction (**239**→**237**; R^1=Me, R^2=Ph).[333,334]

Compound (**237**; R^1 = Ph, R^2 = H) is methylated by methyl iodide affording the salt (**234**; R^1 = Ph, R^2 = H) which is a useful intermediate for preparing 1,2,3-thiadiazolium-5-aminides (**233**) (Section 3.11.3) and 1,2,3-thiadiazolium-5-methylides (**240**) (Section 3.11.5).[295]

3.11.5 1,2,3-Thiadiazolium-5-methylides (240)

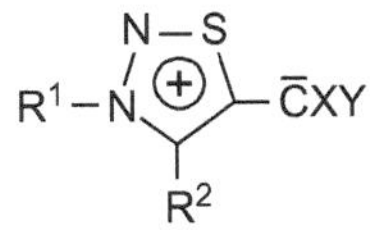

240

Only a single member of this class has been reported. Treatment of the 1,2,3-thiadiazolium salt (**234**; R^1 = Ph, R^2 = H) with malononitrite in the presence of triethylamine yields the orange-yellow derivative (**240**; R^1 = Ph, R^2 = H, X = Y = CN).[295]

3.11.6 1,2,4-Thiadiazoliurn-3-olates (241)

241

242 X = OH
243 X = NH_2
244 X = $NHSO_2 \cdot C_6H_4 \cdot 4Me$

Methylation of 3-hydroxy-5-phenyl-1,2,4-thiadiazole (**242**) using methyl toluene-4-sulfonate yields a variety of methylated products including a small amount (10%) of the meso-ionic 1,2,4-thiadiazolium-3-olate (**241**; R^1 = Ph, R^2 = Me) which has been fully characterized.[335] Alkaline hydrolysis of this compound gives *N*-benzoyl-*N*-methyl urea (12%) and *N*-methylbenzamide (24%).[335]

3.11.7 1,2,4-Thiadiazolium-3-aminides (245)

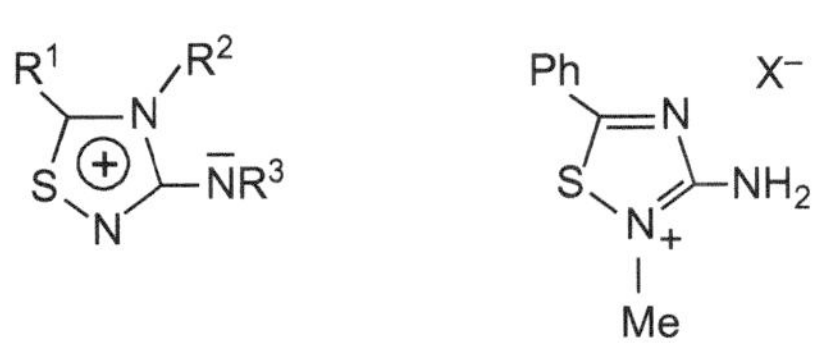

245 **246**

Methylation of the 3-sulfonamido-1,2,4-thiadiazole (**244**) using methyl fluorosulfonate yields all three possible *N*-monomethylated nitrogen compounds including the meso-ionic 1,2,4-thiadiazolium-3-aminide (**245**; R^1=Ph, R^2=Me, R^3=$SO_2 \cdot C_6H_4 \cdot 4Me$) (18.5%).[335]

A report[336] of the synthesis of compound (**245**; R^1=Ph, R^2=Me, R^3=H) by methylation of compound (**243**) and deprotonation has been shown to be incorrect. Methylation, in fact, gives the 2-methyl salt (**246**; X=I) which in the presence of base undergoes a Dimroth rearrangement.[337]

3.11.8 1,3,4-Thiadiazolium-2-olates (*247*)

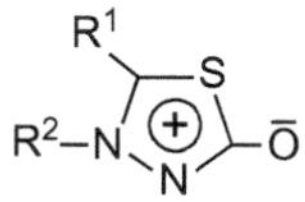

247

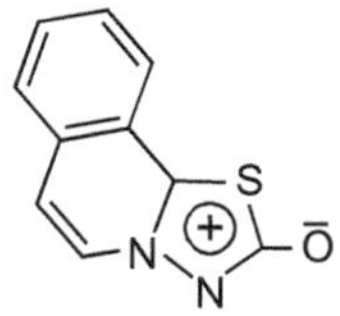

248

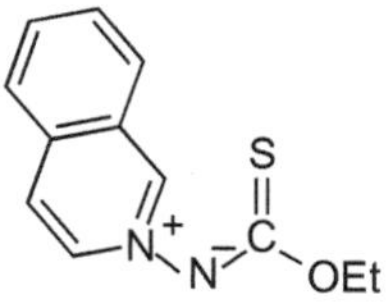

249

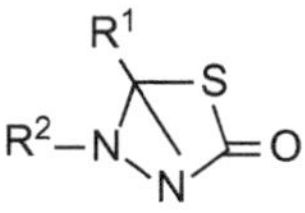

250

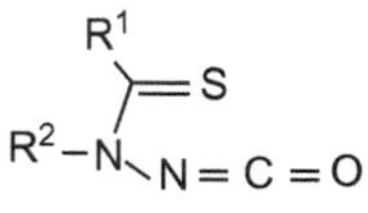

251

A new synthesis of the meso-ionic system (**247**) involves treatment of the corresponding meso-ionic 1,3,4-thiadiazolium-2-thiolate (**254**) (Section 3.11.10) with 2,4,6-trimethylphenyl cyanate.[297] The tricyclic system (**248**) has been obtained by irradiation of the isoquinolinium-*N*-imide derivative (**249**).[338,339] Upon prolonged irradiation in ethanol solution compound (**248**) undergoes further reaction giving 1-ethoxycarbonylaminoisoquinoline.[338]

The photochemical behavior of the 1,3,4-thiadiazolium-2-olates (**247**) has been reinvestigated.[340] Two competing processes each involving valence tautomerism have been identified. Formation of the bicyclic tautomer (**250**) leads to generation of a nitrile-imine and carbon oxysulfide: formation of the acyclic tautomer (**251**) leads to thioamide derivatives.[340]

3.11.9 1,3,4-Thiadiazolium-2-aminides (*252*)

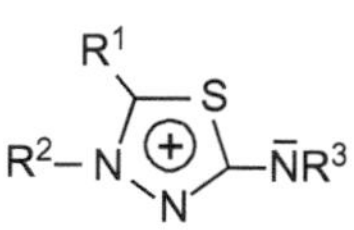

252

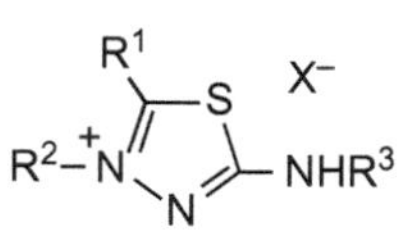

253

Reaction of 2-amino-4,5-dimethyl-1,3,4-thiadiazolium iodide (**253**; $R^1=R^2=Me$, $R^3=H$, X=I) with one equivalent of diethyl benzoylphosphonate [$(EtO)_2PO.COPh$] yields the meso-ionic 1,3,4-thiadiazolium-2-aminide (**252**; $R^1=R^2=Me$, $R^3=COPh$) as the major product (36%).[341] This product was identical with a sample prepared by condensation of *N*-methyl-*N*-thioacetylhydrazine (MeCS. NMe. NH_2) with benzoylisothiocyanate.[341]

1,3,4-Thiadiazolium-2-aminides possessing the novel exocyclic benzylidenehydrazinide grouping (**252**; $R^3=N=CHR$) have been prepared by condensation of dibromodiazabutadienes ($ArCH=N \cdot N=CBr_2$) with *N*-thiobenzoyl hydrazines ($PhCS{\cdot}NR{\cdot}NH_2$) and deprotonation of the resulting 1,3,4-thiadiazolium bromides (**253**; $R^1=Ph$, $R^2=Me$ or Ph, $R^3=N{=}CHAr$).[316] In aqueous ethanolic ammonia solution, rearrangement of the meso-ionic derivative (**252**; $R^1=Ph$, $R^2=Me$, $R^3=N{=}CH \cdot pClC_6H_4$) to the isomeric meso-ionic 1,2,4-triazolium-3-thiolate (**220**; $R^1=Me$, $R^2=Ph$, $R^3=N{=}CH \cdot pClC_6H_4$) is observed.[316] The same derivative (**252**; $R^1=Ph$, $R^2=Me$, $R^3=N{=}CH \cdot pClC_6H_4$) forms a 1, 3-dipolar cycloadduct with 4-phenyl-1,2,4-triazoline-3,5-dione.[316]

3.11.10 1,3,4-Thiadiazolium-2-thiolates (254)

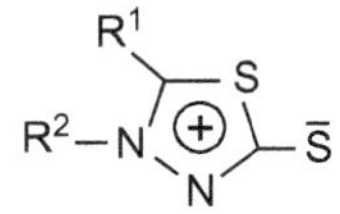

254

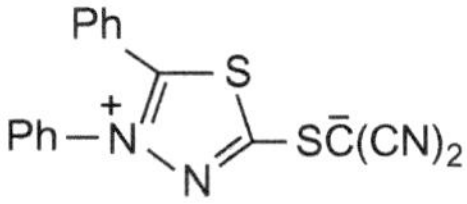

255

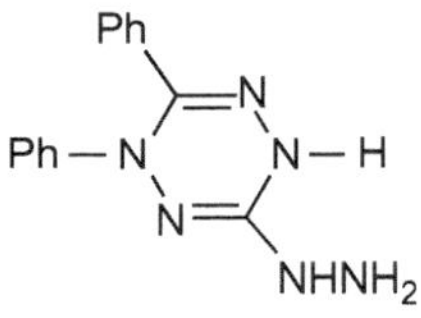

256

Meso-ionic 1,3,4-thiadiazolium-2-thiolates (**254**) are well known. Several new tricyclic derivatives have been prepared.[326] The diphenyl derivative (**254**; $R^1=R^2=Ph$) yields the interesting zwitterion (**255**) with tetracyano-ethylene oxide[342] and the 3-amino-tetrazine (**256**) with hydrazine.[343] Photolysis of compound (**254**; $R^1=Ph$, $R^2=Me$) in moist acetonitrile gives *N*-methylthiobenzamide.[340]

The effect of solvent on the electronic absorption spectra of several systems has been discussed.[344] Some derivatives have been claimed as anticonvulsants.[345,346]

3.11.11 1,3,4-Thiadiazolium-2-methylides (257)

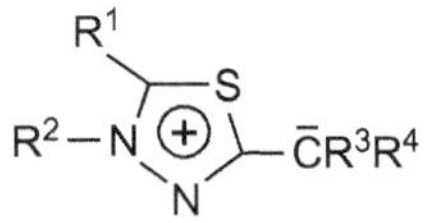

257

Treatment of *N*-methyl-*N*-thiobenzoylhydrazide with phenylthioacetylthioglycolic acid ($PhCH_2CS{\cdot}SCH_2CO_2H$) yields the derivative (**257**; $R^1=R^3=Ph$, $R^2=Me$, $R^4=CS{\cdot}CH_2PH$), the structure of which has been determined by X-ray crystallography.[347]

3.12 Oxatriazoles

3.12.1 1,2,3,4-Oxatriazolium-5-olates (258)

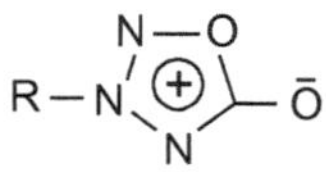

258

An X-ray analysis of the 3-phenyl derivative (**258**; R = Ph) is in agreement with the meso-ionic structure.[348] Further support for the structure of these compounds has been provided by their mass spectra[349] and electric dipole moments.[350]

Irradiation of compound (**258**; R = Ph) yields phenyl azide by a mechanism which involves phenyl migration and not an antiaromatic triazene intermediate. A second, minor, pathway yields phenylisocyanate and nitrous oxide.[351] The kinetics of acid hydrolysis of the 3-methyl (**258**; R = Me) and 3-isopropyl (**258**; R = isoPr) derivatives have been investigated.[352,353]

Some aryl derivatives (**258**; R = Ar) have been claimed to have pesticidal and herbicidal activity.[354]

3.12.2 1,2,3,4-Oxatriazolium-5-aminides (259)

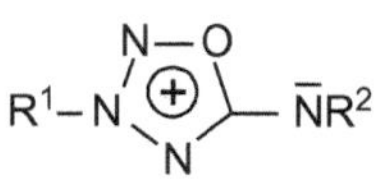

259

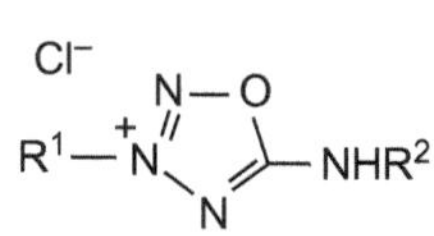

260

Nitrosation of 1,4-diarylthiosemicarbazides yields the 1,2,3,4-oxatriazolium chlorides (**260**) which are deprotonated by dry ammonia affording the meso-ionic 1,2,3,4-oxatriazolium-5-aminides (**259**; R^1 and R^2 = Ar).[355] In hot ethanolic sodium hydroxide solution these diaryl derivatives (**259**; R^1 and R^2 = Ar) rearrange to meso-ionic 1,2,3,4-tetrazolium-5-olates (**263**) (Section 3.13.1). Meso-ionic systems possessing four heteroatoms are usually unreactive toward 1,3-dipolarophiles. It is interesting to note, therefore, that the diphenyl compound (**259**; $R^1=R^2=Ph$) with *p*-chlorophenylisocyanate yields the *p*-chlorophenyl derivative (**259**; R^1=Ph, R_2=*p*·Cl·C_6H_4) (55%)—presumably *via* cycloaddition and cycloreversion.[355]

Recent physical studies on this meso-ionic system (**259**) include: (i) an X-ray investigation of the diphenyl compound (**259**; R^1=R^2=Ph)[356]; (ii) measurements of electric dipole moments of three diaryl derivatives[350]; and (iii) a discussion of their mass spectral fragmentation pattern.[349] Some derivatives have been claimed to possess pesticidal and herbicidal properties.[354]

3.12.3 1,2,3,4-Oxatriazolium-5-tholates (261)

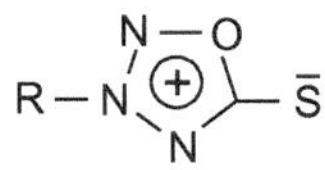

261

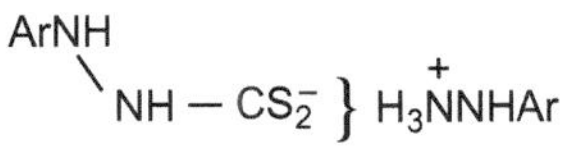

262

3-Aryl-1,2,3,4-oxatriazolium-5-thiolates (**261**) have been prepared by nitrosation of the dithiocarbazinic acid salts (**262**).[357] The meso-ionic structure is supported by their mass spectra[349] and electric dipole moments.[350]

1,2,3,4-Oxatriazolium-5-thiolates (**261**) are isomerized to meso-ionic 1,2,3,4-thiatriazolium-5-olates (**271**) (Section 3.14.1) in hot ethanolic ammonia solution.[357] Pesticidal and herbicidal activity has been claimed for compounds of this type.[354]

3.13 Tetrazoles

3.13.1 1,2,3,4-Tetrazolium-5-olates (263)

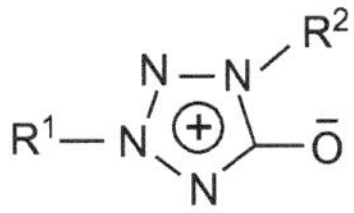

263

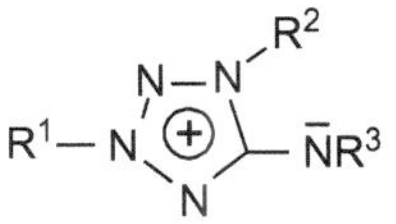

264

Diaryl derivatives (**263**; R^1 and R^2=Ar) have been prepared by rearrangement of meso-ionic 1,2,3,4-oxatriazolium-5-aminides (**259**) (Section 3.12.2) in ethanolic alkali and by treatment of potassium arylhydrazonomethanedisulfonates with aryldiazonium chlorides.[355] The diphenyl compound (**263**; R^1=R^2=Ph) has also been made (45%) by thermal rearrangement of the type B meso-ionic 1,2,3,4-tetrazolium-5-olate (**307**; R^1=R^2=Ph) (Section 4.7.1).[141,358]

The meso-ionic structure (**263**) is supported by mass spectral[349] and electric dipole moment studies.[350] Pesticidal and herbicidal activities have been claimed.[354]

3.13.2 1,2,3,4-Tetrazolium-5-aminides (264)

No new studies on this system (**264**) have been reported since the subject was last reviewed.[6]

3.13.3 1,2,3,4-Tetrazolium-5-thiolates (265)

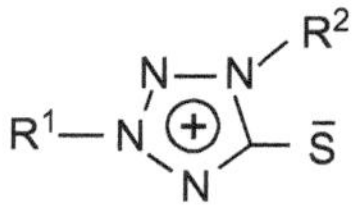

265

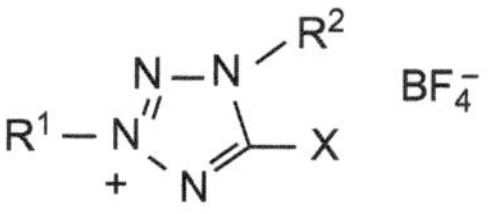

266

Treatment of the tetrazolium tetrafluoroborates (**266**; R^1 and R^2=Ar, X=OEt) with sodium sulfide in dimethylformamide affords the 1,3-diaryl-1,2,3,4-tetrazolium-5-thiolates (**265**; R^1 and R^2=Ar).[359] In a similar fashion, the derivative (**265**; R^1=Et, R^2=Ph) has been prepared by reaction of the salt (**266**; R^1=Et, R^2=Ph, X=SMe) with sodium hydrosulfide in methanol solution.[360,361] The diphenyl derivative (**265**; R^1=R^2=Ph) is obtained in low yield by thermal rearrangement of dehydrodithizone (Section 4.7.3).[141,358]

Recent structural studies on this meso-ionic system (**265**) include: (i) an X-ray investigation of the diphenyl derivative (**265**; R^1=R^2=Ph)[205]; (ii) an analysis of the ^{13}C-NMR spectrum of compound (**265**; R^1=Et, R^2=Ph)[360]; (iii) electric dipole moment measurements[350] and (iv) an analysis of the mass spectral fragmentation pattern.[349]

3.13.4 1,2,3,4-Tetrazolium-5-methylides (267)

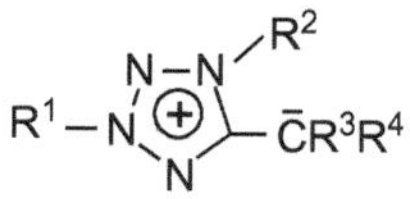

267

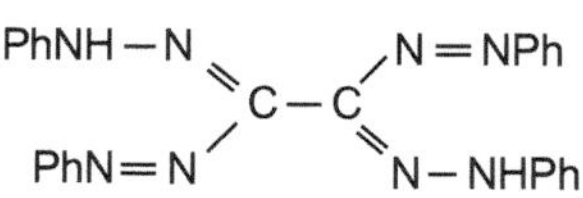

268

Diformazyl, the product obtained by Bamberger and co-workers in 1893 by treatment of the acids [$RCH_2CH_2CO_2H$; R=MeCO, or $HO_2C(CH_2)_2CO$] with phenyldiazonium hydroxide,[362,363] has recently been shown to be the meso-ionic compound (**267**; R^1=R^2=Ph, R^3=R^4=

N_2Ph) and not the biformazan (**268**) as originally formulated.[364] The structure (**267**; $R^1=R^2=Ph$, $R^3=R^4=N_2Ph$) has been confirmed by an X-ray analysis of its hydrochloride salt. Ozonolysis of diformazyl (**267**; $R^1=R^2=Ph$, $R^3=R^4=N_2Ph$) yields 1,3-diphenyl-1,2,3,4-tetrazolium-5-olate (**263**; $R^1=R^2=Ph$) (Section 3.13.1).[364]

Reaction of the tetrazolium tetrafluoroborates (**266**; R^1 and $R^2=Ar$, $X=OEt$) with malononitrile in triethylamine yields the derivatives (**267**; R^1 and $R^2=Ar$, $R^3=R^4=CN$).[365] The electric dipole moment[350] and mass spectrum[349] of compound (**267**; $R^1=R^2=Ph$, $R^3=R^4=CN$) ($\mu=9.54D$) support its meso-ionic structure.

3.13.5 1,2,3,4-Tetrazolium-5-selenolates (269)

This novel selenium containing meso-ionic heterocycle has recently been synthesized. Treatment of the 5-methylthiotetrazolium salt (**270**) with sodium hydrogen selenolate (NaSeH) yields 3-ethyl-l-phenyl-1,2,3,4-tetrazolium-5-selenolate (**269**; $R^1=Et$, $R^2=Ph$).[360,361]

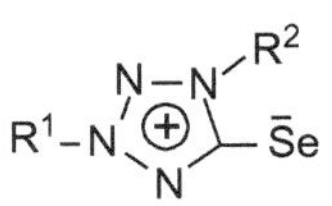

269

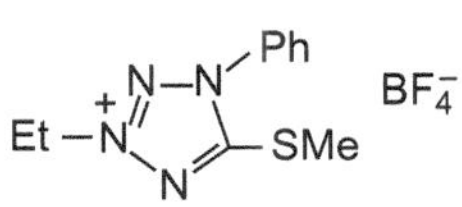

270

3.14 Thiatriazoles

3.14.1 1,2,3,4-Thiatriazolium-5-olates (271)

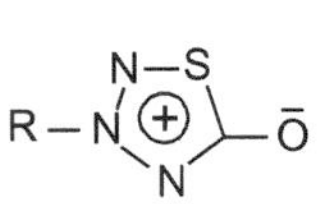

271

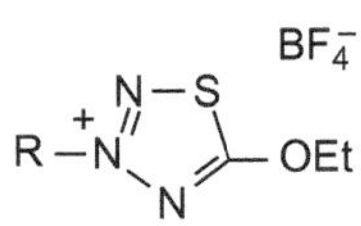

272

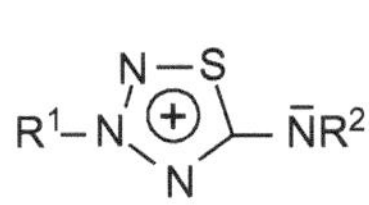

273

Examples of this system (**271**) have been prepared by rearrangement of meso-ionic 1,2,3,4-oxatriazolium-5-thiolates (**261**) (Section 3.12.3) is hot ethanolic ammonia.[357]

Studies on their mass spectral fragmentation patterns[349] and electric dipole moments[350] have been reported. Treatment with triethyloxonium tetrafluoroborate gives the thiatriazolium salts (**272**).[357] Some derivatives of this meso-ionic system (**271**) have been claimed to have useful pesticidal and herbicidal activity.[354]

3.14.2 1,2,3,4-Thiatriazolium-5-aminides (273)

The meso-ionic 1,2,3,4-thiatriazolium-5-aminides (**273**; R^1 and R^2=Ar) have been prepared by reaction of the salts (**272**) with aromatic amines followed by treatment with base.[359]

3.14.3 1,2,3,4-Thiatriazolium-5-thiolates (274)

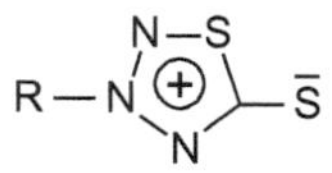

274

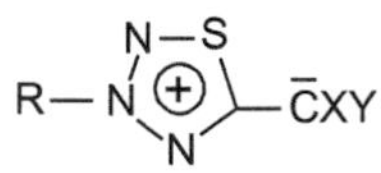

275

Treatment of the 5-ethoxy-1,2,3,4-thiatriazolium tetrafluoroborates (**272**; R=Ar) with sodium sulfide in aqueous dimethyl formamide yields the meso-ionic 1,2,3,4-thiatriazolium-5-thiolates (**274**).[357] The structure (**274**) is supported by mass spectrometry.[349]

3.14.4 1,2,3,4-Thiatriazolium-5-methylides (275)

Treatment of the salts (**272**; R=Ar) with malononitrile in the presence of triethylamine yields the meso-ionic 1,2,3,4-thiatriazolium-5-methylides (**275**; X=Y=CN).[365] Mass spectra[349] and electric dipole moments[350] have been measured.

3.15 Dithiadiazoles

3.15.1 1,3,2,4-Dithiadiazolium-5-olate (276)

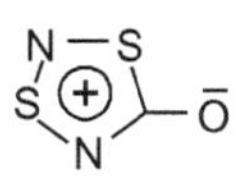

276

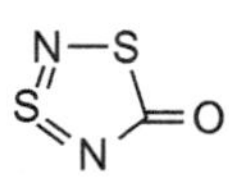

277

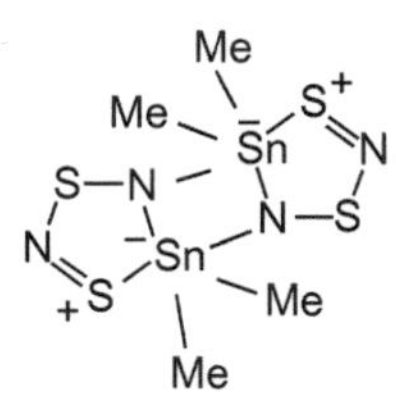

278

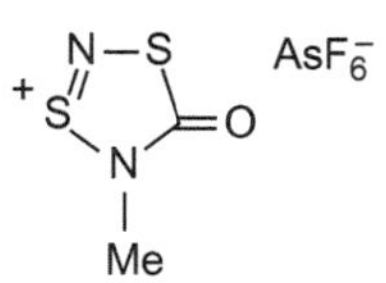

279

This new heterocycle has recently been synthesized by two different methods. Reaction of the dimeric 1,3,2,4,5-dithiadiazastannole (**278**) with excess phosgene or carbonyldifluoride gives the heterocycle (**276**) in moderately good yield (*ca.* 50%).[366] Alternatively, treatment of *N,N'-bis* (trimethylsilyl) sulfur diimide ($Me_3SiN{=}S{=}NSiMe_3$) with l-(chlorothio) formyl chloride (ClCO-SCl) also gives compound (**276**) (46%).[367]

The structure of this meso-ionic compound (**276**) has been confirmed by an X-ray analysis.[368] The extent of the participation of sulfur *d*-orbitals in the bonding of this heterocycle (**276** ↔ **277**) is not clear. With Lewis acids, 1:1 adducts (BF_3, SO_3, AsF_5, $SbCI_5$ SbF_2CI_3) or 2:1 adducts ($SnCI_4$, $TiCI_4$) are formed by complexing through the oxygen atom.[368,369] Interestingly, methylation of this heterocycle (**276**) using $MeOSO^+AsF_6^-$ results in *N*-methylation giving the salt (**279**).[370]

3.15.2 1,3,2,4-Dithiadiazolium-5-aminides (280)

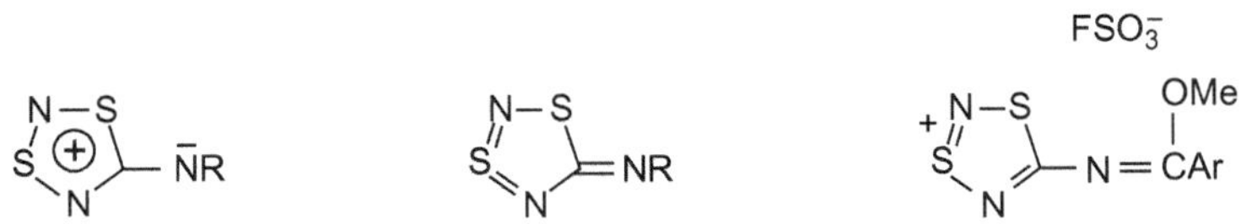

280 **281** **282**

The derivatives (**280**) have been prepared by reaction of *N*-acyl-l-(chlorothio)formimidoyl chlorides (RCO·N=CCl·SC1) with *N,N'-bis* (trimethylsilyl)sulfur diimide ($Me_3SiN{=}S{=}NSiMe_3$).[367,371] The structure of compound (**280**; R = *p* Me·C_6H_4·CO) has been confirmed by an X-ray study.[367,372] The extent of *d*-orbital participation in the *π*-bonding of these molecules (i.e., **280** ↔ **281**) is not clear. Methylation (FSO_3Me) of the derivatives (**280**; R = COAr) results in *O*-methylation to give the salts (**282**).[367,371]

4. The chemistry of meso-ionic compounds of type B (1976–1980)

These systems are discussed in the order given in Table 2. Since the review by Ollis and Ramsden in 1976,[6] derivatives of 4 new type B meso-ionic systems have been reported bringing the total number of known systems up to 13.

4.1 Dioxoles

4.1.1 1,2, Dioxolium-4-olates (283)

283 **284**

Table 2 Known meso-ionic heterocycles of type B (**10**).

Parent system	Heterocycle	Atom or group[a] a	b	c	d	e	f
Oxazoles							
1,2-Oxazolium-4-olates	(**285**)	CR	O	NR	CR	C	O
1,2-Oxazolium-4-aminides	(**287**)	CR	O	NR	CR	C	NR
Diazoles							
1,2-Diazolium-4-olates	(**288**)	CR	NR	NR	CR	C	O
1,2-Diazolium-4-aminides	(**290**)	CR	NR	NR	CR	C	NR
Thiazoles							
1,2-Thiazolium-4-olates	(**291**)	CR	S	NR	CR	C	O
1,2-Thiazolium-4-aminides	(**294**)	CR	S	NR	CR	C	NR
Dithioles							
1,2-Dithiolium-4-olates	(**295**)	CR	S	S	CR	C	O
1,2-Dithiolium-4-thiolates	(**302**)	CR	S	S	CR	C	S
Thiadiazoles							
1,2,5-Thiadiazolium-3-olates	(**304**)	N	S	NR	CR	C	O
Tetrazoles							
1,2,3,4-Tetrazolium-5-olates	(**307**)	N	NR	NR	N	C	O
1,2,3,4-Tetrazolium-5-aminides	(**310**)	N	NR	NR	N	C	NR
1,2,3,4-Tetrazolium-5-thiolates	(**311**)	N	NR	NR	N	C	S
1,2,3,4-Tetrazolium-5-methylides	(**320**)	N	NR	NR	N	C	CXY

[a]The groupings b and c each contribute 2 electrons to the π-electron system of the heterocycle; a, d, e, and f each contribute 1 electron.

In principle, the 1,2,3-propane-triones (**284**) can exist in equilibrium with the meso-ionic valence tautomers (**283**). An X-ray study of the triketone (**284**; $R^1 = R^2 = p \cdot BrC_6H_4$).confirmed the absence of any valence tautomerism (**283** $\rightleftharpoons$ **284**) and attempts to promote reactions of the meso-ionic species (**283**) in solutions of the triketone (**284**; $R^1 = R^2 = Ph$) were also unsuccessful.[373]

4.2 Oxazoles

4.2.1 *1,2-Oxazolium-4-olates (285)*

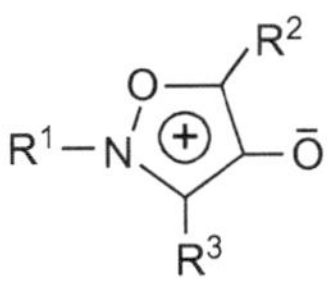

285

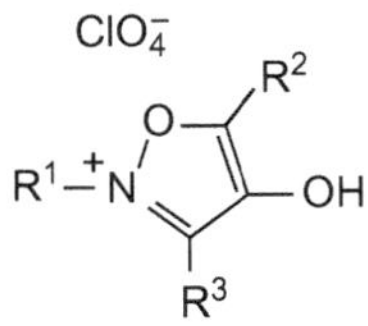

286

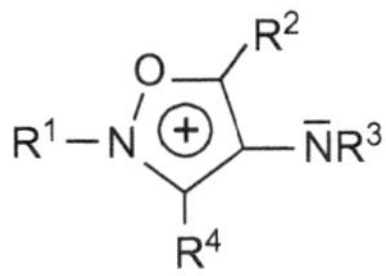

287

Meerwein methylation of 4-hydroxy-3,5-diphenylisoxazole leads to the oxazolium perchlorate (**286**; $R^1 = Me$, $R^2 = R^3 = Ph$) which on treatment with methanolic alkali affords the yellow, unstable 1,2-oxazolium-4-olate (**285**; $R^1 = Me$, $R^2 = R^3 = Ph$) (84%), m.p. 76–78°.[374]

4.2.2 *1,2-Oxazolium-4-aminides (287)*

No new work has been reported on this system since the subject was last reviewed.[6]

4.3 Diazoles

4.3.1 *1,2-Diazolium-4-olates (288)*

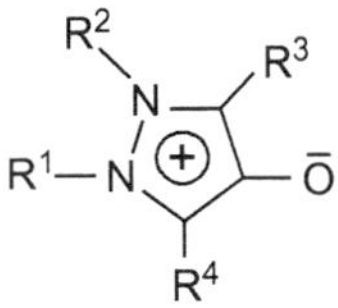

288

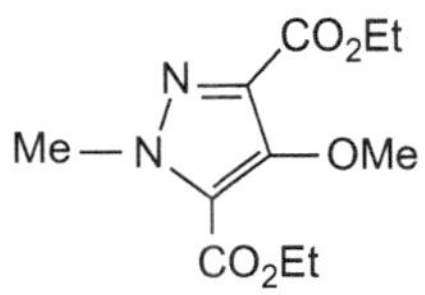

289

A large number of meso-ionic 1,2-diazolium-4-olate derivatives (**288**; $R^1 = Me$, $R^2 = Me$, Ph, R^3 and $R^4 = Me$, Ph, CO_2Et) have been prepared by methylation (Me_2SO_4) of 4-hydroxy pyrazoles. The tetramethyl derivative is obtained by condensation of *N,N*-dimethyl hydrazine with 2-acetoxy-l, 3-dimethylpropane-l,3-dione (Ac_2CHOAc). Compound (**288**; $R^1 = R^2 =$ Me, $R^3 = R^4 = CO_2Et$) is unreactive toward acetylenes but slowly isomerized to the 4-methoxy pyrazole (**289**).[375]

4.3.2 1,2-Diazolium-4-aminides (290)

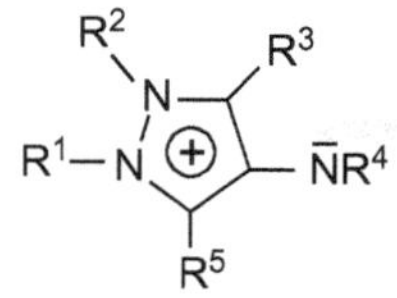

290

An X-ray analysis of the crystal structure of compound (**290**; $R^1=R^2=R^3=R^5=Me$, $R^4=p\text{-}MeC_6H_4{\cdot}SO_2$) has been reported and the results are consistent with the meso-ionic formulation.[376]

4.4 Thiazoles

4.4.1 1,2-Thiazolium-4-olates (291)

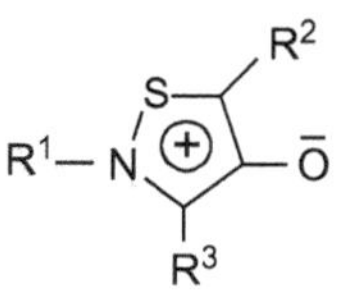

291

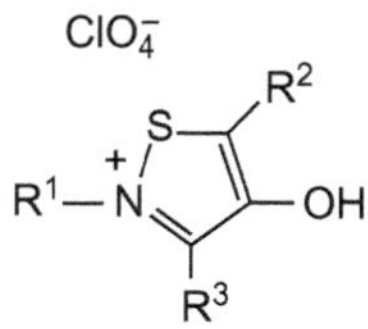

292

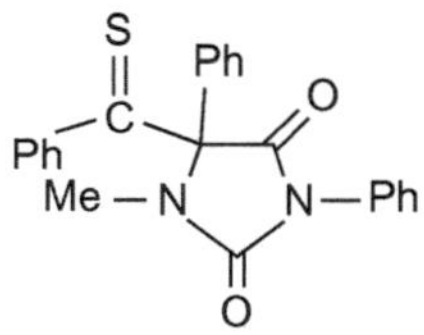

293

Two independent groups have recently achieved the synthesis of derivatives of this novel type B meso-ionic system.[121,377] Alkylation of 3,5-diaryl-4-hydroxy isothiazoles gives the salts (**292**; $R^1=Me$, $R^2=R^3=Ar$) which on treatment with base yield the red 1,2-thiazotium-4-olates (**291**; $R^1=Me$, $R^2=R^3=Ar$).[121,377] These compounds are methylated on the exocyclic oxygen atom by dimethyl sulfate.[377] With phenyl isocyanate in hot xylene the derivative (**291**; $R^1=Me$, $R^2=R^3=Ph$) yields a blue compound which has been formulated as the hydantoin (**293**).[121]

The absorption spectra and dipole moments of 1,2-thiazolium-4-olate derivatives (**291**) have been discussed.[378]

4.4.2 1,2-Thiazolium-4-aminides (294)

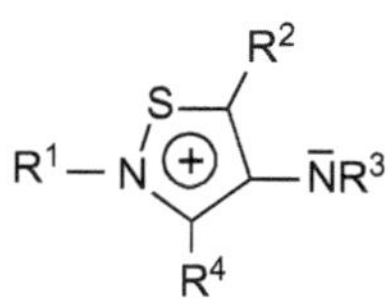

294

No new work on this meso-ionic system (**294**) has been reported since the subject was last reviewed.[6]

4.5 Dithioles

4.5.1 1,2-Dithiolium-4-olates (295)

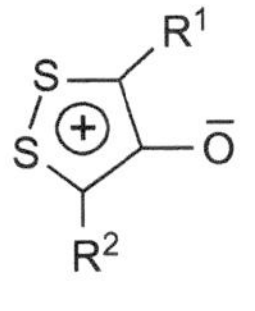

295

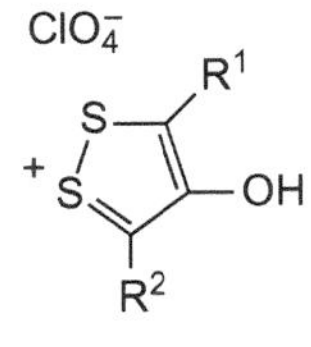

296

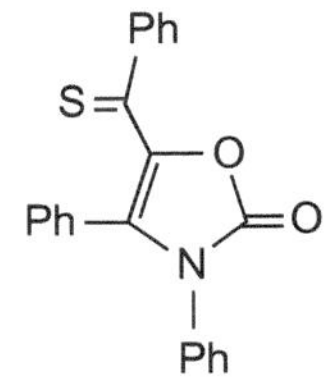

297

A new method of preparing derivatives of this meso-ionic system (**295**) involves treatment of propane-l, 3-dione esters ($(RCO)_2CHOAc$) with hydrogen disulfide or tetraphosphorus decasulfide followed by perchloric acid to give the salts (**296**) which are deprotonated by pyridine. This type of route has been used to prepare alkyl, thioalkyl and aryl derivatives (**295**; R^1 or R^2=Alkyl, S.Alkyl or Aryl).[379–382] Compound (**295**; R^1=Ph, R^2=SMe) is converted into the derivative (**295**; R^1=Ph, R^2=$p\cdot Me_2NC_6H_4$) by treatment with *N,N*-dimethylaniline.[382]

Several interesting transformations of the diphenyl derivative (**295**; R^1=R^2=Ph) have been reported. Reaction with phenyl isocyanate yields the 1,3-oxazol-2-one (**297**) by a mechanism which presumably involves extrusion of sulfur.[121] Treatment with ammonia gives 3,5-diphenyl-4-hydroxyisothiazole[121,377] Reaction of3,5-diphenyl-1,2-dithiolium-4-olate (**298**; R^1=R^2=Ph) with boiling aniline gives the blue benzothiazine (**301**)[121,383] and not 3,4,5-triphenylisothiazole as described by earlier workers.[384] A by-product in this reaction is the maroon indenylamine (**300**) which is probably formed by the sequence shown in Scheme 8.[121] In

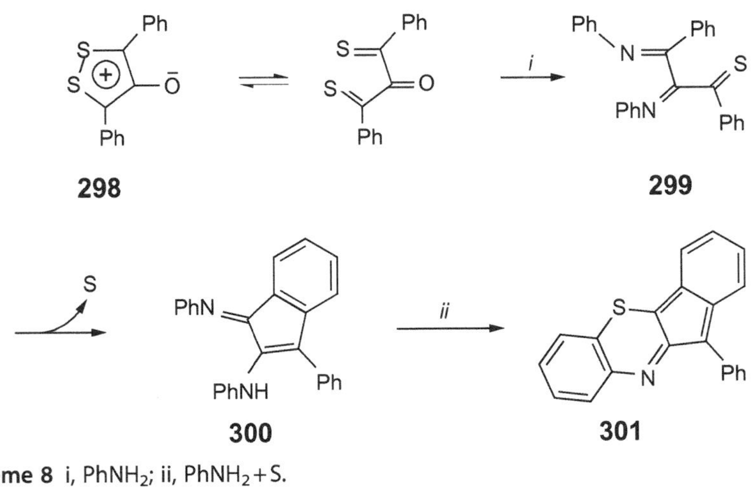

Scheme 8 i, $PhNH_2$; ii, $PhNH_2$+S.

the presence of sulfur, boiling aniline converts compound (**300**) into the blue benzothiazine (**301**) suggesting that the indenylamine (**300**) is an intermediate in the transformation (**298** → **301**) (Scheme 8).[121,383]

Two X-ray investigations have confirmed the meso-ionic structure of compound (**295**; $R^1{=}R^2{=}Ph$).[373,378] Dipole moments and ^{13}C-NMR spectra have also been reported of I, 2-dithiolium-4-olate derivatives (**295**).[378]

4.5.2 1,2-Dithiolium-4-thiolates (302)

Compound (**302**; $R^1{=}CN$, $R^2{=}NHCOPh$), the first example of this class of type B meso-ionic heterocycle, has been prepared from the dithiole derivatives (**303**). The highly functionalized nature of this meso-ionic molecule (**302**; $R^1{=}CN$, $R^2{=}NHCOPh$) leads to a number of interesting transformations including *S*-alkylation and oxidative dimerization.[385]

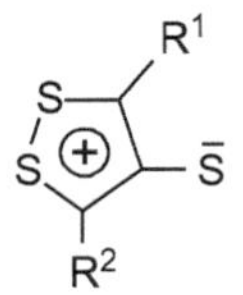

302

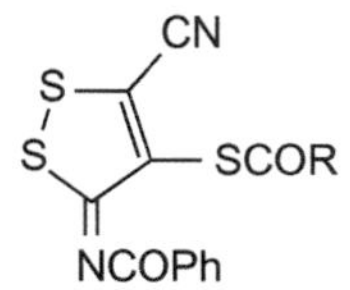

303

4.6 Thiadiazoles

4.6.1 1,2,5-Thiadiazolium-3-olates (304)

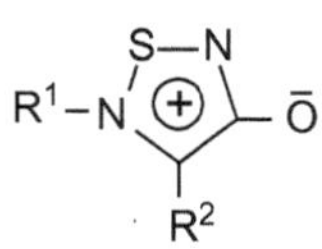

304

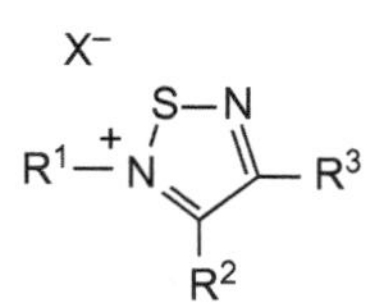

305

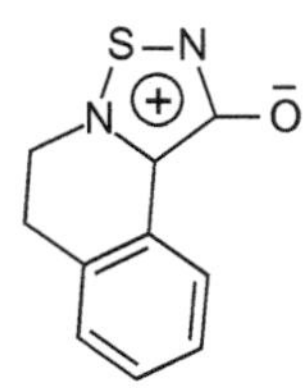

306

Treatment of *α*-amino amides ($R^1NH\text{-}CHR^2\text{-}CONH_2$; $R^1{=}$alkyl, $R^2{=}$aryl) with sulfur monochloride in DMF yields the 1,2,5-thiadiazolium chlorides (**305**; $R^1{=}$alkyl, $R^2{=}$aryl, $R^3{=}OH$, X=CI) which are readily deprotonated giving the meso-ionic derivatives (**304**). The tricyclic derivative (**306**) has been prepared similarly.[386]

Alkaline hydrolysis of compound (**304**; $R^1{=}Me$, $R^2{=}Ph$) yields an imino-amide ($MEN{=}CPh{\cdot}CONH_2$) and sodium borohydride reduction affords the amide ($MeNH\text{-}CHPh{\cdot}CONH_2$). Meerwein alkylation ($Et_3O^+BF_4^-$) of compound (**304**; $R^1{=}Me$, $R^2{=}Ph$) gives the salt (305; $R^1{=}Me$, $R^2{=}Ph$, $R^3{=}OEt$, $X{=}BF_4$) which reacts with methylamine giving the salt (**305**; $R^1{=}Me$, $R^2{=}Ph$, $R^3{=}NHMe$, $X{=}BF_4$).[386]

4.7 Tetrazoles

4.7.1 1,2,3,4-Tetrazolium-5-olates (307)

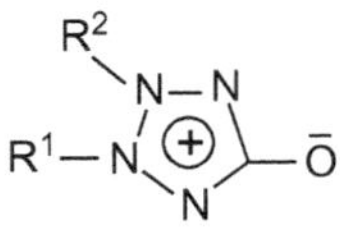

307

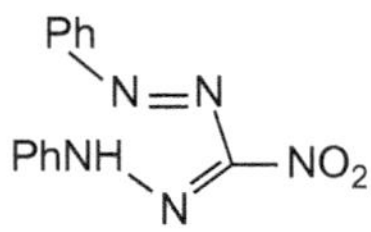

308

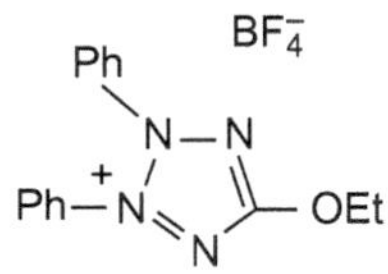

309

Oxidation of 3-nitro-l,5-diphenylformazan (**308**) by pentyl nitrite yields 2,3-diphenyl-l,2,3,4-tetrazolium-5-olate (**307**; $R^1=R^2=Ph$) (79%). Meerwein alkylation ($Et_3O^+BF_4^-$) of this compound affords the 5-ethoxy tetrazolium salt (**309**): attempts to displace the ethoxy group by nucleophiles were unrewarding.[365] Pyrolysis of compound (**307**; $R^1=R^2=Ph$) at 150° results in isomerization to the type A meso-ionic 1,3-diphenyl-1,2,3, 4-tetrazolium-5-olate (**263**; $R^1=R^2=Ph$) (45%) (Section 3.13.1).[141,358]

Two X-ray studies of the crystal structure of compound (**307**; $R^1=R^2=Ph$) have been reported and are consistent with the meso-ionic structure.[205,387]

4.7.2 1,2,3,4-Tetrazolium-5-aminides (310)

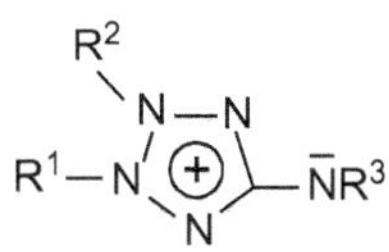

310

No new work on this meso-ionic system has been reported since the subject was last reviewed.[6]

4.7.3 1,2,3,4-Tetrazolium-5-thiolates (311)

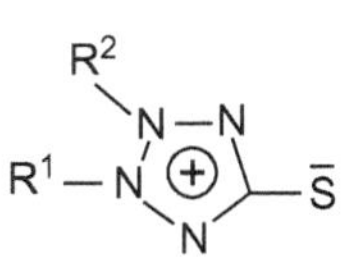

311

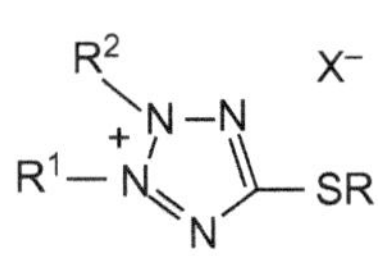

312

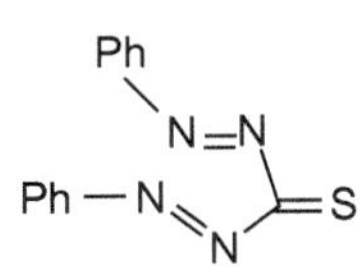

313

Dehydrodithizone (**311**; $R^1=R^2=Ph$) which is normally prepared by oxidation of dithizone ($PhN_2CSNHNHPh$)[1,388] has recently been prepared by reaction of diphenylthiocarbazide [$(PhNHNH)_2CS$] with diethoxydisulfide [$(EtO)_2S_2$].[389] Alkylation of dehydrodithizone (**311**; $R^1=R^2=Ph$) occurs on the exocyclic sulfur atom: treatment with chloroacetic acid yields the tetrazolium chloride (**312**; $R^1=R^2=Ph$, $R=CH_2CO_2H$, $X=Cl$)[388] and Meerwein ethylation gives the tetrafluoroborate (**312**; $R^1=R^2=Ph$, $R=Et$, $X=BF_4$).[365] Attempted displacement of the thioethoxy group in the latter salt yielded 3-ethylthio-1,5-diphenylformazan (87%).[365] Thermolysis of dehydrodithizone (**311**; $R^1=R^2=Ph$) gives a low yield (16%) of the isomeric Type A 1,2,3,4-tetrazolium-5-thiolate (**265**; $R^1=R^2=Ph$) (Section 3.13.3).[141,358]

Many of the reactions of dehydrodithizone (**311**; $R^1=R^2=Ph$) can be interpreted in terms of participation of the acyclic valence tautomer (**313**)[6] (see Scheme 2). Treatment with iron pentacarbonyl [$Fe(CO)_5$] yields the 1,3,4-thiadiazolone (**314**).[141,390] Addition of β-piperidino styrene gives the adduct (**315**) and pyrrolidino cyclohexene affords the adduct (**316**).[391] With the electron rich 1-diethylaminopropyne the major product is the pyrazole (**317**).[391] Reaction with 2-(dialkylamino) indenes gives products with the general structure (**318**).[391,392]

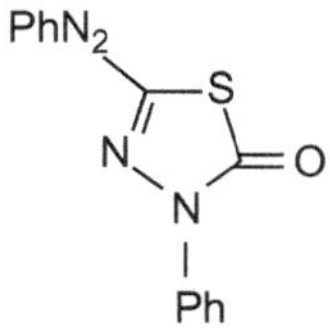

314

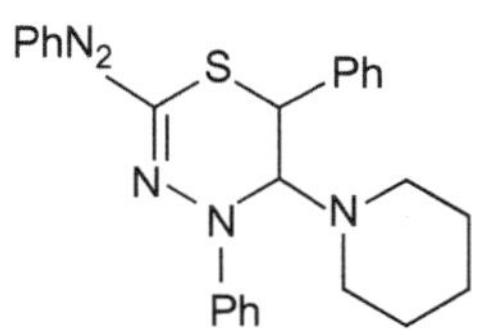

315

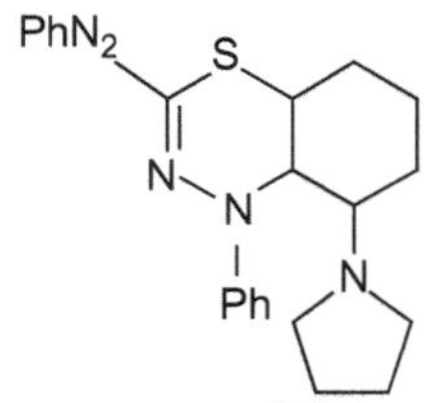

316

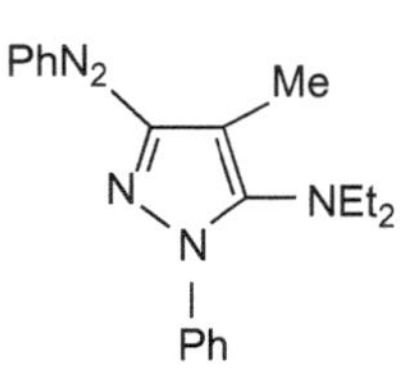

317

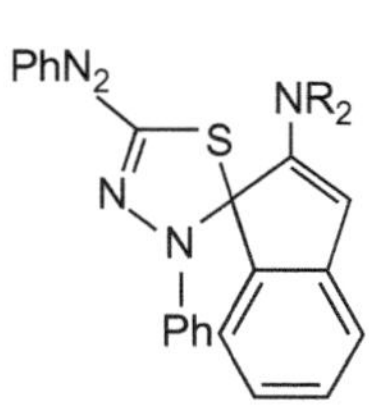

318

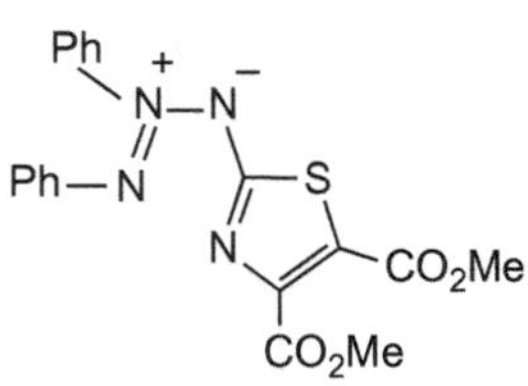

319

Addition of dimethyl acetylenedicarboxylate to dehydrodithizone (**311**; $R^1=R^2=Ph$) gives a product which has been shown by an X-ray analysis to have the dipolar structure (**319**)[393]: reaction with tetraphenylcyclopentadienone gives a similar adduct.[394] Treatment of dehydrodithizone with benzyne yields 2-phenylazobenzothiazole.[393]

Other studies of 1,2,3,4-tetrazolium-5-thiolates (**311**) include a description of their electronic absorption spectra[395] and their polarographic reduction.[396] Several Ni(II) and Cu(II) complexes of dehydrodithizone have been prepared.[397]

4.7.4 1,2,3,4-Tetrazolium-5-methylides (320)

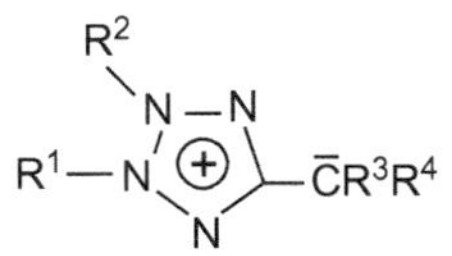

320

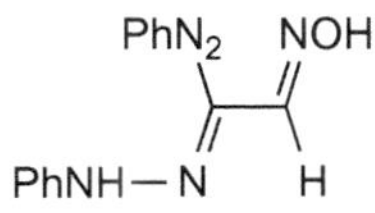

321

Mild oxidation of an authentic sample of the biformazan (**268**) (see also Section 3.13.4) with tetrakis (4-methylphenyl)hydrazine [(p · Me· $C_6H_4)_4N_2$] yields the red-brown meso-ionic 1,2,3,4-tetrazolium-5-methylide (**320**; $R^1=R^2=Ph$, $R^3=R^4=PhN_2$) (40%). Similarly, lead tetra-acetate oxidation of compound (**321**) gives the derivative (**320**; $R^1=R^2=Ph$, $R^3=H$, $R^4=NO$). These compounds are the first reported examples of Type B meso-ionic 1,2,3,4-tetrazolium-5-methylides (**320**).[398]

5. Conclusion

The number of papers published during the period 1976–1980 which describe the chemistry of meso-ionic systems is clear indication of the continuing interest in this class of heterocycles. In an earlier review Ollis and Ramsden have proposed[6] that the term meso-ionic be restricted to five membered heterocycles of types A and B (see Section 2) and this is the definition which has been used to determine the scope of this review. We believe that the restriction of the term meso-ionic to these two classes of heterocycle is essential if the term is to retain its usefulness. This restriction on the use of the term has the clear advantage that meso-ionic compounds are precisely defined and their relationship to other heterocyclic molecules is readily understood.

The modified definition of the term meso-ionic has been welcomed by many chemists. However, some authors continue to describe other classes of dipolar species as meso-ionic. This can only lead to confusion. We strongly urge heterocyclic chemists to adopt the definition of meso-ionic which has been advocated by Ollis and Ramsden. A system of classifying and naming other classes of dipolar heterocyclic will be the subject of a future publication.[399]

References

1. Fischer, E.; Besthorn, E. *Ann.* **1882**, *212*, 316.
2. Baker, W.; Ollis, W. D. *Quart. Rev.* **1957**, *11*, 15.
3. Earl, J. C.; Mackney, A. W. *J. Chem. Soc. D* **1935**, 899.
4. Schönberg, A. *J. Chem. Soc. D* **1938**, 824.

5. Huisgen, R. *Chem. Soc. Spec. Publ.* **1967**, *21*, 51.
6. Ollis, W. D.; Ramsden, C. A. *Adv. Heterocycl. Chem.* **1976**, *19*, 1; Ramsden C. A. In Barton D. H. R., Ollis W. D. Eds.; *Comprehensive Organic Chemistry*, Vol. 4, p. 1171. Pergamon Press, Oxford (1979).
7. Berk, H. C.; Zwikelmaier, K. E.; Franz, J. E. *Synth. Commun.* **1980**, *10*, 707.
8. Hamaguchi, M.; Ibata, T. *Tetrahedron Lett.* **1974**, *4475*.
9. Ibata, T.; Hamaguchi, M.; Kiyohara, H. *Chem. Lett.* **1975**, 21.
10. Potts, K. T.; Marshall, J. L. *J. Organomet. Chem.* **1979**, *44*, 626.
11. Hamaguchi, M.; Ibata, T. *Chem. Lett.* **1975**, 499.
12. Hamaguchi, M. *J. Chem. Soc., Chem. Commun.* **1978**, 247.
13. Friedrichsen, W.; Schroer, W. D.; Debaerdemaeker, T. *Ann.* **1980**, 1850.
14. Hamaguchi, M.; Ibata, T. *J. Chem. Soc., Chem. Commun.* **1980**, 230.
15. Alper, H.; Tanaka, M. *J. Am. Chem. Soc.* **1979**, *101*, 4245.
16. Tighineanu, E.; Chiraleu, F.; Raileanu, D. *Tetrahedron Lett.* **1978**, 1887.
17. Tighineanu, E.; Chiraleu, F.; Raileanu, D. *Tetrahedron* **1980**, *36*, 1385.
18. Höfle, G.; Steglich, W.; Daniel, H. *Chem. Ber.* **1976**, *109*, 2648.
19. Hershenson, F. M. *J. Heterocyclic Chem.* **1979**, *16*, 1093.
20. Hershenson, F. M. *J. Organomet. Chem.* **1975**, *40*, 1260.
21. Rebek, J., Jr.; Gehret, J. C.-E. *Tetrahedron Lett.* **1977**, 3027.
22. Pizznrno, M. T.; Albonico, S. M. *Chem. & Ind.* **1978**, 349.
23. Pizzorno, M. T.; Albonico, S. M. *J. Organomet. Chem.* **1974**, *39*, 731.
24. Pizzorno, M. T.; Albonico, S. M. *J. Organomet. Chem.* **1977**, *42*, 909.
25. Robins, D. J.; Sakdarat, S. *J. Chem. Soc. Commun.* **1979**, 1181. Robins D. J.; Sakdarat, S.; *J. Chem. Soc. Perkin Trans* **1981**, 1, 909.
26. Anderson, W. K.; Corey, P. F. *J. Med. Chem.* **1977**, *20*, 812.
27. Anderson, W. K.; Corey, P. F. *J. Organomet. Chem.* **1977**, *42*, 559.
28. Uchida, T.; Tsubokawa, S.; Harihara, K.; Matsumoto, K. *J. Heterocyclic Chem.* **1978**, *15*, 1303.
29. Potts, K. T.; Yao, S. *J. Organomet. Chem.* **1979**, *44*, 977.
30. Kato, H.; Nakazawa, S.; Kiyosawa, T.; Hirakawa, K. *J. Chem. Soc. Perkin Trans. I* **1976**, 672.
31. Kato, H.; Koshinaka, E.; Ogawa, N.; Yamagishi, K.; Mitani, K.; Kubo, S.; Hanaoka, M. *Chem. Pharm. Bull.* **1980**, *28*, 2194.
32. Kane, J. M. *J. Organomet. Chem.* **1980**, *45*, 5396.
33. Turchi, I. J.; Maryanoff, C. A.; Mastrocola, A. R. *J. Heterocyclic Chem.* **1980**, *17*, 1593.
34. Benages, I. A.; Albonico, S. M. *J. Organomet. Chem.* **1978**, *43*, 4273.
35. Matsukubo, H.; Kato, H. *J. Chem. Soc. Perkin Trans. I* **1976**, 2565.
36. Matsukubo, H.; Kato, H. *Bull. Chem. Soc. Japan* **1976**, *49*, 3314.
37. Myers, J. A.; Wilkerson, W. W.; Council, S. L. *J. Organomet. Chem.* **1975**, *40*, 2875.
38. Lukac, J.; Heimgartner, H.; Schmid, H. *Chimia (Switz.)* **1977**, *31*, 138.
39. Matsukubo, H.; Kato, H. *Bull. Chem. Soc. Japan* **1976**, *49*, 3333.
40. Friedrichsen, W.; Schroer, W. D. *Tetrahedron Lett.* **1977**, 1603.
41. Myers, J. A.; Moor, L. D.; Whitter, W. L.; Council, S. L.; Waldo, R. M.; Lanier, J. L.; Omoji, B. U. *J. Organomet. Chem.* **1980**, *45*, 1202.
42. Padwa, A.; Gingrich, H. L.; Lim, R. *Tetrahedron Lett.* **1980**, *21*, 3419.
43. Potts, K. T.; Baum, J.; Houghton, E. *J. Organomet. Chem.* **1976**, *41*, 818.
44. Matsukubo, H.; Kato, H. *J. Chem. Soc. Perkin Trans. I* **1975**, 632.
45. Huisgen, R.; Schmidt, T. *Ann.* **1978**, 29.
46. Potts, K. T.; Chen, S. J.; Szmuszkovicz, S. *J. Organomet. Chem.* **1977**, *42*, 2525.
47. Potts, K. T.; Baum, J.; Datta, S. K.; Houghton, E. *J. Organomet. Chem.* **1976**, *41*, 813.
48. Friedrichsen, W.; Schwarz, I. *Tetrahedron Lett.* **1977**, 3581.
49. Friedrichsen, W.; Schroer, W. D.; Debaerdemaeker, T. *Ann.* **1980**, 1836.
50. Sain, B.; Thyagarajan, G.; Sandhu, J. S. *Can. J. Chem.* **1980**, *58*, 2034.

51. Lukac, J.; Bieri, J. H.; Heimgartner, H. *Heir. Chim. Acta* **1977**, *60*, 1657.
52. Lukac, J.; Heimgartner, H. *Heir. Chim. Acta* **1979**, *62*, 1236.
53. Gotthardt, H.; Reiter, F. *Ann.* **1979**, 650.
54. Toubro, N. H.; Hansen, B.; Harrit, N.; Holm, A.; Potts, K. T. *Tetrahedron* **1979**, *35*, 229.
55. Kato, H.; Tani, K.; Kurumisawa, H.; Tamura, Y. *Chem. Lett.* **1980**, 717.
56. Dixit, V. M.; Bhat, V.; Trozzolo, A. M.; George, M. V. *J. Organomet. Chem.* **1979**, *44*, 4169.
57. Clerin, D.; Meyer, B.; Fleury, J. P.; Fritz, H. *Tetrahedron* **1976**, *32*, 1055.
58. McEwan, W. E.; Grossi, A. V.; MacDonald, R. J.; Stamegna, A. P. *J. Organomet. Chem.* **1980**, *45*, 1301.
59. Cook, M. J.; Katritzky, M. R.; Page, A. D. *J. Am. Chem. Soc.* **1977**, *99*, 165.
60. McEwan, W. E.; Cabello, C. C.; Calabro, M. A.; Ortega, A. M.; Stott, P. E.; Zapata, A. J.; Zepp, C. M.; Lubinkowski, J. J. *J. Organomet. Chem.* **1979**, *44*, 111.
61. Bhattacharjee, D.; Popp, F. D. *J. Heterocyclic Chem.* **1980**, *17*, 1035.
62. Gotthardt, H.; Weisshuhn, M. C.; Dörhöfer, K. *Angew. Chem. lnt. Ed.* **1975**, *14*, 422.
63. Gotthardt, H.; Weisshuhn, C. M.; Dörhöfer, K. *Chem. Ber.* **1978**, *111*, 3336.
64. Potts, K. T.; Kane, J.; Carnahan, E.; Singh, U. P. *J. Chem. Soc., Chem. Commun.* **1975**, 417.
65. Masuda, K.; Adachi, J.; Nomura, K. *Chem. Pharm. Bull.* **1977**, *25*, 1471.
66. Ports, K. T.; Chen, S. J.; Kane, J.; Marshall, J. L. *J. Organomet. Chem.* **1977**, *42*, 1633.
67. Hamaguchi, M.; Ibata, T. *Chem. Lett.* **1975**, 169.
68. Potts, K. T.; Chen, S. J. *J. Organomet. Chem.* **1977**, *42*, 1639.
69. C. G. Newton, W. D. Ollis, M. L. Podmore and D.E. Wright, to be published.
70. C. G. Newton, W. D. Ollis and D. E. Wright, to be published.
71. Yoshioka, H.; Nakatsu, K.; Hayashi, M.; Mizuno, K. *Tetrahedron Lett.* **1975**, 4031.
72. Mizuno, K.; Tsujino, M.; Takada, M.; Hayashi, M.; Atsumi, K.; Asano, K.; Matsuda, T. *J. Antibiotics* **1974**, *27*, 775.
73. Mizuno, K.; Yaginuma, S.; Hayashi, M.; Takada, M.; Muto, N. *Hakko Kogaku Zasshi* **1975**, *53*, 609. *Chem. Abstr.* **1975**, *83*, 191293.
74. Hayashi, M.; Hirano, T.; Yaso, M.; Mizuno, K.; Ueda, T. *Chem. Pharm. Bull.* **1975**, *23*, 245.
75. Tarumi, Y.; Takebayashi, Y.; Moriguchi, K.; Atsumi, T.; Fukumaru, T.; Yamamoto, H. *J. Heterocyclic Chem.* **1980**, *17*, 1425.
76. Atsumi, T.; Takebayashi, Y.; Tarumi, Y.; Yamamoto, H. *(Sumitomo Chemical KK) Belgian Pat. 856403*; 1977.
77. Begtrup, M. *J. Chem. Soc. Perkin Trans. I* **1979**, 521.
78. Begtrup, M. *J. Chem. Soc. Perkin Trans.* **1979**, 1132.
79. Talukdar, P. B.; Sengupta, S. K.; Datta, A. K. *Ind. J. Chem.* **1976**, *14B*, 382.
80. Fischer, E.; Rembarz, G.; Wollin, K. M. *J. Prakt. Chem.* **1980**, *322*, 375.
81. J. T. Edward and R. H. Sheffler, ACS/CSJ Chemical Congress. Abstract ORGN-168. Honolulu, Hawaii (April 1979).
82. Potts, K. T.; Choudhury, D. R. *J. Organomet. Chem.* **1978**, *43*, 2697.
83. Talukdar, P. B.; Sengupta, S. K.; Datta, A. K.; Roy, T. K. *Ind. J. Chem.* **1977**, *15B*, 41. Talukdar P. B.; Sengupta S. K.; Datta A. K. *Ind. J. Chem.* **1980**, *19B*, 638.
84. Barton, D. H. R.; Buschmann, E.; Haüsler, J.; Holzapfel, C. W.; Sheradsky, T.; Taylor, D. A. *J. Chem. Soc. Perkin Trans. I* **1977**, 1107.
85. Potts, K. T.; Choudhury, D. R. *J. Organomet. Chem.* **1978**, *43*, 2700.
86. Potts, K. T.; Kanemasa, S. *J. Organomet. Chem.* **1979**, *44*, 3803.
87. Potts, K. T.; Kanemasa, S. *J. Organomet. Chem.* **1979**, *44*, 3808.
88. Singh, H. K.; Gandhi, C. S. *Ind. J. Chem.* **1978**, *16B*, 331.

89. Singh, H.; Ahuja, A. S.; Gandhi, C. S. *J. Chem. Res. (M)* **1979**, 2935. *J. Chem. Res. (S)* **1979**, 264.
90. Baudy, M.; Robert, A. *J. Chem. Soc., Chem. Commun.* **1976**, 23.
91. Baudy, M.; Robert, A.; Foucaud, A. *J. Organomet. Chem.* **1978**, *43*, 3732.
92. Baudy, M.; Robert, A. *Tetrahedron Lett.* **1980**, *21*, 2517.
93. Baudy, M.; Robert, A. *Comptes Rend. Acad. Sci.* **1980**, *290*, 287.
94. Gorb, L. T.; Sych, E. D.; Tolmachev, A. I.; Shpileva, I. S. *Khim. Geterotsikl. Soedin.* **1979**, 1066. *Chem. Abstr.* **1979**, *91*, 212554.
95. Tavernier, D.; Jamoulle, J. C.; Lapiere, C. L. *Can. J. Chem.* **1978**, *56*, 2109.
96. Gorb, L. T.; Romanov, N. N.; Tolmachev, A. I. *Khim. Geterotsikl. Soedin.* **1979**, 1343. *Chem. Abstr.* **1980**, *92*, 128789.
97. Gorb, L. T.; Romanov, N. N.; Sych, E. D.; Tolmachev, A. I. *Dopov. Akad. Nauk. Ukr. RSR, Ser. B: Geol. Khim. Biol. Nauki.* **1978**, 894. *Chem. Abstr.* **1979**, *90*, 38831.
98. Sych, E. D.; Gorb, L. T. *Khim. Geterotsikl. Soedin.* **1980**, 312. *Chem. Abstr.* **1980**, *92*, 128789.
99. Gorb, L. T.; Romanov, N. N.; Tolmachev, A. I. *Khim. Geterotsikl. Soedin.* **1979**, 989. *Chem. Abstr.* **1979**, *91*, 211314.
100. Gorb, L. T.; Kachkovskii, A. T.; Romanov, N. N.; Shpileva, I. S.; Tolmachev, A. I. *Khim. Geterotsikl. Soedin.* **1980**, 621. *Chem. Abstr.* **1980**, *93*, 132416.
101. Robert, A.; Ferrey, M.; Foucaud, A. *Tetrahedron Lett.* **1975**, 1377.
102. Robert, A.; Ferrey, M.; Le Marechal, A. *Tetrahedron* **1980**, *36*, 1571.
103. Matsukubo, H.; Kato, H. *J. Chem. Soc., Perkin Trans. I* **1976**, 2562.
104. Robert, A.; Baudy, M.; Foucaud, A.; Golic, L.; Stanovnik, B. *Tetrahedron* **1978**, *34*, 3525.
105. Sheradsky, T.; Zbaida, D. *J. Organomet. Chem.* **1980**, *45*, 4850.
106. Sheradsky, T.; Zbaida, D. *Tetrahedron Lett.* **1978**, 2037.
107. Sheradsky, T.; Zbaida, D. *J. Organomet. Chem.* **1980**, *45*, 2165.
108. Talukdar, P. B.; Sengupta, S. K.; Datta, A. K. *Ind. J. Chem.* **1978**, *16B*, 678. *Chem. Abstr.* **1979**, *90*, 54901.
109. Talukdar, P. B.; Sengupta, S. K.; Datta, A. K. *Ind. J. Chem.* **1979**, *18B*, 39. *Chem. Abstr.* **1979**, *91*, 210691.
110. Talukdar, P. B.; Chakraborty, A. *Ind. J. Chem.* **1978**, *16B*, 310. *Chem. Abstr.* **1978**, *89*, 107293.
111. Elgendy, M.; Eldawy, M. A. *Egypt. J. Pharm. Sci.* **1976**, *17*, 35. *Chem. Abstr.* **90**, 152061 (1979).
112. Barrett, G. C.; Walker, R. *Tetrahedron* **1976**, *32*, 571.
113. Barrett, G. C.; Walker, R. *Tetrahedron* **1976**, *32*, 579.
114. Duncan, J. A.; Bosse, M. L.; Masnovi, J. M. *J. Organomet. Chem.* **1980**, *45*, 3176.
115. Barrett, G. C.; Walker, R. *Tetrahedron* **1976**, *32*, 583.
116. Cava, M. P.; Saris, L. E. *J. Chem. Soc., Chem. Commun.* **1975**, 617.
117. Potts, K. T.; Huang, F.; Khattak, R. K. *J. Organomet. Chem.* **1977**, *42*, 1644.
118. Gotthardt, H.; Weisshuhn, M. C.; Christi, B. *Chem. Ber.* **1976**, *109*, 740.
119. Potts, K. T.; Choudhury, D. R.; Elliott, A. J.; Singh, U. P. *J. Organomet. Chem.* **1976**, *41*, 1724.
120. Ohta, M.; Sugiyama, M. *Bull. Chem. Soc. Japan* **1995**, *38*, 596.
121. G. P. Rowson, PhD. Thesis, University of Sheffield, 1976.
122. Gotthardt, H.; Weisshuhn, C. M.; Huss, O. M.; Brauer, D. J. *Tetrahedron Lett.* **1978**, 671.
123. Gotthardt, H.; Huss, O. M.; Weisshuhn, C. M. *Chem. Ber.* **1979**, *112*, 1650.
124. Gotthardt, H.; Weisshuhn, M. C.; Christl, B. *Chem. Ber.* **1976**, *109*, 753.
125. Gotthardt, H.; Weisshuhn, C. M. *Chem. Ber.* **1978**, *111*, 2021.

126. Gotthardt, H.; Weisshuhn, C. M. *Chem. Ber.* **1978**, *111*, 2028.
127. Gotthardt, H.; Weisshuhn, C. M.; Christl, B. *Ann.* **1979**, 360.
128. Gotthardt, H.; Christl, B. *Chem. Ber.* **1978**, *111*, 3029.
129. Gotthardt, H.; Weisshuhn, C. M.; Christl, B. *Chem. Ber.* **1978**, *111*, 3037.
130. Gotthardt, H.; Weisshuhn, C. M. *Chem. Ber.* **1978**, *111*, 3171.
131. Gotthardt, H.; Weisshuhn, C. M. *Chem. Ber.* **1978**, *111*, 3178.
132. Tezuka, H.; Shiba, T.; Aoki, N.; Iijima, K.; Kato, H. *Kokagaku Toronkai Koen Yoshishu* **1979**, 8. *Chem. Abstr.* **1980**, *92*, 197661.
133. Fridman, A. L.; Yufareva, E. G.; Kolobov, N. A. *Khim. Prom-st., Set.: Reakt. Osobo Chist. Veshchestva* **1979**, 14. *Chem. Abstr.* **1979**, *91*, 123684.
134. Tien, H. J.; Yeh, M. Y.; Lin, S. T.; Ohta, M. *Hua Hsueh* **1978**, 7. *Chem. Abstr.* **1980**, *92*, 94316.
135. Gotthardt, H.; Reiter, F. *Ann.* **1979**, 63.
136. Tien, H. J.; Nonaka, T.; Sekine, T. *Chem. Lett.* **1979**, 283.
137. Greco, C. V.; Mehta, J. R. *J. Heterocyclic Chem.* **1979**, *16*, 1059.
138. Baker, W.; Ollis, W. D.; Poole, V. D. *J. Chem. Soc. D* **1950**, 1542.
139. Weintraub, P. M.; Bambury, R. E. *Tetrahedron Lett.* **1969**, 579.
140. Tien, H. J.; Nonaka, T.; Fuchigami, T.; Sekine, T.; Ohta, M. *Nippon Kagaku Kaishi* **1978**, 1310. *Chem. Abstr.* **1979**, *90*, 72115.
141. Preston, P. N.; Turnbull, K. *J. Chem. Soc. Perkin Trans. I* **1977**, 1229.
142. Tien, H. J. *Hua Hsueh* **1977**, 8. *Chem. Abstr.* **1980**, *92*, 110930.
143. Fridman, A. L.; Yufareva, E. G.; Kolobov, N. A.; Novikov, S. S. *Khim. Geterotsikl. Soedin.* **1977**, 701. *Chem. Abstr.* **1977**, *87*, 117814.
144. Fridman, A. L.; Yufareva, E. G.; Gabitov, F. A.; Andreichikov, Y. S. *Khim. Geterotsikl. Soedin.* **1979**, 1425. *Chem. Abstr.* **1980**, *92*, 110929.
145. Ugarkar, B. G.; Badami, B. V.; Puranik, G. S. *Arch. Pharm.* **1979**, *312*, 977. *Chem. Abstr.* **92**, 181092 (1980).
146. Neumann, B. M.; Henning, H. G.; Gloyna, D.; Bandlow, M. *J. Prakt. Chem.* **1976**, *318*, 823.
147. Sorm, M.; Nespurek, S. *Czech. Patent 165174*; 1976. *Chem. Abstr.* **1977**, *87*, 23269.
148. Sorm, M.; Nespurek, S. *Czech. Patent 167719*; 1977. *Chem. Abstr.* **1977**, *87*, 152217.
149. Nespurek, S.; Sorm, M. *Czech. Patent 167730*; 1977. *Chem. Abstr.* **1977**, *87*, 152218.
150. Greco, C. V.; Mehta, J. R. *J. Chem. Soc. Perkin Trans. I* **1980**, 20.
151. Yen, M. Y.; Tien, H. J. *Cheng-kung Ta Hsueh Hsueh Pao* **1979**, *14*, 23. *Chem. Abstr.* **1980**, *92*, 181091.
152. Tien, H. J.; Yen, M. Y.; Lian, C. K. *J. Chin. Chem. Soc.* **1977**, *24*, 49. *Chem. Abstr.* **1977**, *87*, 135216.
153. Kozinsky, V. A.; Burmistrov, S. I.; Mikheeva, O. V. *Zh. Org. Khim.* **1976**, *12*, 1572. *Chem. Abstr.* **1976**, *85*, 159992.
154. Tien, H. J.; Yeh, M. Y. *J. Chin. Chem. Soc.* **1977**, *24*, 123. *Chem. Abstr.* **1978**, *88*, 50744.
155. Fridman, A. L.; Yufareva, E. G.; Kolobov, N. A. *Khim. Geterotsikl. Soedin.* **1977**, 1962. *Chem. Abstr.* **1978**, *88*, 121061.
156. Legrand, J. J.; Renault, C. *(Marpha Societe d'Etude et d'Exploitation de Marques) Belg. Patent 843959*; 1977. *Chem. Abstr.* **1977**, *87*, 152198.
157. Badami, B. V.; Puranik, G. S. *Rev. Roam. Chim.* **1975**, *20*, 981. *Chem. Abstr.* **1976**, *84*, 4864.
158. Kishimoto, S.; Noguchi, S.; Masuda, K. *Chem. Pharm. Bull.* **1976**, *24*, 3001.
159. Gotthardt, H.; Reiter, F. *Chem. Ber.* **1979**, *112*, 1193.
160. Gotthardt, H.; Reiter, F. *Chem. Ber.* **1979**, *112*, 1206.
161. Meier, H.; Heimgartner, H.; Schmid, H. *Heir. Chim. Acta* **1977**, *60*, 1087.
162. Bieri, J. H.; Meier, H.; Heimgartner, H. *Heir. Chim. Acta* **1978**, *61*, 1091.

163. Jones, R. A. J.; Marriott, M. T. P. *Heterocycles* **1980**, *14*, 185.
164. Sasaki, T.; Manabe, T.; Nishida, S. *J. Organomet. Chem.* **1980**, *45*, 479.
165. Haneda, A.; Imagawa, T.; Kawanisi, M. *Bull. Chem. Soc. Japan* **1976**, *49*, 748.
166. Franz, J. E. *(Monsanto Co.) Ger. Offen. 28059*; 1978. *Chem. Abstr.* **1979**, *90*, 22560.
167. Berk, H. C.; Franz, J. E. *J. Organomet. Chem.* **1979**, *44*, 2395.
168. Berk, H. C.; Fraraz, J. E. *Synth. Commun.* **1980**, *10*, 189.
169. Kato, H.; Shiba, T.; Kitajima, E.; Kiyosawa, T.; Yamada, F.; Nishiyama, T. *J. Chem. Soc. Perkin Trans. I* **1976**, *863*.
170. Märky, M.; Meier, H.; Wunderli, A.; Heimgartner, H.; Schmid, H.; Hansen, H. J. *Heir. Chim. Acta* **1978**, *61*, 1477.
171. Gotthardt, H.; Reiter, F. *Chem. Ber.* **1979**, *112*, 1635.
172. Pfoertner, K. H.; Foricher, J. *Heir. Chim. Acta* **1980**, *63*, 653.
173. Kar, S. K. *Ind. J. Chem.* **1977**, *15B*, 184. *Chem. Abstr.* **1977**, *87*, 117810.
174. Neumann, B. M.; Henning, H. G.; Gloyna, D.; Sauer, J. *J. Prakt. Chem.* **1978**, *320*, 81.
175. Henning, H. G.; Neumann, B. M.; Alder, L. *Z. Chem.* **1978**, *18*, 262.
176. Bauschke, E.; Tomaschewski, G. *J. Prakt. Chem.* **1978**, *302*, 206.
177. Gotthardt, H.; Reiter, F.; Gieren, A.; Lamm, V. *Tetrahedron Lett.* **1978**, 2331.
178. Gieren, A.; Lamm, V. *Acta Cryst.* **1978**, *B34*, 3248.
179. Bhat, V.; Dixit, V. M.; Ugarker, B. G.; Trozzolo, A. M.; George, M. V. *J. Organomet. Chem.* **1979**, *44*, 2957.
180. Kopecky, K. R.; Pope, P. M.; Sastre, J. A. L. *Can. J. Chem.* **1976**, *54*, 2639.
181. McLean, C. D.; Middleton, S. *Austral. J. Chem.* **1978**, *31*, 2659.
182. Slyusarenko, I. S.; Ogorodnikova, V. V.; Yashunskii, V. G. *Khim. Geterotsikl. Soedin.* **1976**, 450. *Chem. Abstr.* **1976**, *85*, 63001.
183. Paetzotd, P. I.; Bohm, P.; Richter, A.; Scholl, E. *Z Naturforsch.* **1976**, *31B*, 754.
184. Paetzold, P. I.; Schimmel, G. *Z. Naturforsch.* **1980**, *35B*, 568.
185. Dickopp, H. *Chem. Ber.* **1980**, *113*, 1830.
186. Hasek, J.; Obrda, J.; Huml, K.; Nespurek, S.; Chojnacki, H.; Sorm, M. *Acta. Cryst.* **1978**, *B34*, 2756.
187. Hasek, J.; Obrda, J.; Huml, K.; Nespurek, S.; Sorm, M. *Acta. Cryst.* **1979**, *B35*, 437.
188. Hasek, J.; Obrda, J.; Huml, K.; Nespurek, S.; Sorm, M. *Acta. Cryst.* **1979**, *B35*, 2449.
189. Stefaniak, L. *Tetrahedron* **1977**, *33*, 2571.
190. Stefaniak, L.; Witanowski, M.; Kamienski, B.; Webb, G. A. *Org. Magn. Reson.* **1980**, *13*, 274.
191. Huang, M. S.; Lu, S. L.; Lin, L. C.; Tien, H. J. *Hua Hsueh* **1977**, 83. *Chem. Abstr.* **1980**, *92*, 110113.
192. Henning, H. G.; Neumann, B. M.; Gloyna, D.; Sauer, J. *J. Prakt. Chem.* **1978**, *320*, 71.
193. Eber, G.; Schneider, S.; Doerr, F. *Ber. Bunsenges. Phys. Chem.* **1980**, *84*, 281. *Chem. Abstr.* **1980**, *93*, 113392.
194. Inoue, S.; Asai, N.; Yasuda, G.; Hori, Y. *Bull. Chem. Soc. Japan* **1977**, *50*, 3268.
195. Sorm, M.; Nespurek, S. *Collect. Czech. Chem. Commun.* **1975**, *40*, 1534.
196. Sorm, M.; Nespurek, S. *Collect. Czech. Chem. Commun.* **1975**, *40*, 3459.
197. Nespurek, S.; Sorm, M. *Collect. Czech. Chem. Commun.* **1977**, *42*, 811.
198. Trozzolo, A. M.; Leslie, Y. M.; Sarpotdar, A. S.; Small, R. D.; Ferraudi, G. J.; Do, M. Y.; Hartless, R. L. *Pure Appl. Chem.* **1979**, *51*, 261. *Chem. Abstr.* **1979**, *91*, 107414.
199. Nespurek, S.; Sorm, M. *Czech. Patent. 168397*; 1977. *Chem. Abstr.* **1978**, *88*, 56987.
200. Nespurek, S.; Ulbert, K.; Sorm, M. *Czech. Patent 172689*; 1978. *Chem. Abstr.* **1978**, 89, 224007.
201. Uhlmann, D. R.; Snitzer, E.; Hovey, R. J.; Chu, N. Y. C.; Fournier, J. I. *U.S. Pat. 4012232*; 1977. *Chem. Abstr.* **1977**, *86*, 157168.
202. Lemire, R. J.; Sears, P. G. *J. Chem. Eng. Data* **1977**, *22*, 376. *Chem. Abstr.* **1977**, *87*, 141446.

203. Lemire, R. J.; Sears, P. G. *J. Solution Chem.* **1980**, *9*, 553. *Chem. Abstr.* **1980**, *93*, 246506.
204. Fridman, A. L.; Gomza, L. D.; Legotkina, G. I.; Yufareva, E. G.; Andreichikov, Y. S. *Zh. Obshch. Khim.* **1979**, *49*, 2337. *Chem. Abstr.* **1980**, *92*, 75529.
205. King, T. J.; Preston, P. N.; Suffolk, J. S.; Turnbull, K. *J. Chem. Soc. Perkin Trans. II* **1979**, 1751.
206. Wen, C.-M.; Tien, H.-J. *J. Chin. Chem. Soc.* **1975**, *22*, 265. *Chem. Abstr.* **1976**, *84*, 30203.
207. Sauer, J.; Bauschke, E.; Tomaschewski, G. *J. Prakt. Chem.* **1977**, *319*, 83.
208. Sauer, J. *Chem. Phys. Lett.* **1978**, *55*, 119. *Chem. Abstr.* **1978**, *88*, 200109.
209. Kier, L. B.; Roche, E. B. *J. Pharm. Sci.* **1967**, *56*, 149.
210. Hill, J. B.; Ray, R. E.; Wagner, H.; Aspinall, R. L. *J. Med. Chem.* **1975**, *18*, 50.
211. Ray, R. E.; Wagner, H. A. *(G. D. Searle and Co.) U.S. Pat. 4020079*; 1977. *Chem. Abstr.* **1977**, *87*, 53315.
212. Hill, J. B. *(G. D. Searle and Co.), Br. Pat. 1407540*; 1975. *Chem. Abstr.***1977**, *87*, 135343.
213. Padhya, K. G. U.; Badami, B. V.; Puranik, G. S.; Biradar, V. N.; Naojappa, S. *Arch. Pharm.* **1980**, *313*, 684. *Chem. Abstr.* **1981**, *94*, 47195.
214. Albert, L.; Tokes, B.; Czegeni, L.; Domokos, L. *Rev. Med. (Tirgu-Mures, Rom.)* **1978**, *24*, 171. *Chem. Abstr.* **1980**, *92*, 70162.
215. Peters, W.; Davies, E. E.; Robinson, B. L. *Ann. Trop. Med. Parasitol.* **1975**, *69*, 311. *Chem. Abstr.* **1976**, *84*, 12314.
216. Ugarkar, B. G.; Badami, B. V.; Puranik, G. S.; Bhat, K. G. S. *Arch. Pharm.* **1978**, *311*, 109. *Chem. Abstr.* **1978**, *89*, 43251.
217. Bulger, W. H.; Wells, P. R.; Roche, E. B. *J. Pharm. Sci.* **1976**, *65*, 109.
218. Havanur, S. B.; Badami, B. V.; Puranik, G. S. *J. Heterocyclic Chem.* **1980**, *17*, 1049.
219. Fridman, A. T.; Yufareva, E. G.; Kozhevnikov, Y. A.; Zalesov, V. S.; Kolobov, N. A. *Khim.-Farm. Zh.* **1976**, *10*, 53. *Chem. Abstr.* **1976**, *85*, 5557.
220. Finland, M.; Garner, C.; Wilcox, C.; Sabath, D. L. *Antimicrob. Agents Chemother.* **1976**, *9*, 11. *Chem. Abstr.* **1976**, *84*, 130823.
221. Finland, M.; Garner, C.; Wilcox, C.; Sabath, L. D. *J. Infect. Diseases* **1976**, *134*, 75. *Chem. Abstr.* **1976**, *85*, 117296.
222. Ryan, C. W.; Blanchard, W. B. *(Eli Lilly and Co.) U.S. Pat. 3943126*; 1976. *Chem. Abstr.* **1976**, *85*, 21407.
223. Montavon, M.; Reiner, R. *(F. Hoffman-La Roche and Co., A.G.), Ger. Offen. 2527291*; 1976. *Chem. Abstr.* **1976**, *84*, 150646.
224. Shimizu, B.; Saito, A.; Kaneko, M.; Yanagisawa, H.; Nakao, H. *(Sankyo Co. Ltd.) Get. Offen. 2552861*; 1976. *Chem. Abstr.* **1976**, *85*, 160133.
225. Shimizu, B.; Saito, A.; Taneko, M. *(Sankyo Co. Ltd.) Japan Kokai 7652191*; 1976. *Chem. Abstr.* **1976**, *85*, 192748.
226. Berges, D. A. *(Smithkline Corp.) Ger. Offen. 2558022*; 1976. *Chem. Abstr.* **1977**, *86*, 72667.
227. Kariyone, K.; Tanaka, K.; Yazawa, H. *(Fujisawa Pharmaceutical Co., Ltd.) Austrian Patent 333951*; 1976. *Chem. Abstr.* **1977**, *87*, 23306.
228. Berges, D. A. *(Smithkline Corp.) U.S. Pat. 4093723*; 1978. *Chem. Abstr.* **1978**, *89*, 180025.
229. Beeby, P. J. *[Syntex (U.S.A.) Inc.] U.S. Pat. 4139618*; 1979. *Chem. Abstr.* **1979**, *90*, 204121.
230. Berges, D. A. *(Smithkline Corp.) U.S. Pat. 4159373*; 1979. *Chem. Abstr.* **1979**, *91*, 175374.
231. Yashunskii, V. G.; Kholodov, L. E. *Russ. Chem. Rev.* **1980**, *49*, 1.
232. Shaikh, A. A. *Vidya, B* **1976**, *19*, 137. *Chem. Abstr.* **1978**, *88*, 136524.
233. Shaikh, A. A. *J. Inst. Chem.* **1976**, *48*, 186. *Chem. Abstr.* **1977**, *86*, 155581.

234. Olovyanishnikova, Z. A.; Ivanova, T. M.; Sviridova, V. E.; Yashunskii, V. G. *Khim. Geterotsikl. Soedin.* **1978**, 170. *Chem. Abstr.* **1978**, *89*, 6284.
235. Ogorodnikova, V. V.; Kholodov, L. E.; Yashunskii, V. G. *Khim. Geterotsikl. Soedin.* **1977**, 39. *Chem. Abstr.* **1977**, *86*, 170328.
236. Borisov, E. V. *Zh. Fiz. Khim.* **1980**, *54*, 1523. *Chem. Abstr.* **1980**, *93*, 185337.
237. Olovyanishnikova, Z. A.; Galkin, O. M.; Yashunskii, V. G.; Davidovich, Y. A. *Khlm. Get. Soedin.* **1978**, 1335. *Chem. Abstr.* **1979**, *90*, 138165.
238. Ivanova, T. M.; Olovyanishnikova, Z. A.; Yashunskii, V. G. *Khim. Geterotsikl. Soedin.* **1978**, 1193. *Chem. Abstr.* **1979**, *90*, 22177.
239. Kholodov, L. E.; Yashunskii, V. G.; Al'tshuler, R. A.; Mashkovskii, M. D.; Ogorodnikova, V. V.; Olovyanishnikova, Z. A.; Vitvitskaya, A. S.; Parshin, V. A.; Kelekhsaeva, E. A. *(Ordzhonikidze, S., All-Union Scientific Research Chemical Pharmaceutical Institute), Get. Offen. 2738022*; 1978. *Chem. Abstr.* **1978**, *89*, 129511.
240. Yashunskii, V. G.; Olovyanishnikova, Z. A.; Ogorodnikova, V. V.; Kholodov, L. E. *Khim. Geterotsikl. Soedin.* **1980**, 117. *Chem. Abstr.* **1978**, *92*, 215360.
241. Olovyanishnikova, Z. A.; Bryantsev, B. I.; Slyusarenko, I. S.; Yashunskii, V. G. *Khim. Geterotsikl. Soedin.* **1975**, 1198. *Chem. Abstr.* **1976**, *84*, 30971.
242. Slyusarenko, I. S.; Ogorodnikova, V. V.; Yashunskii, V. G. *Khim. Geterotsikl. Soedin.* **1976**, 450. *Chem. Abstr.* **1976**, *85*, 63001.
243. Borisov, E. V. *Zh. Fiz. Khim.* **1979**, *53*, 1708. Chem. Abstr. **1979**, *91*, 192374.
244. Polievktov, M. K.; Motoricheva, T. A.; Kholodov, L. E. *Khim. Geterotsikl. Soedin.* **1977**, 33. *Chem. Abstr.* **1977**, *86*, 170453.
245. Borisov, E. V.; Ogorodnikova, V. V.; Lebedeva, N. D.; Yashunskii, V. G. *Khim. Geterotsikl. Soedin.* **1976**, 53. *Chem. Abstr.* **1976**, *84*, 163945.
246. Temkov, I.; Tomov, T.; Kamenov, G. *Nevrol., Psikhiatr. Neurokhir.* **1975**, *14*, 179. *Chem. Abstr.* **1976**, *84*, 38935.
247. Lyubimov, B. I.; Ostrovskaya, G. Z. *Farmakol. Toksikol.* **1977**, *40*, 133. *Chem. Abstr.* **1977**, *86*, 183247.
248. Prokudin, V. N. *Dtsch. Gesundheitswes.* **1977**, *32*, 805. *Chem. Abstr.* **1977**, *87*, 47836).
249. Andreev, B. V.; Patkina, N. A. *Psikhofarmakol. Emotsionarnogo Stressa Zoosotsiarnogo Vzaimodeistviya* **1975**, 87. *Chem. Abstr.* **1977**, *87*, 96057.
250. Arushanyan, E. B.; Baturin, V. A. *Byull. Eksp. Biol. Med.* **1977**, *84*, 303. *Chem. Abstr.* **1977**, *87*, 194036.
251. Al'tshuler, R. A.; Roshchina, L. F.; Mashkovskii, M. D. *Farmakol. Toksikol.* **1976**, *39*, 9. *Chem. Abstr.* **1976**, *84*, 99398.
252. Al'tshuler, R. A.; Kovalenko, T. A.; Kholodov, L. E.; Polievktov, M. K. *Khim.-Farm. Zh.* **1976**, *10*, 20. *Chem. Abstr.* **1976**, *84*, 159481.
253. Tat'yanenko, L. V.; Lavretskaya, E. F.; Raikhman, L. M.; Moshkovskii, Y. S.; Lebedeva, O. I. *Vopr. Med. Khim.* **1976**, *22*, 406. *Chem. Abstr.* **1976**, *85*, 72100.
254. Mashkovskii, M. D.; Raevskii, K. S.; Maisov, N. I.; Sandalov, Y. G.; Andreeva, N. I. *Khim.-Farm. Zh.* **1976**, *10*, 3. *Chem. Abstr.* **1977**, *86*, 50521.
255. Al'tshuler, R. A.; Kholodov, L. E.; Meirena, D. V.; Seredenin, S. B.; Sultanov, I. A.; Stelletskaya, N. V.; Blednov, Y. A. *Khim.-Farm. Zh.* **1977**, *11*, 16. *Chem. Abstr.* **1977**, *86*, 133316.
256. Al'tshuler, R. A.; Kholodov, L. E.; Gilev, A. P.; Meirena, D.; Odynets, A. G. *Khim.-Farm. Zh.* **1977**, *11*, 12. *Chem. Abstr.* **1977**, *86*, 182899.
257. Lavretskaya, E. F.; Tat'yanenko, L. V.; Lebedeva, O. I. *Farmakol. Toksikol.* **1977**, *40*, 12. *Chem. Abstr.* **1977**, *87*, 62383.
258. Rosdy, B.; Kiss, B.; Szporny, L. *Proc. Hung. Annu. Meet. Biochem.* **1977**, *17*, 89. *Chem. Abstr.* **1978**, *88*, 15896.
259. Erdo, S. L. *Proc. Hung. Annu. Meet. Biochem.* **1978**, *18*, 153. *Chem. Abstr.* **1979**, *90*, 33846.

260. Parshin, V. A. *Khim.-Farm. Zh.* **1979**, *13*, 42. *Chem. Abstr.* **1979**, *91*, 13458.
261. Kholodov, L. E.; Lil'in, E. T. *Mater Resp. Rasshir. Konf. Farmakol. Gruz.* **1977**, 84. *Chem. Abstr.* **1979**, *91*, 49198.
262. Kulinskii, V. I.; Kononova, L. I.; Savorenko, N. S. *Vopr. Med. Khim.* **1979**, *25*, 387. *Chem. Abstr.* **1979**, *91*, 106235.
263. Veksler, I. G.; Ryabukha, V. N.; Balitskii, K. P.; Artshuler, R. A.; Mashkovskii, M. D. *Farmakol. Toksikol.* **1980**, *43*, 349. *Chem. Abstr.* **1977**, *93*, 37419.
264. Erdo, S. L.; Kiss, B.; Rosdy, B. *Proc. Hung. Annu. Meet. Biochem.* **1979**, *19*, 163. *Chem. Abstr.* **1980**, *92*, 34248.
265. Medvedev, O. S.; Stepochkina, N. A. *Farmakol. Toksikol.* **1977**, *40*, 50. *Chem. Abstr.* **1977**, *86*, 100989.
266. Petryaevskaya, N. V.; Nesterovich, N. A. *Psikho armakol. Emotsional'nogo Stressa Zoosotsial'nogo Vzaimodeistviya* **1975**, 41. *Chem. Abstr.* **1977**, *87*, 96052.
267. Katkova, E. B.; Bystritskii, A. S.; Martynikhin, A. V. *Psikho armakol. Emotsional'nogo Stressa Zoosotsial'nogo Vzaimodeistviya* **1975**, *91*. *Chem. Abstr.* **1977**, *87*, 111697.
268. Baturin, V. A. *Farmakol. Toksikol.* **1977**, *40*, 691. *Chem. Abstr.* **1978**, *88*, 45028.
269. Belozertsev, Y. A.; Fomin, S. N. *Farmakol. Toksikol.* **1978**, *41*, 5. *Chem. Abstr.* **1978**, *88*, 99190.
270. Schchekoldina, T. G.; Vatolkina, O. E.; Libinzon, R. E. *Vopr. Med. Khim.* **1978**, *24*, 166. *Chem. Abstr.* **1978**, *89*, 264.
271. Soredenin, S. B.; Vedernikov, A. A. *ByuU. Eksp. Biol. Med.* **1979**, *88*, 38. *Chem. Abstr.* **1979**, *91*, 117460.
272. Shuvalova, M. E.; Liberman, S. S.; Egorova, E. F. *Khim.-Farm. Zh.* **1976**, *10*, 70. *Chem. Abstr.* **1977**, *86*, 83724.
273. Rapoport, I. A.; Efremova, G. I. *Khim. Mutagen. Gibrid.* **1978**, 182. *Chem. Abstr.* **1979**, *91*, 49294.
274. Seredin, S. B.; Vedernikov, A. A.; Durnev, A. D. *Deposited Doc., VINITI 1889*; 1979. *Chem. Abstr.* **1980**, *93*, 125650.
275. Czira, C.; Polgar, M.; Tamas, J.; Holly, S.; Radics, L.; Vereczkey, L. *Acta. Pharm. Hung.* **1978**, *48*, 23. *Acta. Pharm. Hung.* **1978**, *48*, 25.
276. Bulat, A. D.; Kholodov, L. G.; Usaevich, Y. Y.; Kivokurtseva, L. N. *Khim.-Farm. Zh.* **1975**, *9*, 9. *Chem. Abstr.* **1976**, *85*, 21221).
277. Baumann, D. *Pharm. Prax.* **1976**, 36. Chem. Abstr. **1976**, *85*, 10324.
278. Hirata, M.; Oku, Y.; Tanabe, M.; Kikuchi, K. *Takeda Kenkyusho Ho.* **1975**, *34*, 139. *Chem. Abstr.* **1976**, *84*, 252.
279. Masuda, K.; Toga, T.; Hayashi, N. *J. Labelled Compounds* **1975**, *11*, 301. *Chem. Abstr.* **1976**, *84*, 121730.
280. Grund, E.; Müller-Ruchholtz, E. R.; Lapp, E. R.; Lösch, H.-M.; Lochner, W. *Arnzeim-Forsch.* **1978**, *28*, 1624.
281. Holtz, J.; Bassenge, E.; Kolin, A. *Basic Res. Cardiol.* **1978**, *73*, 469. *Chem. Abstr.* **1976**, *90*, 115207.
282. Ciaeeri, G.; Attaguile, G.; Marini, P. P. *Boll, Sedute Accad. Gioenia Sci. Nat. Catania* **1975**, *12*, 94. *Chem. Abstr.* **1978**, *89*, 140497.
283. Ciaceri, G.; Marini, P. P.; Attaguile, G. *Boll, Sedute Accad. Gioenia Sci. Nat. Catania* **1975**, *12*, 105. *Chem. Abstr.* **1978**, *89*, 140498.
284. Scholtholt, J.; Fiedler, V. B.; Keil, M. *Arzneim-Forsch.* **1978**, *28*, 1612.
285. Scholtholt, J.; Fiedler, V. B.; Keil, M. *Arzneim-Forsch.* **1978**, *28*, 1619.
286. Berdeaux, A.; Tato, F.; Ho, S.; Boissier, J. R.; Giudicelli, J. F. *J. Pharmacol.* **1978**, *9*, 219. *Chem. Abstr.* **1979**, *90*, 81015.
287. Fiedler, V. B.; Scholtholt, J. *Arzneim-Forsch.* **1978**, *28*, 1605.
288. Dell, D.; Chamberlain, J. *J. Chromatogr.* **1978**, *146*, 465. *Chem. Abstr.* **1979**, *90*, 114758.

289. W. Lochner and F. Bender "Molsidomine" New Aspects in the Therapy of Ischemic Heart Disease, Ist Molsidomine Symposium, Munich, 1978), *Urban and Schwartzenburg, Munich* **1979**; *Chem. Abstr.* **1979**, *91*, 151536.
290. Glushkov, R. G.; Dronova, L. N.; Nikolaeva, L. A.; Medvedev, B. A.; Mashkovskii, M. D.; Solov'eva, N. P.; Turchin, K. F.; Persianova, I. V. *Khim.-Farm. Zh.* **1978**, *12*, 59. *Chem. Abstr.* **1978**, *89*, 109286.
291. Al'tshuler, R. A.; Parshin, V. A. *FarmakoL Toksikol.* **1980**, *43*, 153. *Chem. Abstr.* **1980**, *93*, 36827.
292. Dobrokhotov, Z. M.; Rakov, A. P.; Blokhin, Y. I.; Makarov, N. P. *USSR Patent 502910*; 1976. *Chem. Abstr.* **1976**, *85*, 6687.
293. Dobrokhotov, Z. M.; Makarov, N. P.; Blokhin, Y. I.; Samoilovskikh, N. A. *Fosfororgan. Soedineniya i Polimery* **1976**, 97. *Chem. Abstr.* **1977**, *87*, 40173.
294. Dobrokhotov, Z. M.; Makarov, N. P.; Blokhin, Y. I.; Samoilovskikh, N. A. *Izv. Vyssh. Uchebn. Zaved., Khim. Tekhnol.* **1977**, *20*, 148. *Chem. Abstr.* **1977**, *86*, 140865.
295. Masuda, K.; Adachi, J.; Nomura, K. *J. Chem. Soc. Perkin Trans I* **1979**, 956.
296. Alajarin, M.; Molina, P. *Tetrahedron Lett.* **1980**, *21*, 4025.
297. Grashey, R.; Weidner, M.; Schroll, G. *Chem. -Ztg.* **1976**, *100*, 497.
298. Isukul, E. A.; Tillett, J. G. *J. Chem. Soc. Perkin Trans. II* **1978**, 908.
299. Grashey, R.; Janchen, E. *Chem. -Ztg.* **1980**, *104*, 240.
300. Rousseau, D.; Taurins, A. *Can. J. Chem.* **1977**, *55*, 3736.
301. Dunkin, I. R.; Poliakoff, M.; Turner, J. J.; Harrit, N.; Holm, A. *Tetrahedron Lett.* **1976**, 873.
302. Holm, A.; Harrit, N.; Toubro, N. H. *Tetrahedron* **1976**, *32*, 2559.
303. Holm, A.; Harrit, N.; Toubro, N. H. *J. Am. Chem. Soc.* **1975**, *97*, 6197.
304. Holm, A.; Harrit, N.; Trabjerg, I. *J. Chem. Soc. Perkin Trans. I* **1978**, 746.
305. Holm, A.; Christiansen, J. J.; Lohse, C. *J. Chem. Soc. Perkin Trans. I* **1979**, 960.
306. Abu-EI-Haj, M. J.; McFarland, J. W. *(Pfizer Inc.) U.S. Patent 3933843*; 1976.
307. Abu-EI-Haj, M. J.; McFarland, J. W. *(Hoechst A.G.) U.S. Patent 4002636*; 1977. *Chem. Abstr.* **1977**, *86*, 155663.
308. Nielsen, K. *Acta Chem. Scand.* **1975**, *29 (A)*, 647.
309. Kubota, S.; Uda, M. *Chem. Pharm. Bull.* **1976**, *24*, 1336.
310. Saito, A.; Shimizu, B. *Bull. Chem. Soc. Japan* **1977**, *50*, 1596.
311. Brown, D. J.; Dunlap, W. C.; Grigg, G. W.; Danckwerts, L.; Nagamatsu, T. *Austral. J. Chem.* **1978**, *31*, 397.
312. Ziman, S. D. *J. Heterocyclic Chem.* **1980**, *17*, 1319.
313. Treuner, U. D.; Breuer, H. *(E. R. Squibb & Sons, Inc.) U.S. Patent. 4124764*; 1978. *Chem. Abstr.* **1976**, *90*, 87508.
314. Takano, T.; Kamiya, T. *(Fujisawa Pharmaceutical Co. Ltd) Japan Tokyo Koho, 7916514*; 1979. *Chem. Abstr.* **1979**, *91*, 175336.
315. Cannon, J. R.; Raston, C. L.; White, A. H. *Austral. J. Chem.* **1980**, *33*, 2237.
316. Cawkill, E.; Ollis, W. D.; Ramsden, C. A.; Rowson, G. P. *J. Chem. Soc., Chem. Commun.* **1976**, 439. *J. Chem. Soc. Perkin Trans I*, **1979**, *724*.
317. Lazaris, A. Y.; Shmuilovich, S. M.; Egorochkin, A. N. *Khim. Geterotsikl. Soedin.* **1976**, 424. *Chem. Abstr.* **1976**, *85*, 21235.
318. Masuda, K.; Adachi, J.; Shibata, T.; Nomura, K. *Chem. Pharm. Bull.* **1979**, *27*, 1683.
319. Ollis, W. D.; Rowson, G. P. *J. Chem. Soc., Chem. Commun.* **1976**, 440.
320. Lazaris, A. Y.; Shmuilovich, S. M.; Egorochkinl Dokl, A. N. *Akad. Nauk SSR* **1976**, *226*, 843. *Chem. Abstr.* **1976**, *85*, 32924.
321. Masuda, K.; Adachi, J.; Shibata, T.; Nomura, K. *Chem. Pharm. Bull.* **1979**, *27*, 1688.
322. Lazaris, A. Y.; Shmuilovich, S. M.; Egorochkin, A. N. *lzv. Adak. Nauk SSSR, Ser. Khim.* **1977**, 1827. *Chem. Abstr.* **1977**, *87*, 184467.

323. Mokrushina, G. A.; Postovskii, I. Y.; Kotovskaya, S. K. *Khim. Geterotsikl. Soedin.* **1977**, 411. *Chem. Abstr.* **1977**, *87*, 53170.
324. Postovsky, I. Y.; Kotovskaya, S. K.; Mokrushina, G. A.; Ilenko, V. I.; Platonov, V. G.; Polonyankin, A. V. *Khim.-Farm. Zh.* **1980**, *14*, 50.
325. Talukdar, P. B.; Sengupta, S. K.; Datta, A. K. *Ind. J. Chem.* **1976**, *14B*, 176. *Chem. Abstr.* **1976**, *85*, 123820.
326. Talukdar, P. B.; Sengupta, S. K.; Datta, A. K. *Ind. J. Chem.* **1977**, *15B*, 1110. *Chem. Abstr.* **1978**, *89*, 6298.
327. Lazaris, A. Y.; Kalinin, A. I.; Egorochkin, A. N. *Izv. Akad. Nauk SSSR, Ser. Khim.* **1979**, 2717. *Chem. Abstr.* **1980**, *92*, 198326.
328. Lazaris, A. Y.; Egorochkin, A. N. *Izv. Akad. Nauk SSSR, Ser. Khim.* **1976**, 1191. *Chem. Abstr.* **1976**, *85*, 108584.
329. Lazaris, A. Y.; Egorochkin, A. N. *Izv. Akad. Nauk SSSR, Ser. Khim.* **1978**, 1401. *Chem. Abstr.* **1978**, *89*, 146844.
330. Masuda, K.; Adachi, J.; Nomura, K. *J. Chem. Soc. Perkin Trans. I* **1979**, 2349.
331. Burmistrov, S. I.; Kozinskii, V. A. *Zh. Org. Khim.* **1974**, *10*, 891. *Chem. Abstr.* **1974**, *81*, 13444.
332. Brückner, S.; Fronza, G.; Giunchi, L. M.; Kozinski, V. A.; Zelenskaja, O. V. *Tetrahedron Lett.* **1980**, *21*, 2101.
333. Auricchio, S.; Bruckner, S.; Giunchi, L. M.; Kozinsky, V. A.; Zelenskaja, O. V. *J. Heterocyclic Chem.* **1980**, *17*, 1217.
334. Auricchio, S.; Bruckner, S.; Giunchi, L. M.; Kozinsky, V. A. *Heterocycles* **1980**, *14*, 1757.
335. C. G. Newton, W. D. Ollis and D. E. Wright, to be published.
336. Goerdeler, J.; Roth, W. *Chem. Ber.* **1963**, *96*, 534.
337. Goerdeler, J.; Mertens, P. *Chem. Ber.* **1970**, *103*, 1805.
338. Lempert-Sreter, M.; Lempert, K. *J. Chem. Res. (S)* **1979**, 377.
339. Lempert-Sreter, M.; Lempert, K.; Tamas, J.; Vekey, K. *Acta Chim. Acad. Sci. Hung.* **1980**, *103*, 259. *Chem. Abstr.* **1981**, *94*, 47070.
340. Holm, A.; Toubro, N. H.; Harrit, N. *Tetrahedron Lett.* **1976**, 1909.
341. Takamizawa, A.; Matsushita, Y.; Harada, H. *Chem. Pharm. Bull.* **1980**, *28*, 447.
342. Lazaris, A. Y.; Egorochkin, A. N. *lzv. Akad. Nauk SSSR, Set. Khim.* **1976**, 1191. *Chem. Abstr.* **1976**, *85*, 108585.
343. Lazaris, A. Y.; Shmuilovich, S. M.; Egorochkin, A. N. *Khim. Geterotsikl. Soedin.* **1976**, 713. *Chem. Abstr.* **1976**, *85*, 94325.
344. Kiwan, A. M.; Marafie, H. M. *J. Heterocyclie Chem.* **1976**, *13*, 1273.
345. J. Wood, R. F. Jones and J. H. Davies (Shell International Research), *Br. Pat.* 2015878 (1979).
346. Amann, A.; Koenig, H.; Thieme, P. C.; Giertz, H.; Kretzschmar, R. *(BASF A.-G.), Ger. Often. 2408288*; 1975. *Chem. Abstr.* **1975**, *83*, 193341.
347. Moriarty, R. M.; Mukherjee, R.; Flippen, J. L.; Karle, J. *J. Chem. Soc., Chem. Commun.* **1971**, 1436. Mukherjee R.; Moriarty R. M.; *Ind. J. Chem.* **1977**, *15B*, 499; *Chem. Abstr.* **1978**, *88*, 37707.
348. Ottersen, T. *Acta Chem. Scand.* **1975**, *29A*, 799.
349. Hanley, R. N.; Ollis, W. D.; Ramsden, C. A. *J. Chem. Soc. Perkin Trans. I* **1979**, 747.
350. Hanley, R. N.; Ollis, W. D.; Ramsden, C. A.; Rowlands, G.; Sutton, L. E. *J. Chem. Soc. Perkin Trans. I* **1978**, 600.
351. Bjerre, C.; Christophersen, C.; Hansen, B.; Harrit, N.; Nicolaisen, F. M.; Holm, A. *Tetrahedron* **1979**, *35*, 409.
352. Isukul, E. A.; Ranson, R.; Tillett, J. G. *J. Organomet. Chem.* **1976**, *41*, 3040.
353. Said, Z.; Tillett, J. G. *J. Chem. Soc. Perkin Trans. II* **1979**, 533.

354. J. Wood, R. F. Jones and J. H. Davies (Shell International Research), *Brit. Pat.* 2015878 (1979).
355. Hanley, R. N.; Ollis, W. D.; Ramsden, C. A. *J. Chem. Soc. Perkin Trans. I* **1979**, 736.
356. Ottersen, T.; Christophersen, C.; Treppendahl, S. *Acta Chem. Scand.* **1975**, *29A*, 45.
357. Hanley, R. N.; Ollis, W. D.; Ramsden, C. A. *J. Chem. Soc. Commun.* **1976**, 306. Hanley, R. N.; Ollis, W. D.; Ramsden C. A.; *Z. Chem. Soc. Perkin Trans I*, **1979**, *732*.
358. Preston, P. N.; Tiwari, K. K.; Turnbull, K.; King, T. J. *J. Chem. Soc. Chem. Commun* **1976**, 343.
359. Hanley, R. N.; Ollis, W. D.; Ramsden, C. A. *J. Chem. Soc. Chem. Commun.* **1976**, 307. Hanley, R. N.; Ollis W. D.; Ramsden, C. A.; *J. Chem. Soc. Perkin Trans I*, **1979**, *741*.
360. Bartels-Keith, J. R.; Burgess, M. T.; Stevenson, J. M. *J. Organomet. Chem.* **1977**, *42*, 3725.
361. Bartels-Keith, J. R.; Burgess, M. T. *(Polaroid Corp.), U.S. Patent 4003910*; 1977. *Chem. Abstr.* **1977**, *86*, 171459.
362. Bamberger, E.; Kuhlemann, F. *Ber.* **1893**, *26*, 2978.
363. Bamberger, E.; Müller, J. *J. Prakt. Chem.* **1901**, *64*, 199.
364. Neugebauer, F. A.; Fischer, H.; Krieger, C. *Chem. Ber.* **1979**, *112*, 2369.
365. Hanley, R. N.; OUis, W. D.; Ramsden, C. A. *J. Chem. Soc., Chem. Commun.* **1976**, 307. Hanley, R. N.; Ollis, W. D.; Ramsden, C. A.; Smith, I. S.; *J. Chem. Soc. Perkin Trans I*, **1979**, *744*.
366. Roesky, H. W.; Wehner, B. *Angew. Chem. Int. Edn.* **1975**, *14*, 498.
367. Neidlein, R.; Leinberger, P.; Gieren, A.; Dederer, B. *Chem. Ber.* **1978**, *111*, 698.
368. Roesky, H. W.; Wehner, E.; Zehnder, E. J.; Deiseroth, H. J.; Simon, A. *Chem. Ber.* **1978**, *111*, 1670.
369. Gieren, A.; Dederer, B.; Martin, R.; Schanda, F.; Roesky, H. W.; Eiser, M. *Chem. Ber.* **1980**, *113*, 3904.
370. Roesky, H. W.; Müller, T.; Wehner, E.; Rodek, E. *Chem. Ber.* **1980**, *113*, 2802.
371. Neidlein, R.; Lehr, W. *Chem. Ztg.* **1980**, *104*, 200. *Chem. Abstr.* **1980**, *93*, 204558.
372. Gieren, A.; Dederer, B. *Acta. Cryst.* **1978**, *B34*, 2046.
373. Cannon, J. R.; Potts, K. T.; Raston, C. L.; Sierakowski, A. F.; White, A. H. *Austral. J. Chem.* **1978**, *31*, 297.
374. Friedrichsen, W.; Schmidt, R. *Ann.* **1978**, 1540.
375. Fagan, P. J.; Neidert, E. E.; Nye, M. J.; O'Hare, M. J.; Tang, W.-P. *Can. J. Chem.* **1979**, *57*, 904.
376. Lindley, P. F.; Mahmoud, M. M.; Dodd, C.; Smith, C. H.; Boyd, G. V.; Norris, T. *Acta. Cryst.* **1977**, *B33*, 2160.
377. Barillier, D. *Bull. Soc. Chim. Fr.* **1979**, II-26.
378. Barillier, D. *Phosphorus, Sulphur and Related Elements* **1980**, *8*, 79.
379. Barillier, D.; Rioult, P.; Vialle, J. *Bull. Soc. Chim. Fr.* **1973**, 3031.
380. Barillier, D.; Rioult, P.; Vialle, J. *Bull. Soc. Chim. Fr.* **1976**, 444.
381. Barillier, D.; Rioult, P.; Vialle, J. *Bull. Soc. Chim. Fr.* **1977**, 659.
382. Barillier, D. *Phosphorus, Sulphur and Related Elements* **1978**, *5*, 251.
383. C. G. Newton, M. Sc. Thesis, University of Sheffield (1977).
384. Chinone, A.; Inouye, K.; Ohta, M. *Bull. Chem. Soc. Japan* **1972**, *45*, 213.
385. Kibbel, H. U.; Teller, J.; Hansen, P.; Reiter, A. *J. Prakt. Chem.* **1980**, *322*, 769.
386. Masuda, K.; Adachi, J.; Nomura, K. *J. Chem. Soc., Chem. Commun.* **1979**, 331. *J. Chem. Soc. Perkin Trans. I*, **1981**, 1033.
387. Blaton, N.; Peeters, O.; Willems, G.; De Ranter, C. *Cryst. Struct. Commun.* **1977**, *6*, 603. *Chem. Abstr.* **1977**, *87*, 160281.
388. Hutton, A. T.; Irving, H. M. N. H. *J. Chem. Soc. Perkin Trans. II* **1980**, 139.
389. Kagami, H.; Motoki, S. *Bull. Chem. Soc. Japan* **1979**, *52*, 3463.

390. Preston, P. N.; Robinson, N. J.; Turnbull, K.; King, T. J. *J. Chem. Soc., Chem. Commun.* **1974**, 998.
391. Boyd, G. V.; Norris, T.; Lindley, P. F. *J. Chem. Soc. Perkin Trans. I* **1976**, 1673.
392. Boyd, G. V.; Norris, T.; Lindley, P. F. *J. Chem. Soc., Chem. Commun.* **1975**, 100.
393. Boyd, G. V.; Norris, T.; Lindley, P. F.; Mahmoud, M. M. *J. Chem. Soc. Perkin Trans. I* **1977**, 1612.
394. Boyd, G. V.; Norris, T.; Lindley, P. F. *J. Chem. Soc. Perkin Trans. I* **1977**, 965.
395. Kiwan, A. M.; Kassim, A. Y. *J. Heterocyclic Chem.* **1978**, *15*, 133.
396. Abd-El-Nabey, B. A.; Kiwan, A. M. *J. Electroanal. Chem. Interfacial Electrochem.* **1979**, *105*, 365. *Chem. Abstr.* **1980**, *92*, 101428.
397. Peyronel, G.; Fabretti, A. C. *J. Coord. Chem.* **1977**, 7, 119. *Chem. Abstr.* **1978**, *88*, 181658.
398. Neugebauer, F. A.; Fischer, H. *Chem. Ber.* **1980**, *113*, 1226.
399. W. D. Ollis, C. A. Ramsden and S. P. Stanforth, to be published.

Index

Note: Page numbers followed by "*f*" indicate figures, "*t*" indicate tables, and "*s*" indicate schemes.

N

P

R

S

Kolibri 360 GmbH
Ferdinand-Jühlke-Straße 7
99095 Erfurt, Deutschland
produktsicherheit@kolibri360.de